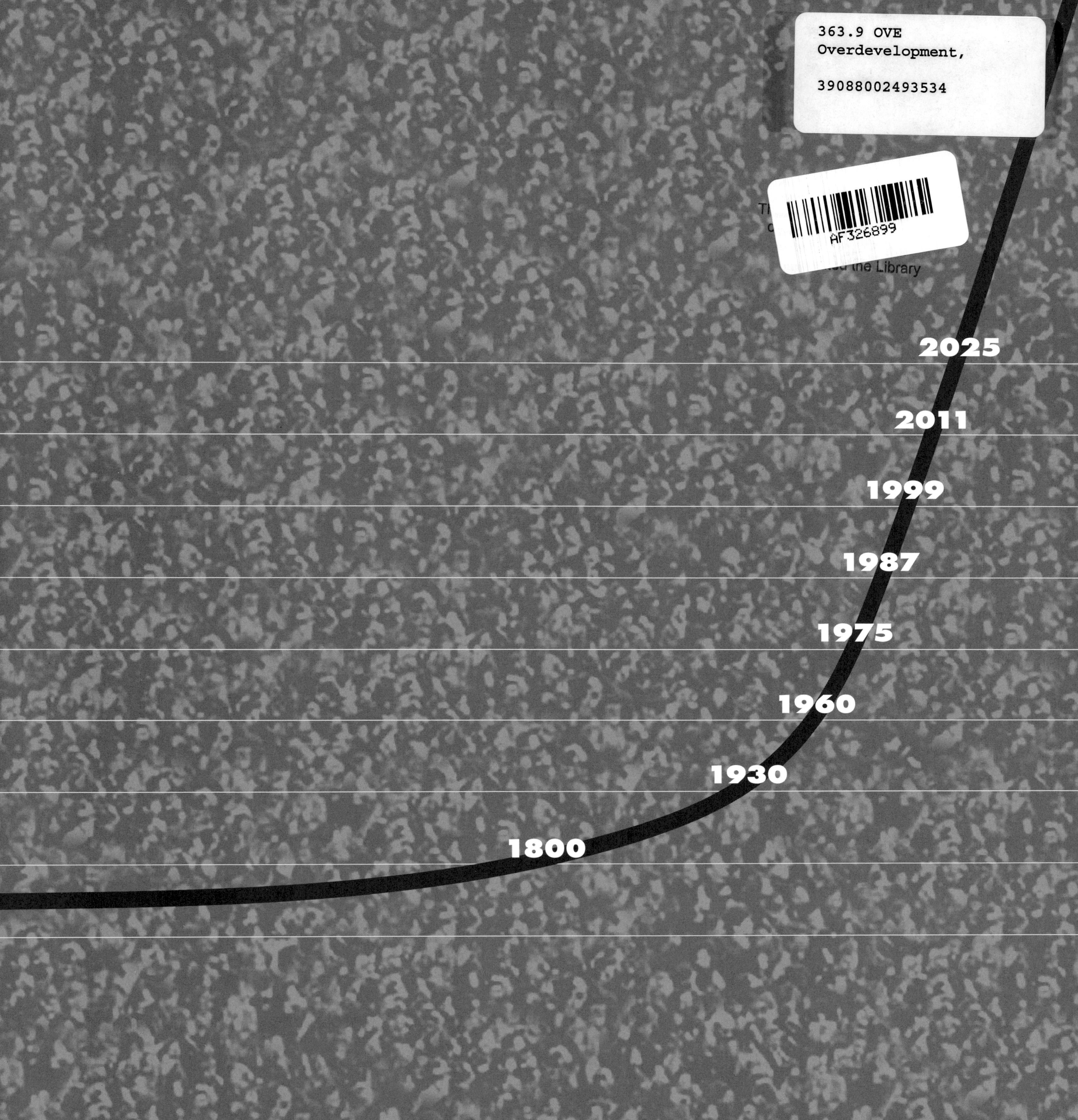
2025
2011
1999
1987
1975
1960
1930
1800

OVER DEVELOPMENT
OVER POPULATION
OVER SHOOT

Edited by Tom Butler

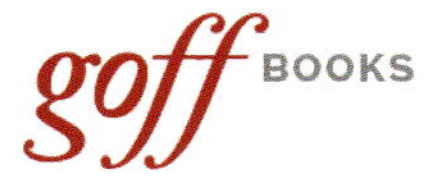

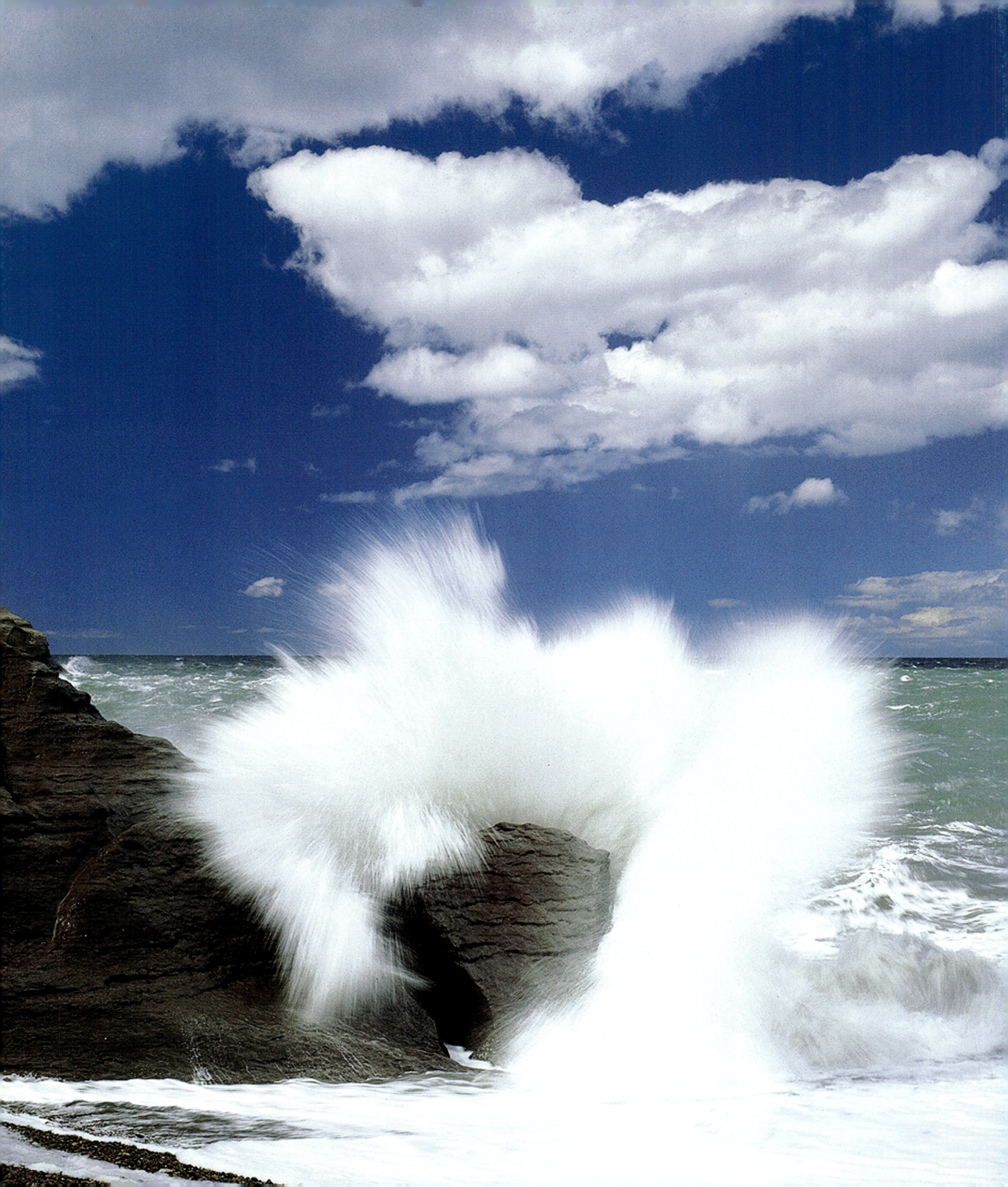

To the wild beauty, ecological richness,

and cultural diversity being swept away

by the rising tide of humanity . . .

and for William R. Catton Jr.,

peerless teacher on the perils of overshoot.

Can you think of any problem in any area of human
endeavor on any scale, from microscopic to global,
whose long-term solution is in any demonstrable way
aided, assisted, or advanced by further increases in
population, locally, nationally, or globally?

—Albert Bartlett

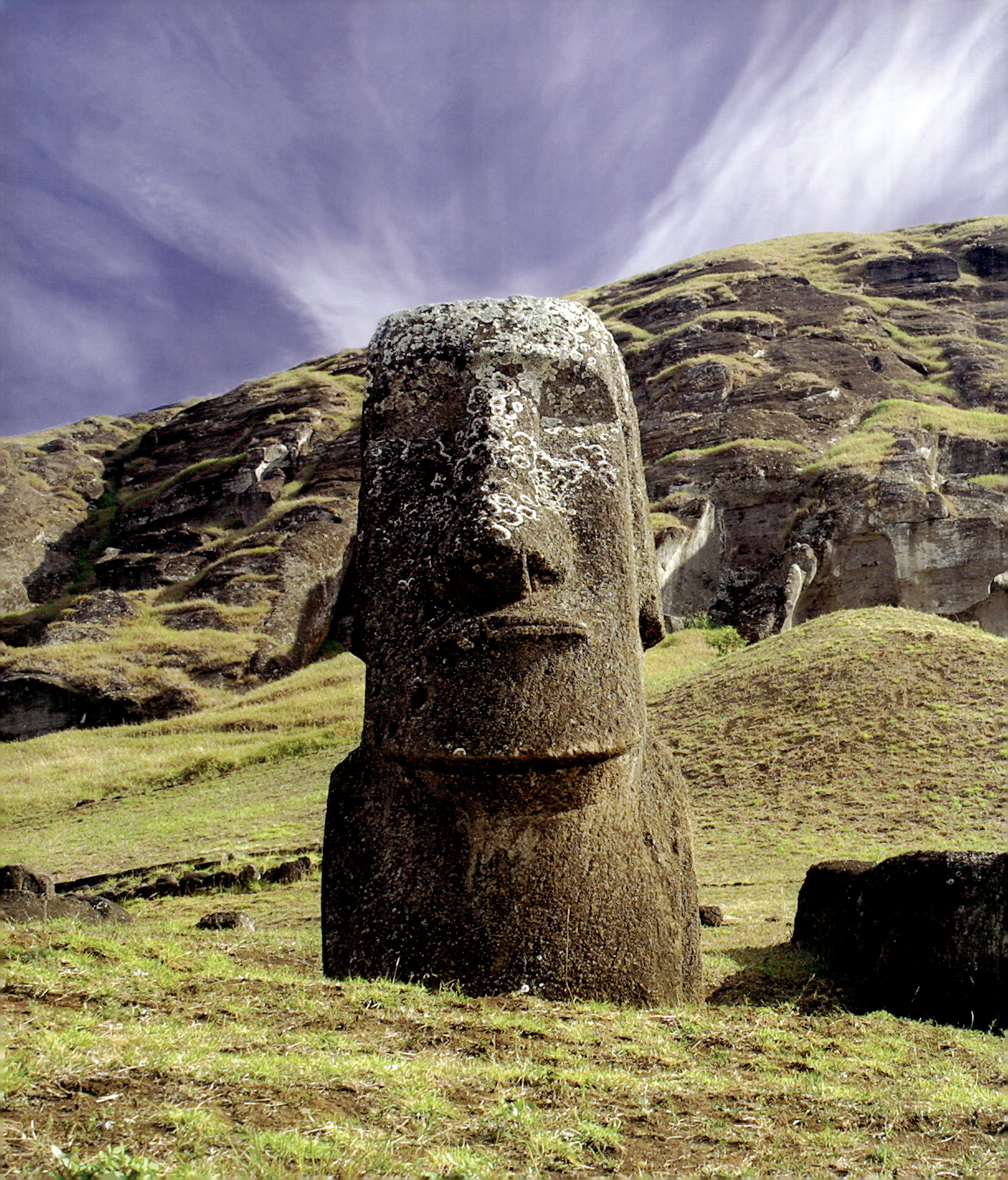

CONTENTS

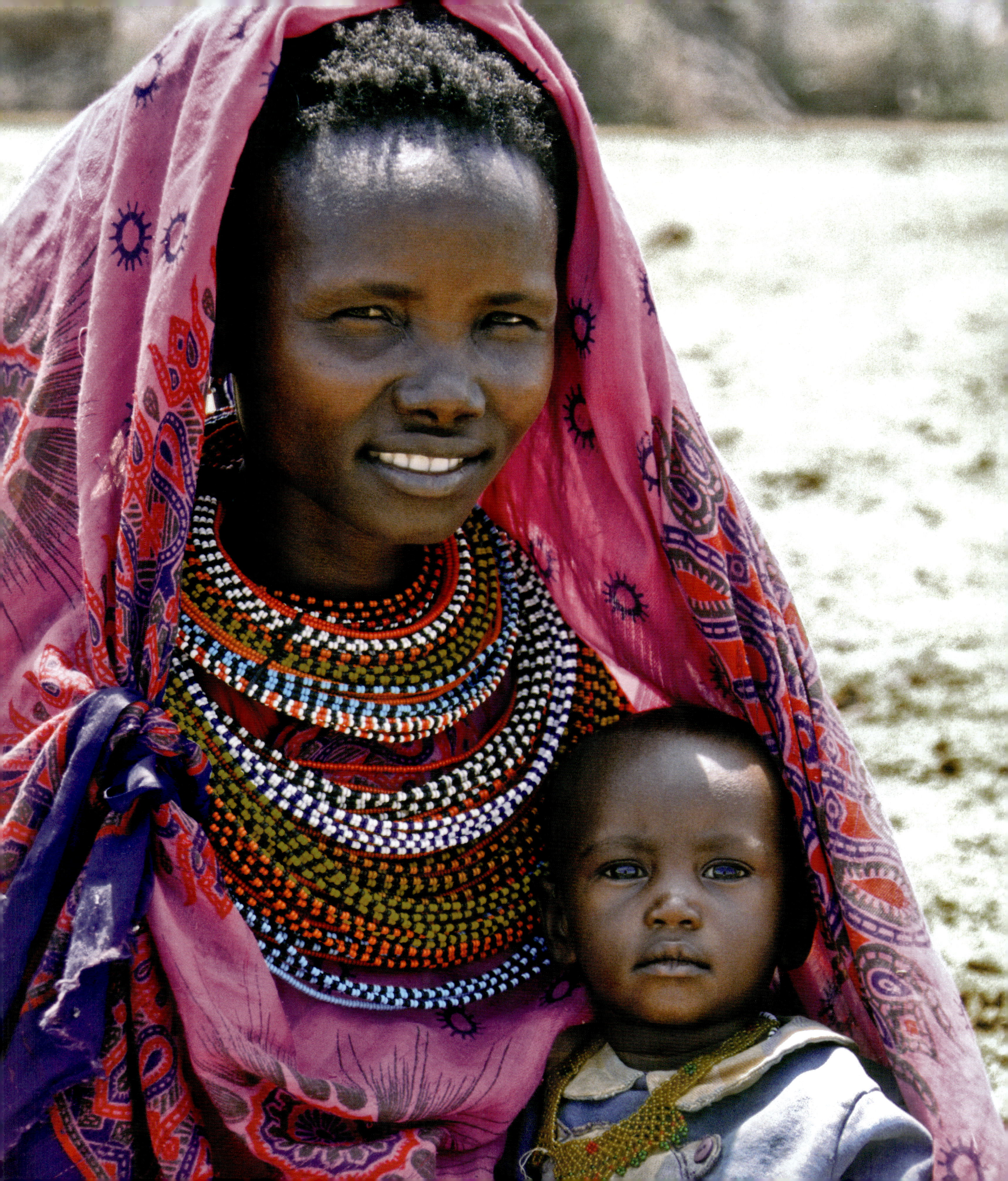

Musimbi Kanyoro

WE ARE ONE HUMAN RACE living on one planet. We aspire for the same things: food, water, good health, and most of all, dignity and loving relationships. We yearn for opportunity, voice, and resources to develop our potential. We want to raise our children in a safe and healthy environment. We want to experience the Earth's beauty and natural bounty.

Realizing our common humanity invites us to embrace common responsibility and to care for one another and the planet on which we live. The emergence of such grave global challenges as biodiversity loss and climate change demands our urgent and undivided attention. The health of the oceans, the air, the water, and the land affects human health. The size of the human family and the way that we live influence the quality of life for people today as well as for future generations. Moreover, our numbers and behavior profoundly affect nonhuman species, all of the creatures with which we share this beautiful but finite planet. The web of life that these species create is what makes the Earth habitable and lovely.

We know that rapid population growth exacerbates social, economic, and ecological problems—whether in rich or poor countries, north or south. Most important, rapid population growth is a fundamental driver of individual as well as societal problems that deny dignity, especially to women who bear the burden of reproduction and caretaking of communities. We have the knowledge to reduce these burdens thoughtfully by using rights-based, culturally appropriate ways to slow population growth while enhancing human dignity and thoughtful development. Taking action in this way is important for my country, Kenya, as it is for all other nations. This is what the world needs to do *today* and not tomorrow.

This urgency strikes home when looking through the images in this powerful book. Who can say, with an honest heart, that the suffering of the Earth and millions of her children is not linked to the exponential growth in human numbers?

I have devoted most of my professional life to advocating for and advancing the universality of human rights, the rights of women and girls, and the rights of poor people. I am not naïve about either the complexity of factors affecting public policy, or about the imbalance of power, voice, and resources across nations, genders, generations, and cultures. Yet, I sincerely believe that family planning is a human right that yields multiple benefits for women, children, and poor people—ultimately for all humanity. It helps sustain a mother's health and gives women choices beyond childbearing. Well-spaced children are healthier, and fewer children per family help their parents to better support their growth and development. All these step-by-step and one-person-at-a-time actions add up to immense social good when implemented on a large scale.

The core ethic that unites all of us in relation to family planning is a respect for individual autonomy. Family planning is not about telling people what to do but about listening to what they want. Over 200 million couples around the world want to limit the number of children they have, but are not using contraception, and every woman wants and deserves a safe delivery. A safe and legitimate way to reduce population growth is to make family planning information and services and access to safe motherhood universally available in a human rights framework.

While the complexities and challenges of achieving this are quite real, the problem of rapid population growth requires that leaders gather collective political will and implement effective policies with the speed and commitment of resources commensurate with the urgency and immensity of the problem. This is the right thing to do, and it is our responsibility to future generations. We owe our moral will to this action.

Some good practices are happening. There is global momentum to promote and invest in the education of girls and to protect them from harmful practices such as child marriage. Despite pockets of distractors, such as the Boko Haram in Northern Nigeria or the Taliban in Pakistan that want to hinder this progress, we must commit to giving the next generation the opportunities to fulfill their dreams. Moving toward that future is a shared responsibility and one that cannot be limited by geography or politics.

All of these things are possible when individuals, families, governments, and international development organizations work cooperatively and quickly to make family planning education and services universally available, moving toward ensuring total equality of opportunity for girls and women, and when everyone works toward narrowing the economic gap between nations.

These times call for an unprecedented level of cooperation and common purpose among genders, cultures, peoples, and nations. The legacy for future generations and so much of Earth's living heritage depends on we who are living today. Our problem is not ignorance. We are drowning in knowledge. Perhaps we lack discernment, or else we are simply are too selfish to care for the future of those to come after us! But indeed, the world community can act—and act quickly—when faced with major threats. Now is the time to take the problem of overshoot seriously and to act while we still have an opportunity to ensure that future generations inherit a sustainable world.

Today we are faced with a challenge that calls for a shift in our thinking, so that humanity stops threatening its life-support system. We are called to assist the Earth to heal her wounds and in the process heal our own—indeed to embrace the whole of creation in all its diversity, beauty, and wonder.

—Wangari Maathai

LORD MAN

A PARABLE . . .

In the beginning the world was whole, and beauty prevailed . . .

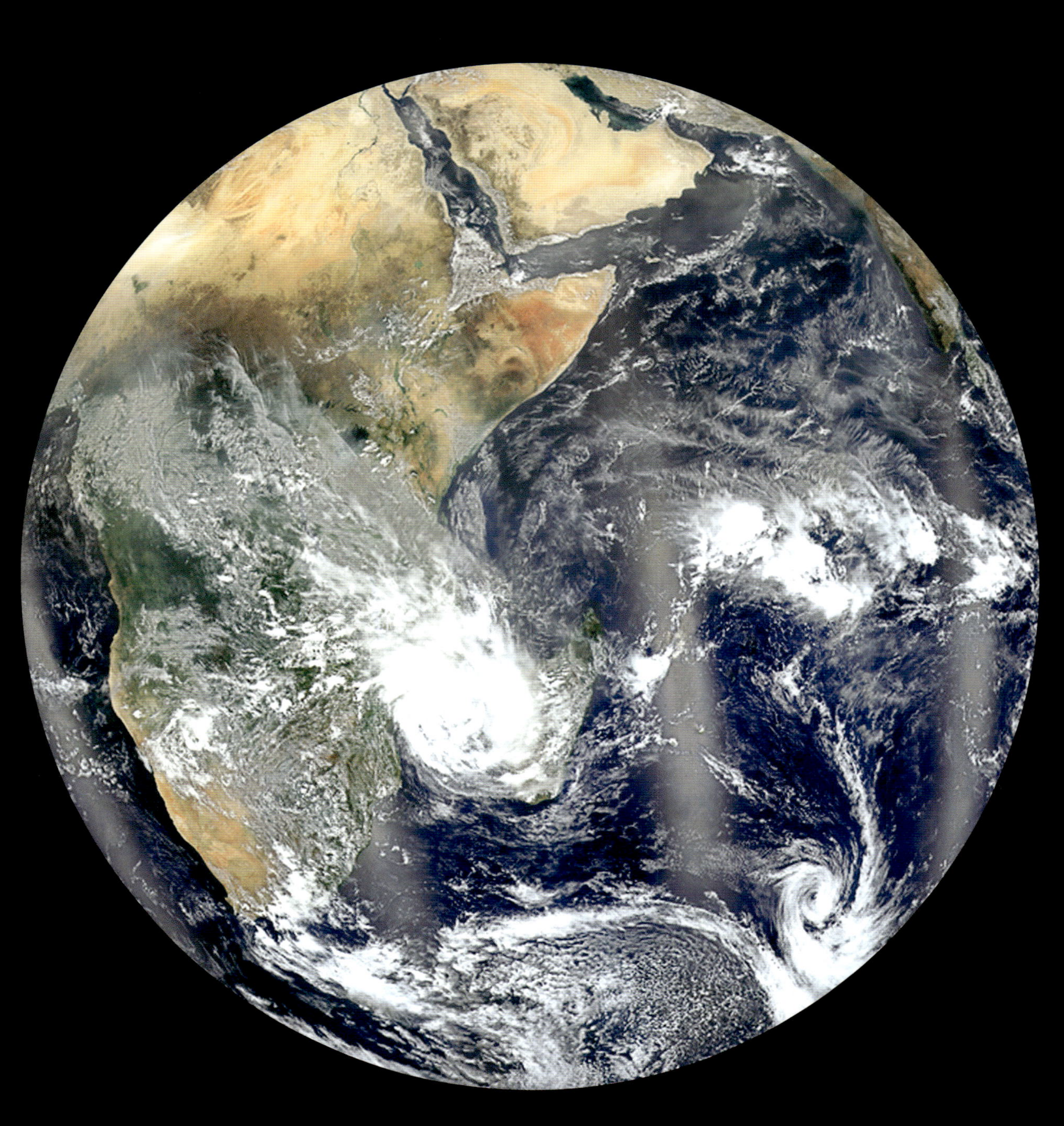

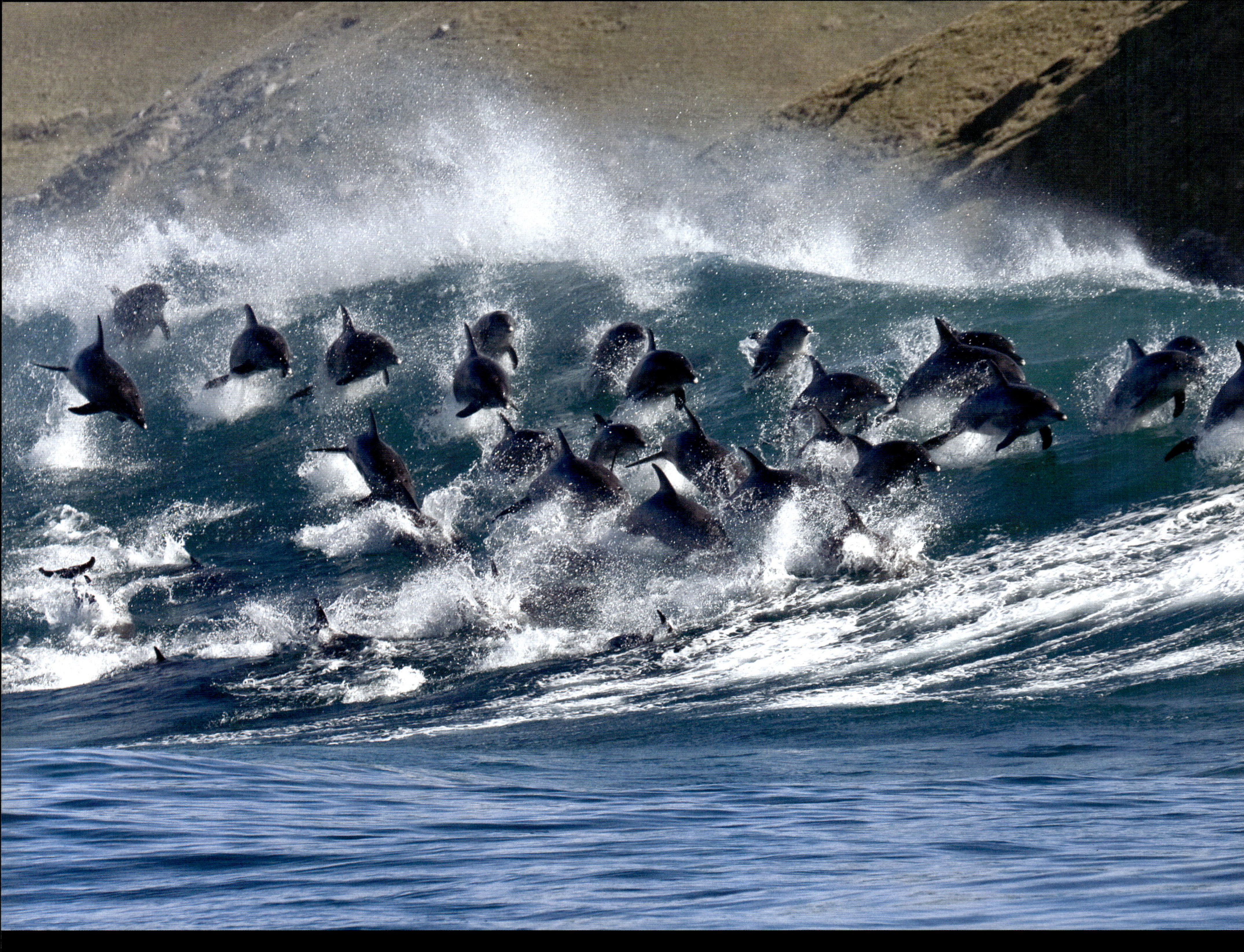

Life begetting life, until the waters,

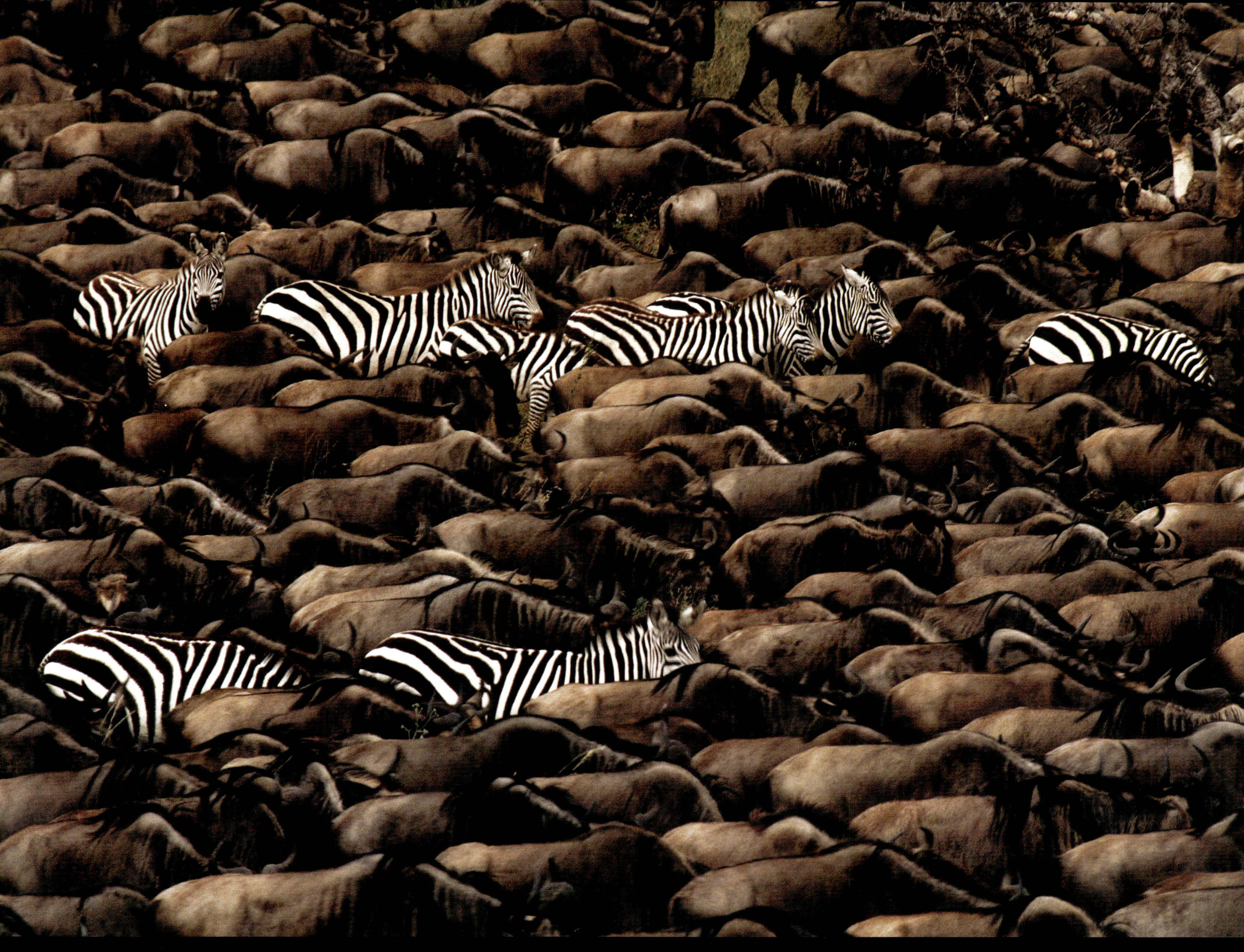

then the lands, were filled with creatures.

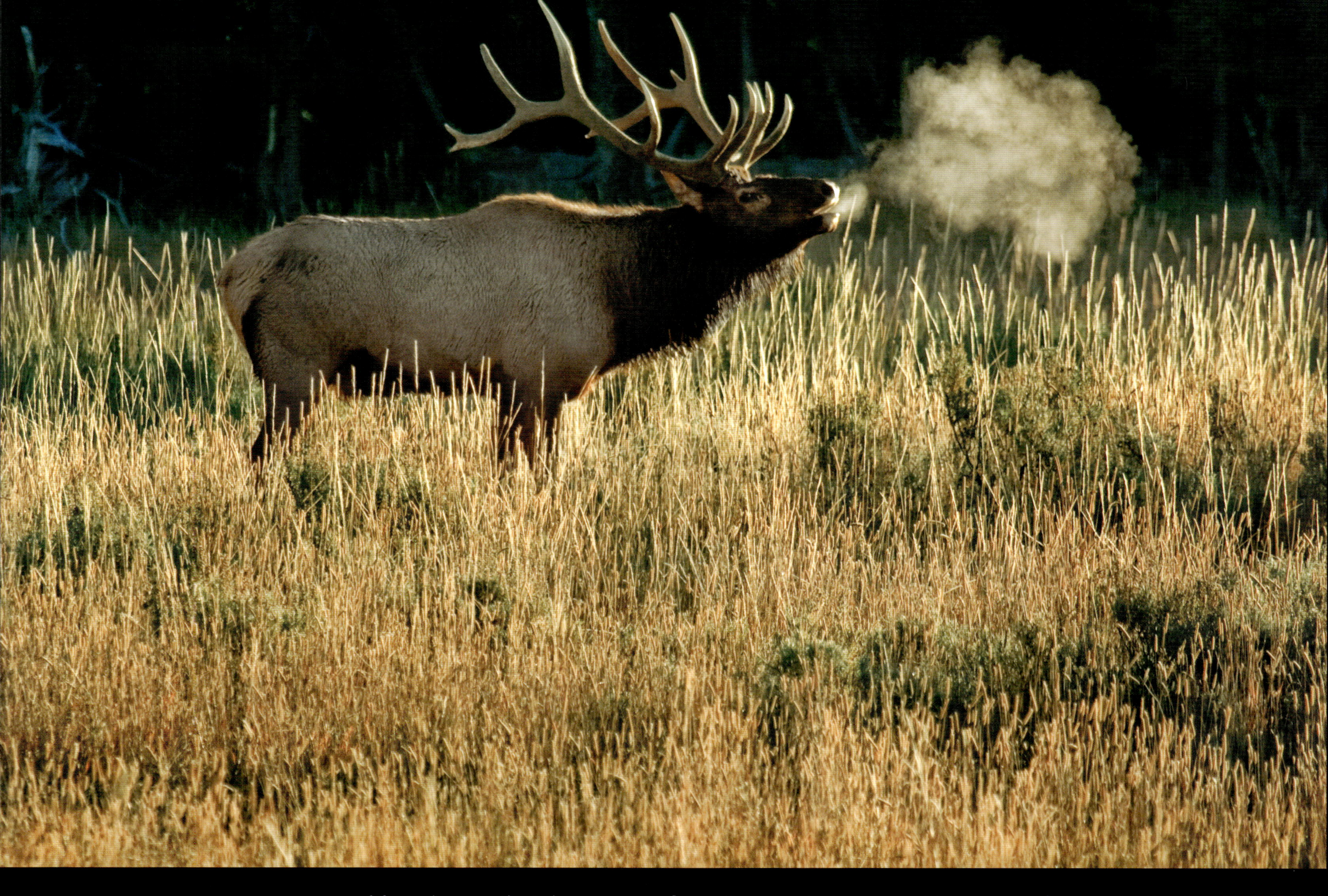

Myriad were their languages, from the nearly imperceptible song of moss to the bugling of elk.

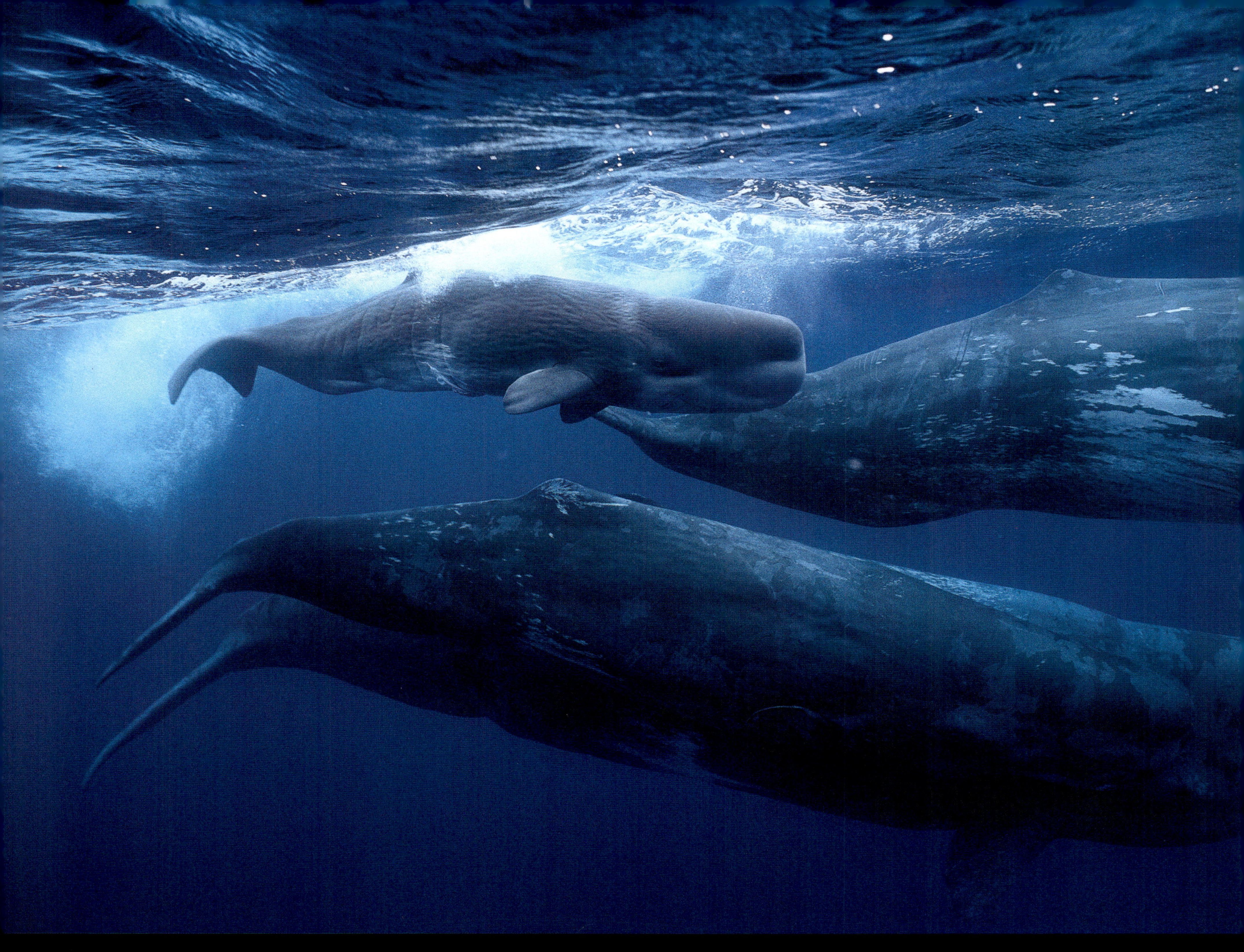

Whales performed their symphonies in the deep. The sounds of life were everywhere.

ife pulsed and contracted and flourished through the ages.

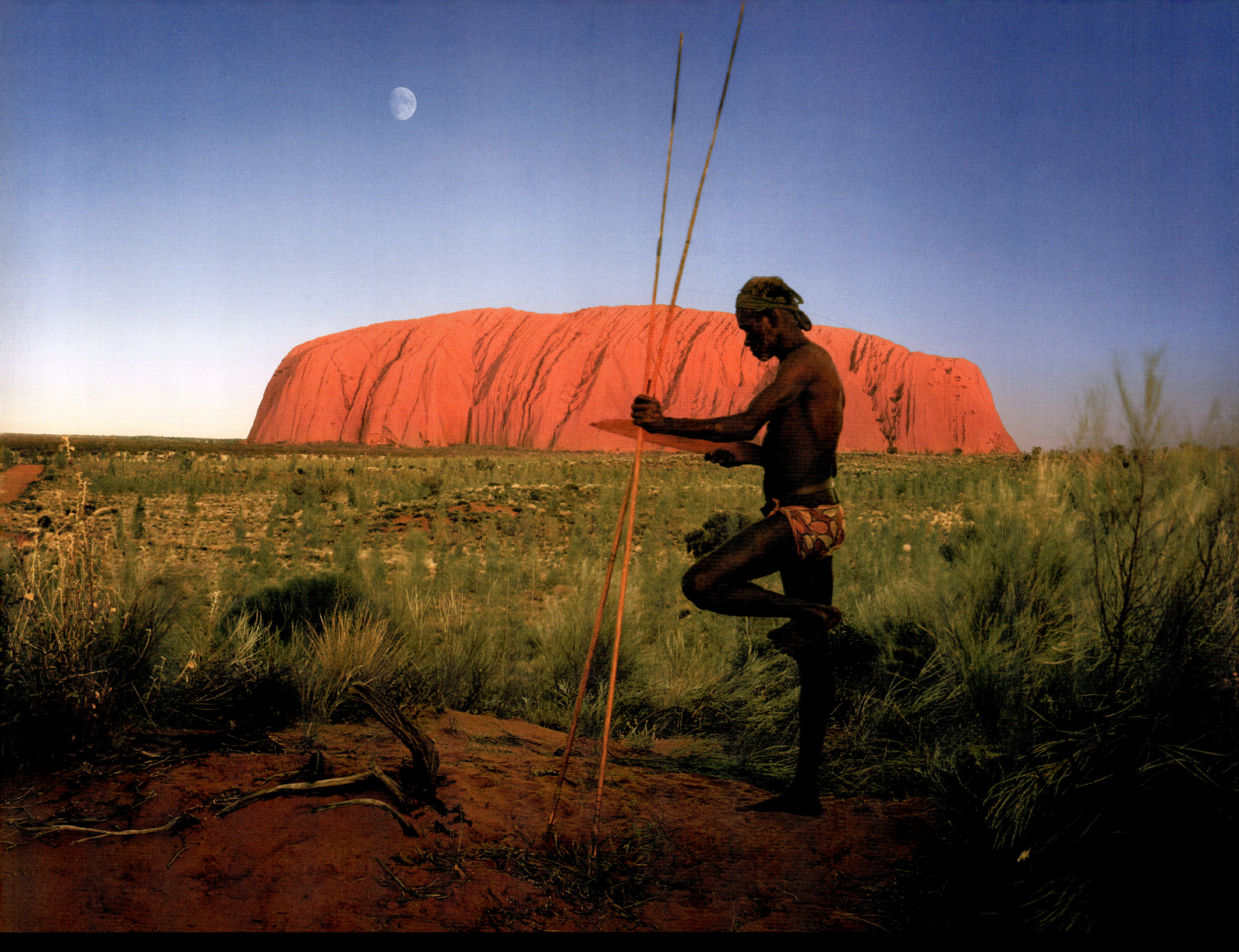

Eventually, a being appeared who learned to speak and count. For millennia he lived well among his wild kin

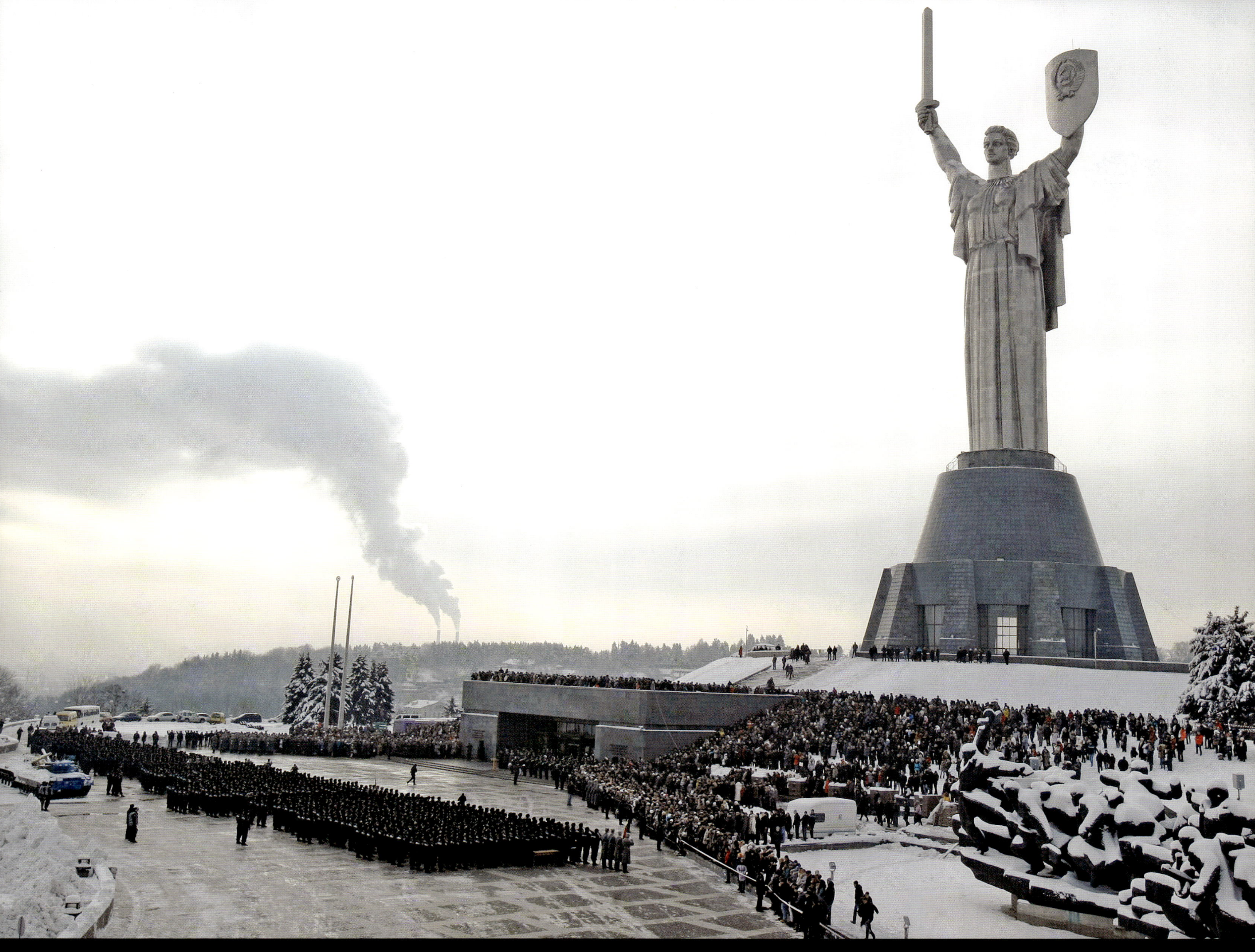

But as his cleverness grew, so did his ambitions, until the day he declared himself ruler of all.

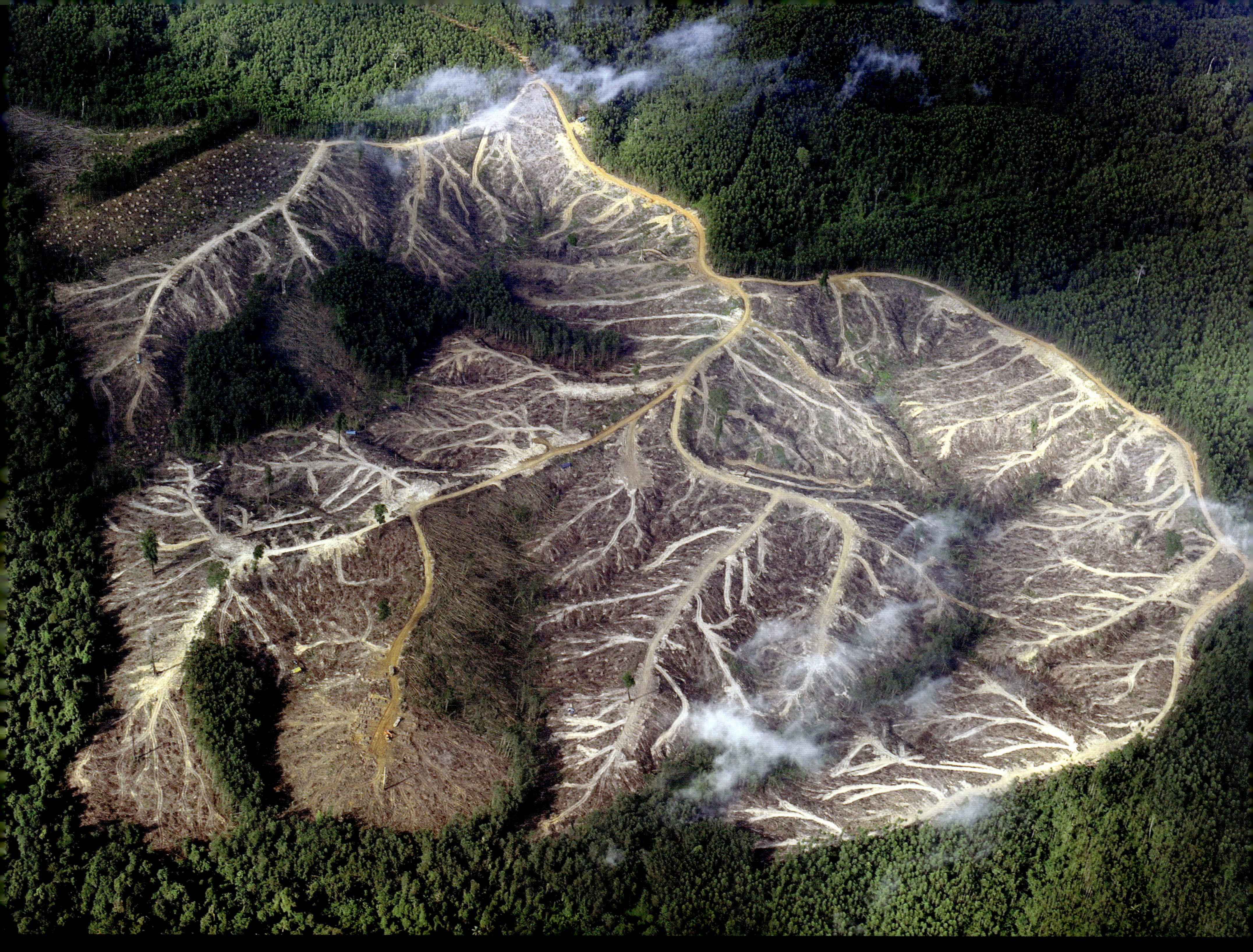

Believing the self-deception that his kind was sovereign over the others, he
taught his children that the Earth had been made for Man's use and profit.

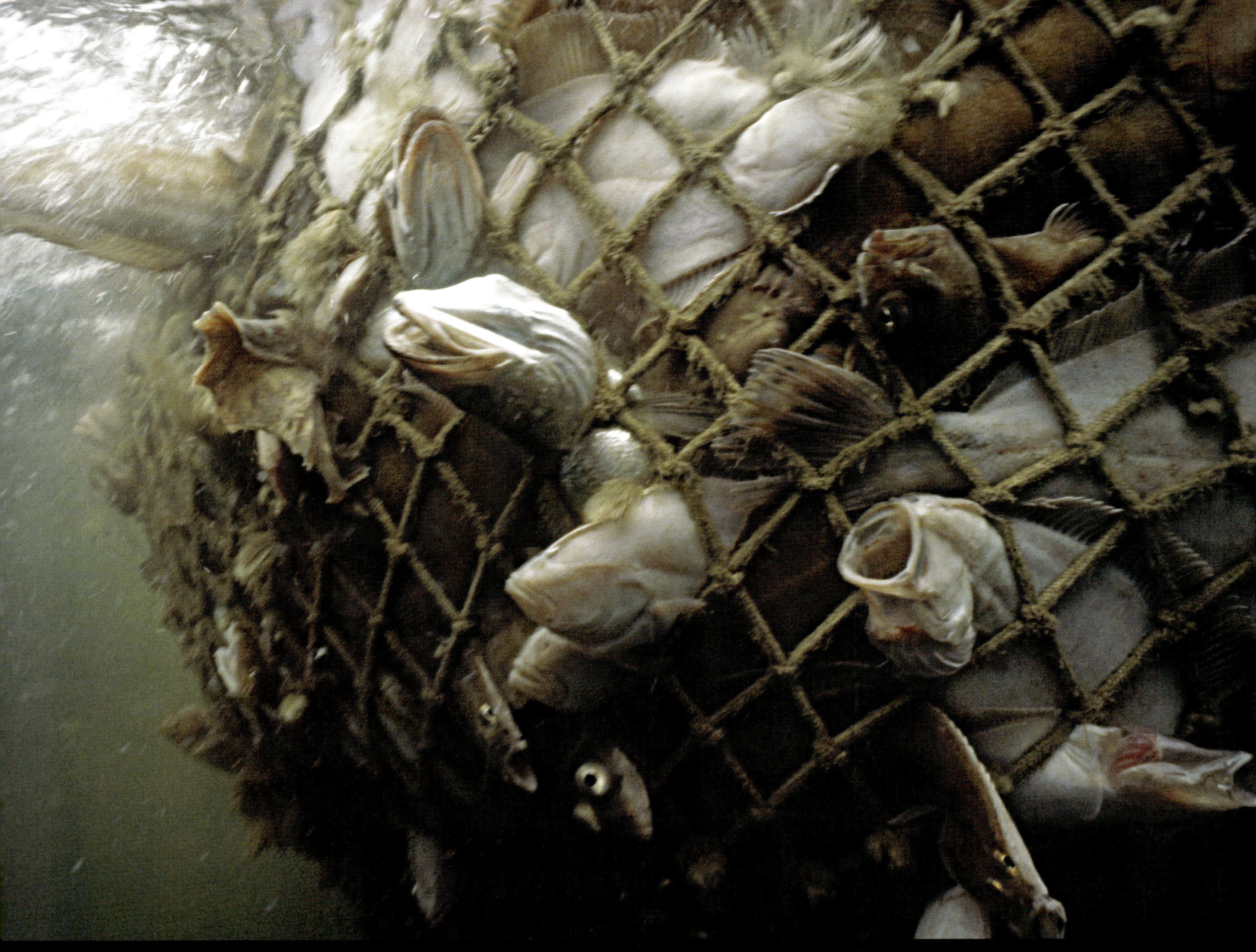

He no longer recognized his neighbors in the community of life, instead calling them "natural resources."

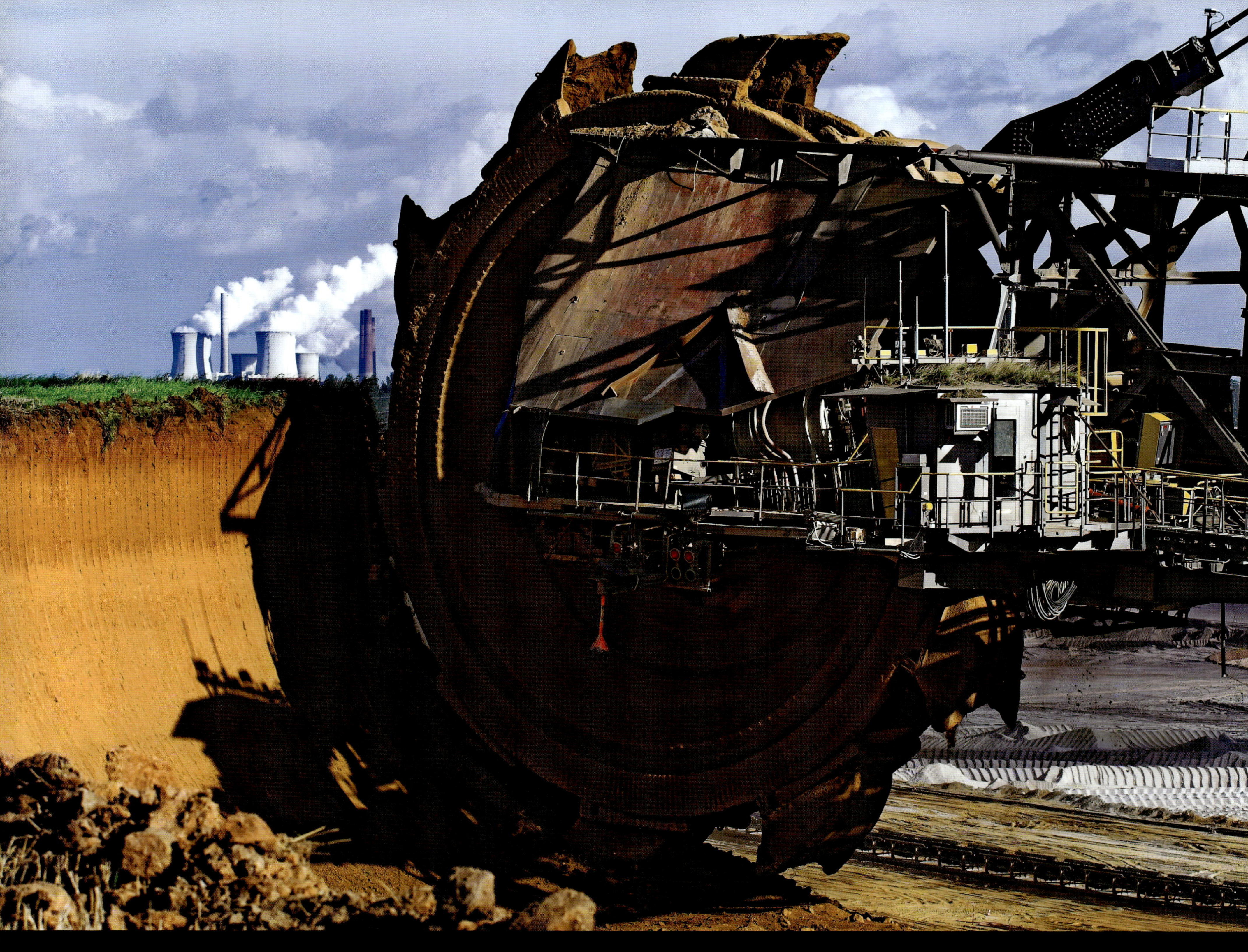

His work he named "progress."

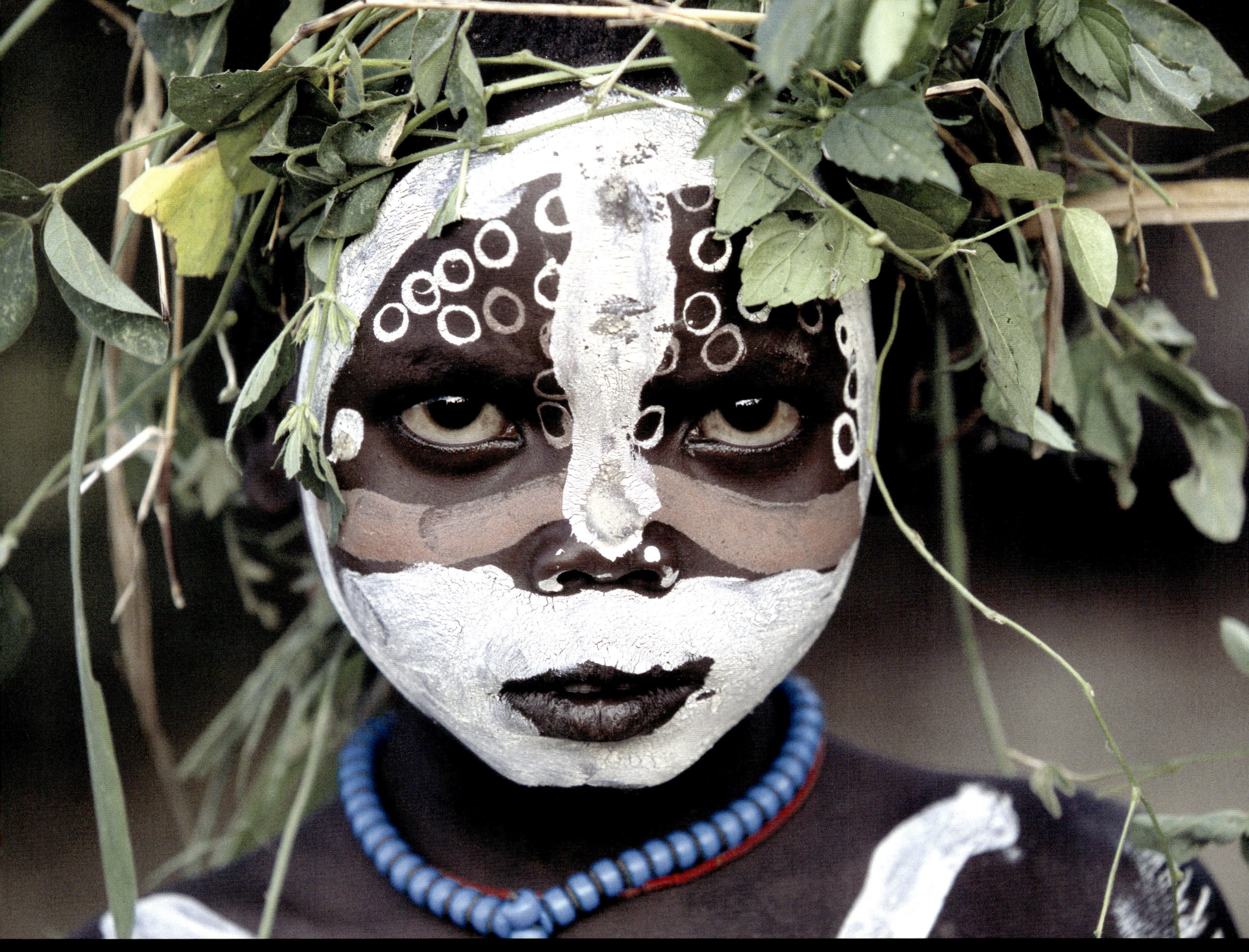

The old religions, which had long tied the human tribe to the other creatures in a circle of reciprocity, were forgotten.

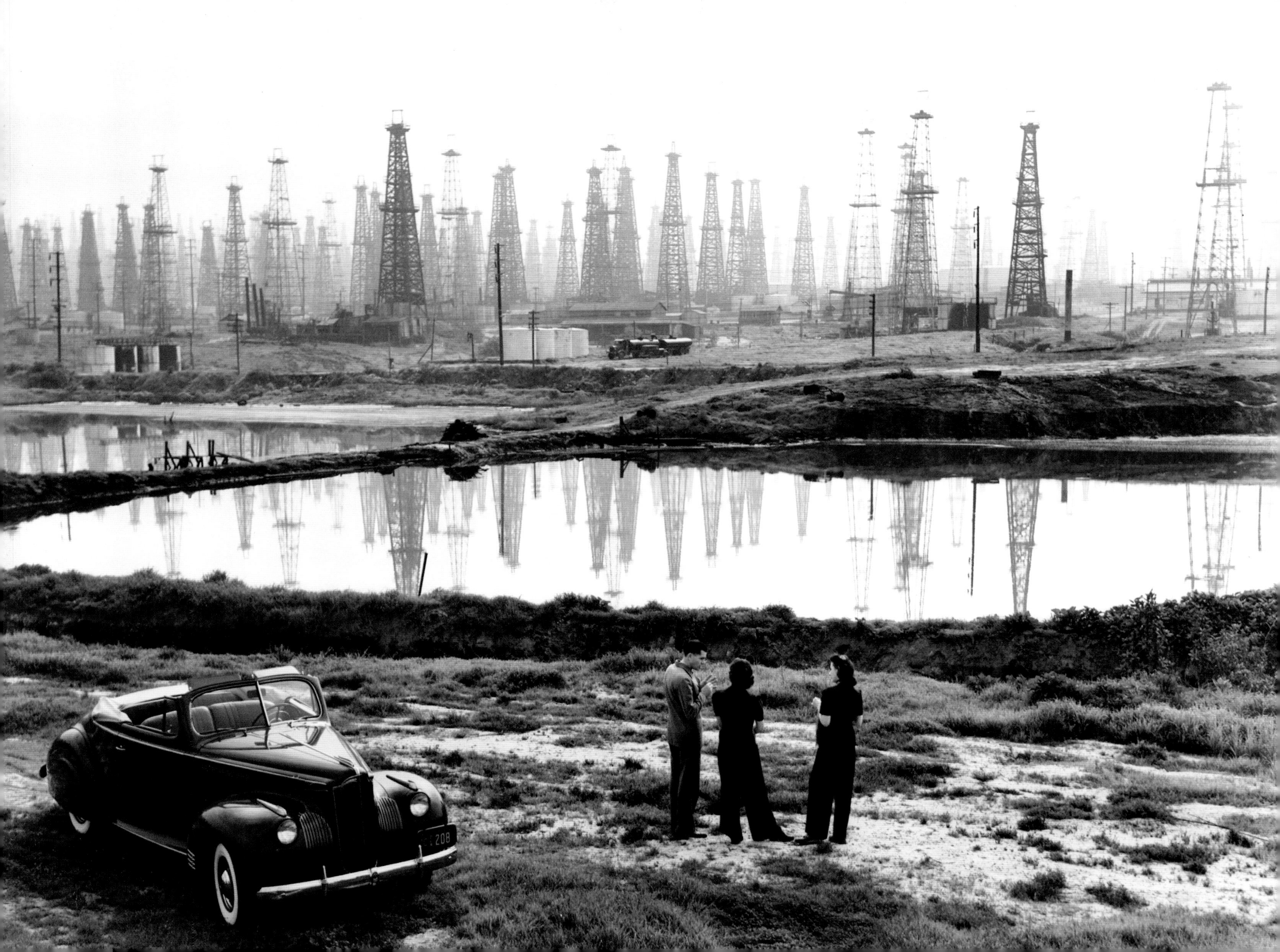

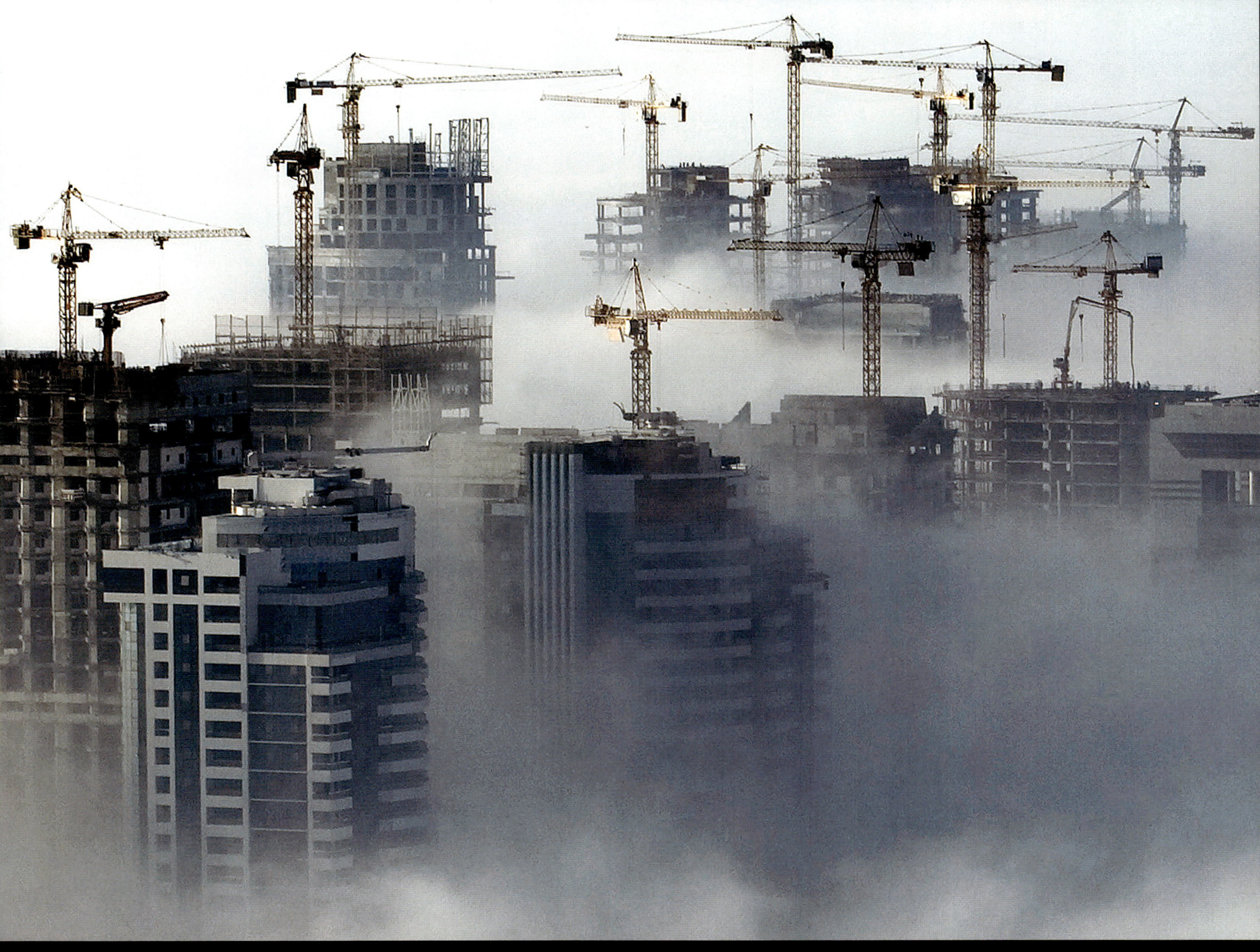

Praise was sung incessantly to the new god: Growth.

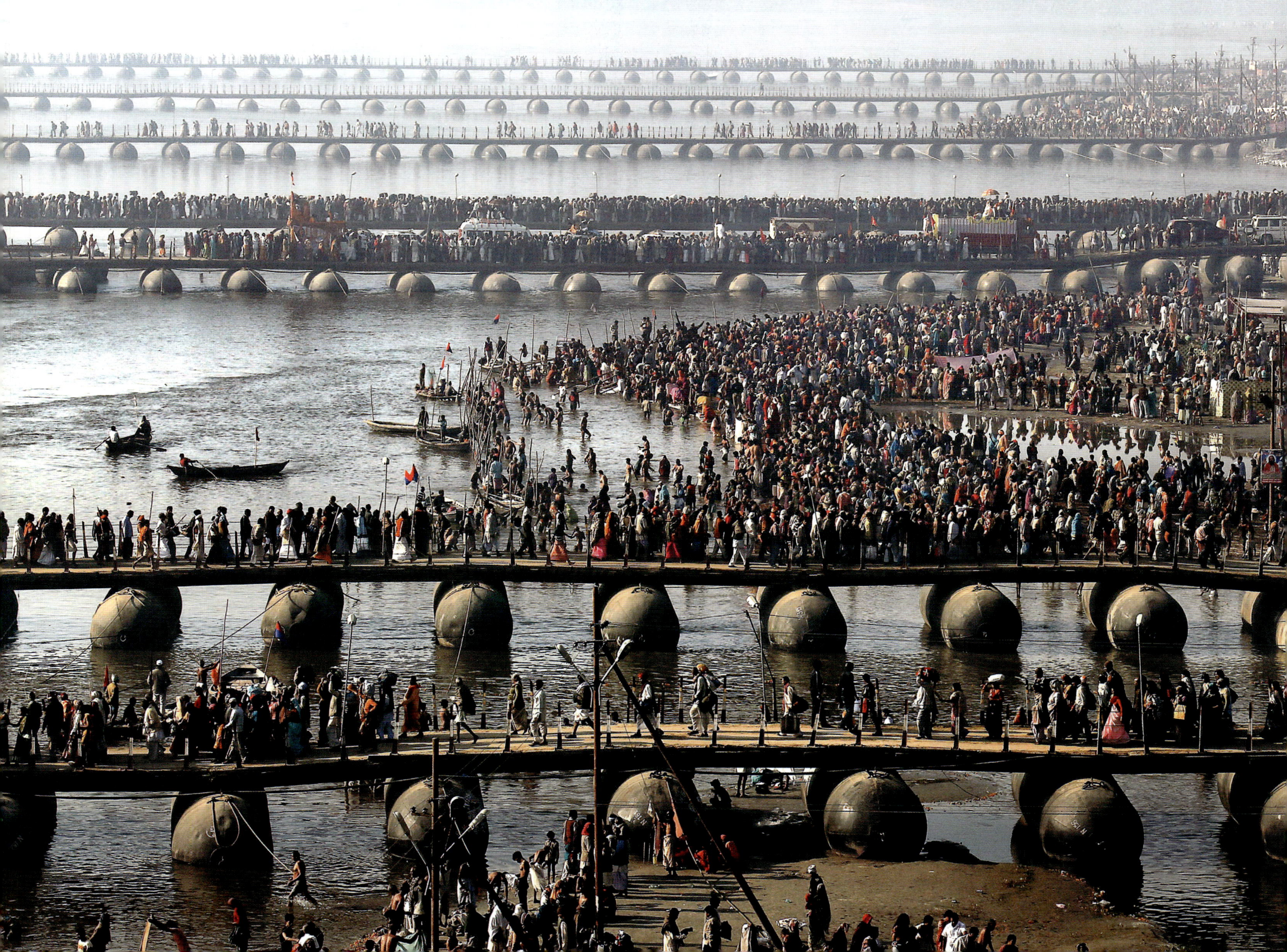

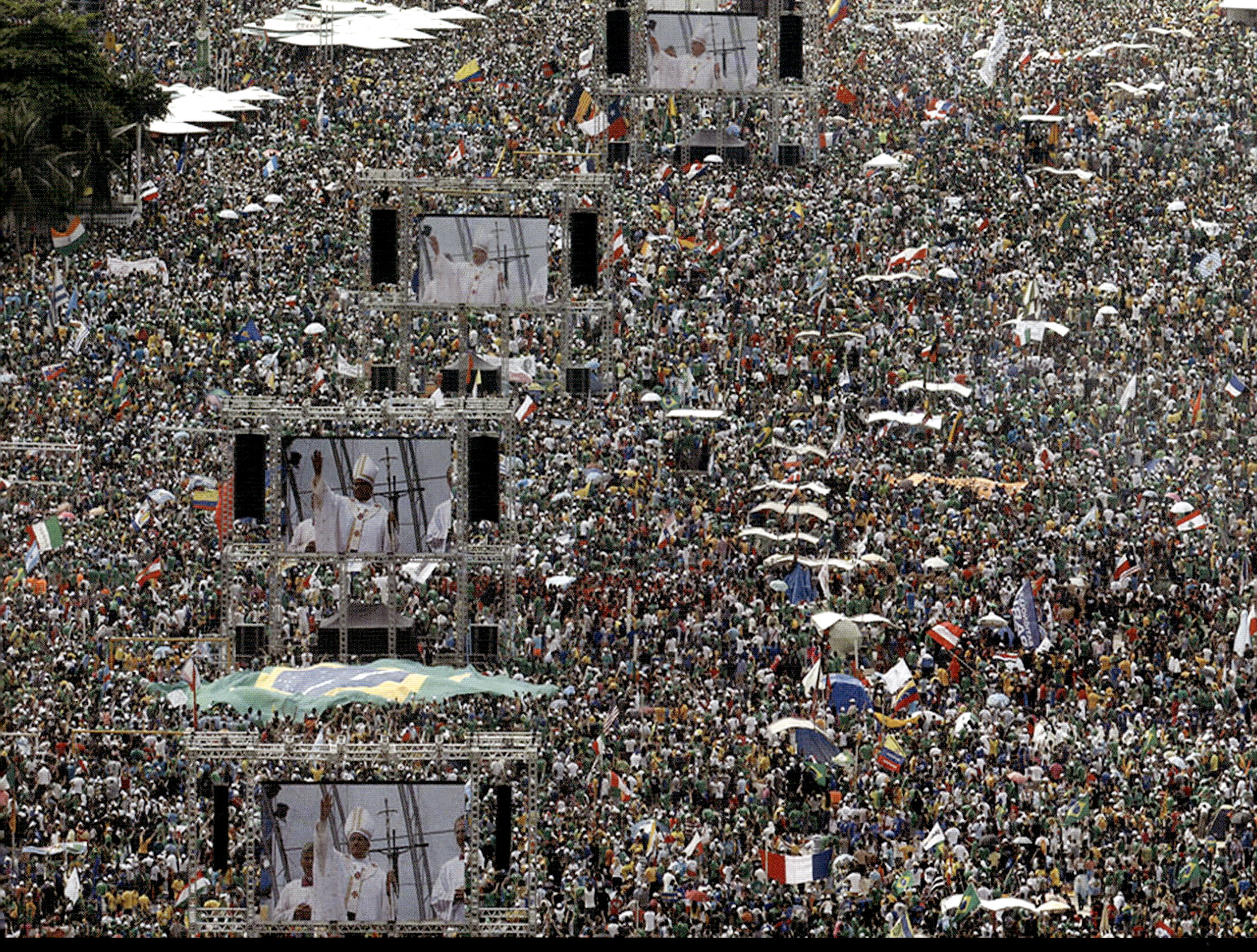

As the multitudes spread across the face of the Earth, the songs of the other creatures grew fewer and fainter.

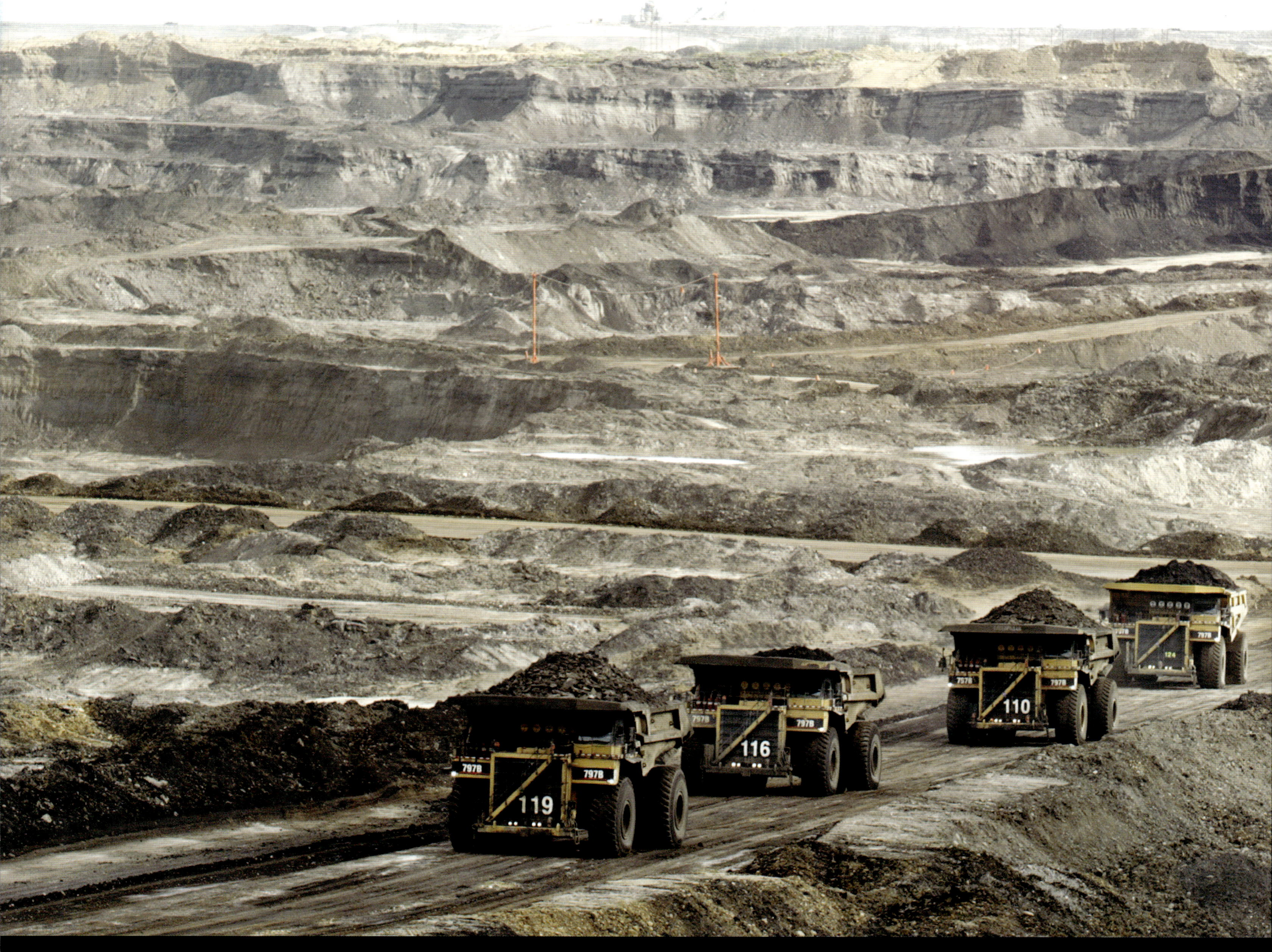
Many voices went permanently quiet, replaced by the sounds of machines—digging, churning, scarring the land,

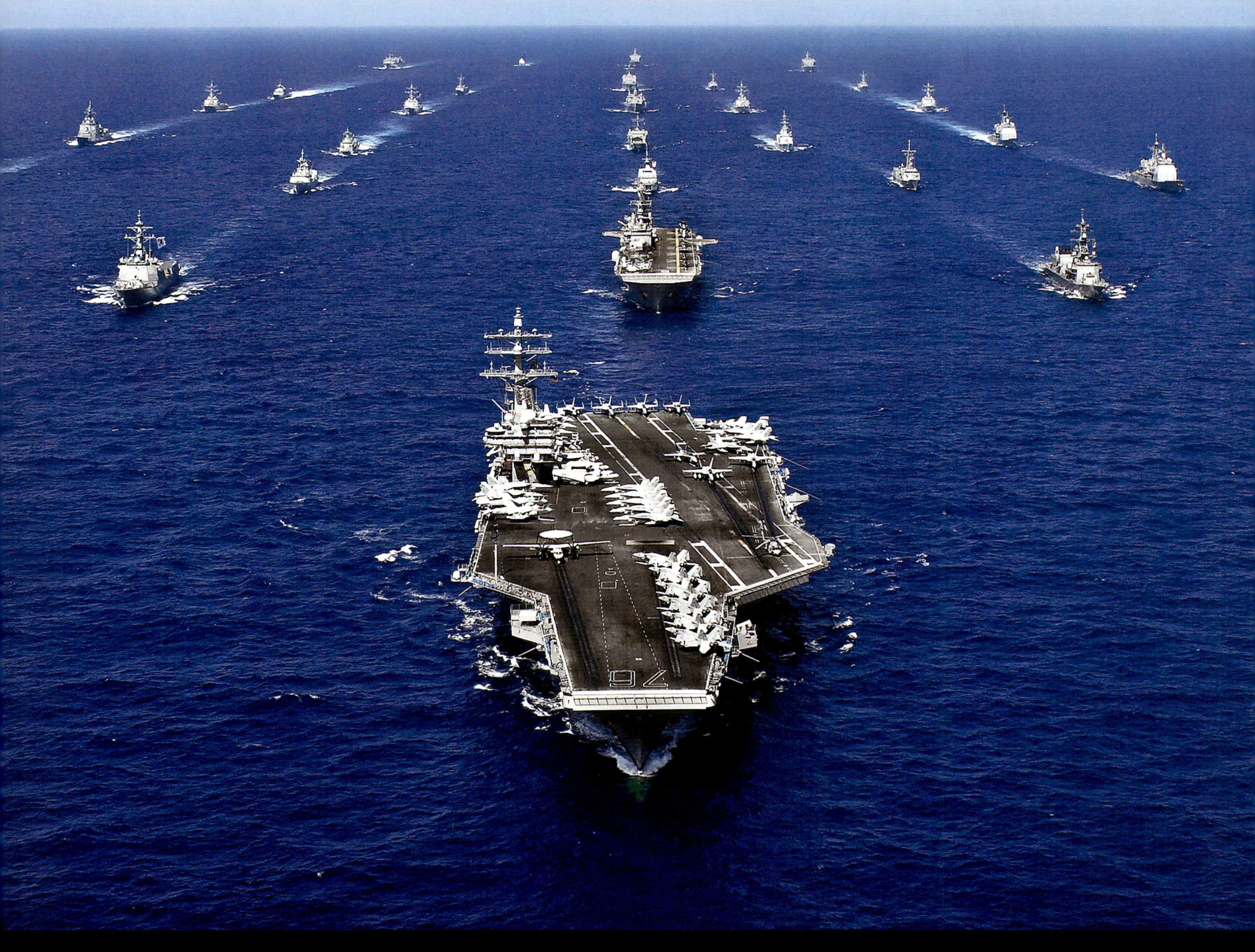

and driving the whales crazy with their noise.

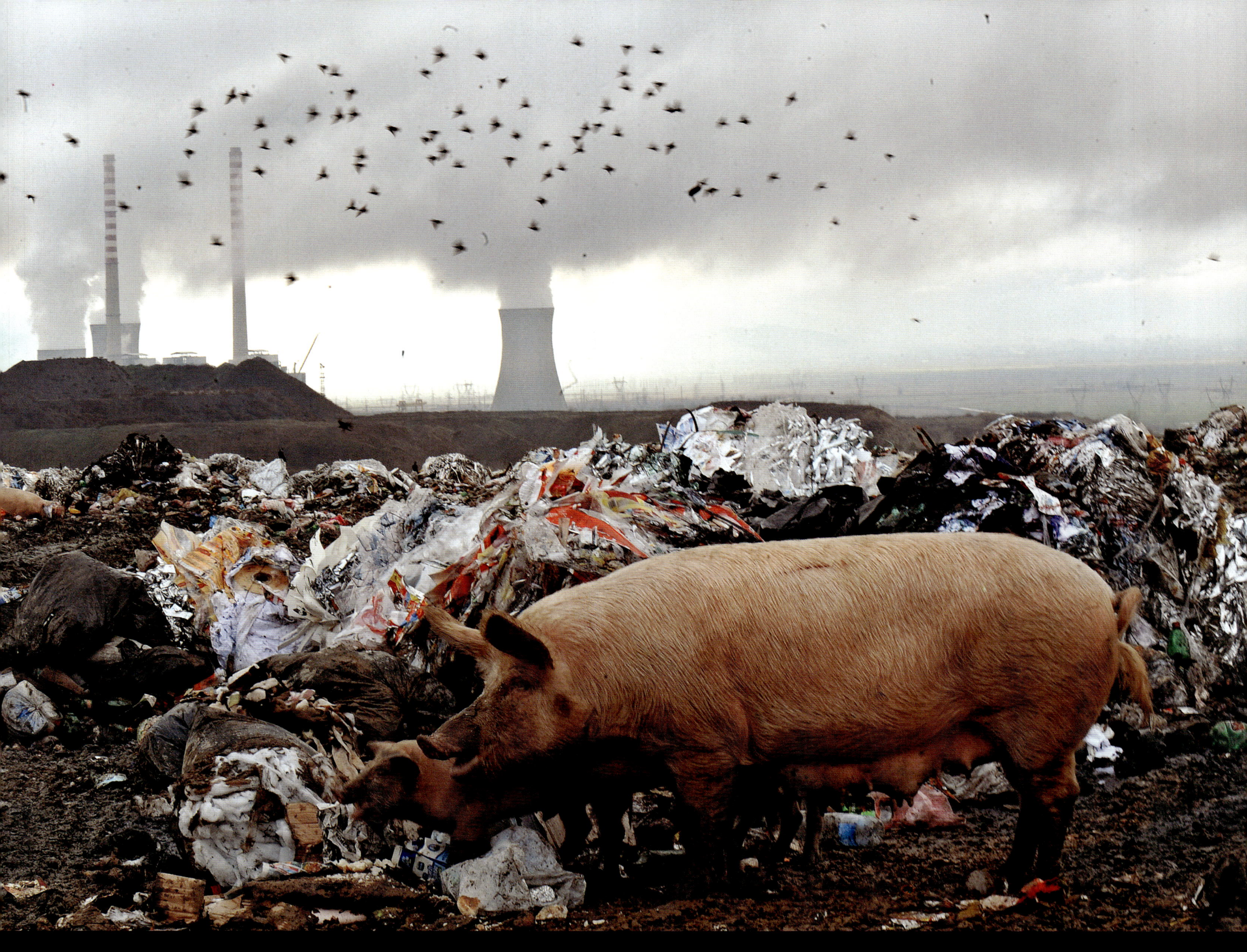

Every day the Earth grew poorer—transformed, bit by bit, by Lord Man's numbers and actions

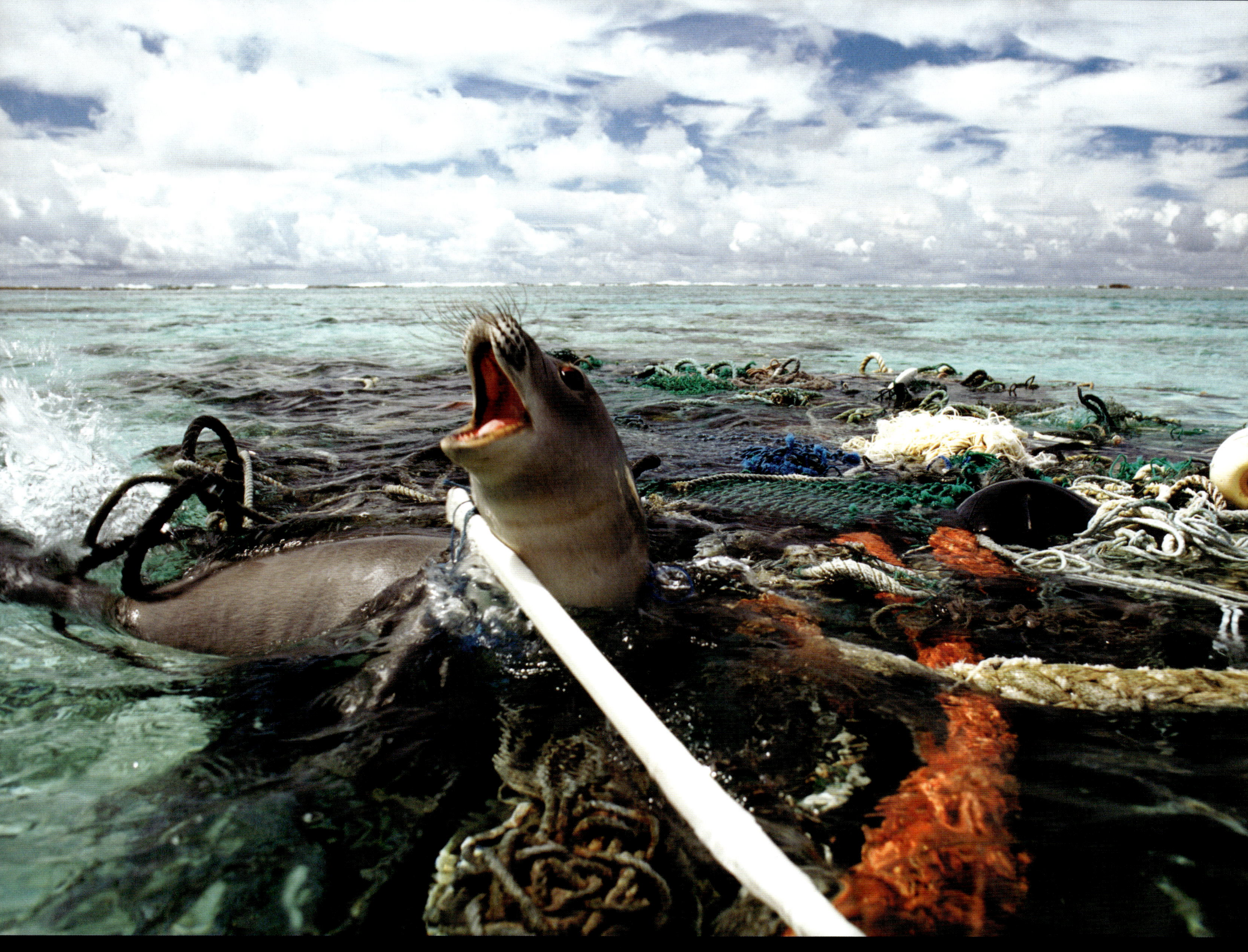

The seas were emptied of fish and filled with garbage.

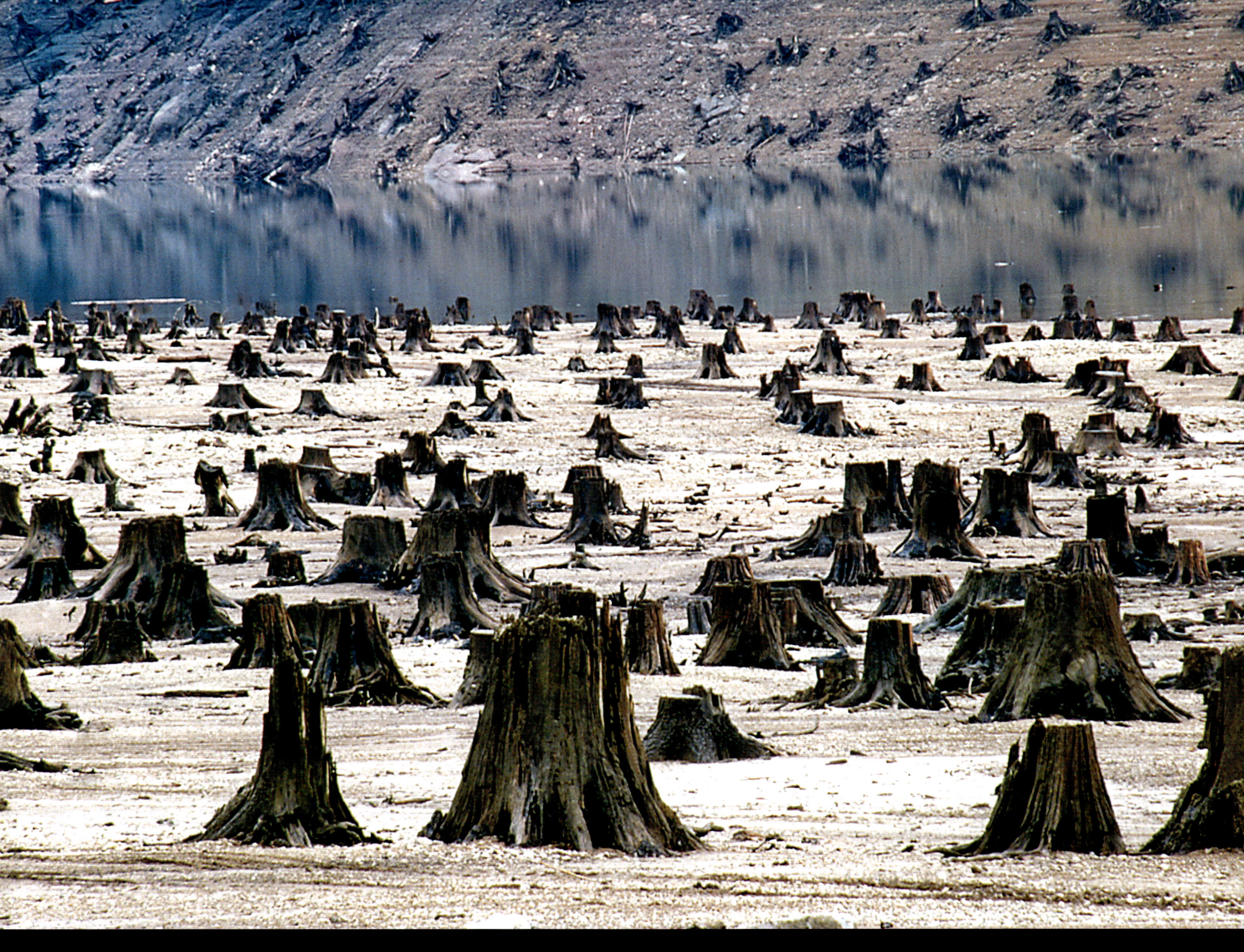

The trees were replaced by bleeding stumps.

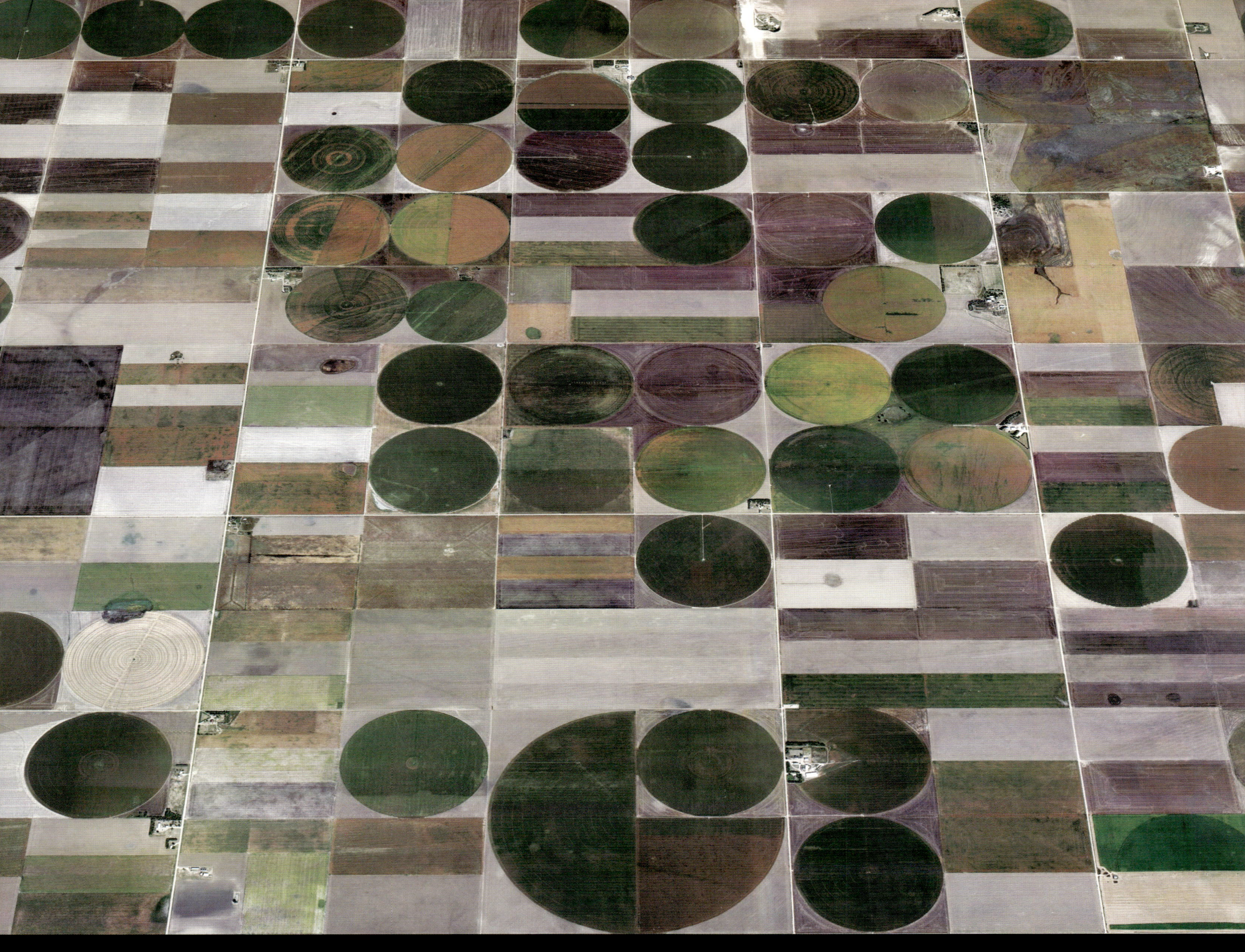

The prairies were transformed into feeding factories for the ever-expanding human masses.

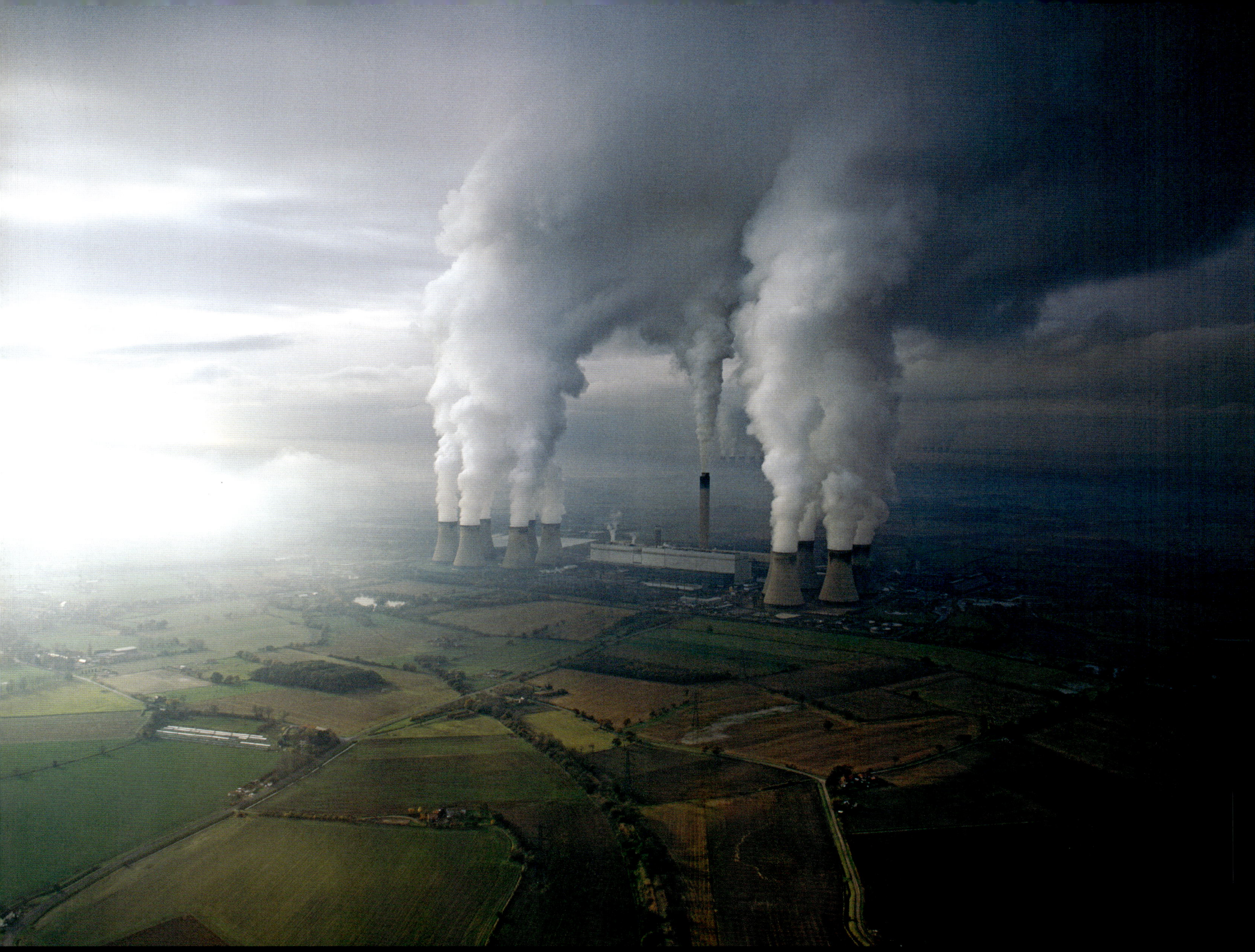

Smokestacks darkened the skies

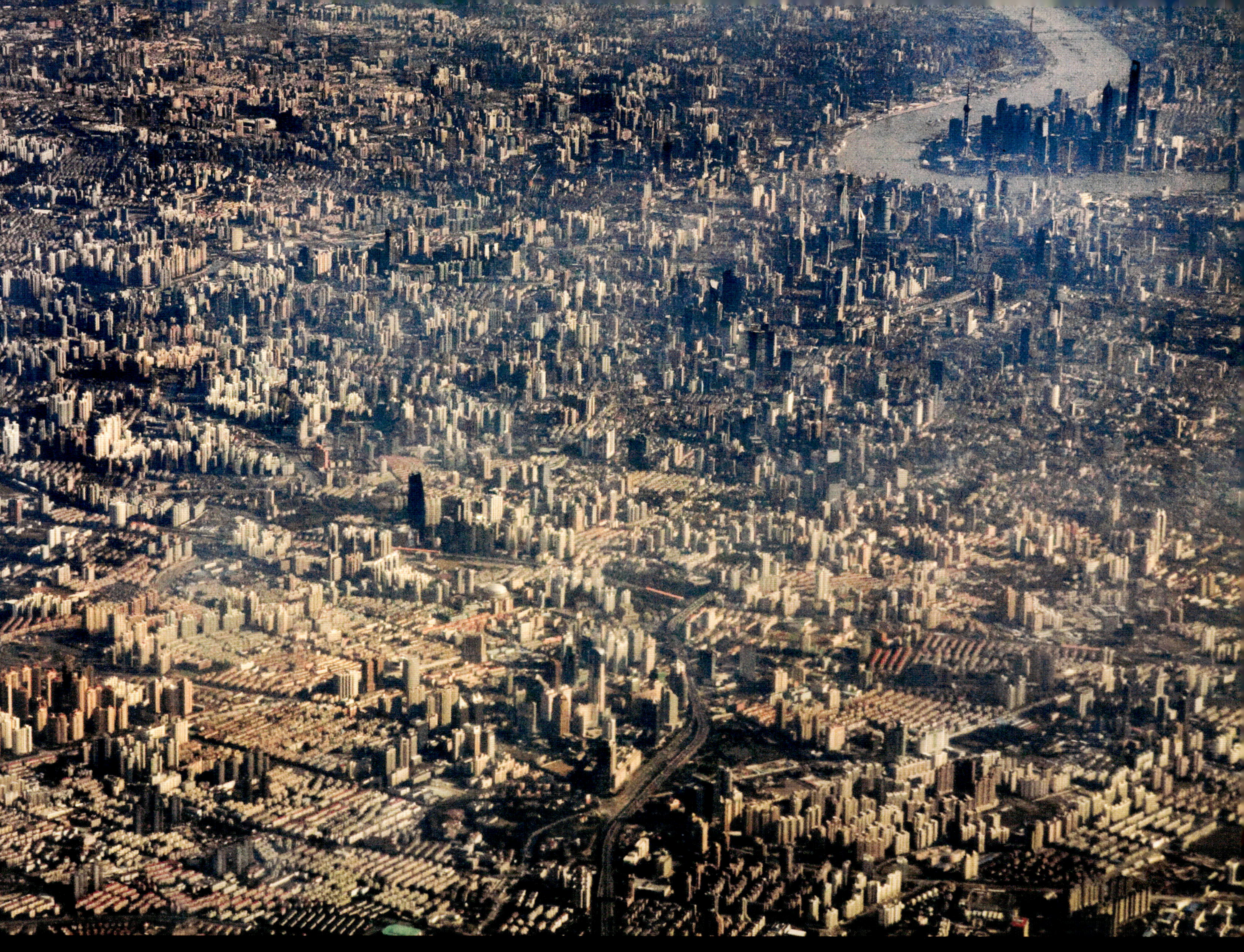

No place was sacred, no landscape safe from the insatiable creature's thirst for more energy to serve his God of Growth.

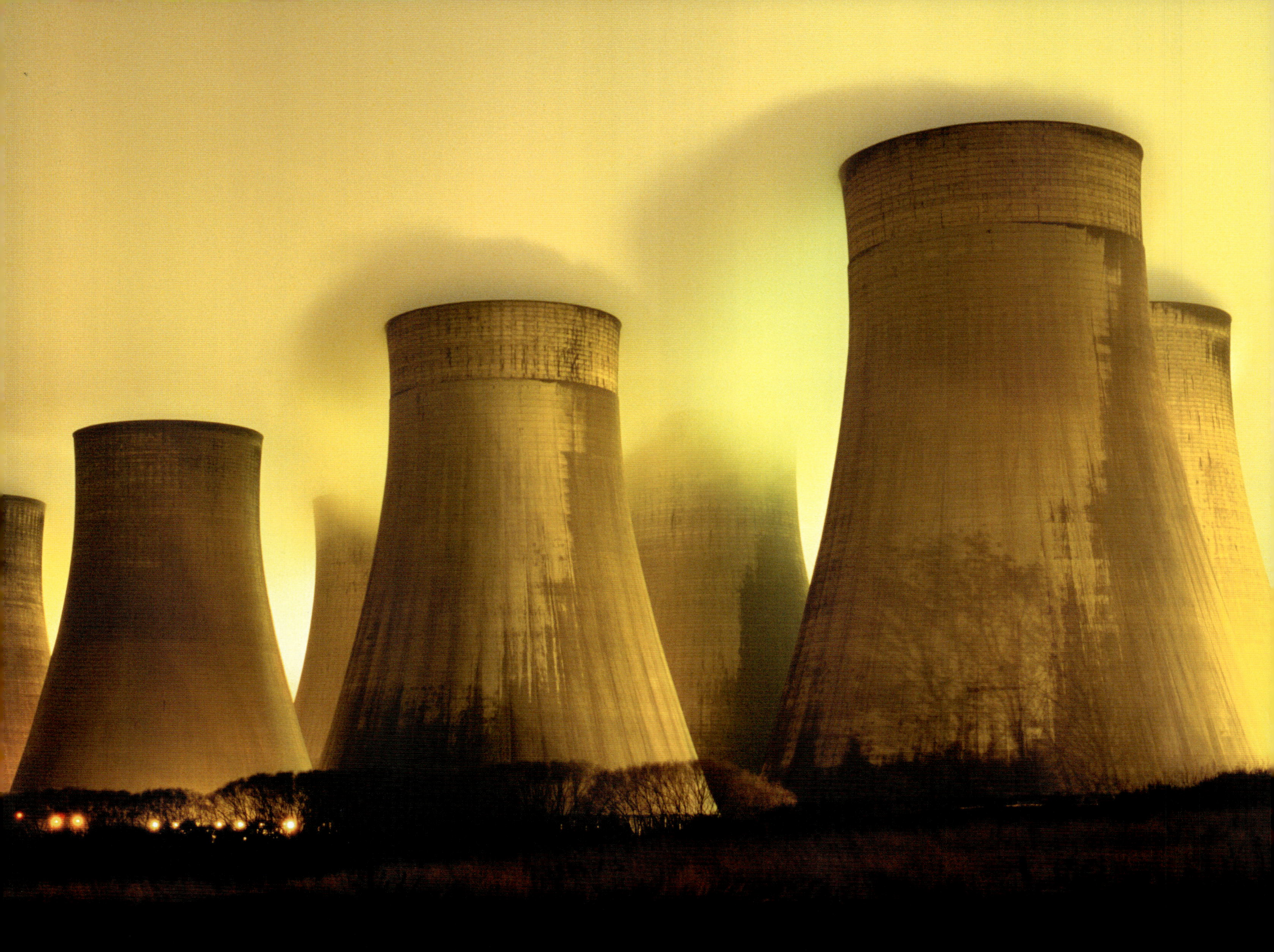
Lord Man tamed rivers, split atoms

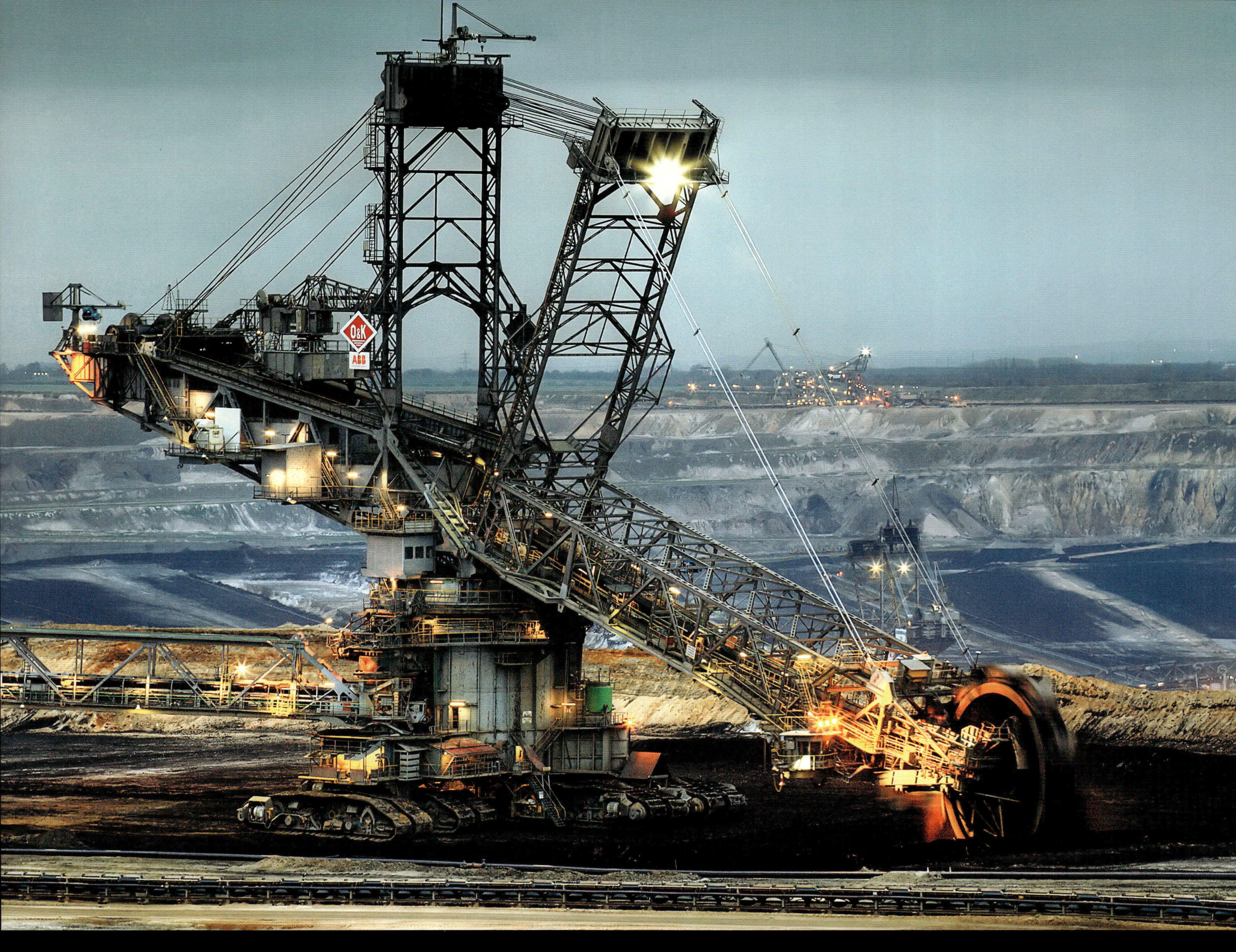

decapitated mountains, and stabbed the Earth everywhere he thought a vein of fuel lay hidden.

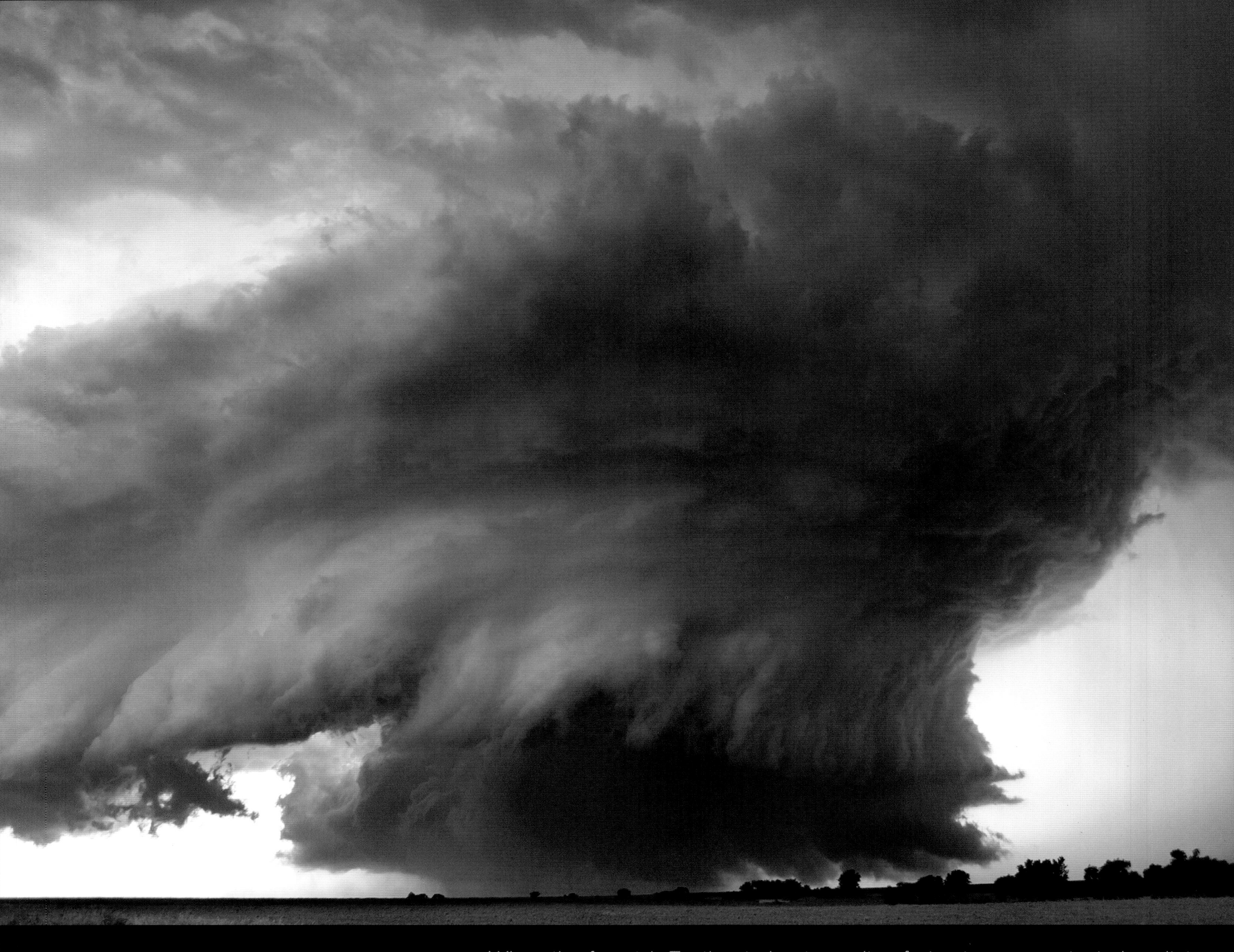

When the feverish Earth cried out, sending furies to communicate her distress.

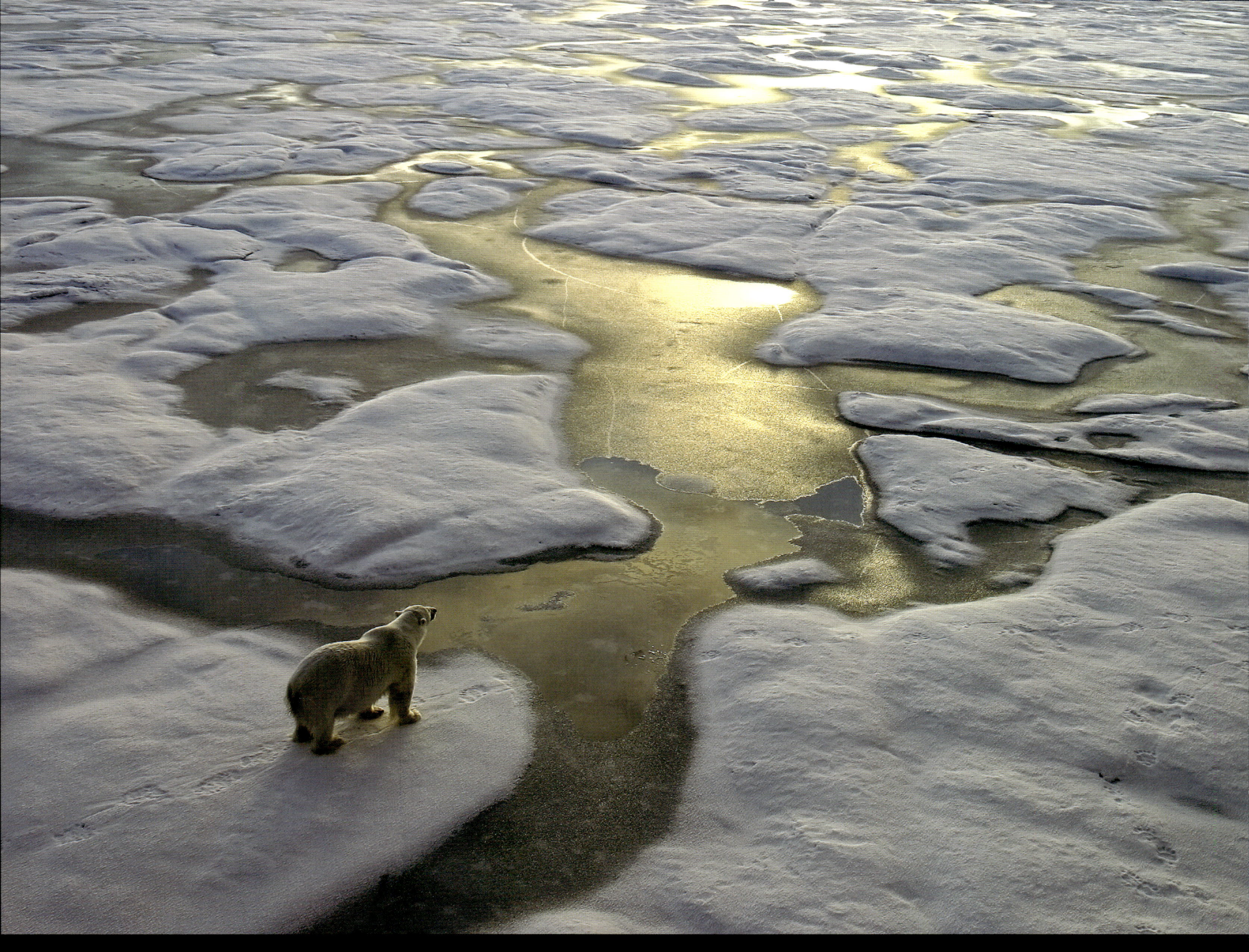

Lord Man ignored the growing sickness until it could no longer be denied.

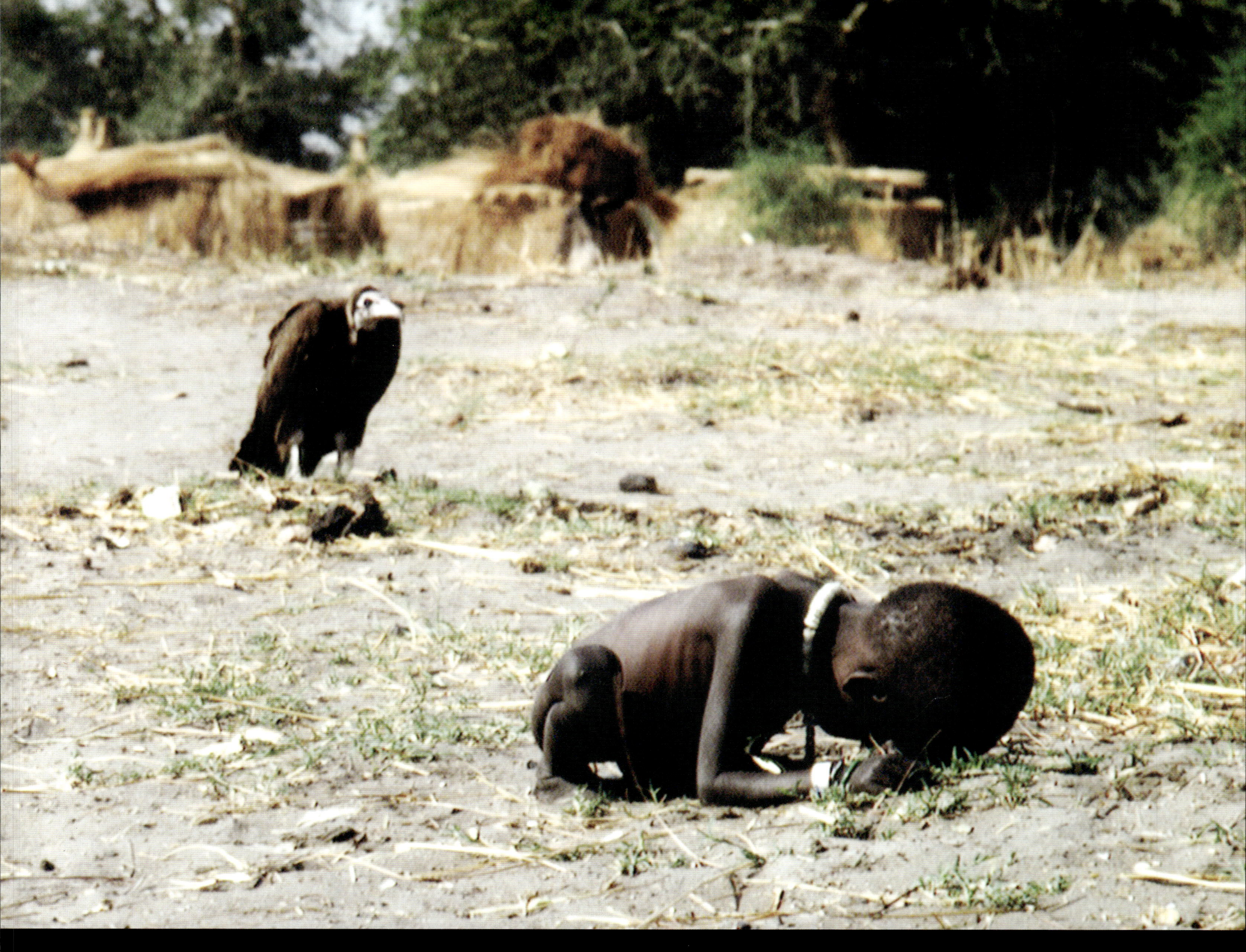

Slowly, the scales began to fall from his eyes when he saw famine ravage the land.

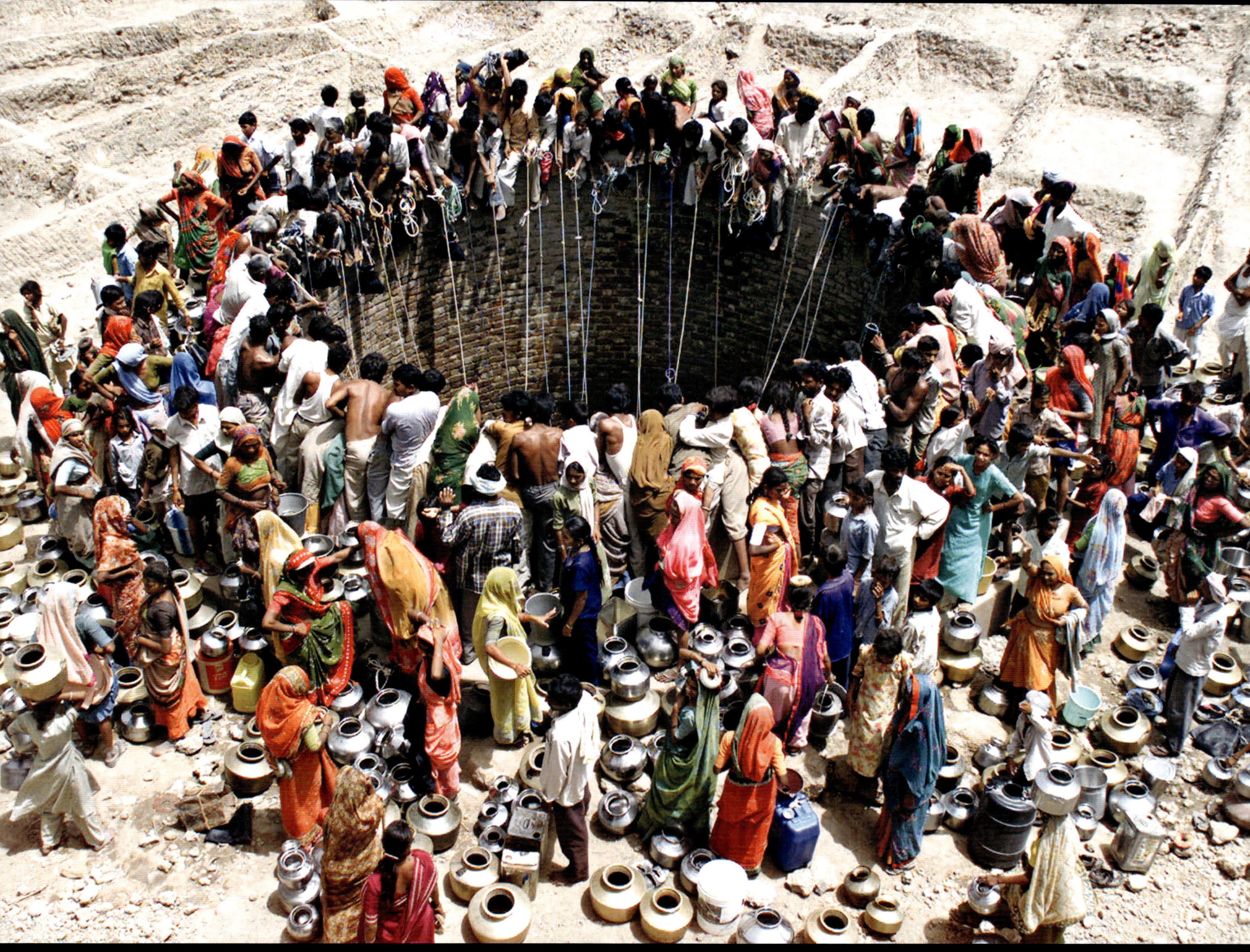

When he saw precious sources of freshwater disappear.

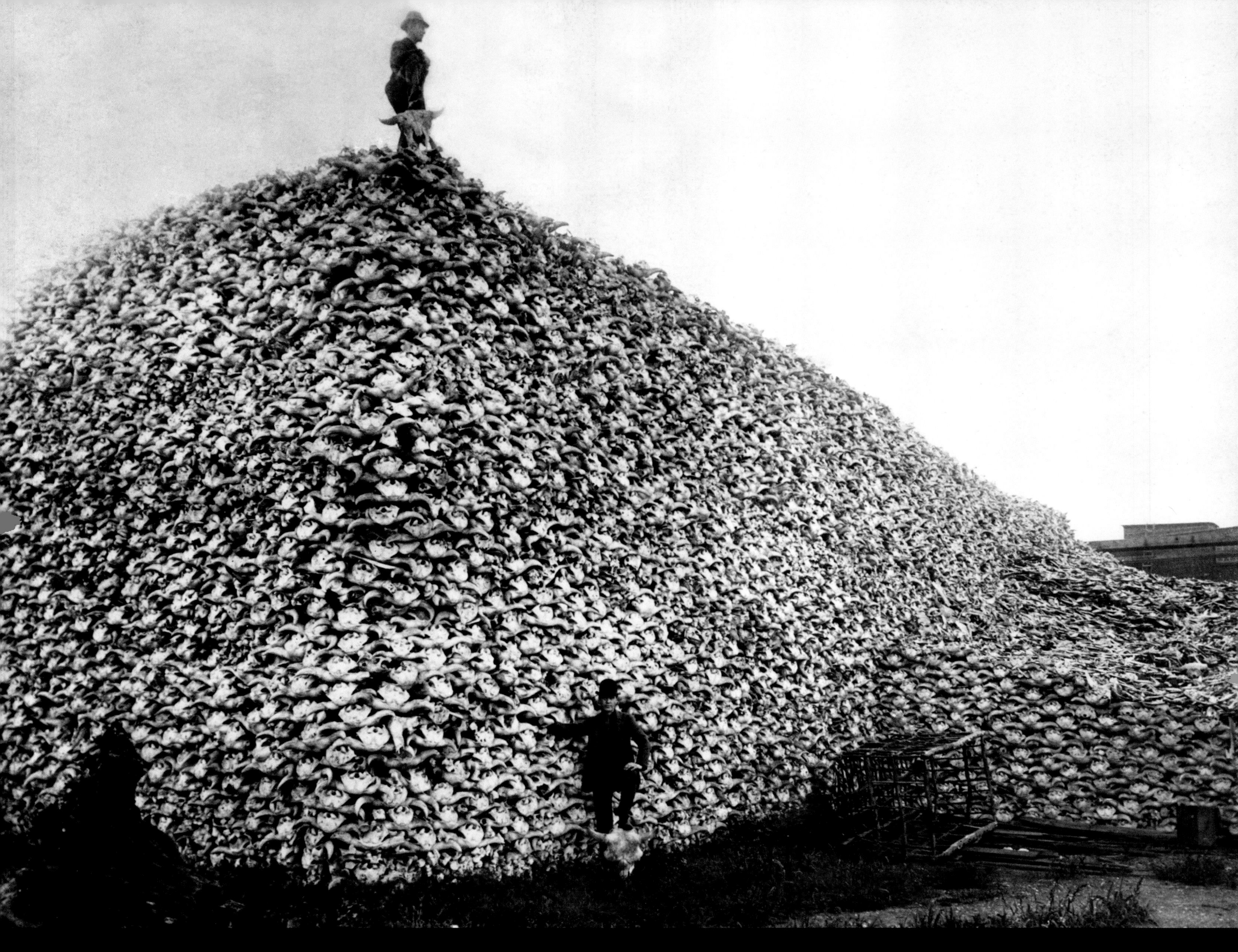
When the longing that gnawed on his spirit made him recall so many creatures that had passed into oblivion

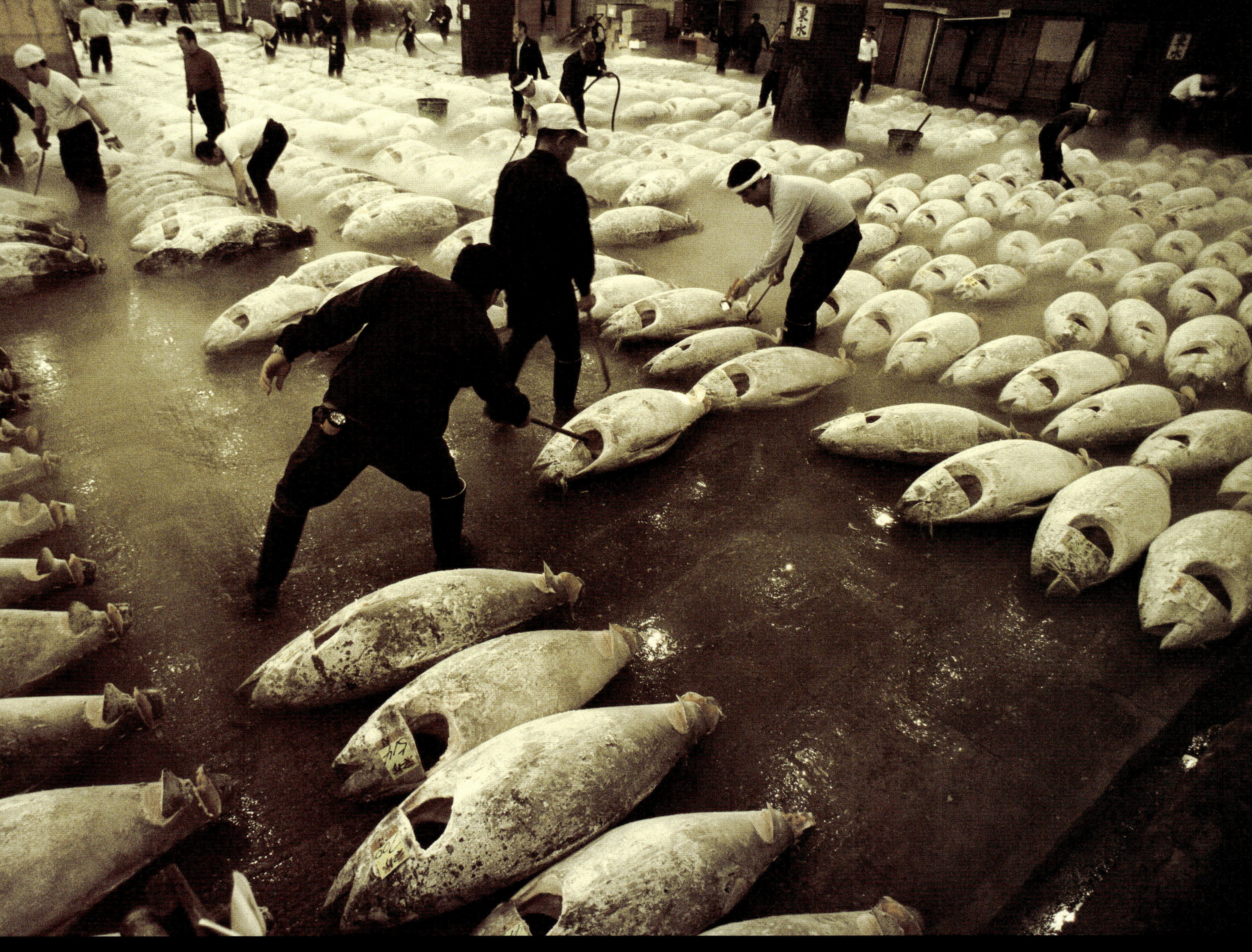

Seeing the effects of his hubris, he began to wonder if his empire was secure.

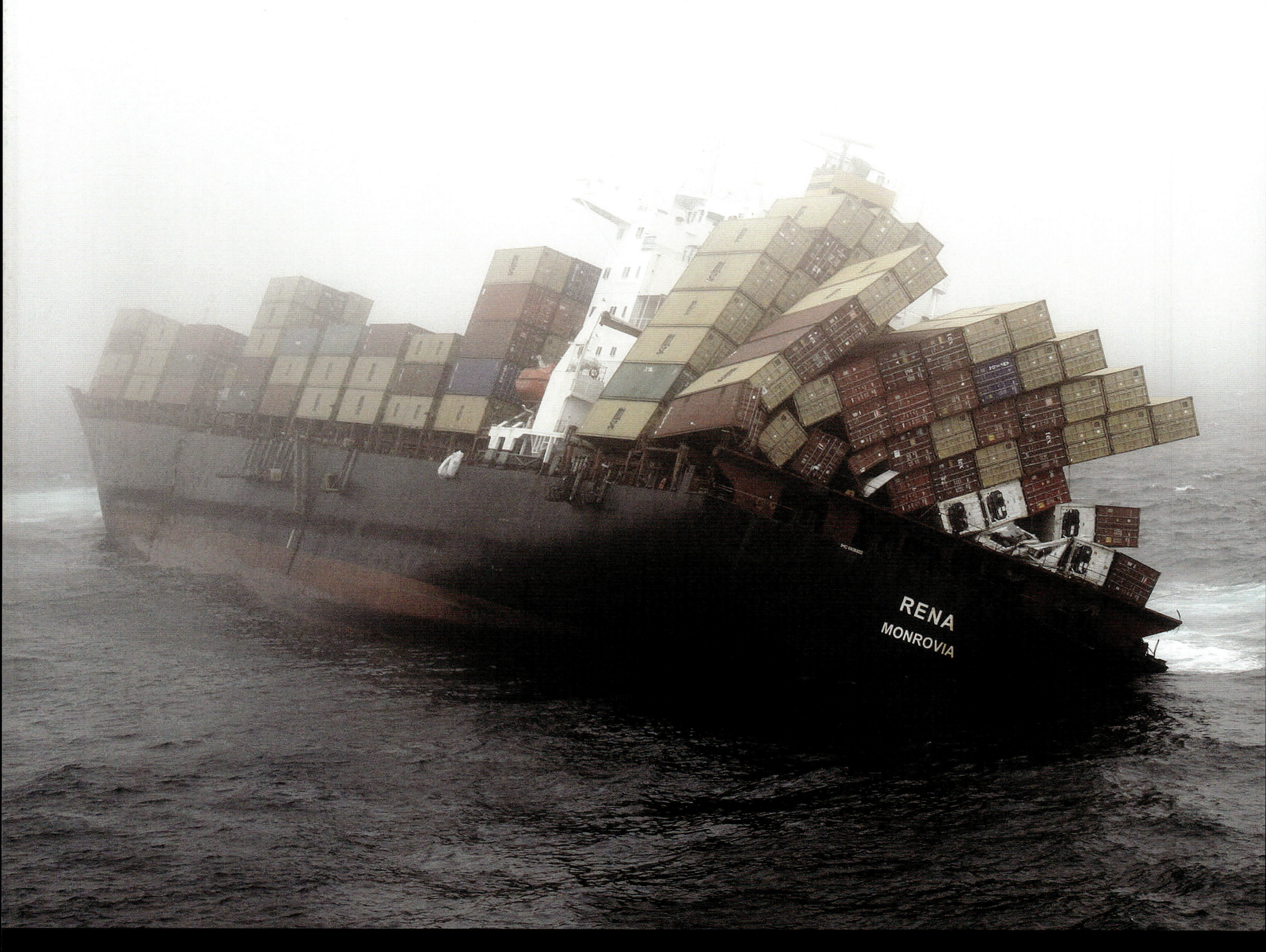

His delusion weakened just enough to reveal the choice before him:

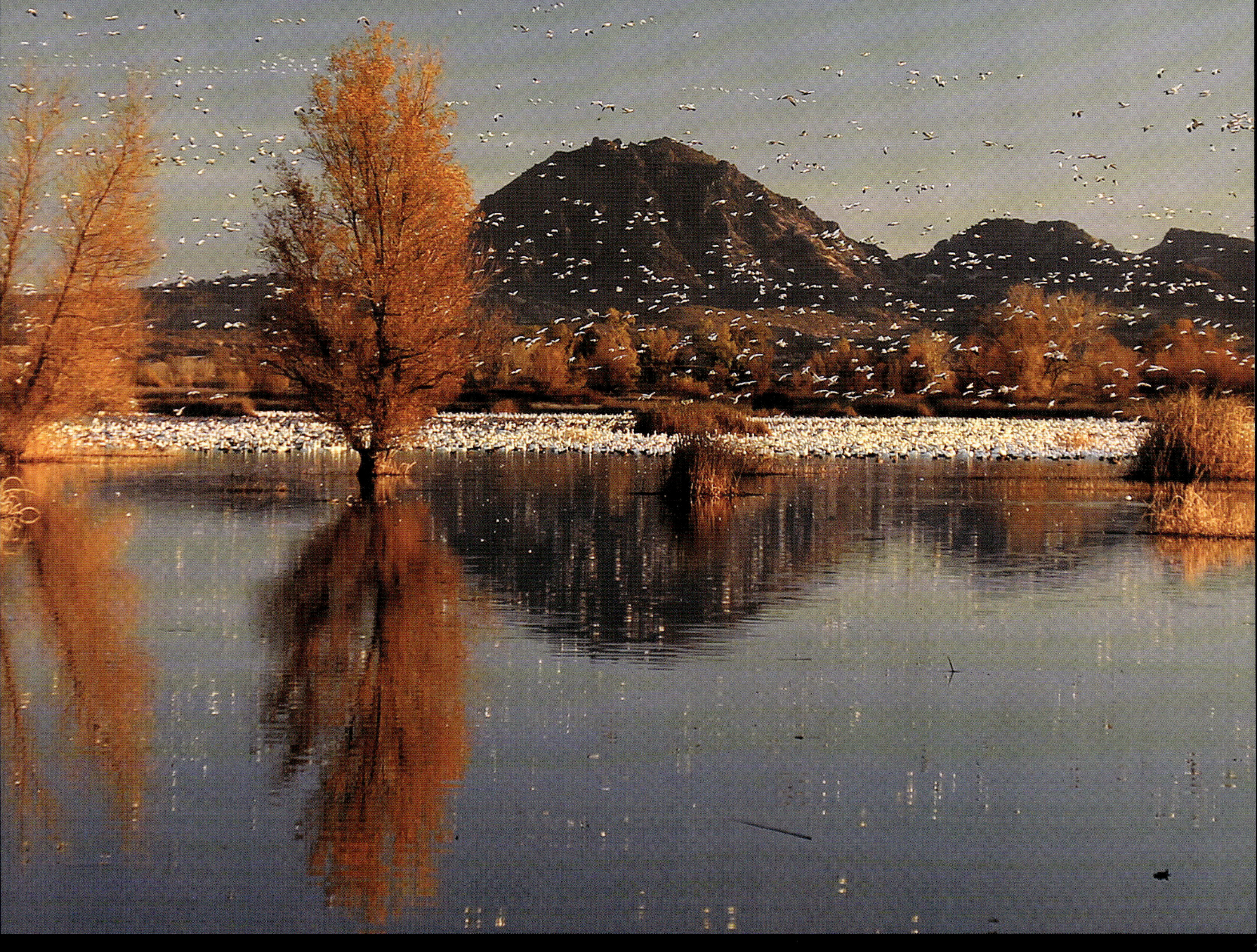

Two paths, one leading to an abundant Earth, filled with birdsong;

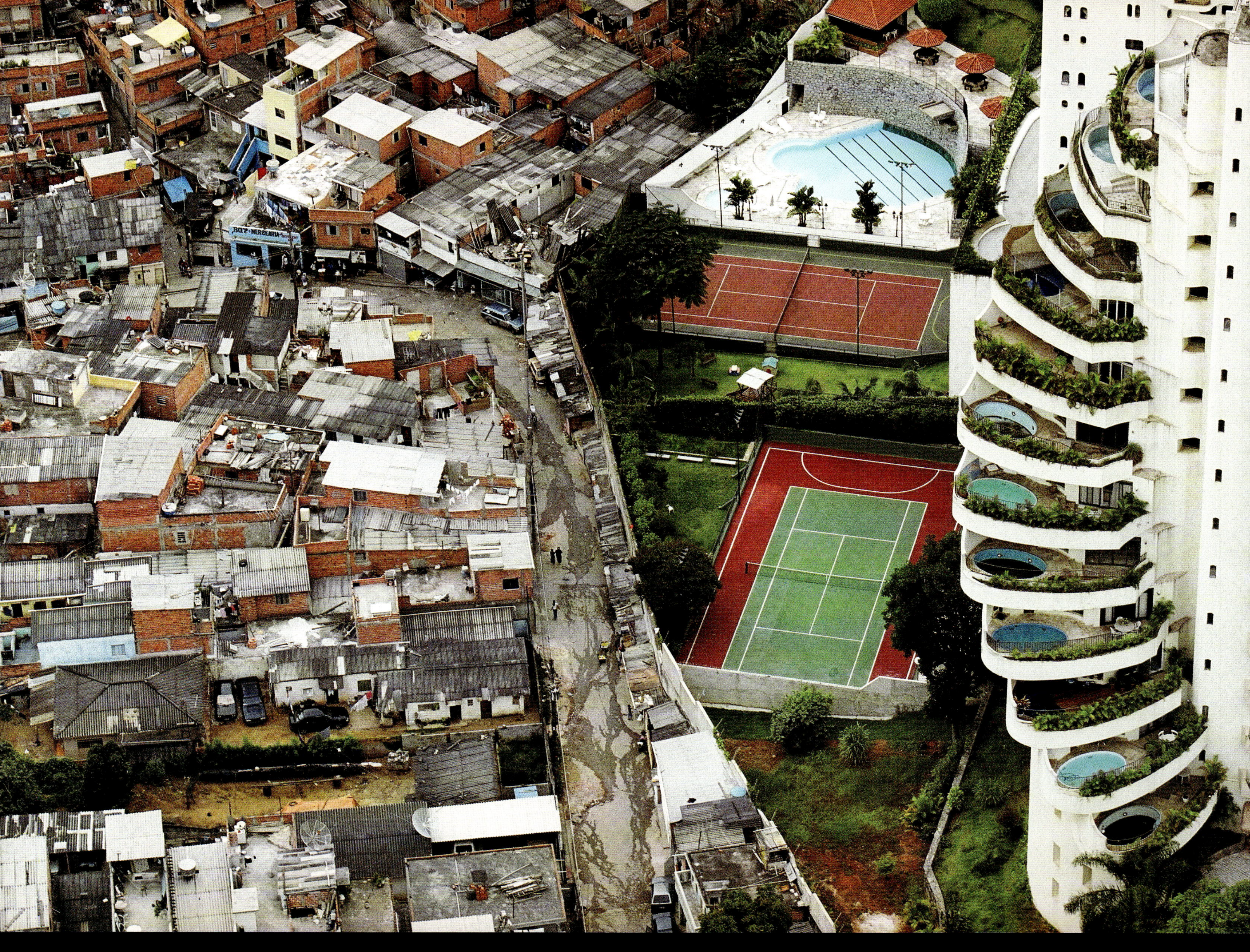

the other—the way of Growth—offered riches for some, misery for many, and ultimate destruction for all his tribe

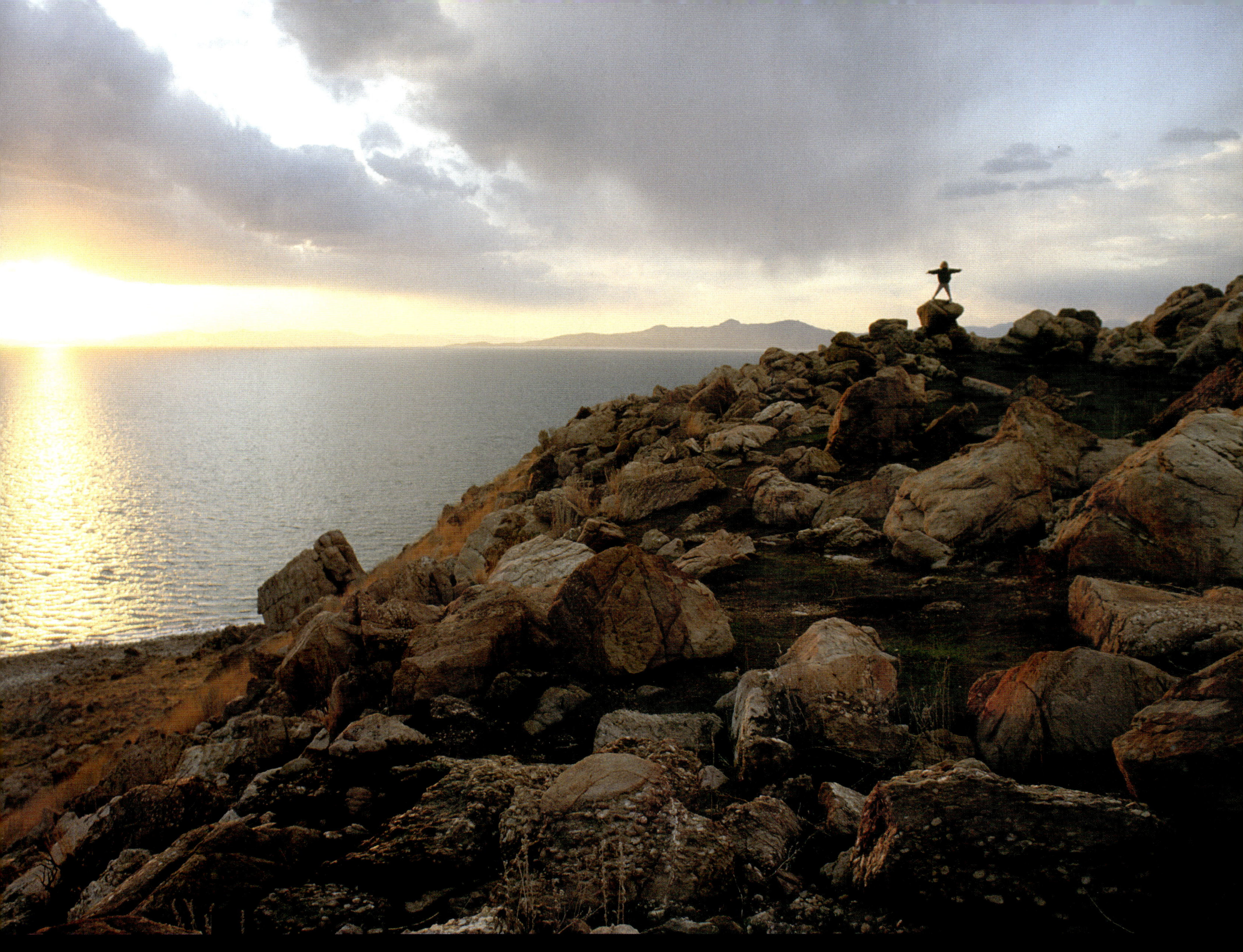

Would he restrain his numbers and rejoin the community of life as plain member and citizen?

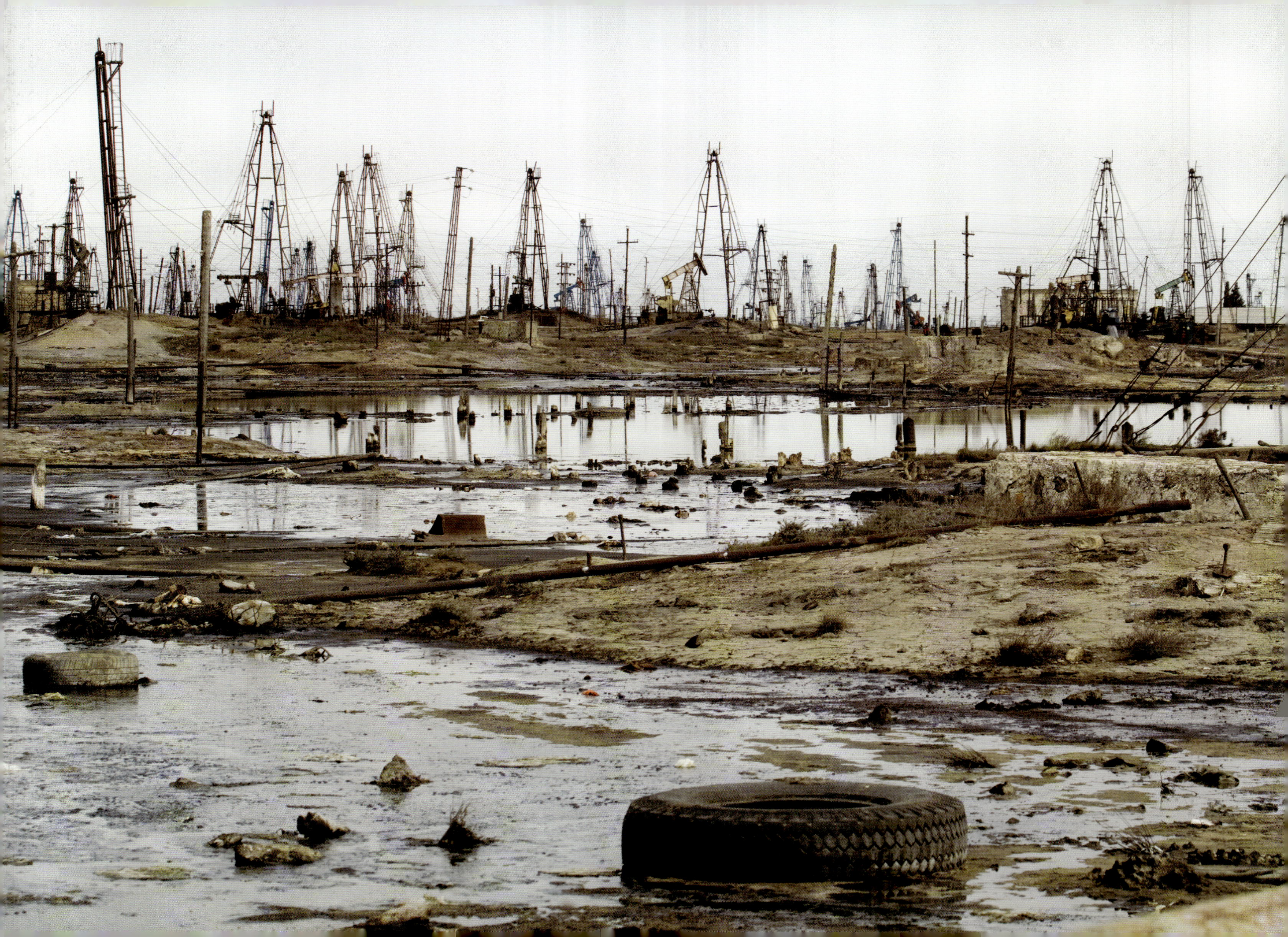

Or attempt to engineer all the Earth to his will, heeding only the call of *more*?

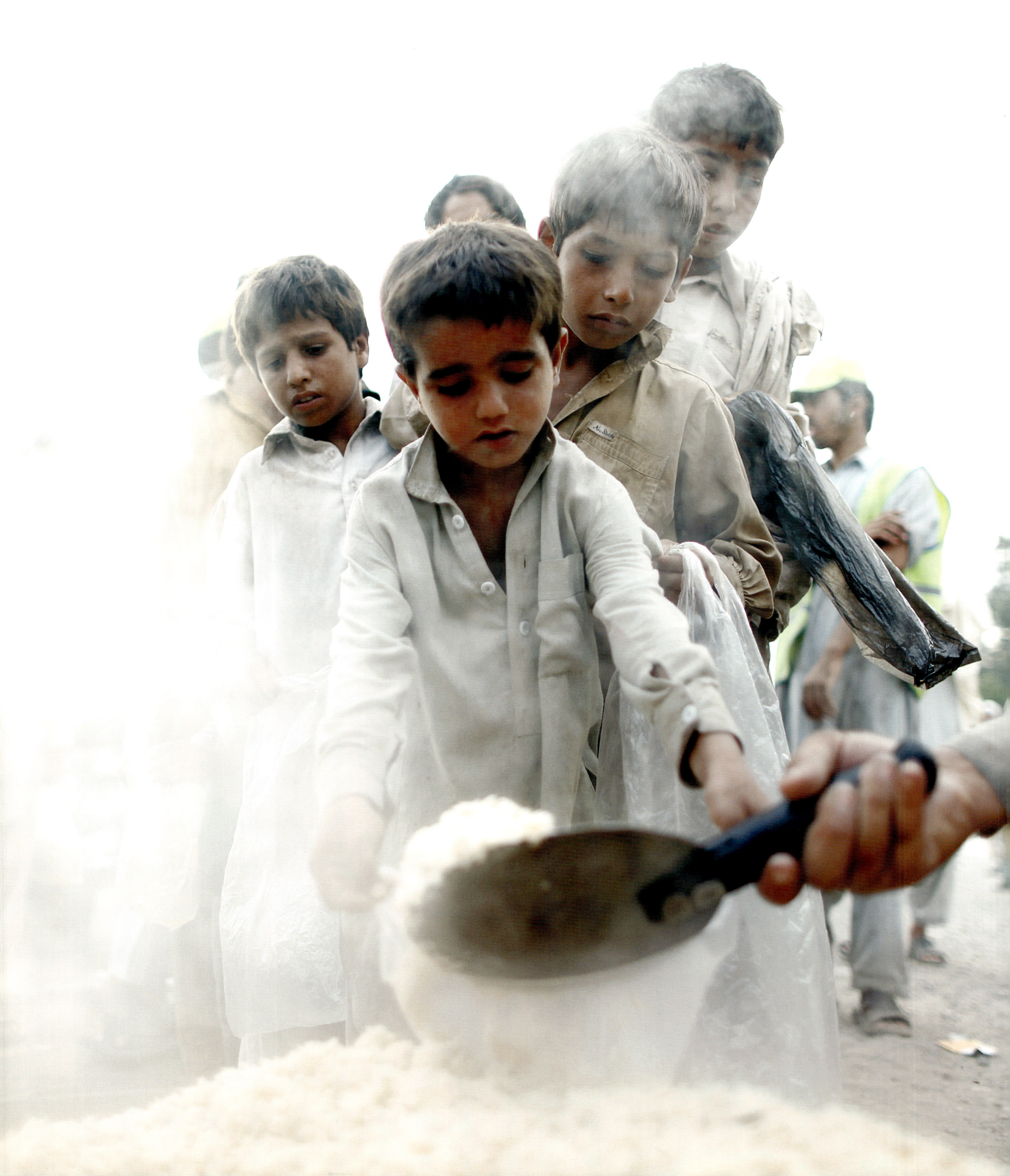

William Ryerson

Most conversations about population begin with statistics—demographic data, fertility rates in this or that region, the latest reports on malnutrition, deforestation, biodiversity loss, climate change, and so on. Such data, while useful, fails to generate mass concern about the fundamental issue affecting the future of the Earth.

In reality, every discussion about population involves people, the world that our children and grandchildren will live to see, and the health of the planet that supports all life. In my roles as president of Population Media Center and CEO of the Population Institute, I spend most of my time in developing countries, where many of my friends and acquaintances are educated and prospering. But I also know individuals who are homeless, unemployed, or hungry. The vast majority of people in these societies, regardless of their current status, do not enjoy a safety net. They live from day to day in hopes that their economic circumstances will improve. Abstract statistics on poverty are irrelevant to families struggling to secure the food, water, and resources needed to sustain a decent life.

Those who blithely dismiss the challenges posed by population growth like to say that we could physically squeeze 7 billion people into an area the size of Texas. They don't stop to consider the suffering already caused by overpopulation. The population debate is not about the maximum number of people that could be packed onto the planet. The crucial question is: How many people can the Earth sustain, at a reasonable standard of living, while leaving room for the diversity of life to flourish? There is no precise answer to this question, but the facts overwhelmingly support one conclusion: We cannot go on the way we are going. We are already doing severe and irreparable harm to the planet. Something has to give.

If we cannot live sustainably with 7.2 billion people, how are we going to support billions more by the end of this century? The United Nations' latest "medium-variant" projection indicates that we could have 10.9 billion people by 2100, but that may be an underestimation. Fertility rates in many parts of the world are not falling as fast as previously anticipated. In some countries, both developed and developing, fertility rates are actually on the rise again. In 2014 the global total fertility rate—the average number of children born to each woman during her lifetime—was 2.5. If this rate were to remain unchanged, demographers suggest that we could have 27 billion people on the planet by the end of the century. Given our limited inheritance of soil, water, and arable land, sustaining a global population of that size is not even remotely possible.

As vividly illustrated by this book, human numbers and activity are already destroying the planet's ecological integrity—running roughshod over myriad other species. But it's not just the environmental damage we're inflicting that should concern us. Equally appalling is how our actions threaten humanity's future prospects. We have passed a crucial tipping point. Our quest for greater and greater material prosperity is now impoverishing future generations. The Global Footprint Network estimates that humans already use 150 percent of the Earth's renewable capacity annually, and it estimates further that by 2030 we will need "two planets" to sustain us. Further growth simply deepens the crisis of ecological "overshoot" as we draw down Earth's carrying capacity, and it comes at the direct expense of our own children and grandchildren. Is that any kind of way to behave?

If you care about people, you must care about what we are doing to the planet. If you care about what we are doing to the planet, you must also care about human numbers. Given a planet with infinite space and resources, population growth could, arguably, be a blessing. We do not live on such a planet. However, there was a time when the Earth's resources appeared boundless. Some people still adhere to that anachronistic belief. If nothing else, the photographs in this book should shatter that illusion.

Many of us today do recognize that the Earth and its resources are limited, yet too many people still cling to the notion that modern science and technology will enable us to defy physical limits. In the Middle Ages, alchemists sought in vain for a "philosopher's stone" that would convert base metals into gold. They never succeeded. Why? Because what they were looking for did not, and could not, exist, because its existence would have violated the physical laws governing the universe.

Modern-day alchemists are trying to find ways of sustaining perpetual growth in a finite and increasingly resource-constrained world, searching for a scientific or technological breakthrough that will enable us to keep growing indefinitely. Like the philosopher's stone, it does not exist. Our faith in breakthroughs is misplaced, as amply demonstrated by the past three hundred years of scientific and technological advances that have accelerated, not slowed, the degradation of the natural world. Even if scientists were to develop a relatively cheap, abundant, and clean form of energy that powered continuous economic and population growth, it would only accelerate the rate at which humanity is destroying the ecological systems that make the planet habitable. In the meantime, while we are waiting for magical breakthroughs, we are in a headlong race to extract and consume fossil fuels at whatever the cost to the Earth. Scientists warn that we will fry the planet if we burn all the world's known reserves of coal, gas, and oil, but that concern has not slowed the relentless exploration for more fossil fuels. An ever-expanding human population and rising demand for products and services make humanity's hunger for fossil fuels utterly insatiable.

Some cling to the notion that we can achieve sustainability by reducing consumption in the overdeveloped world. As meritorious as that idea may be, it has no critical mass of support. A growing number of political leaders are supporting the idea of "greener" or "smarter" growth, but there is not a single politician of significant stature in the world calling for slower economic growth, no growth (a steady-state economy), or de-growth. Yes, there are individuals who are trying to reduce their carbon and ecological "footprints," but their numbers, for the moment, are dwarfed by the growing numbers of people who want to expand their ecological footprint through additional consumption.

Much of humanity, of course, desperately needs a larger share of Earth's resources. More than 2 billion people in the world live on less than $2 per day. Nearly a billion people go to bed hungry every night. About half the people in the world do not have access to toilets or other means of modern sanitation. I do not know of anyone who would deny these people a better quality of life, but if world population continues to grow as currently projected, many, if not most, of these people will never have their most basic needs realized, let alone fulfill their aspirations. The world is not that bountiful. I wish it were, but it is not.

If we have any hope of bringing about a genuine balance between what humans demand of nature and what nature can reasonably provide for humanity, we must take crucial steps. Starting with the first step, we must devote more resources to preventing unplanned pregnancies through expanded access to contraceptives. Women everywhere should have the means to time, limit, or space their pregnancies. But greater access to contraceptives alone will not suffice. In those countries where population growth is most rapid today, girls and women lack reproductive choice; they live in traditionally male-dominated societies where large families are still the norm. Large-family norms, misinformation, and cultural barriers account for most decisions to not use contraception. If we do not enable girls to remain in school and delay marriage until adulthood, provide accurate information, and empower women in the developing world, then we will have failed countless individuals. Moreover, in the face of this humanitarian failure, fertility rate declines may continue only very slowly, or not at all—but certainly not fast enough to avoid the kind of human suffering that results when countries are overpopulated.

In many parts of the world, child marriage is still prevalent. It is estimated that some 14,000 girls become child brides each day. In some areas, particularly poor rural communities, parents require their daughters—who have not yet reached puberty—to wed men who are twice or three times their age. Child brides do not enjoy reproductive choice in any meaningful sense. Most are condemned, if they survive childbirth, to having many children, and their families are condemned, in turn, to a life of continued poverty and deprivation.

As important as it is to reduce unplanned pregnancies in the developing world, it is just as important to do so in the overdeveloped world, where the per capita consumption of resources is so much greater. Nearly half of all pregnancies in the United States are unplanned, and while America's teenage pregnancy rate is declining, it remains the highest among industrialized nations. Shockingly, several state legislatures in recent years have slashed support for family planning, resulting in dozens of clinics having to either close their doors or limit services.

These individual and community-level actions, in aggregate, have global consequences. The leading scientists of the world are concerned that we are approaching as many as nine planetary tipping points, which, if surpassed, would cause irreparable harm to the environment and the well-being of future generations. We have already crossed one boundary in terms of greenhouse gas emissions; the climate is changing, and we have already inflicted incalculable harm on posterity as a result.

Because of population growth and changing diets, the world's demand for food is projected to rise by 70–100 percent over the next forty years. No one knows how we will meet that demand. Cultivated farmlands already occupy a land mass the size of South America, and ranchlands used for livestock grazing occupy a land mass the size of Africa. There's very little arable land left; most of it is in the form of tropical forests, which if cut down to expand agriculture would accelerate biodiversity loss and further complicate efforts to rein in greenhouse gas emissions.

Water scarcity in many parts of the world has already reached crisis proportions. Demand for water is expected to outstrip supply by 40 percent within the next twenty years. As one research organization put it, we will need the equivalent of 20

Nile Rivers—which we do not have—to meet demand. By 2030, an estimated 3.9 billion people, nearly half the world's population, will be living in areas of high water stress.

We live today in a "Catch 22" world, where addressing one urgent problem often exacerbates another. If we double food production to feed a growing world, we expand greenhouse gas emissions. If we discover and exploit more fossil fuels, we fry the planet. If we reduce our water consumption, we curtail our food production. If we grow the world's middle class, we increase the pressure on Earth's natural ecosystems.

There is, however, one exception to this predicament, and that concerns population. Viewed from almost any angle, addressing population is a win-win proposition. By empowering girls and women in the developing world and expanding family planning services and information everywhere, we produce a world of good: Fertility rates decline; maternal and child health improve; food security increases; poverty decreases; education and economic opportunities expand; and degradation of the environment is curtailed.

In discussions about family planning and its many benefits, the health of nature is often an afterthought. Far too often it is overlooked entirely. We tend to see the well-being of people as somehow distinct from the well-being of the Earth. Some even see the environment as being in "competition" with humans. The obvious truth, although unacknowledged by some, is that we are not separate or distinct from nature. Our hopes and our fate are inextricably linked to the fate of the natural world. We are part of a complex web of interdependent life, and our welfare depends upon the health of the whole. When life took hold on this planet it produced millions of species that have lived and evolved and produced both wondrous beauty and diversity. We modern humans are both products and beneficiaries of that evolutionary process.

We are, however, acting as ungrateful beneficiaries. Scientists tell us that we are exterminating our fellow plant and animal species at a rate that is a hundred or even a thousand times faster than the natural rate of extinction. Leading biologists now warn that human numbers and activity are triggering the "sixth mass extinction," the largest since the dinosaurs were wiped out 65 million years ago.

As a young man, after earning undergraduate and graduate degrees in biology with a specialization in ecology and evolution, my interest in moths and butterflies was so strong that I seriously considered becoming a lepidopterist. Many of the species that piqued my interest as a college student are now in danger of becoming extinct. Even the common *Danaus plexippus*, otherwise known as the monarch butterfly, is fast approaching endangered status. Its winter habitat in Mexico has shrunk dramatically. Biologists warn that herbicide use is decreasing the availability of milkweed plants, limiting a primary food source for monarchs and thus diminishing their numbers.

But it's not just the monarch butterfly that is imperiled. Every year there are fresh reports about the senseless slaughter of elephants, rhinos, lions, tigers, and other "megafauna." Some of their population decline is attributable to poachers seeking to harvest ivory or other body parts, but much of the dramatic decline has been caused by an ever-increasing loss of habitat. Many of these animals live in areas, like sub-Saharan Africa, where human fertility rates equate to a doubling of the human population every thirty or forty years.

In my college days, we were taught that, since the end of the last Ice Age about 12,000 years ago, humans have been living in the Holocene Epoch, but our impact upon the planet and its environment has become so great that some geologists today suggest we change the epoch's name to the "Anthropocene," or "Age of Man." To most scientists, that development is a frightening prospect; it means that we are changing the planet—for the worse—on a global scale. Some scientists, though a distinct minority, insist that we can "manage" this change; that we can strike a balance with nature that will allow us to feed, clothe, and meet the economic aspirations of an additional 3 or 4 billion

people moving forward. As well illustrated by the photographs in this book, that line of thought reflects the worst kind of wishful thinking. Our 7.2 billion on the planet are already doing grave harm to the biosphere. Several decades ago, a cartoon character named "Pogo" made popular the oft-quoted saying: "We have met the enemy and he is us." We might say this today in regards to the challenge the world faces, only it's not a comic matter. If we are to reduce severe poverty, defeat hunger, and bring about a sustainable world, we must achieve change on a global scale, beyond just our consumption habits, and that change must begin with us. This conviction led me to work for the Population Institute more than forty years ago and subsequently spurred me to establish the Population Media Center fifteen years ago.

Despite the widespread belief that simply making contraceptives more widely available can stabilize world population, there are other reasons why women in the developing world end up having more children than they might otherwise desire, as revealed through the Demographic and Health Surveys supported by USAID (United States Agency for International Development). In reality, many of these women have no reproductive choice. Child brides often have nothing to say about how many of their own children they will have or when. Some women abstain from using contraceptives because of misinformation or blatant lies about the possible side effects or risks of using modern methods of contraception. Still other women have more children than they want because of fatalism, or religious teachings, or insistent in-laws who want more grandchildren.

At the Population Media Center (PMC) we create long-running serial dramas (soap operas) that serve to educate women about their contraceptive choices. Using a methodology based upon the "social learning" theories of the great Stanford psychologist Albert Bandura and the programs developed by Miguel Sabido, former vice president of Televisa in Mexico, we work with in-country teams to develop long-running dramas, generally broadcast via radio, that provide positive role models for men and women in the developing world. Our listening audiences learn from popular "transitional" characters who are torn between good and bad influences. In the process the characters and the listening audience discover the benefits of family planning and small family norms.

Our programs also address the deeper social stereotypes that demean women and effectively deny them reproductive choice. When girls are educated, women are empowered, and gender equity is achieved, women tend to have smaller, healthier families. By changing attitudes and behavior toward girls and women we can improve their lives, the well-being of their families, and prospects for the planet and our posterity.

At PMC we also use the "Sabido methodology," as it is now known, to achieve positive social change with respect to environmental conservation. In Rwanda, our radio programs have encouraged farmers to participate in reforestation programs aimed at restoring natural habitats and preserving the land for future generations. Similarly, we can use our programs to alter harmful consumption patterns or promote sustainable agricultural practices. The potential is enormous.

WHILE THE OBSTACLES before humanity are real, we should be careful not to overestimate the difficulty of following the path of the United Nations' lowest population projections, which show a possible global stabilization as soon as the year 2050. Achieving this stabilization is a challenge, but it is far from an insurmountable one. The United Nations estimates that it would cost an additional $3.5 billion per year to provide contraceptive information and services to the more than 220 million women in the developing world who want to avoid a pregnancy but who are not using a modern method of contraception. (This is less than 4 percent of what Americans spend on beer each year.) That's a very small price to pay for a more sustainable world. Combine that investment with efforts through entertainment mass media and other means to change attitudes and behavior toward girls and women in the developing world, and we can stabilize world population at 8.3 billion and then begin a gradual reduction in the total number of humans on the planet as soon as 2050.

If we can hew to the United Nations' low-variant demographic projection, by 2100 global population would be back down to 6.7 billion—more than 4 billion fewer than can be expected in the business-as-usual, medium-variant projection of the human population trajectory. Such numbers may seem incomprehensible but the reality is that these two possible futures—one of 6 billion versus one of 10 billion humans to feed, clothe, educate, and employ—is the difference between a world of scarcity and nightmarish suffering for much of humanity and a world in which it may be possible to balance the needs of people and nature. Put another way, a population difference of 4 billion—the result of either staying complacent or working hard to share family planning tools and information around the globe—is 46 percent more than the current combined populations of North America, Central America, South America, Oceania, Europe, and Africa (roughly 2.7 billion)!

While I am deeply concerned about the future of humanity and the planet, I'm not a pessimist. It's not too late. There are things that we can do to achieve a harmonious world and many of the steps that are required, like PMC's radio programs, do not require an enormous investment of resources. Time, however, is beginning to run out.

Given the central role that population dynamics will play in determining the welfare of future generations, what the world needs today is a wake-up call. This book is that wake-up call. The photographs to follow are emotionally jarring. The thoughts expressed herein are not reassuring; they are deeply provocative. But that is the nature of wake-up calls. The way that human numbers and behavior are transforming the Earth, undermining its ability to support the human family and the rest of life, is apparent for all to see. The reality of this urgent moment calls us to think, to care, and to act.

DEMOGRAPHIC EXPLOSION

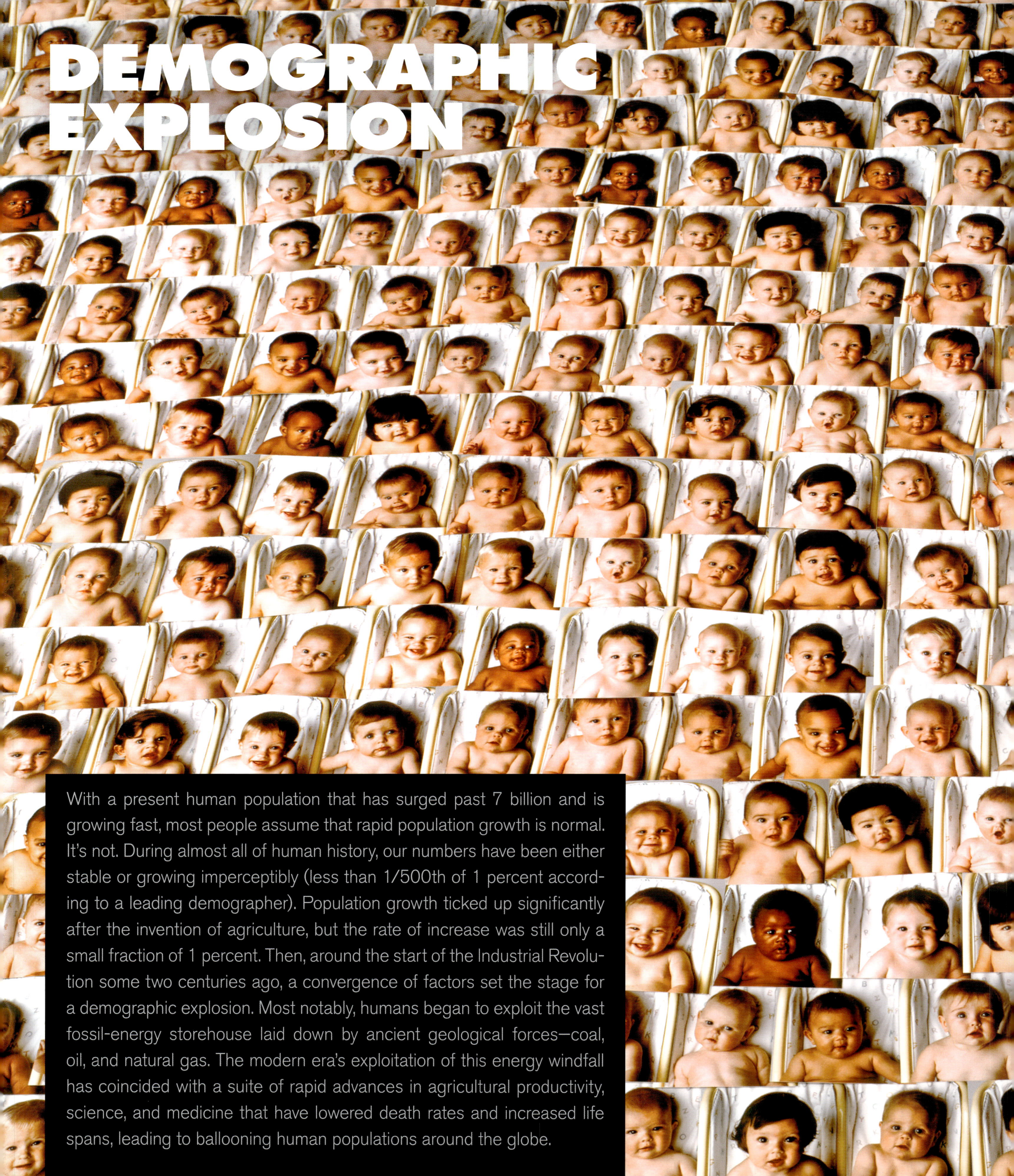

With a present human population that has surged past 7 billion and is growing fast, most people assume that rapid population growth is normal. It's not. During almost all of human history, our numbers have been either stable or growing imperceptibly (less than 1/500th of 1 percent according to a leading demographer). Population growth ticked up significantly after the invention of agriculture, but the rate of increase was still only a small fraction of 1 percent. Then, around the start of the Industrial Revolution some two centuries ago, a convergence of factors set the stage for a demographic explosion. Most notably, humans began to exploit the vast fossil-energy storehouse laid down by ancient geological forces—coal, oil, and natural gas. The modern era's exploitation of this energy windfall has coincided with a suite of rapid advances in agricultural productivity, science, and medicine that have lowered death rates and increased life spans, leading to ballooning human populations around the globe.

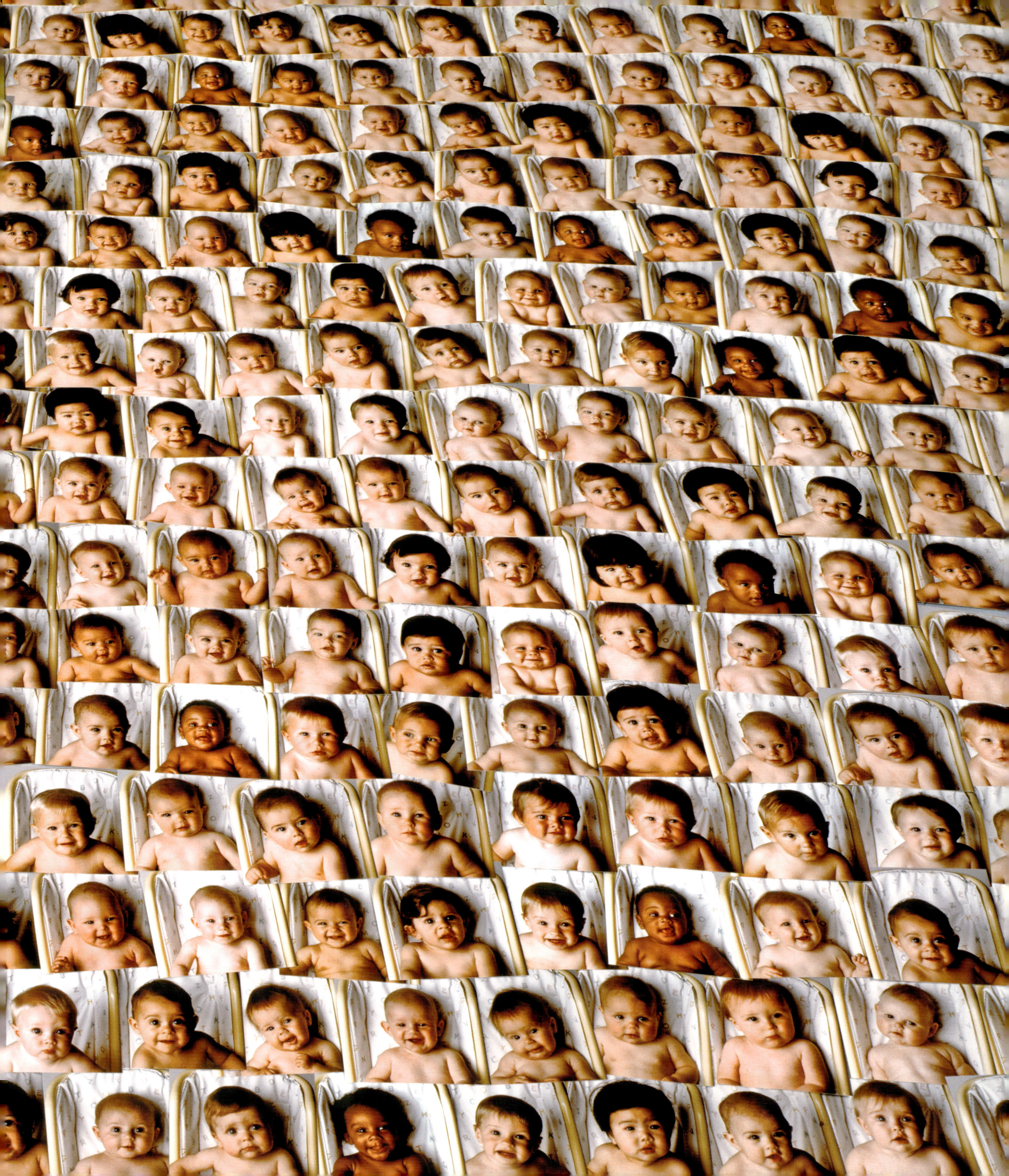

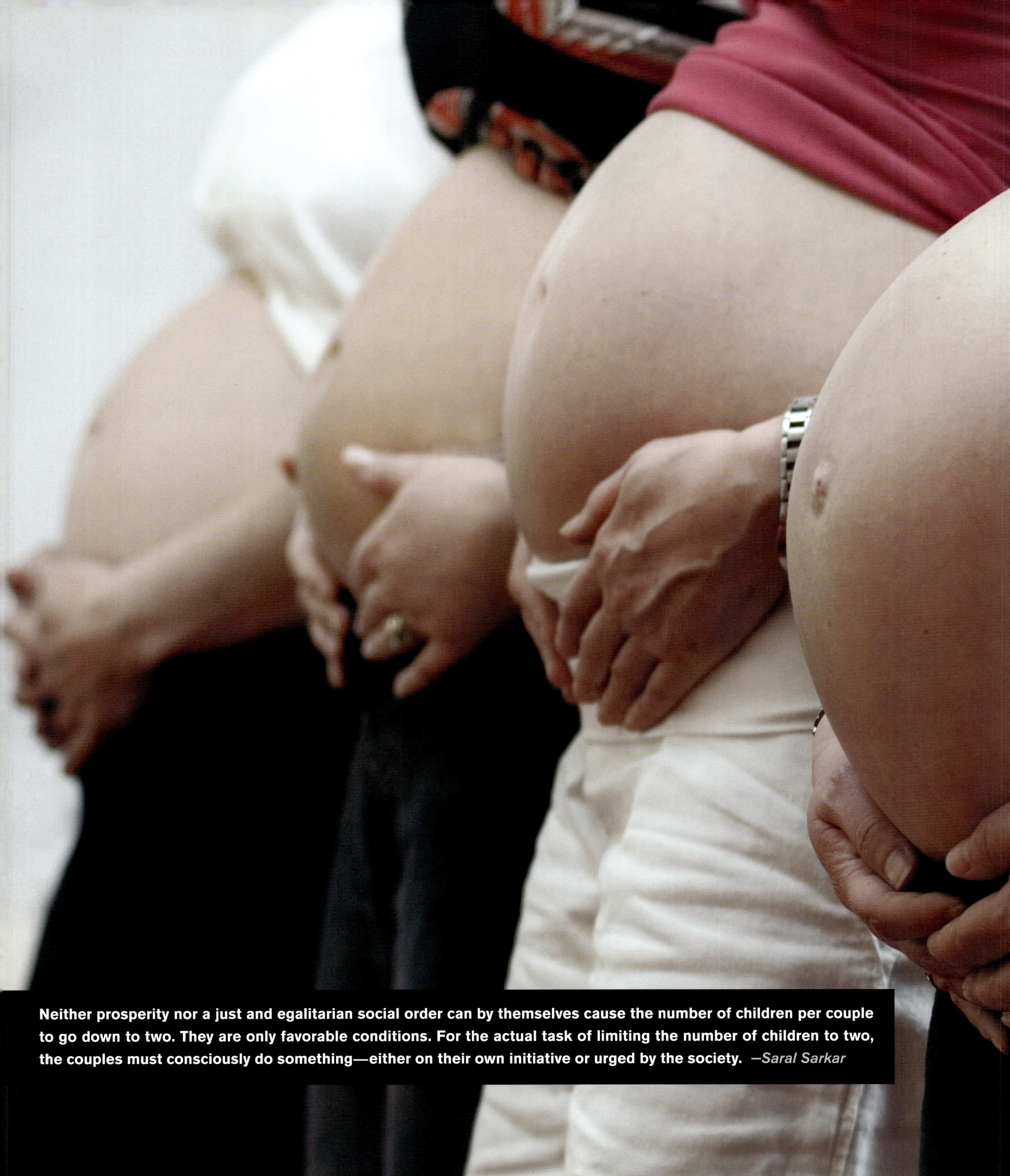

Neither prosperity nor a just and egalitarian social order can by themselves cause the number of children per couple to go down to two. They are only favorable conditions. For the actual task of limiting the number of children to two, the couples must consciously do something—either on their own initiative or urged by the society. —*Saral Sarkar*

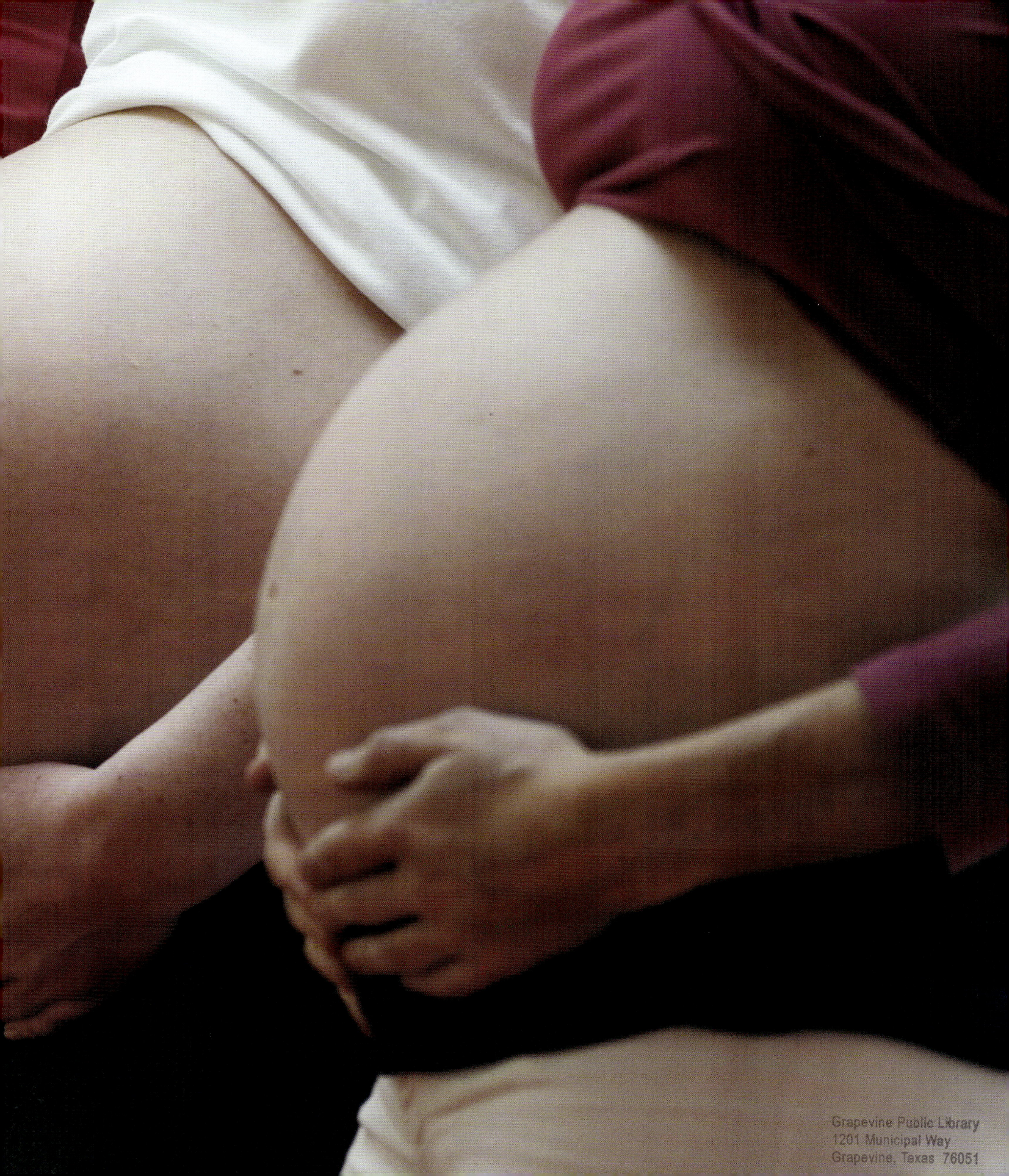
Grapevine Public Library
1201 Municipal Way
Grapevine, Texas 76051

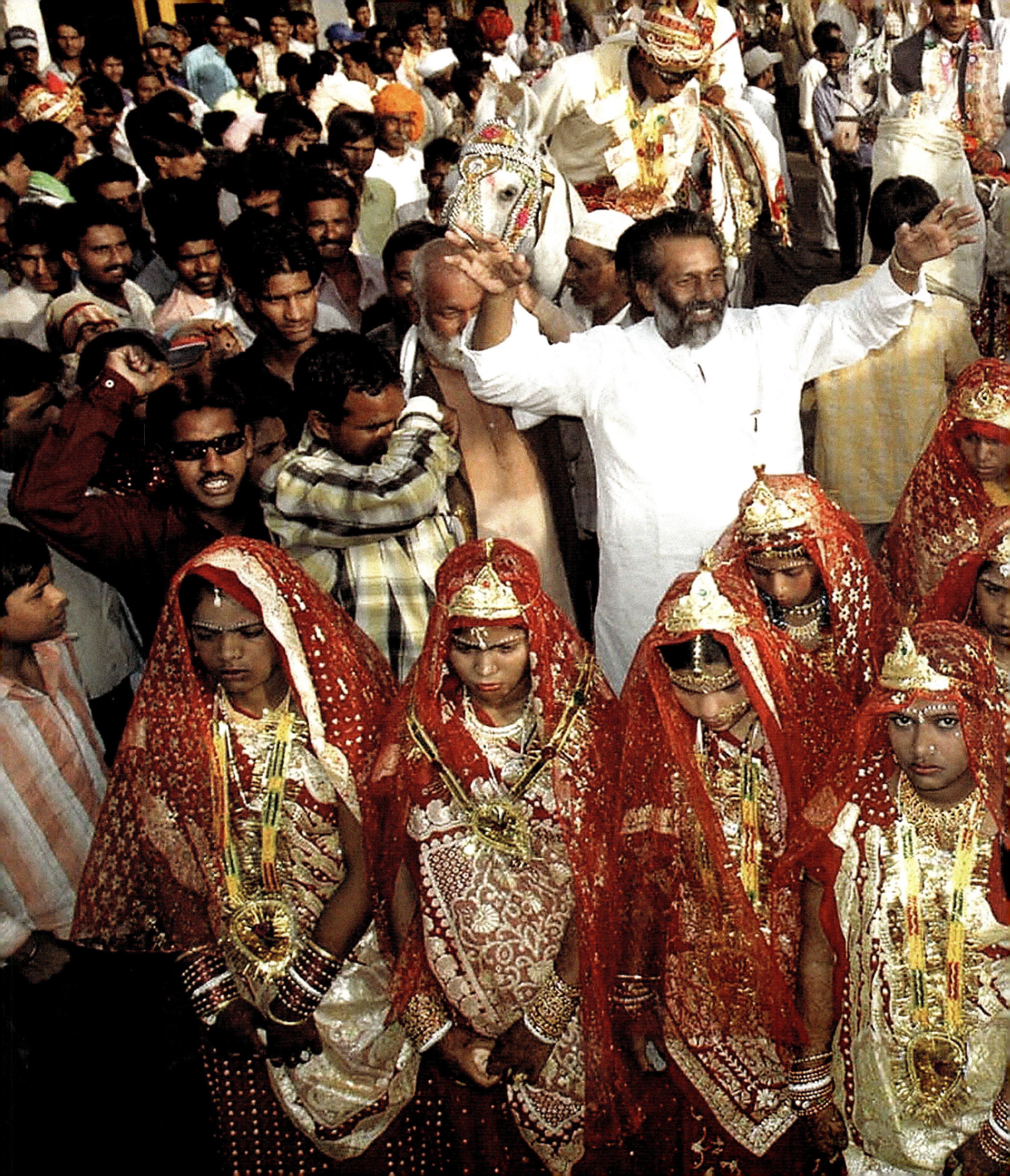

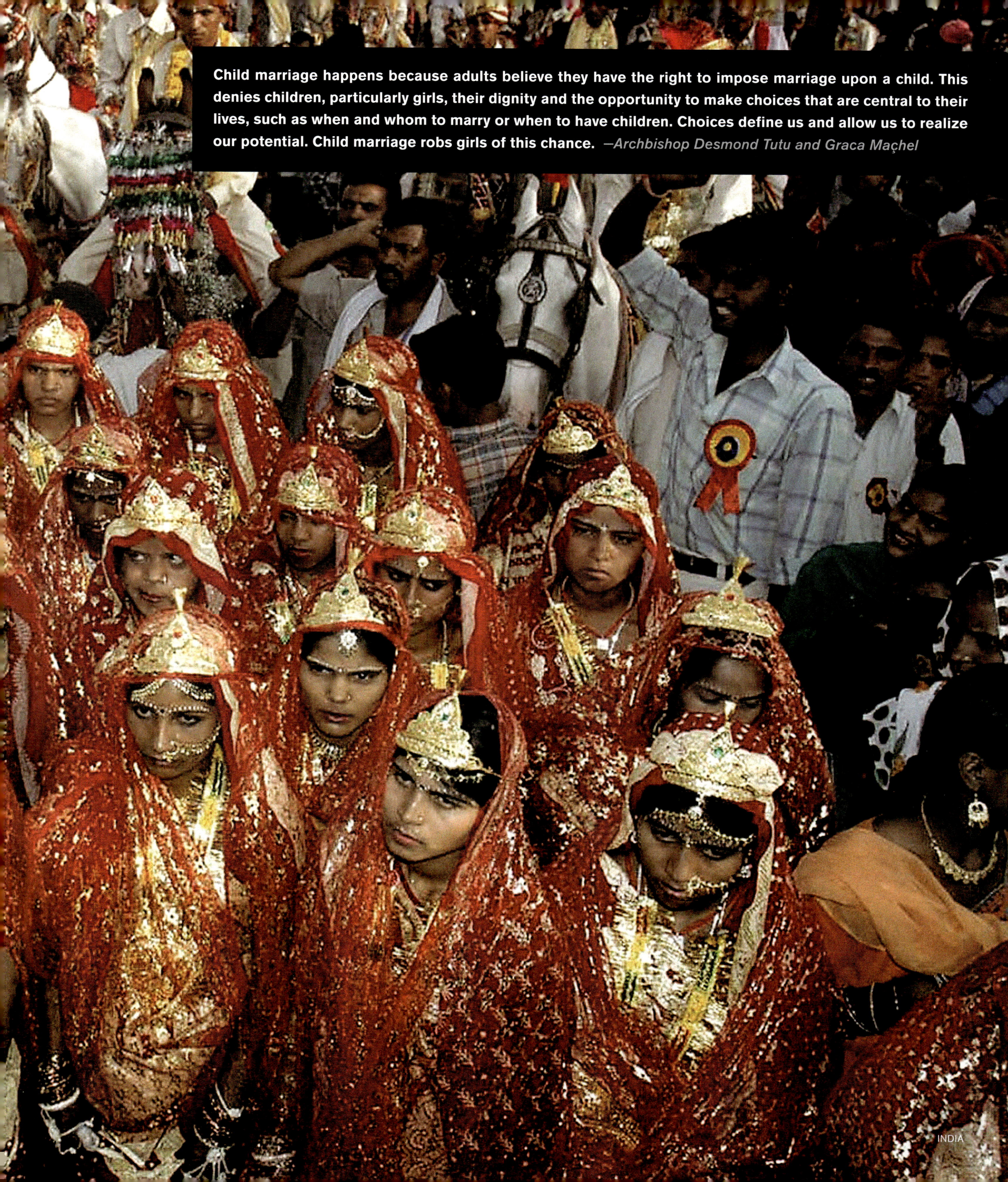

Child marriage happens because adults believe they have the right to impose marriage upon a child. This denies children, particularly girls, their dignity and the opportunity to make choices that are central to their lives, such as when and whom to marry or when to have children. Choices define us and allow us to realize our potential. Child marriage robs girls of this chance. —Archbishop Desmond Tutu and Graca Maçhel
INDIA

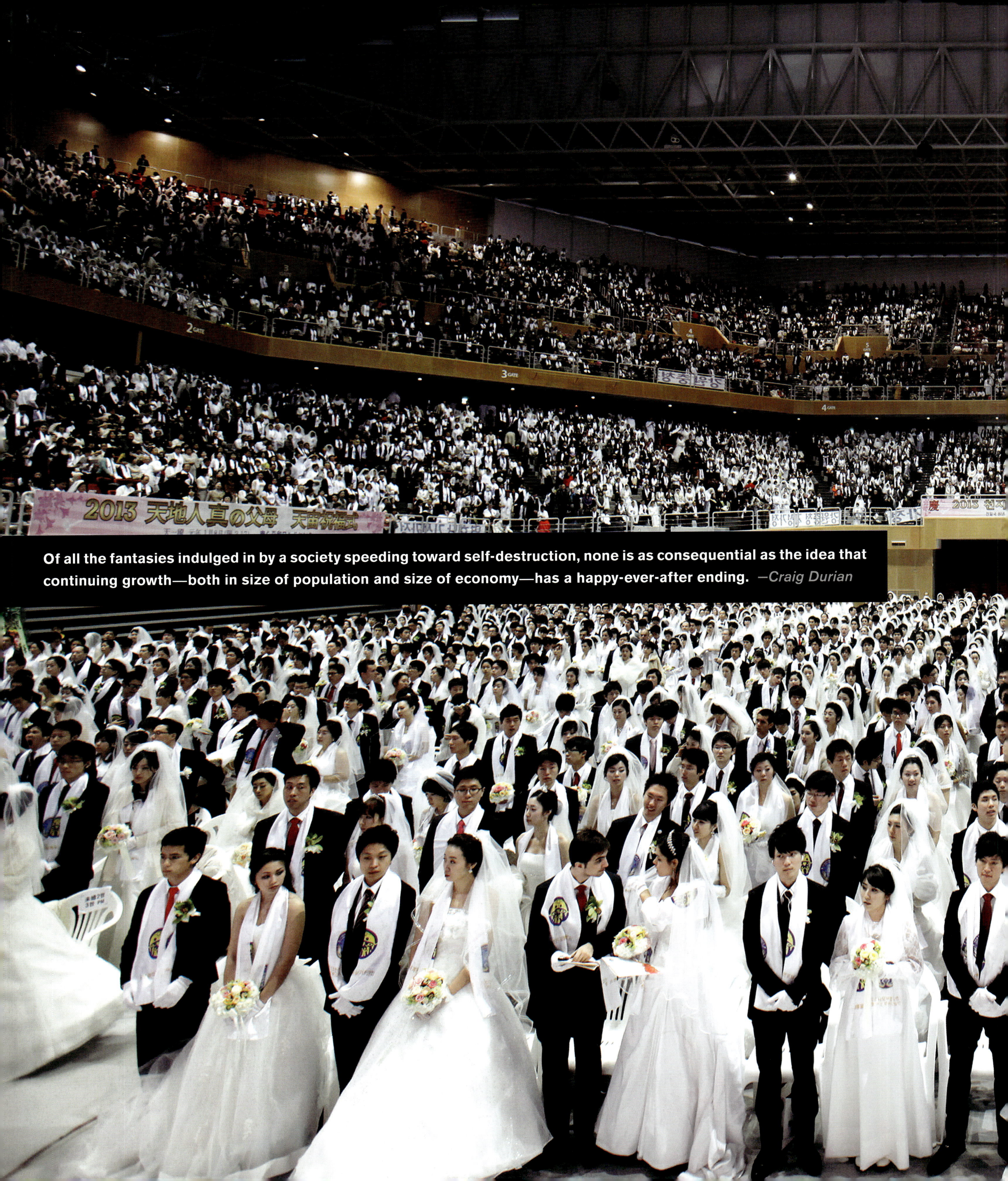

Of all the fantasies indulged in by a society speeding toward self-destruction, none is as consequential as the idea that continuing growth—both in size of population and size of economy—has a happy-ever-after ending. *—Craig Durian*

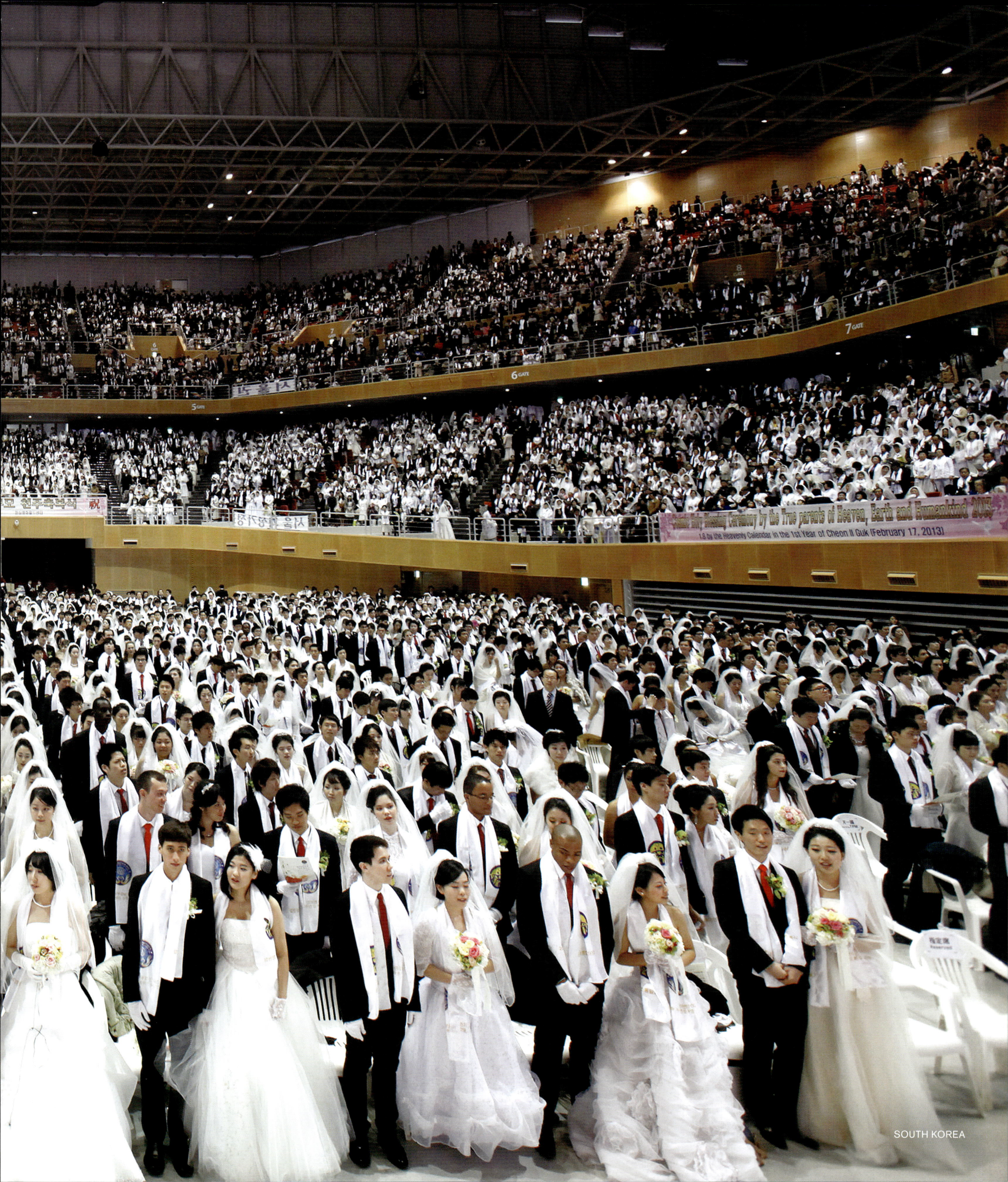
SOUTH KOREA

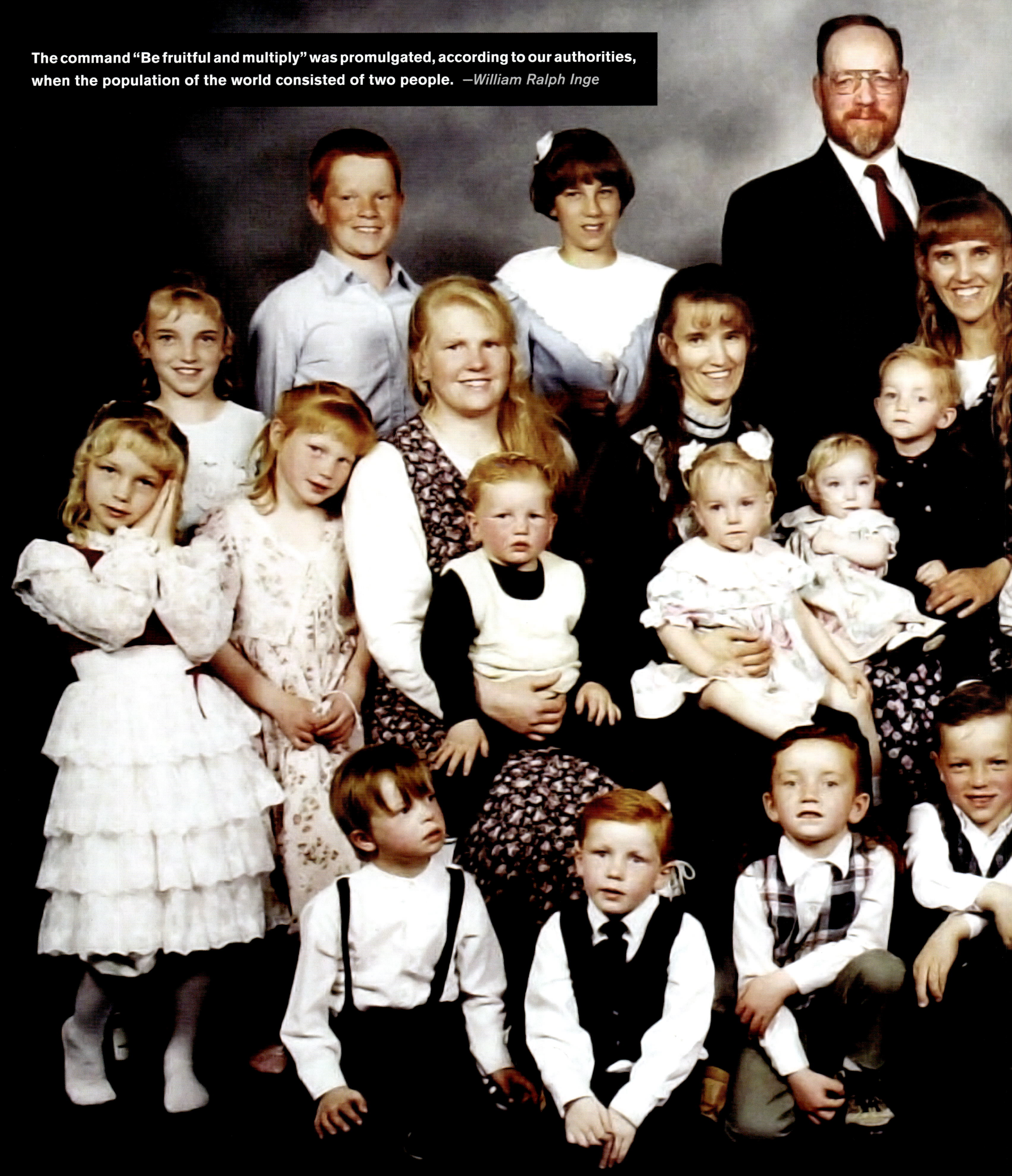

The command "Be fruitful and multiply" was promulgated, according to our authorities, when the population of the world consisted of two people. —William Ralph Inge

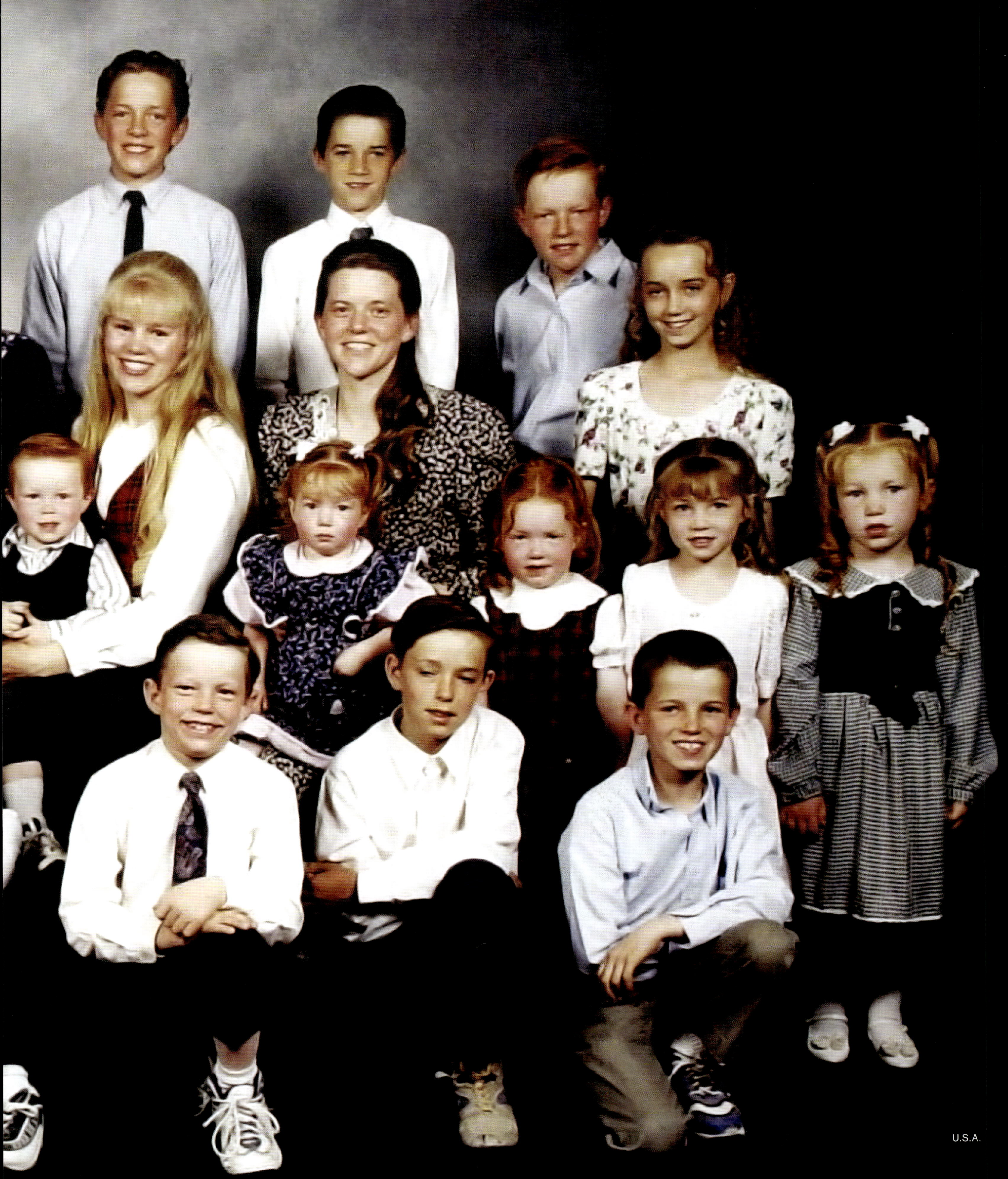

U.S.A.

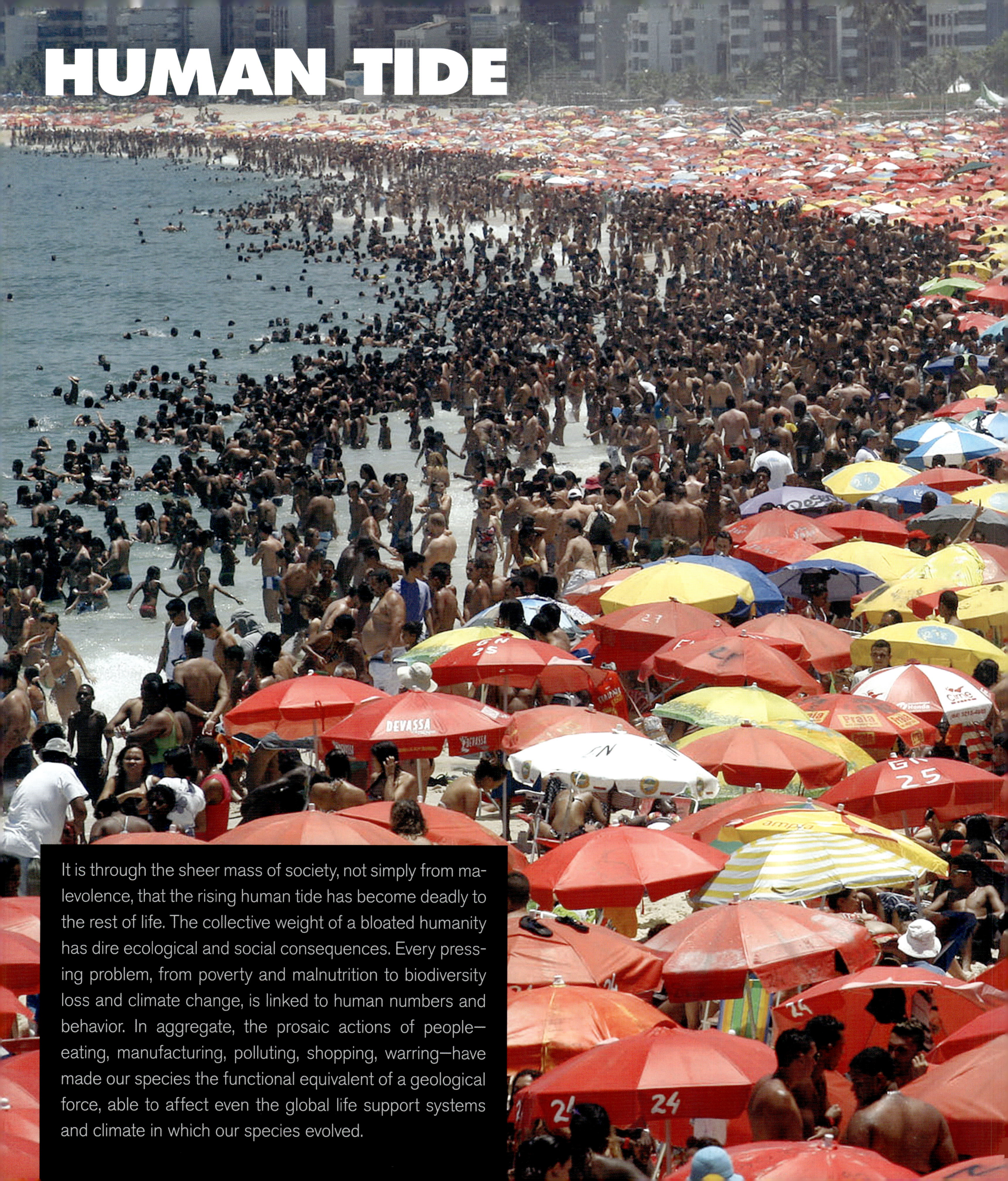

It is through the sheer mass of society, not simply from malevolence, that the rising human tide has become deadly to the rest of life. The collective weight of a bloated humanity has dire ecological and social consequences. Every pressing problem, from poverty and malnutrition to biodiversity loss and climate change, is linked to human numbers and behavior. In aggregate, the prosaic actions of people—eating, manufacturing, polluting, shopping, warring—have made our species the functional equivalent of a geological force, able to affect even the global life support systems and climate in which our species evolved.

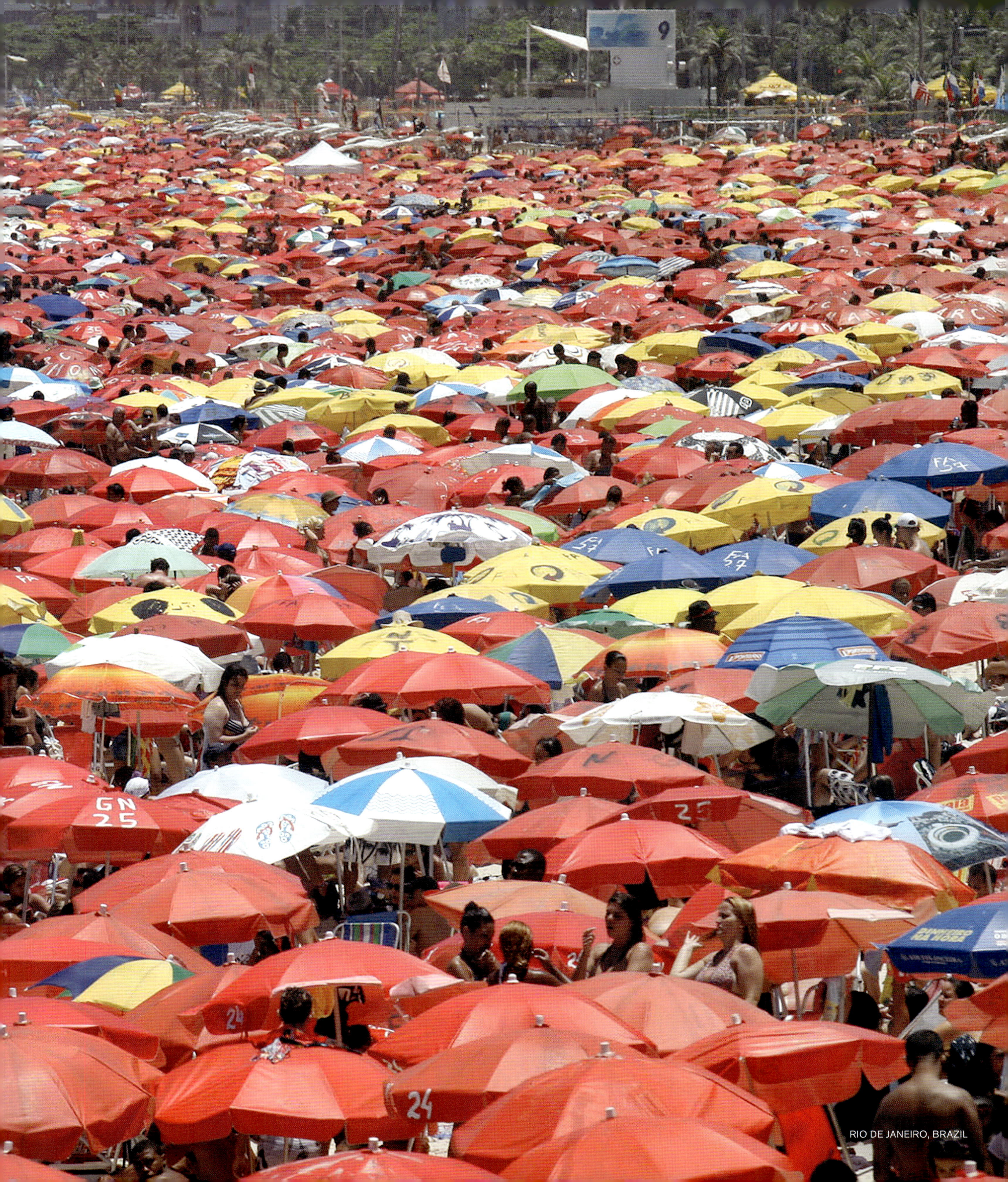
RIO DE JANEIRO, BRAZIL

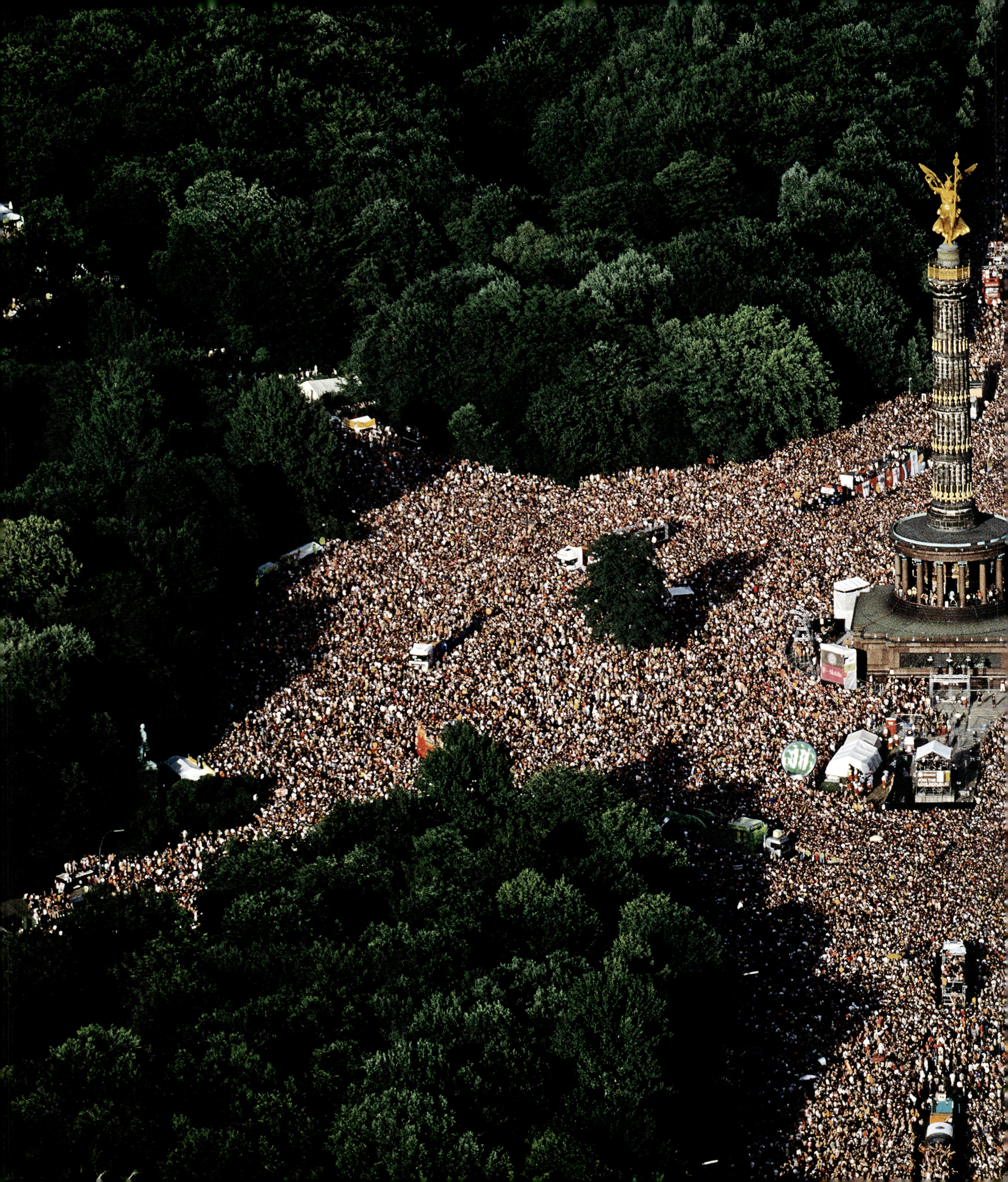

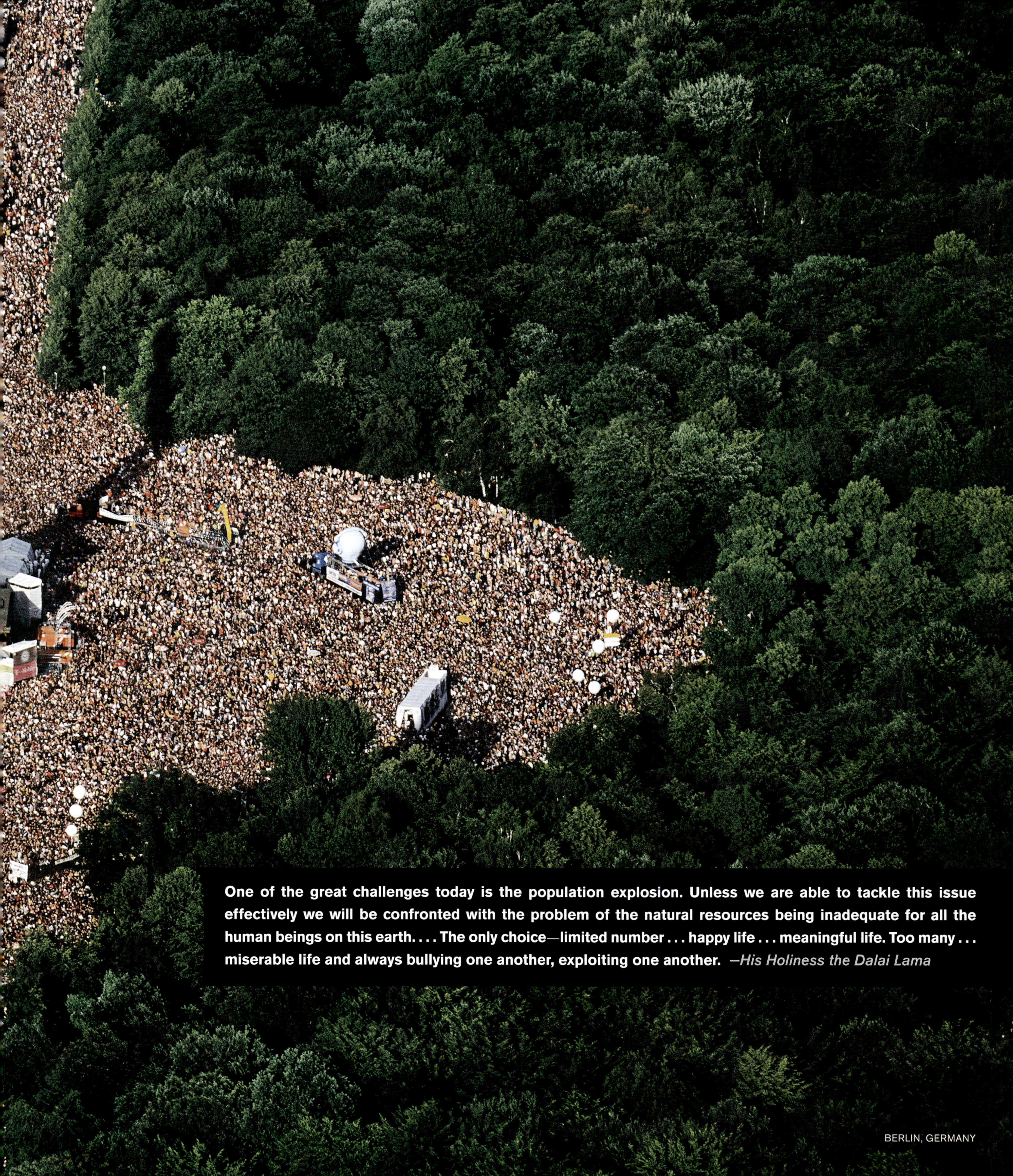

One of the great challenges today is the population explosion. Unless we are able to tackle this issue effectively we will be confronted with the problem of the natural resources being inadequate for all the human beings on this earth. . . . The only choice—limited number . . . happy life . . . meaningful life. Too many . . . miserable life and always bullying one another, exploiting one another. —His Holiness the Dalai Lama

BERLIN, GERMANY

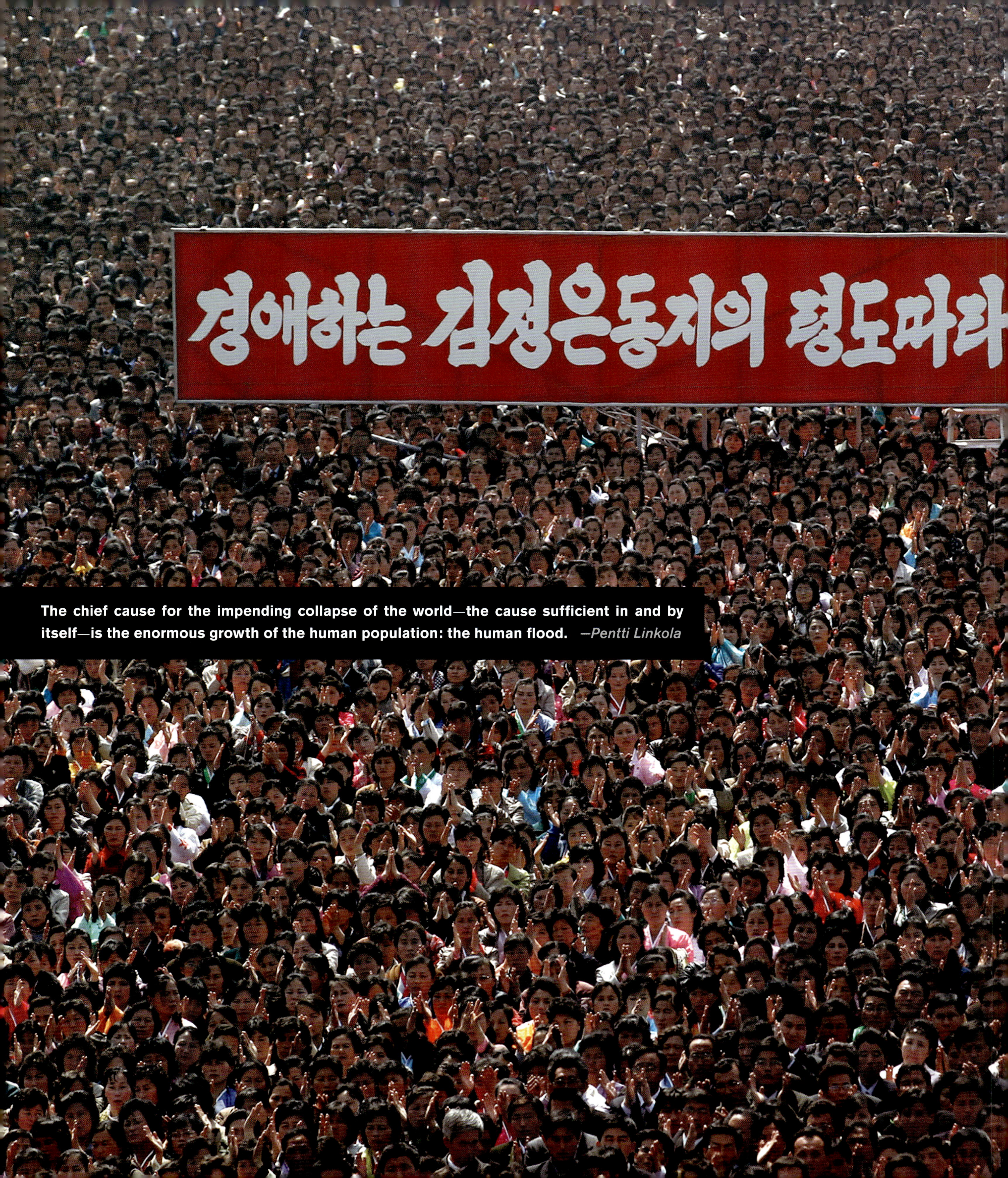

The chief cause for the impending collapse of the world—the cause sufficient in and by itself—is the enormous growth of the human population: the human flood. —*Pentti Linkola*

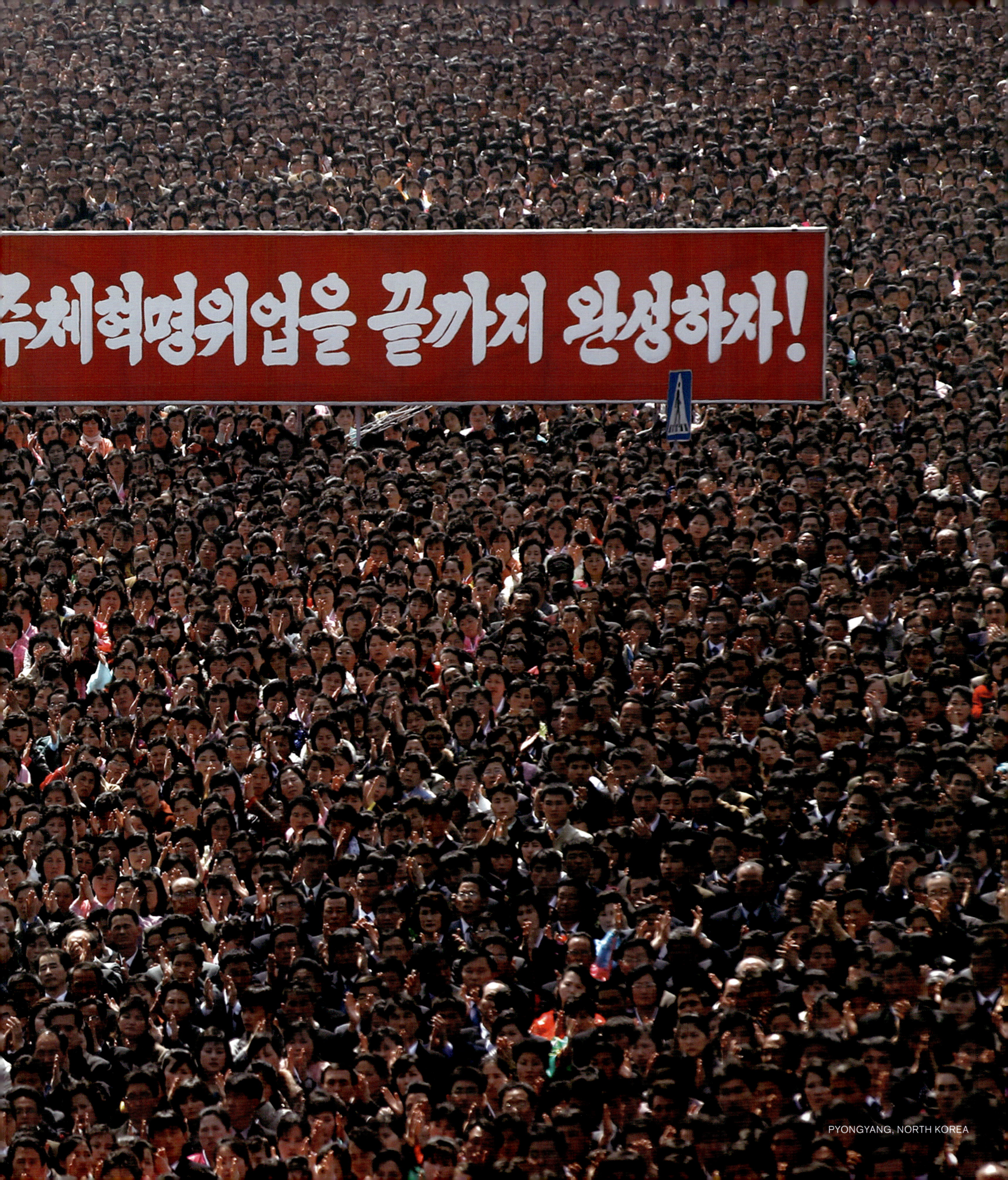

PYONGYANG, NORTH KOREA

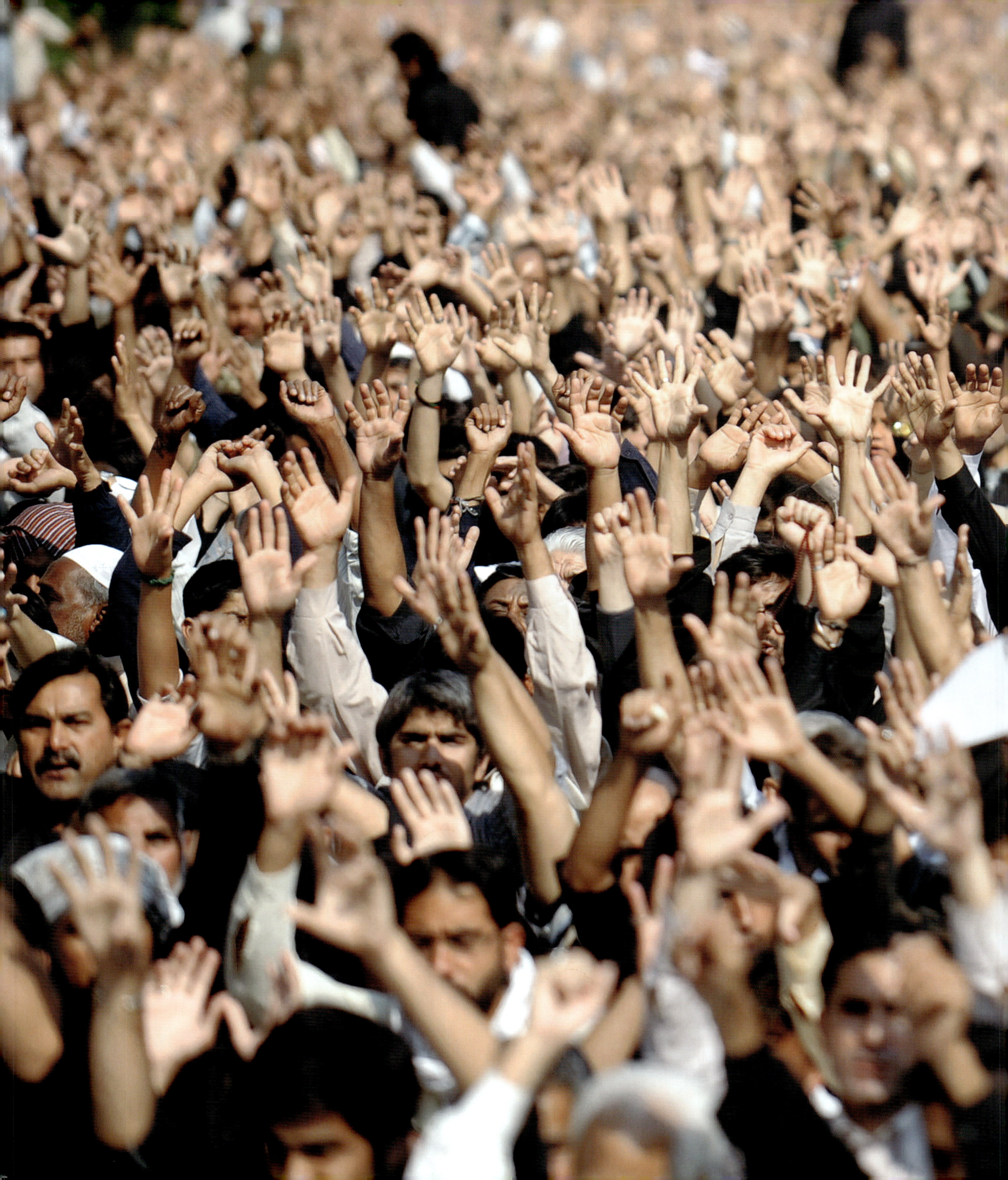

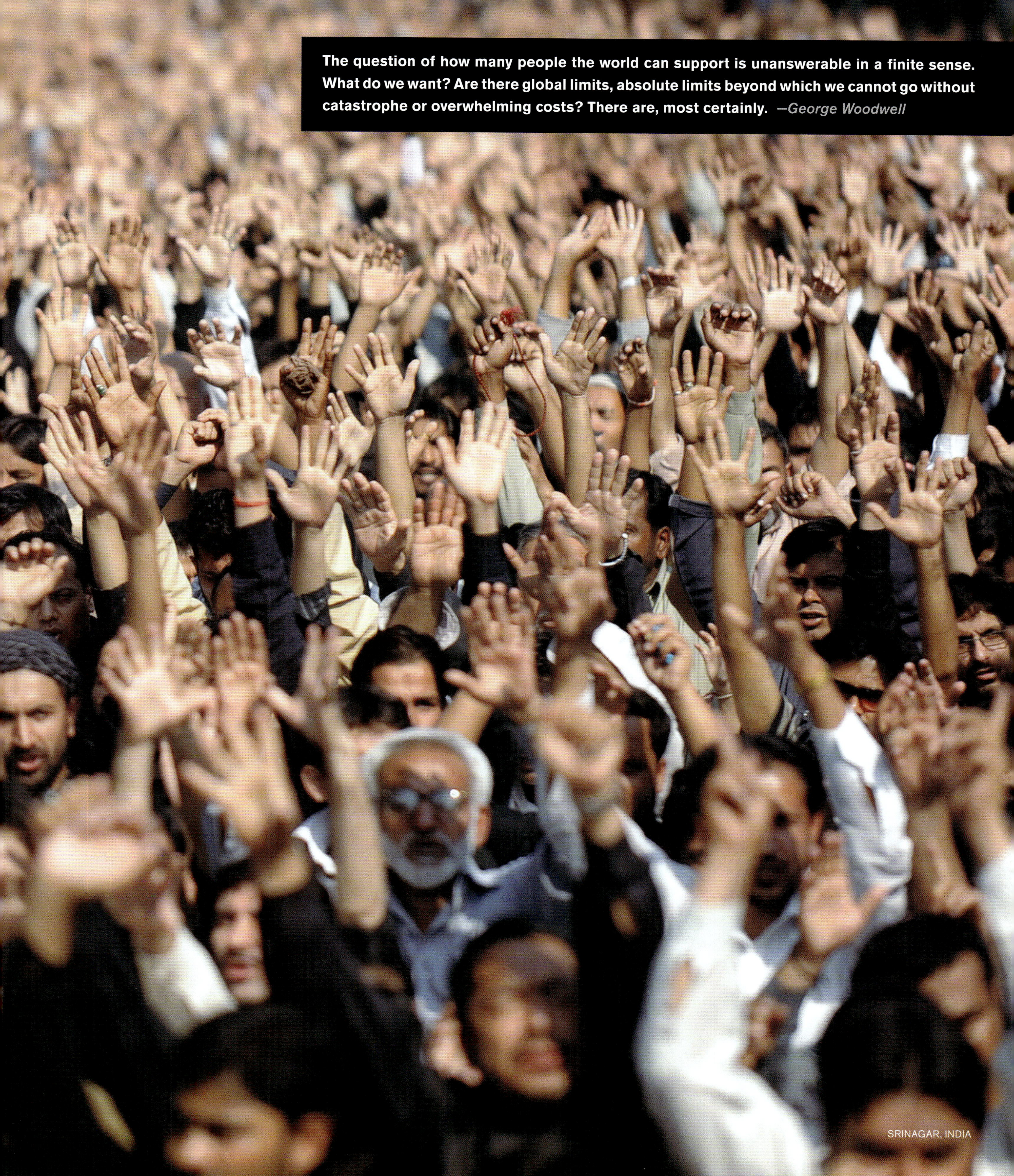
The question of how many people the world can support is unanswerable in a finite sense. What do we want? Are there global limits, absolute limits beyond which we cannot go without catastrophe or overwhelming costs? There are, most certainly. —George Woodwell
SRINAGAR, INDIA

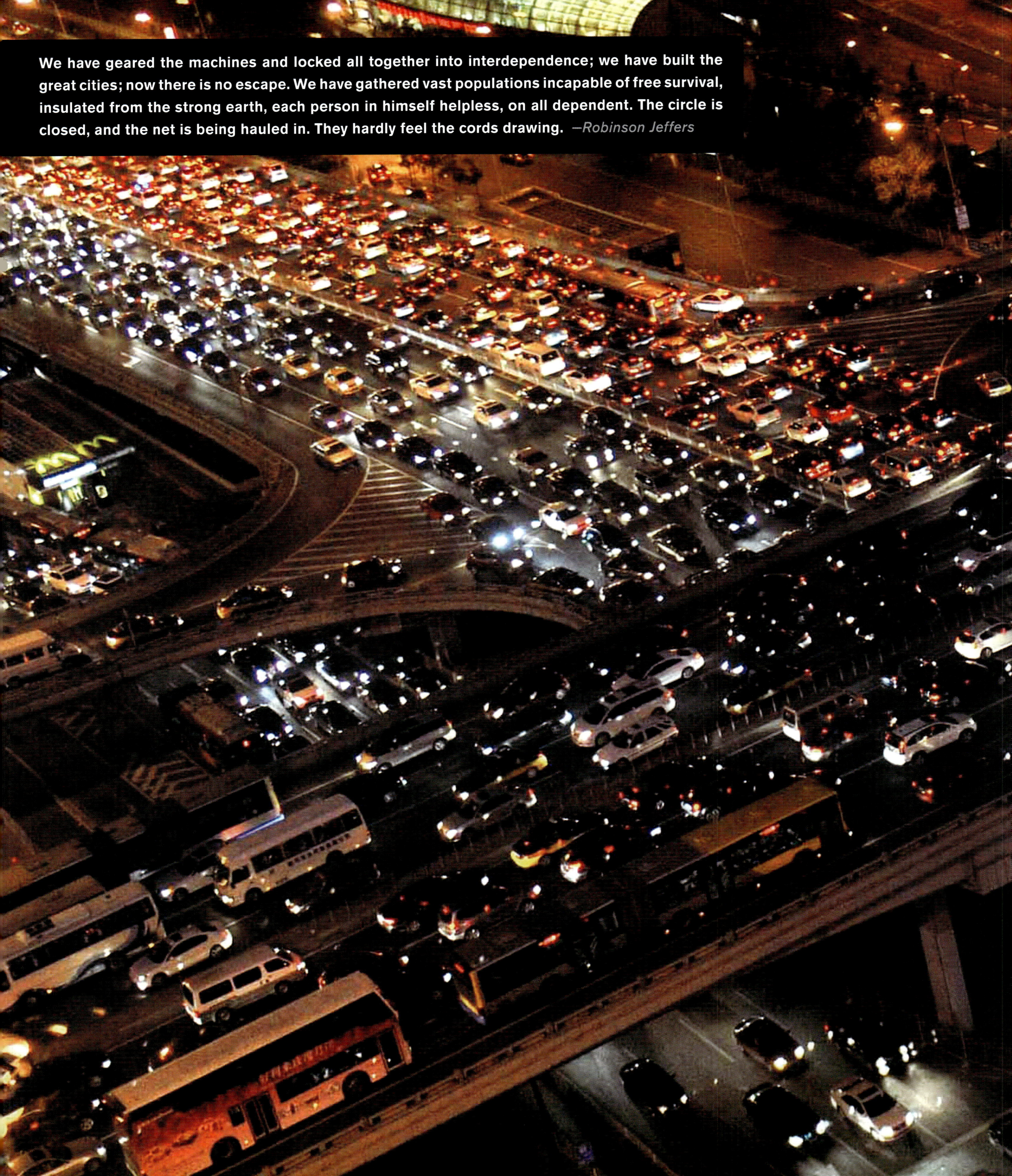

We have geared the machines and locked all together into interdependence; we have built the great cities; now there is no escape. We have gathered vast populations incapable of free survival, insulated from the strong earth, each person in himself helpless, on all dependent. The circle is closed, and the net is being hauled in. They hardly feel the cords drawing. —Robinson Jeffers

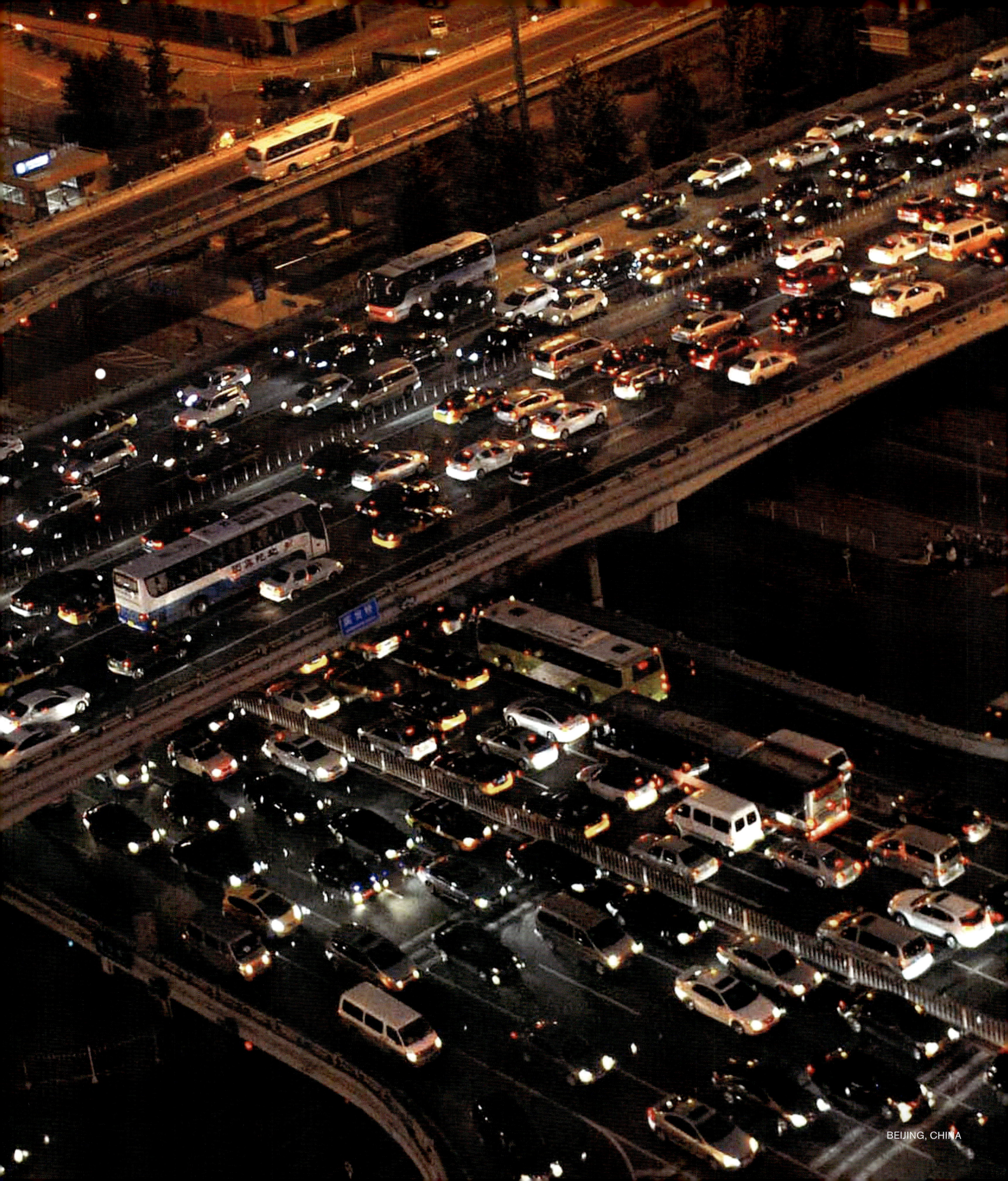

BEIJING, CHINA

URBAN ANIMAL

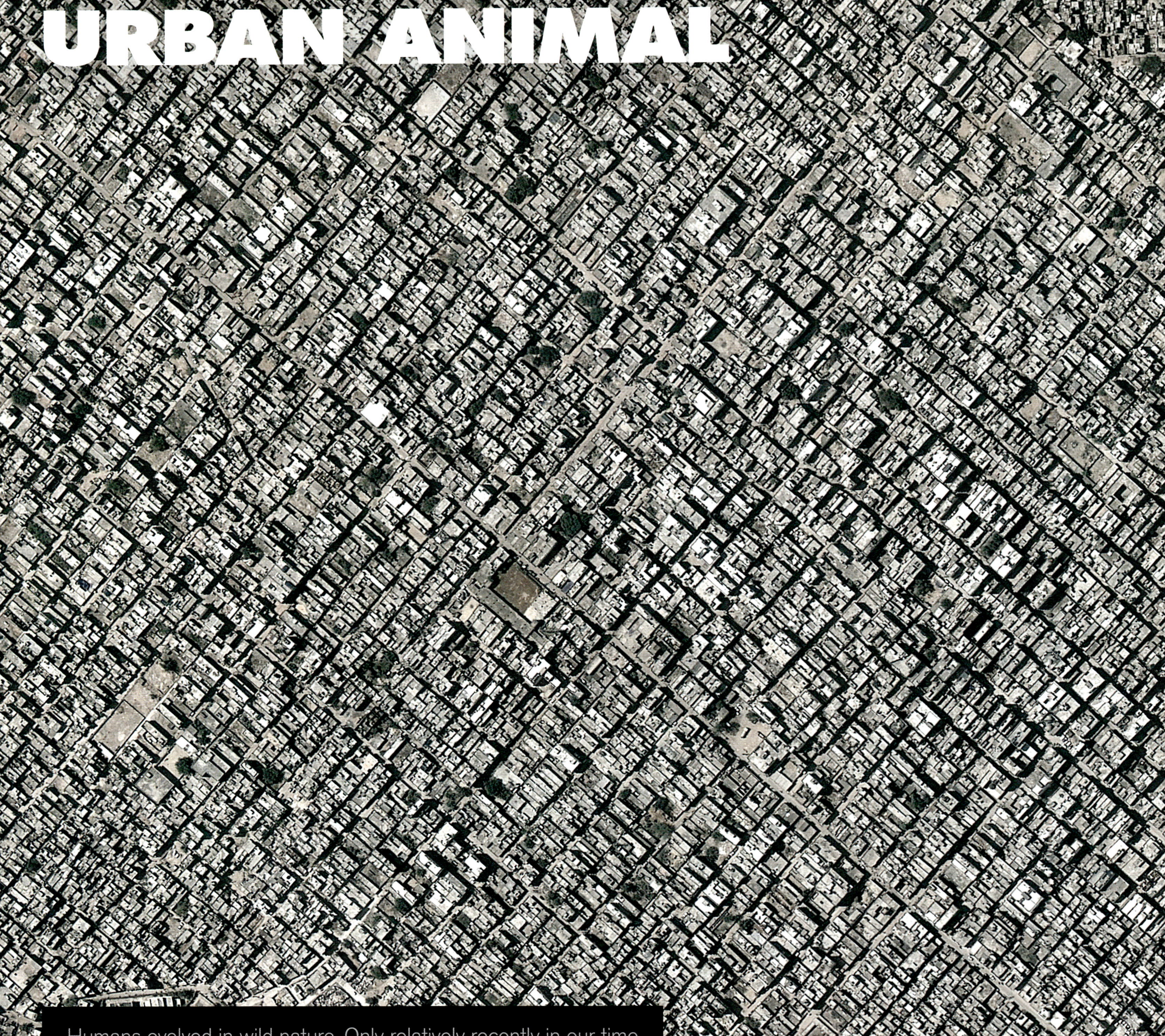

Humans evolved in wild nature. Only relatively recently in our time on Earth, roughly ten to twelve millennia ago, did we begin to domesticate other species—and ourselves. That first agricultural revolution set humanity on a trajectory of population growth and settlement-based land use. Increased social organization and the invention of cities went hand in hand to allow development of increasingly complex economic and political systems. In 2008, for the first time in history, the majority of humans on Earth lived in cities. We had become, at least superficially, urban animals.

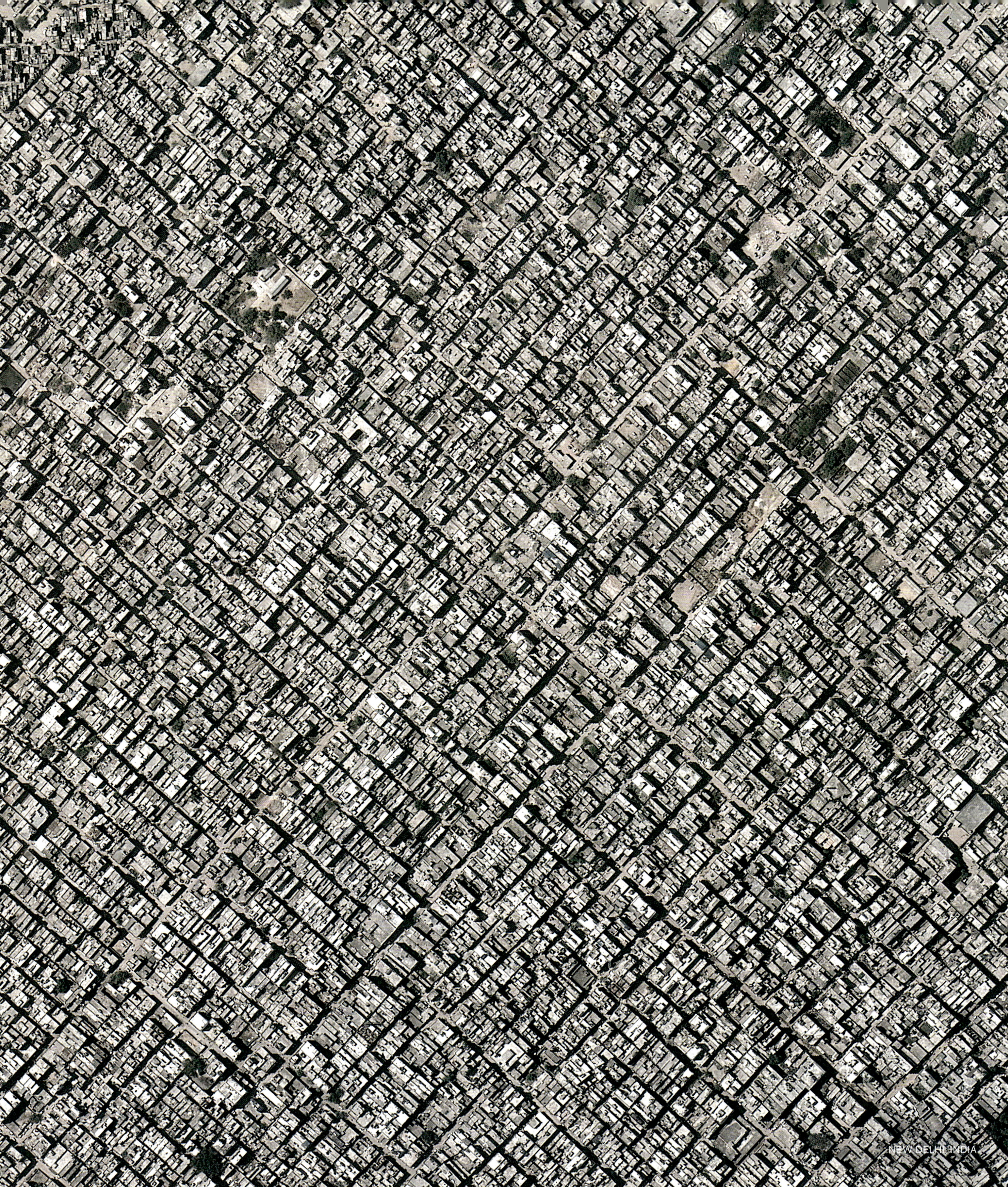
NEW DELHI, INDIA

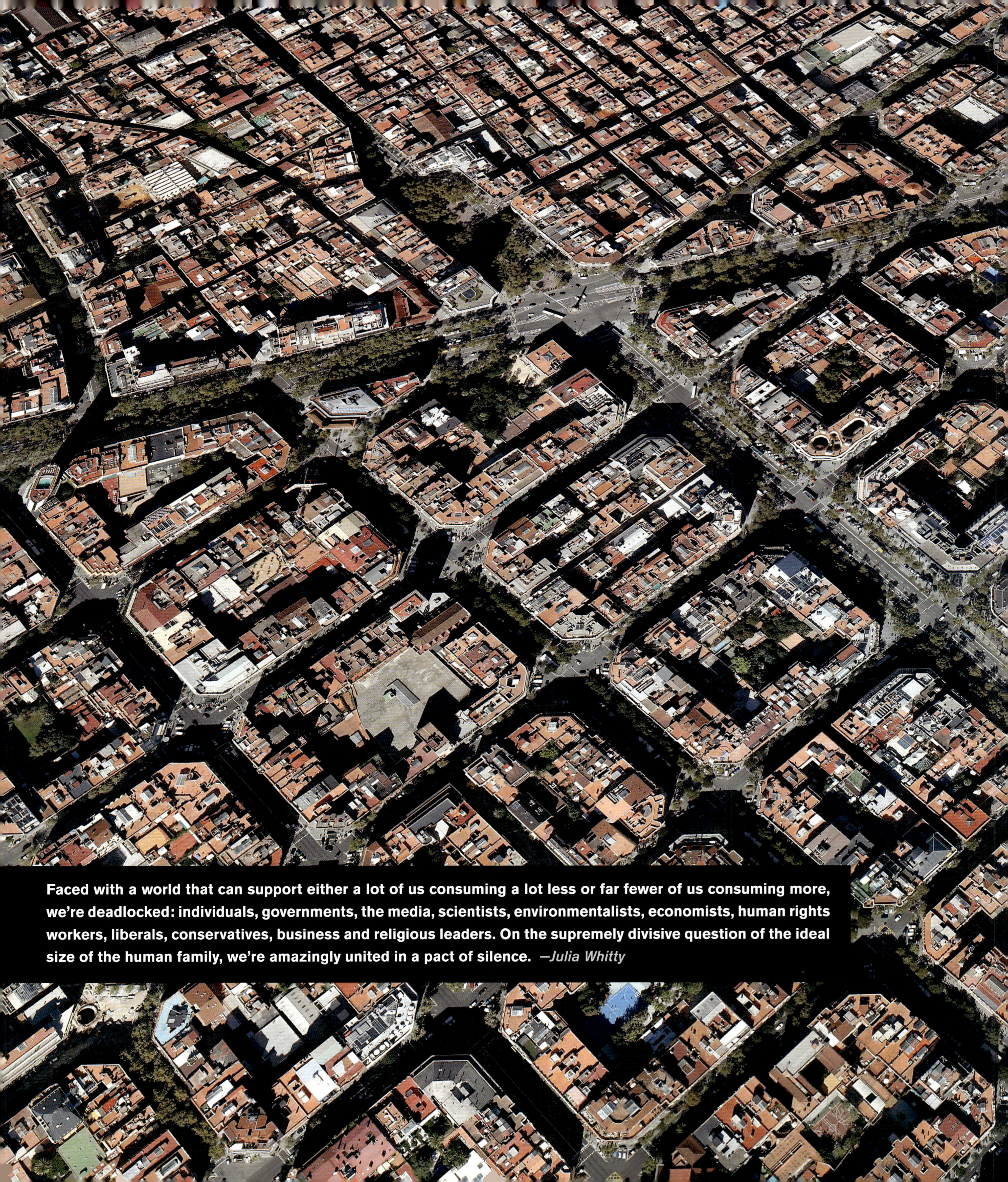

Faced with a world that can support either a lot of us consuming a lot less or far fewer of us consuming more, we're deadlocked: individuals, governments, the media, scientists, environmentalists, economists, human rights workers, liberals, conservatives, business and religious leaders. On the supremely divisive question of the ideal size of the human family, we're amazingly united in a pact of silence. —Julia Whitty

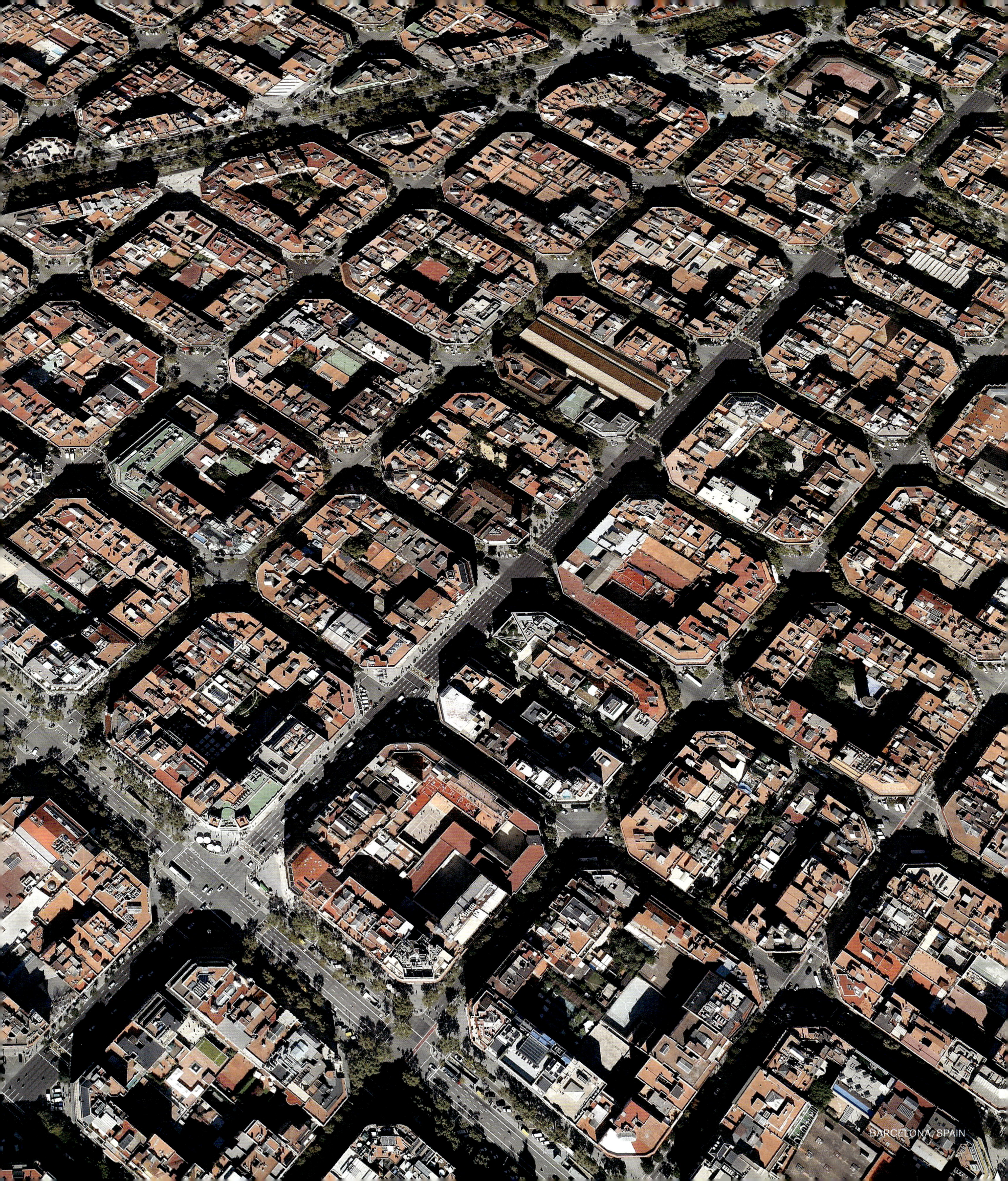
BARCELONA, SPAIN

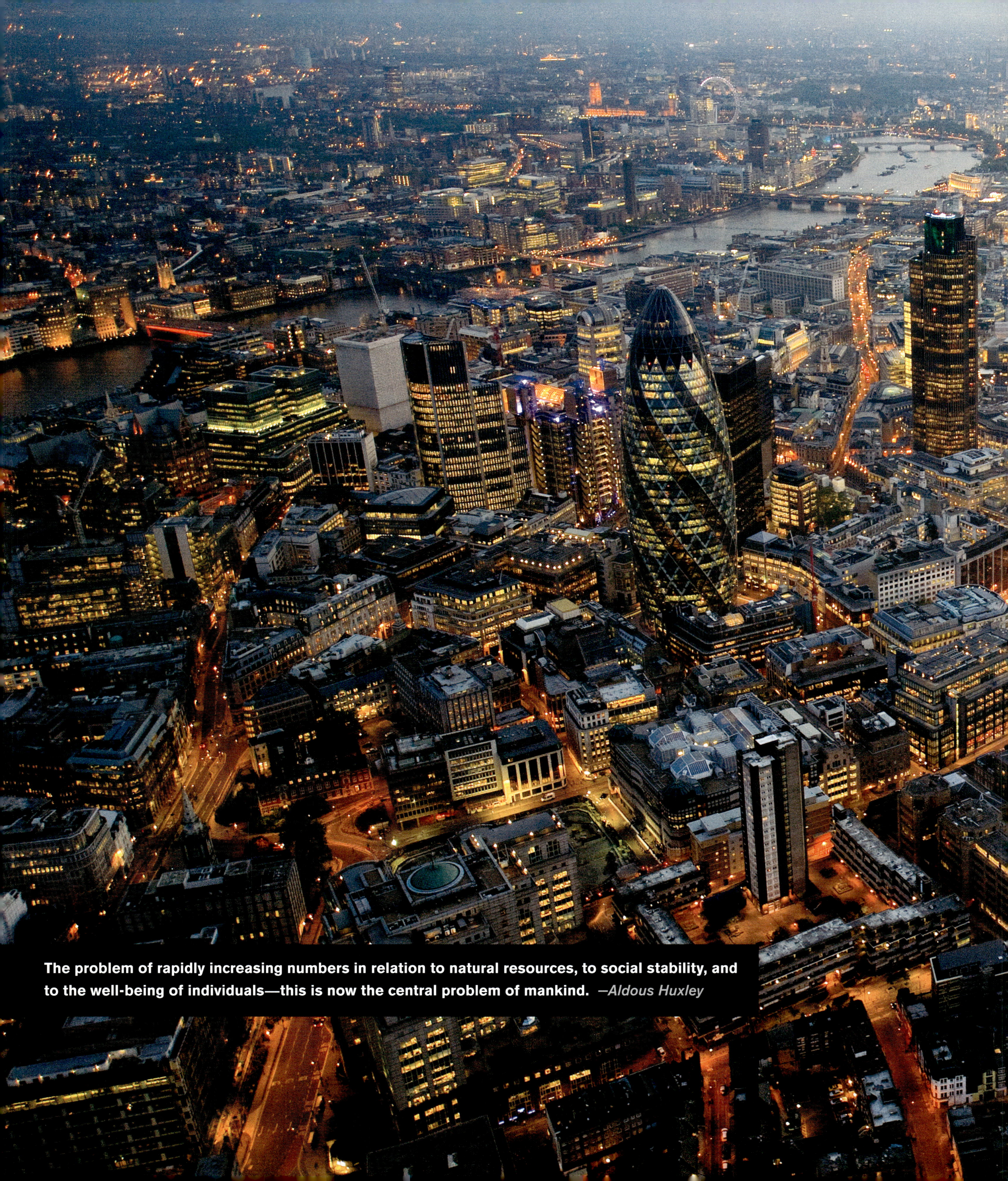

The problem of rapidly increasing numbers in relation to natural resources, to social stability, and to the well-being of individuals—this is now the central problem of mankind. —*Aldous Huxley*

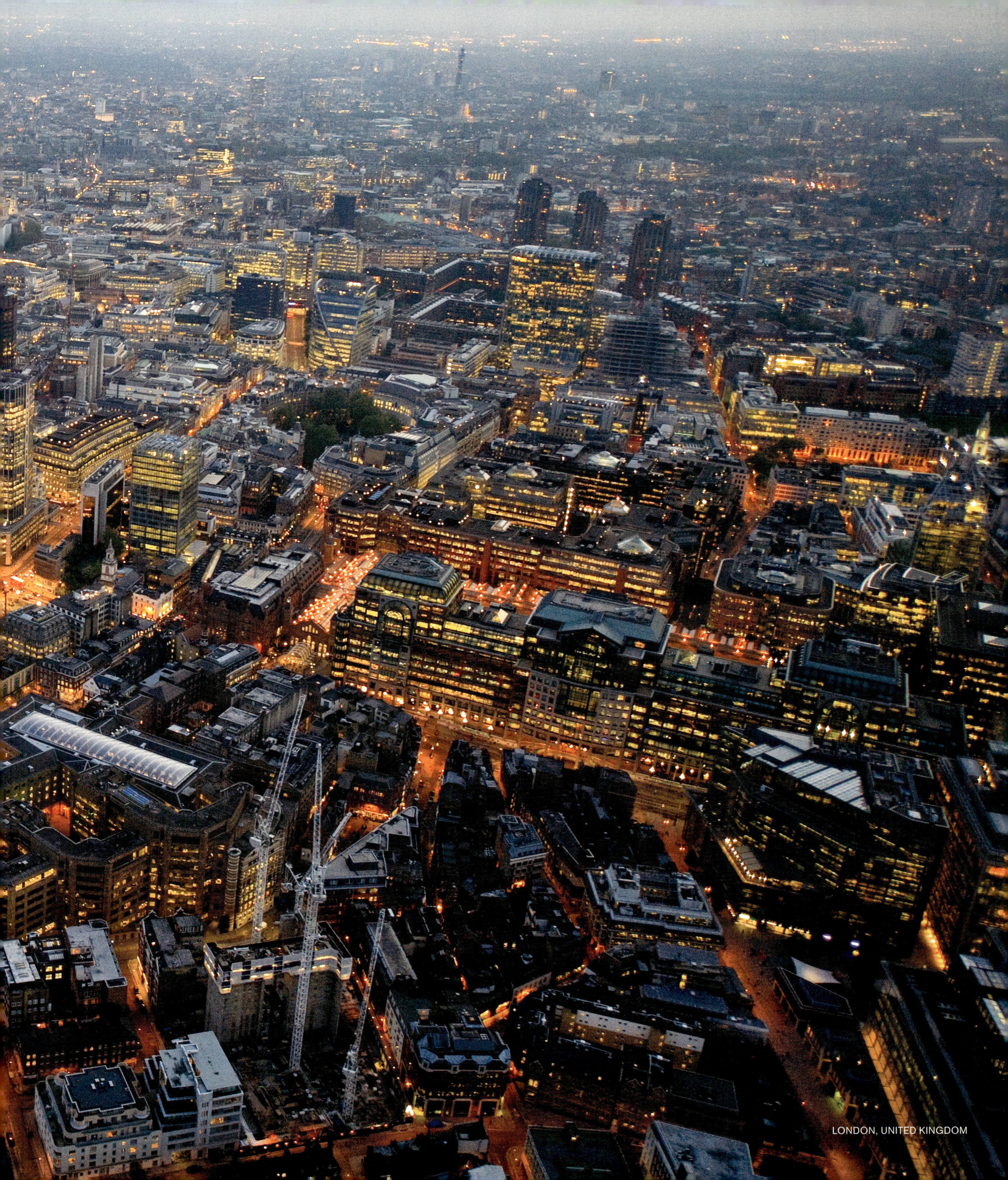

LONDON, UNITED KINGDOM

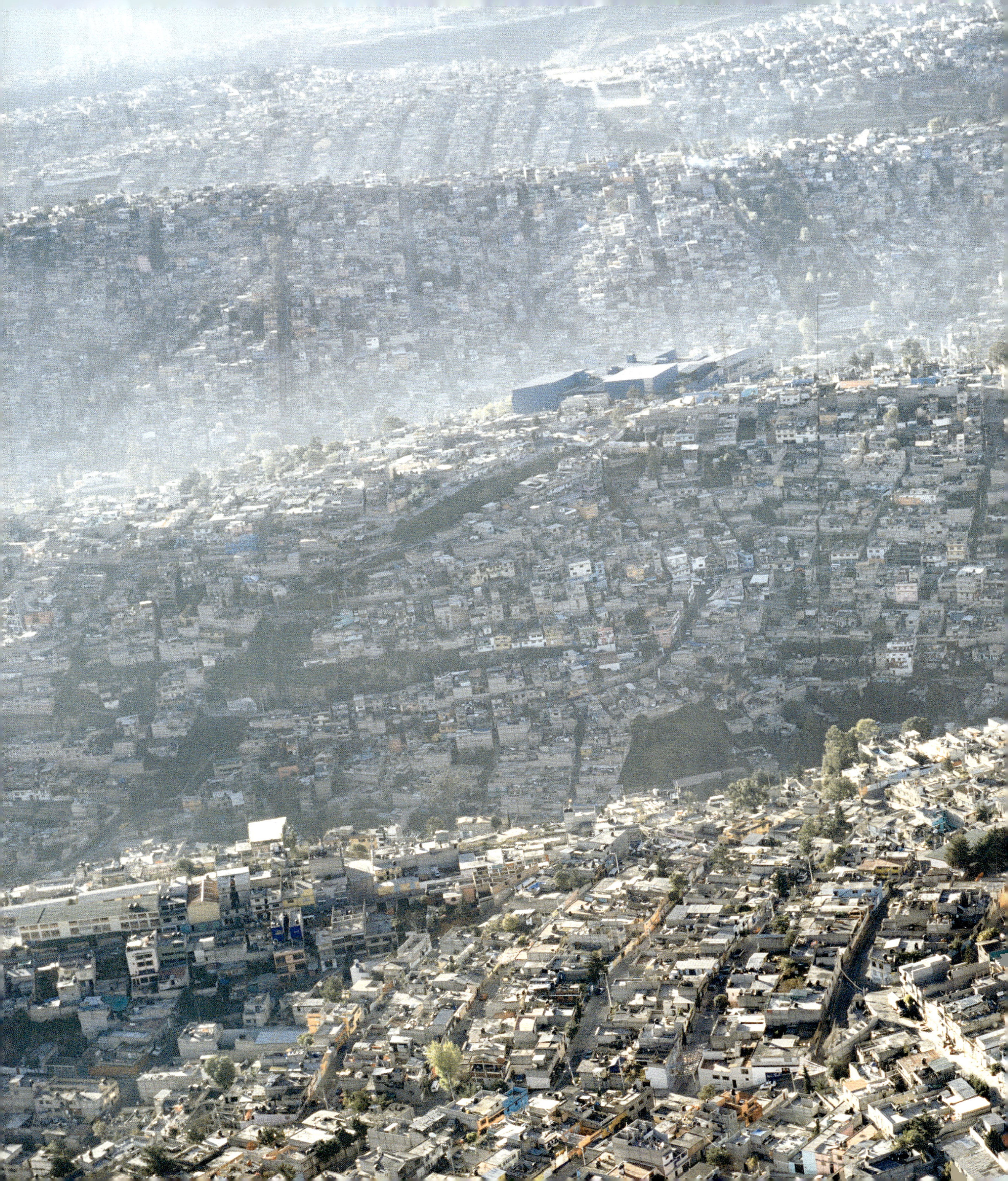

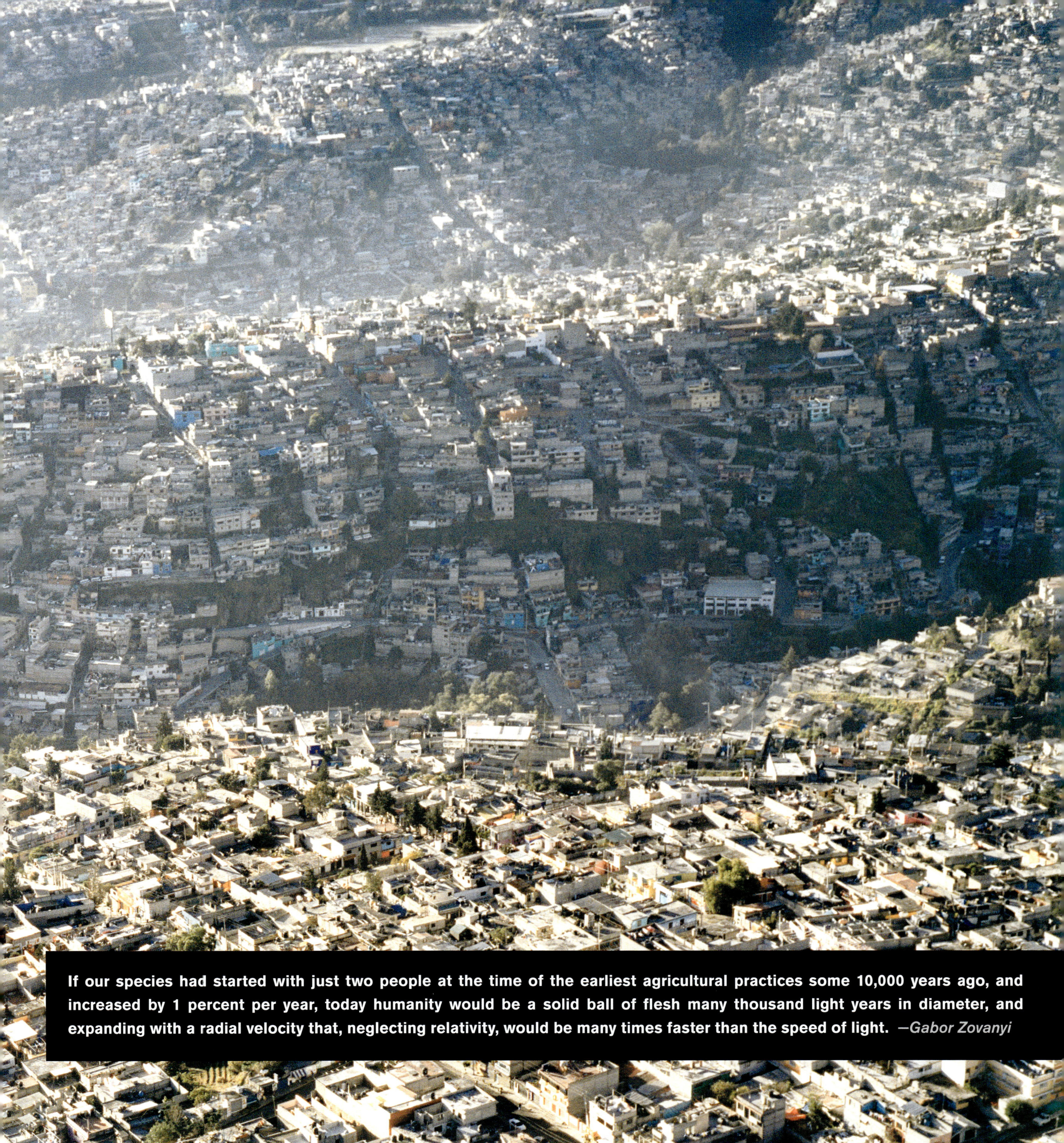

If our species had started with just two people at the time of the earliest agricultural practices some 10,000 years ago, and increased by 1 percent per year, today humanity would be a solid ball of flesh many thousand light years in diameter, and expanding with a radial velocity that, neglecting relativity, would be many times faster than the speed of light. —Gabor Zovanyi
MEXICO CITY, MEXICO

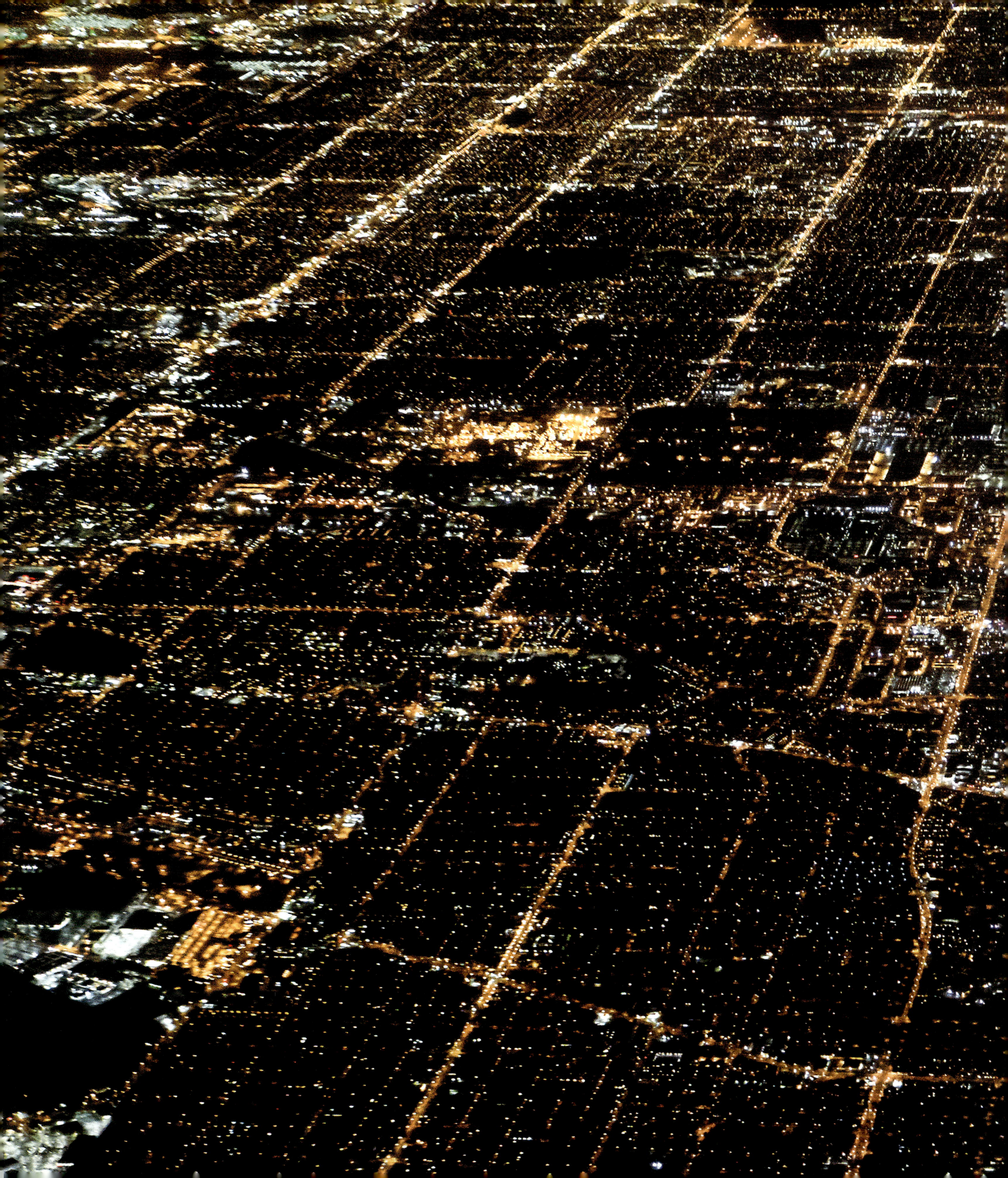

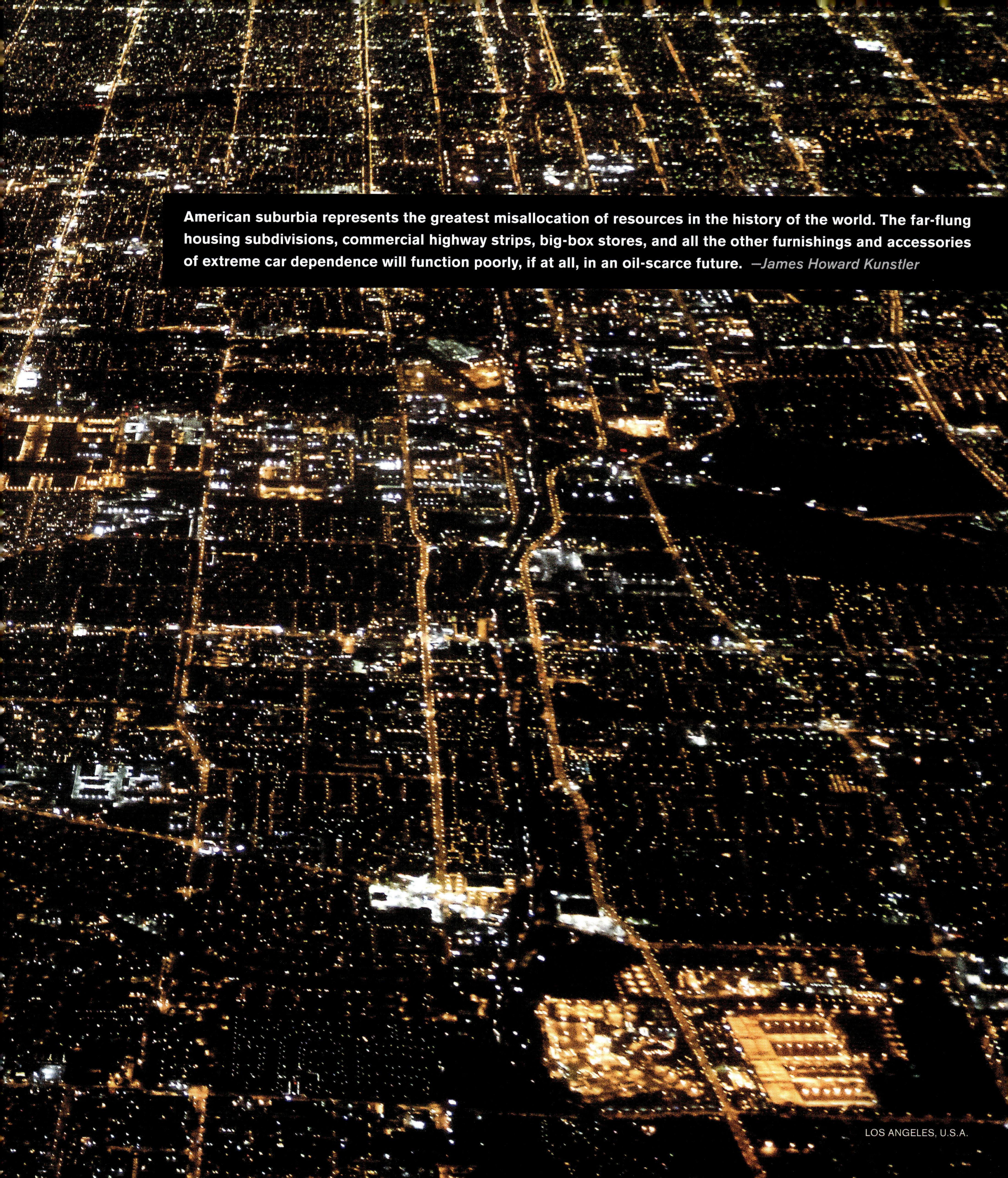

American suburbia represents the greatest misallocation of resources in the history of the world. The far-flung housing subdivisions, commercial highway strips, big-box stores, and all the other furnishings and accessories of extreme car dependence will function poorly, if at all, in an oil-scarce future. —James Howard Kunstler

LOS ANGELES, U.S.A.

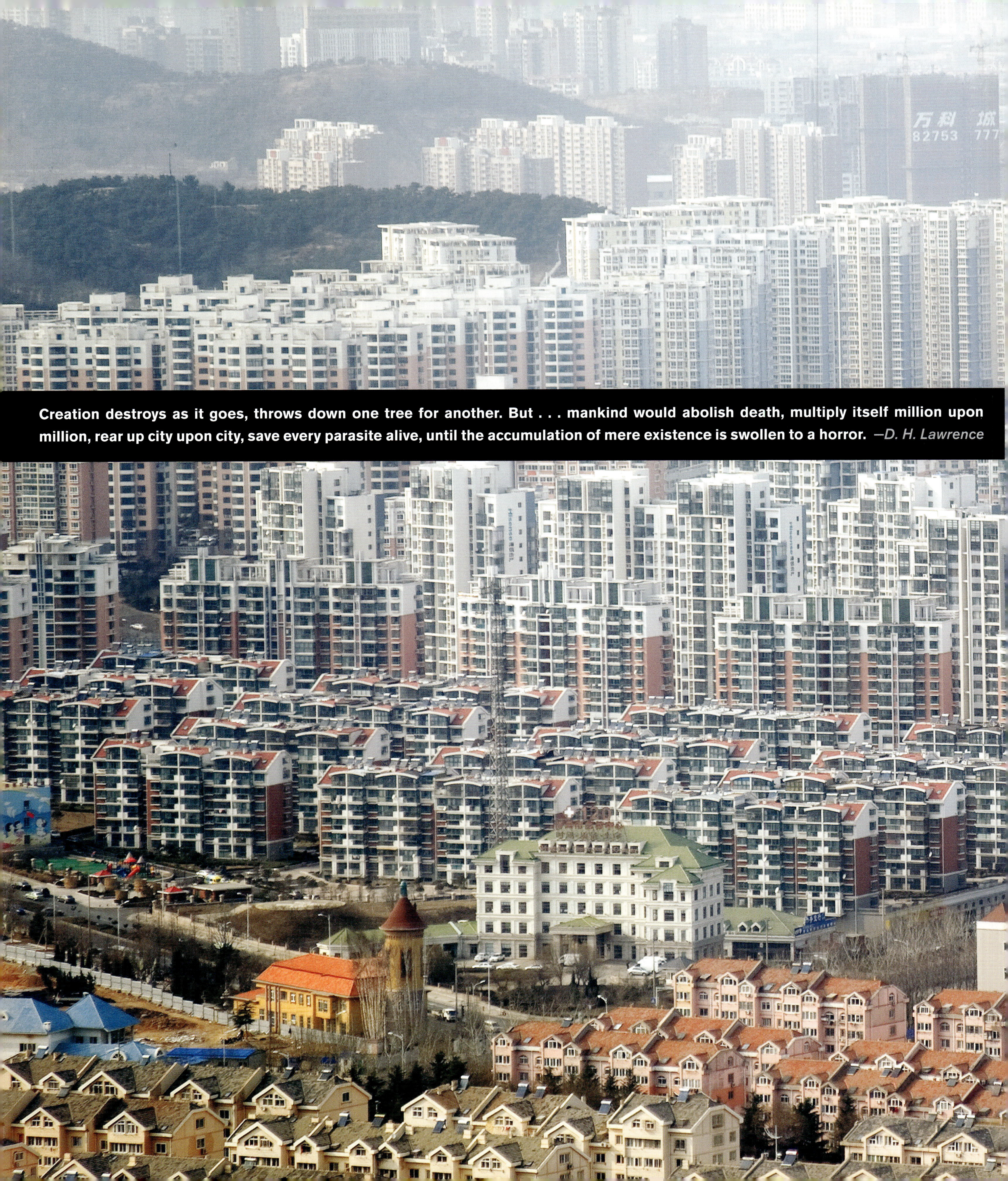

Creation destroys as it goes, throws down one tree for another. But . . . mankind would abolish death, multiply itself million upon million, rear up city upon city, save every parasite alive, until the accumulation of mere existence is swollen to a horror. —D. H. Lawrence

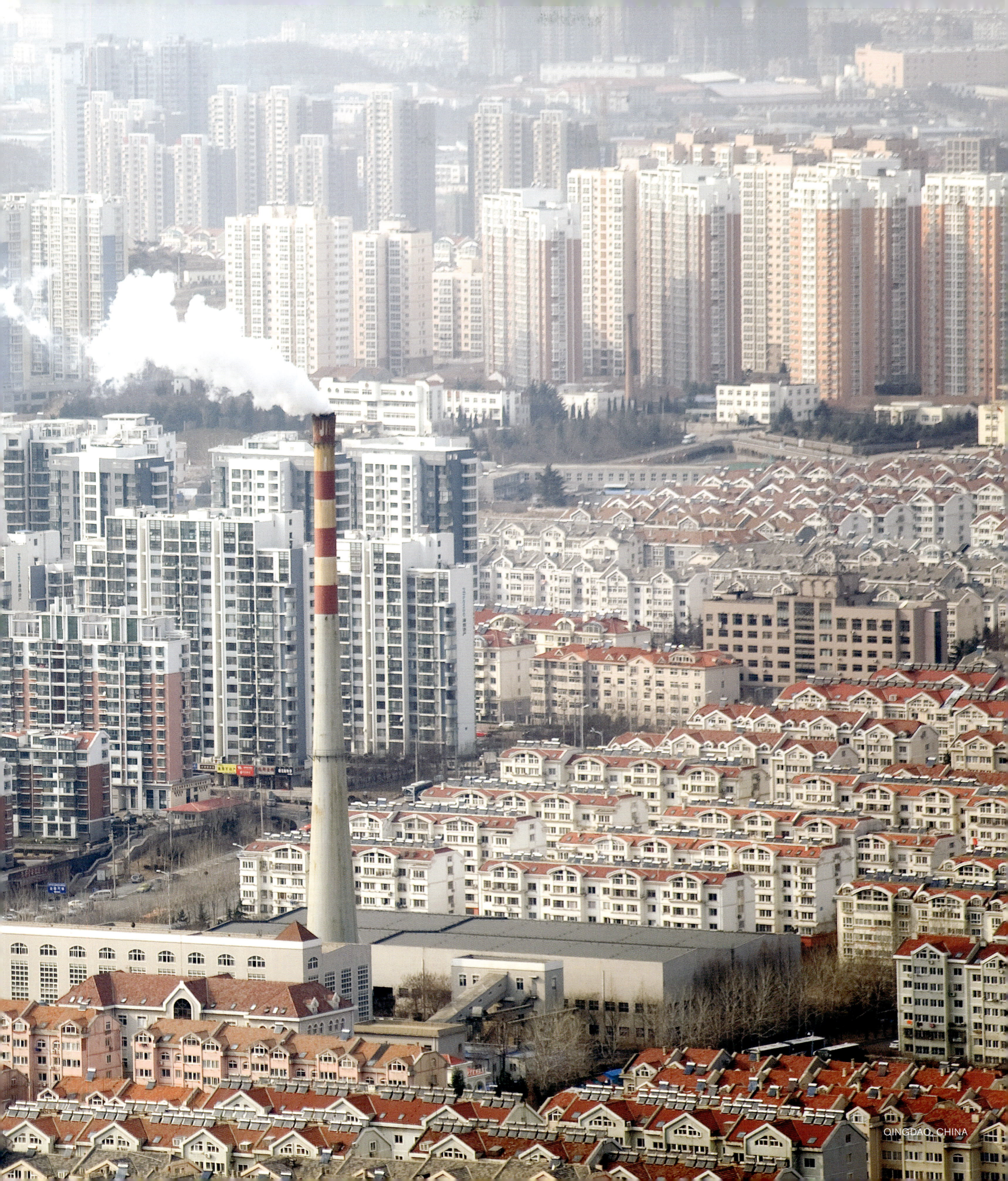

QINGDAO, CHINA

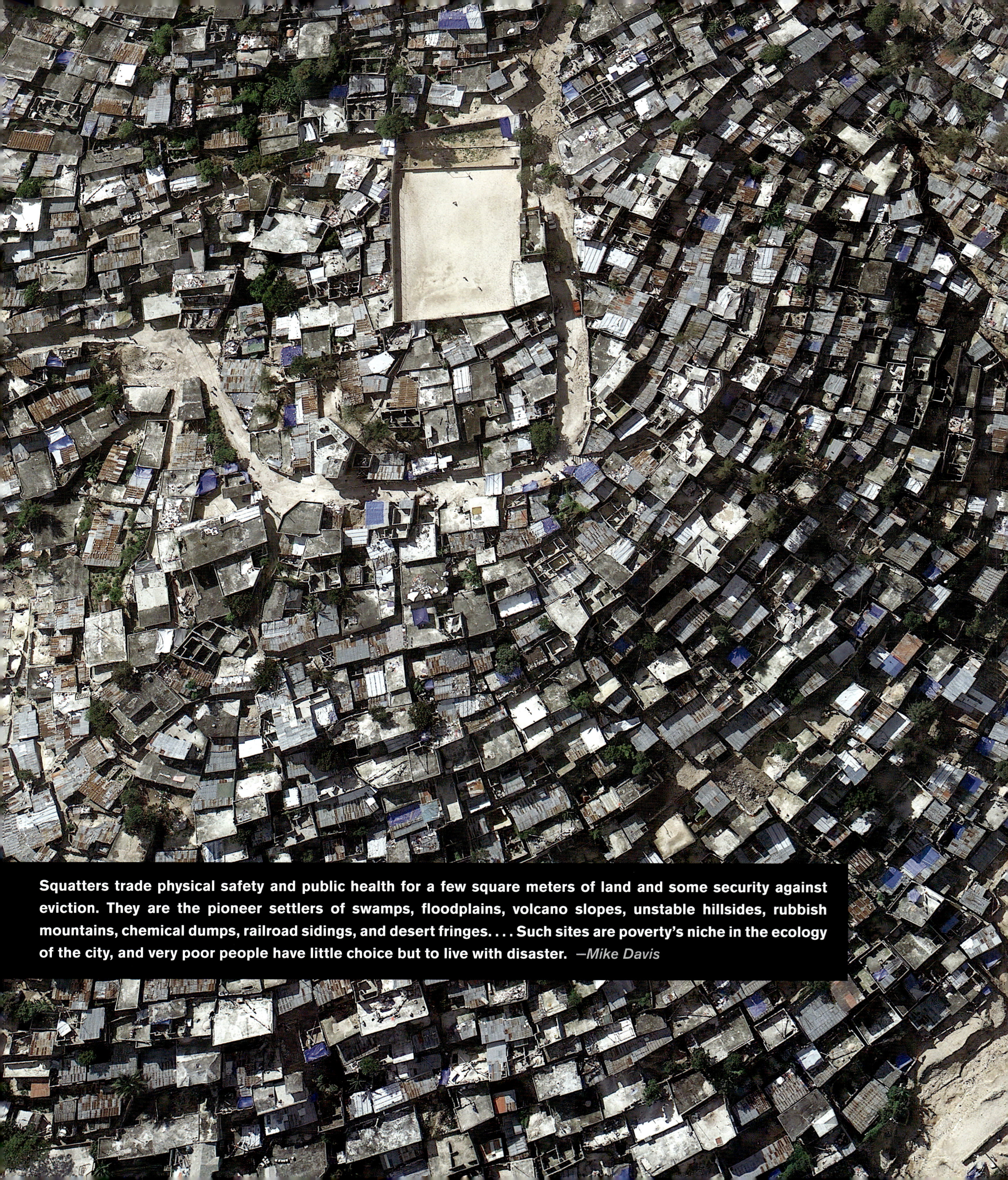

Squatters trade physical safety and public health for a few square meters of land and some security against eviction. They are the pioneer settlers of swamps, floodplains, volcano slopes, unstable hillsides, rubbish mountains, chemical dumps, railroad sidings, and desert fringes. . . . Such sites are poverty's niche in the ecology of the city, and very poor people have little choice but to live with disaster. —*Mike Davis*

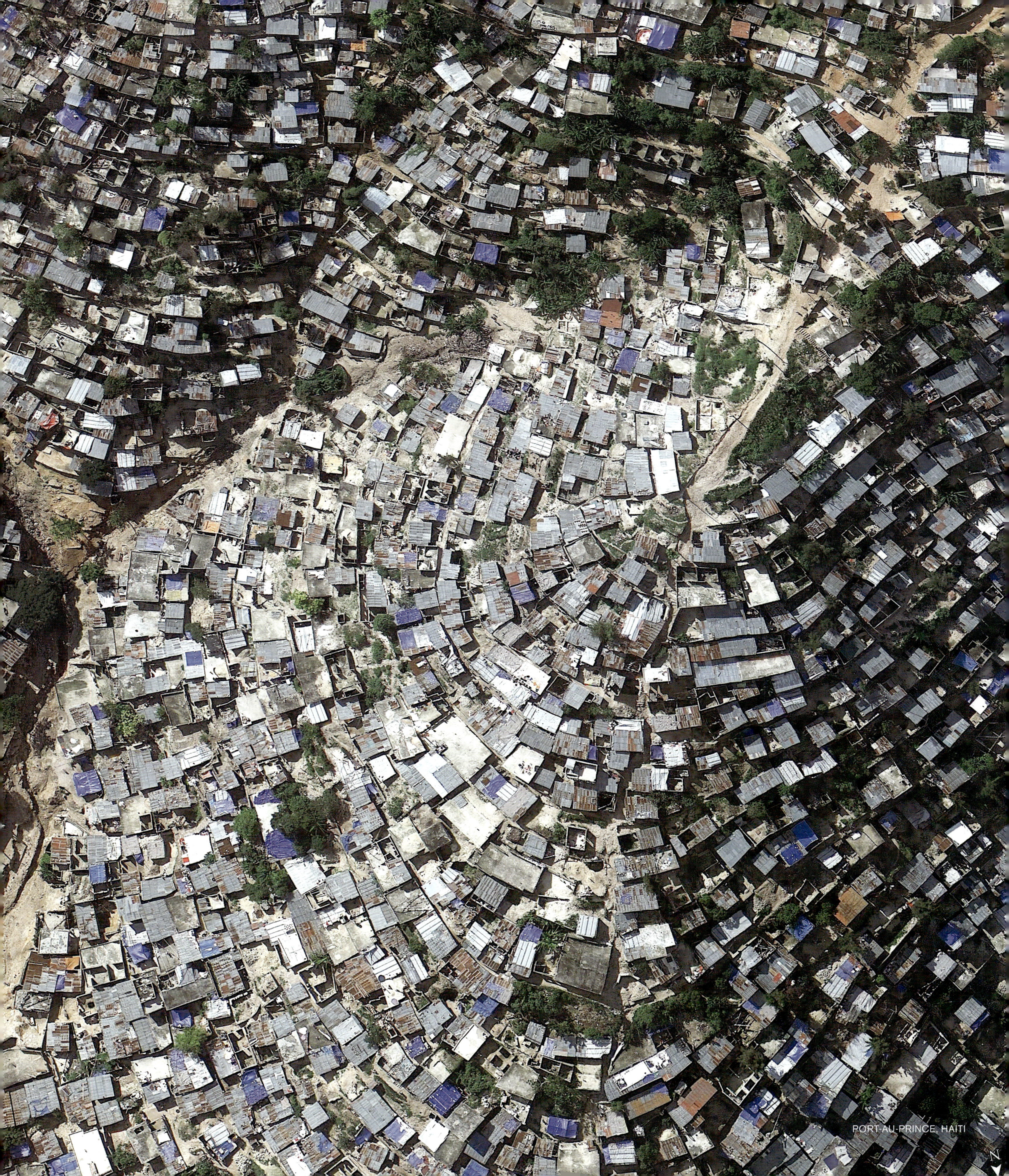
PORT-AU-PRINCE, HAITI

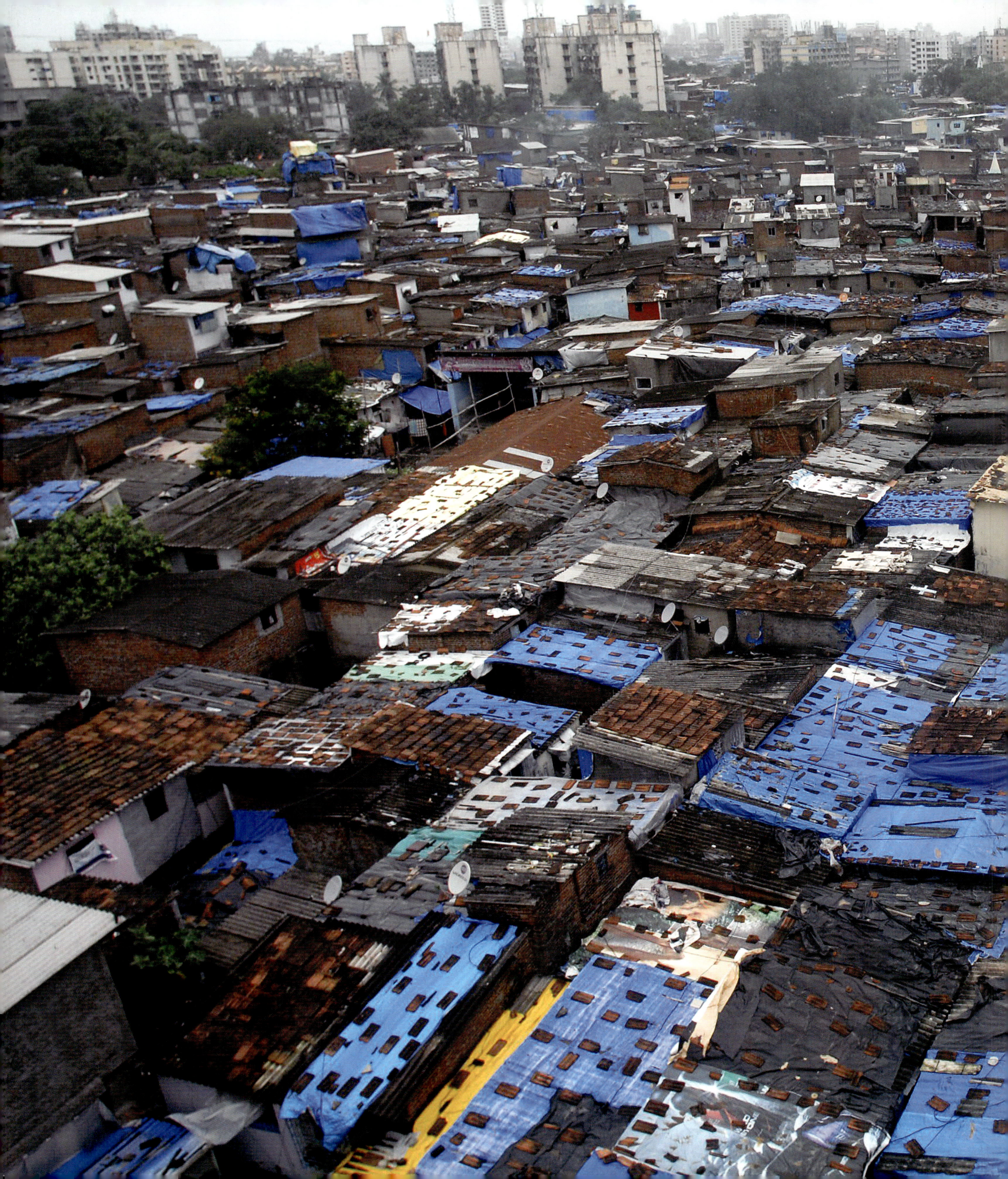

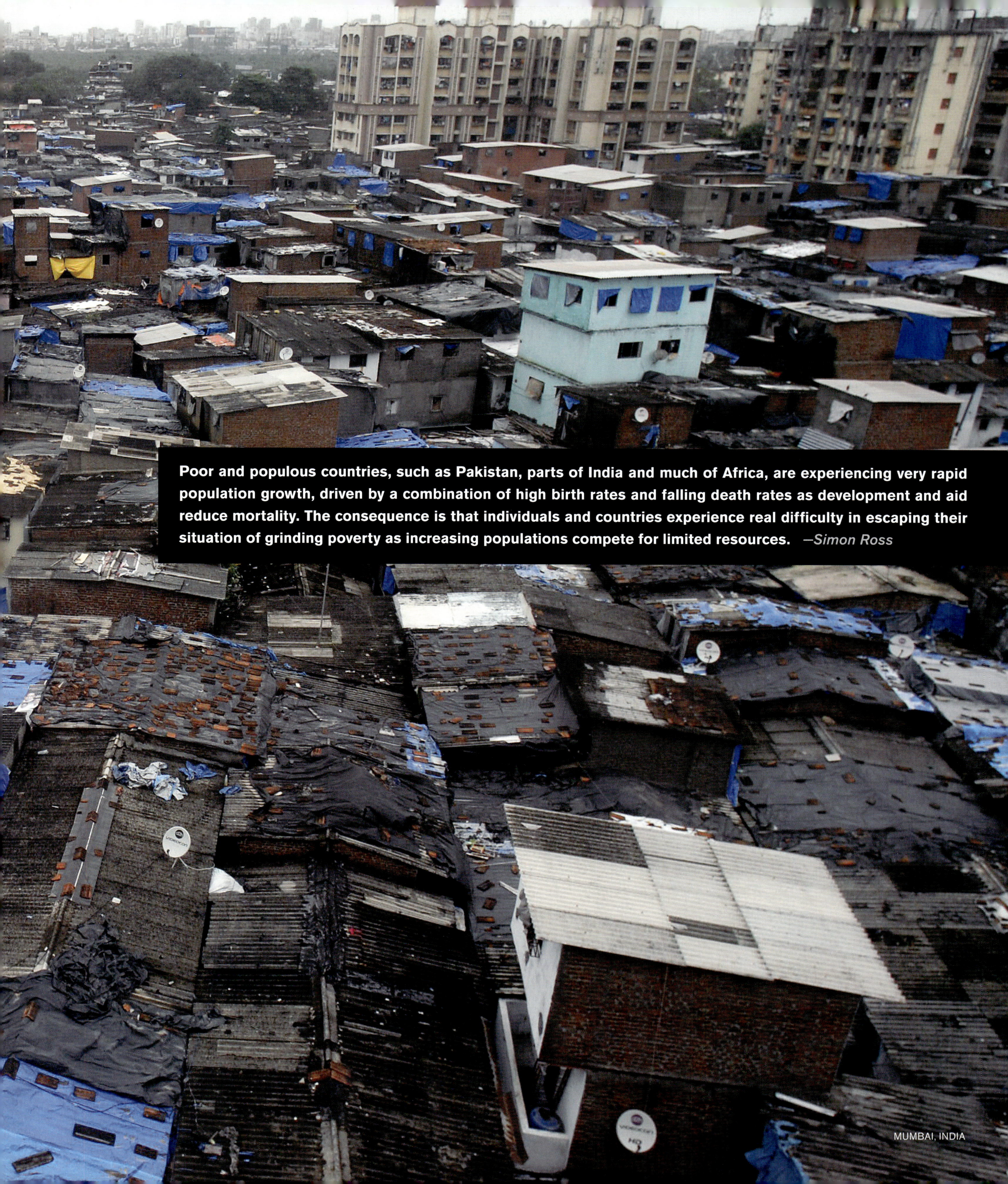
Poor and populous countries, such as Pakistan, parts of India and much of Africa, are experiencing very rapid population growth, driven by a combination of high birth rates and falling death rates as development and aid reduce mortality. The consequence is that individuals and countries experience real difficulty in escaping their situation of grinding poverty as increasing populations compete for limited resources.   —Simon Ross
MUMBAI, INDIA

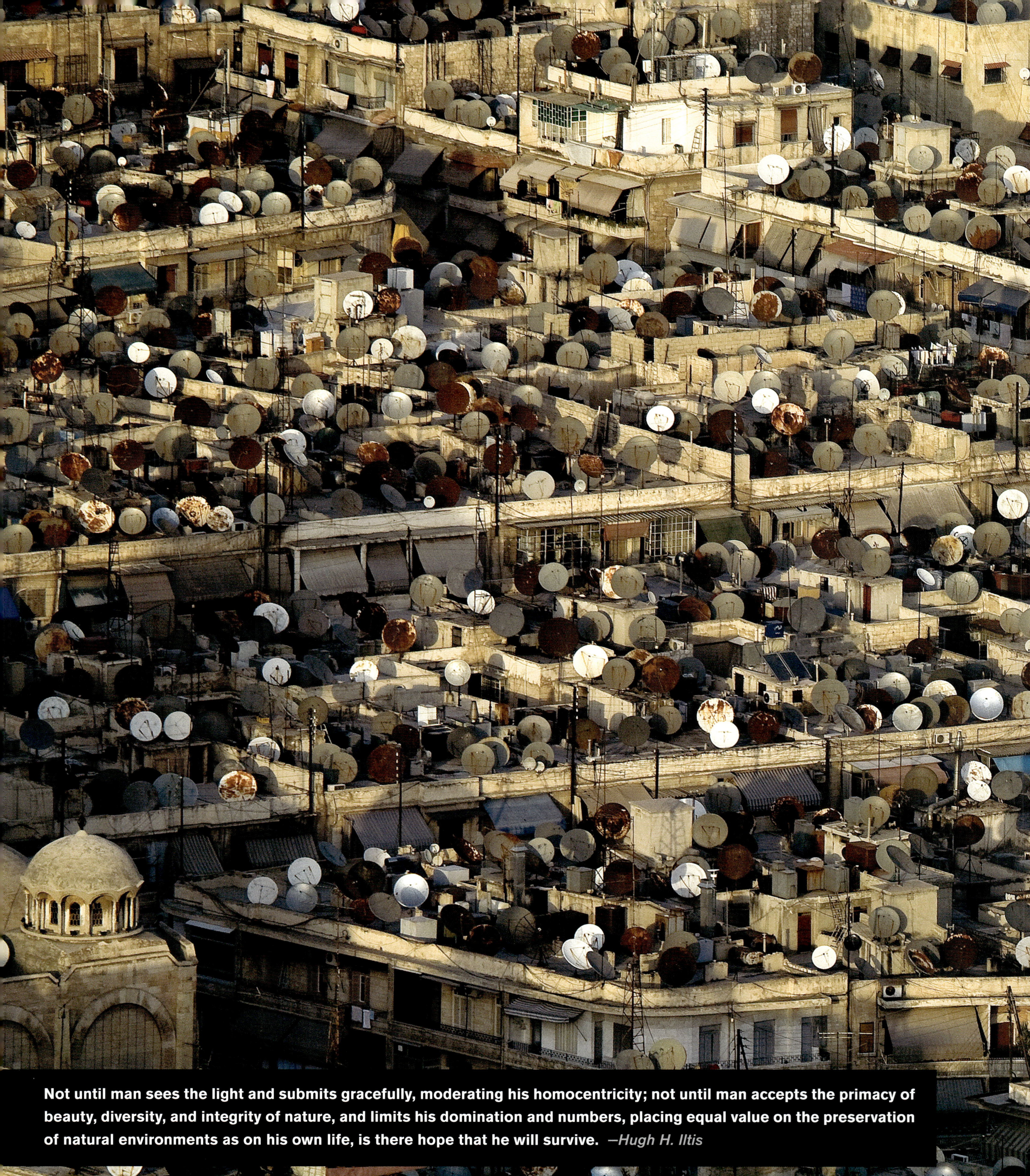

Not until man sees the light and submits gracefully, moderating his homocentricity; not until man accepts the primacy of beauty, diversity, and integrity of nature, and limits his domination and numbers, placing equal value on the preservation of natural environments as on his own life, is there hope that he will survive. —*Hugh H. Iltis*

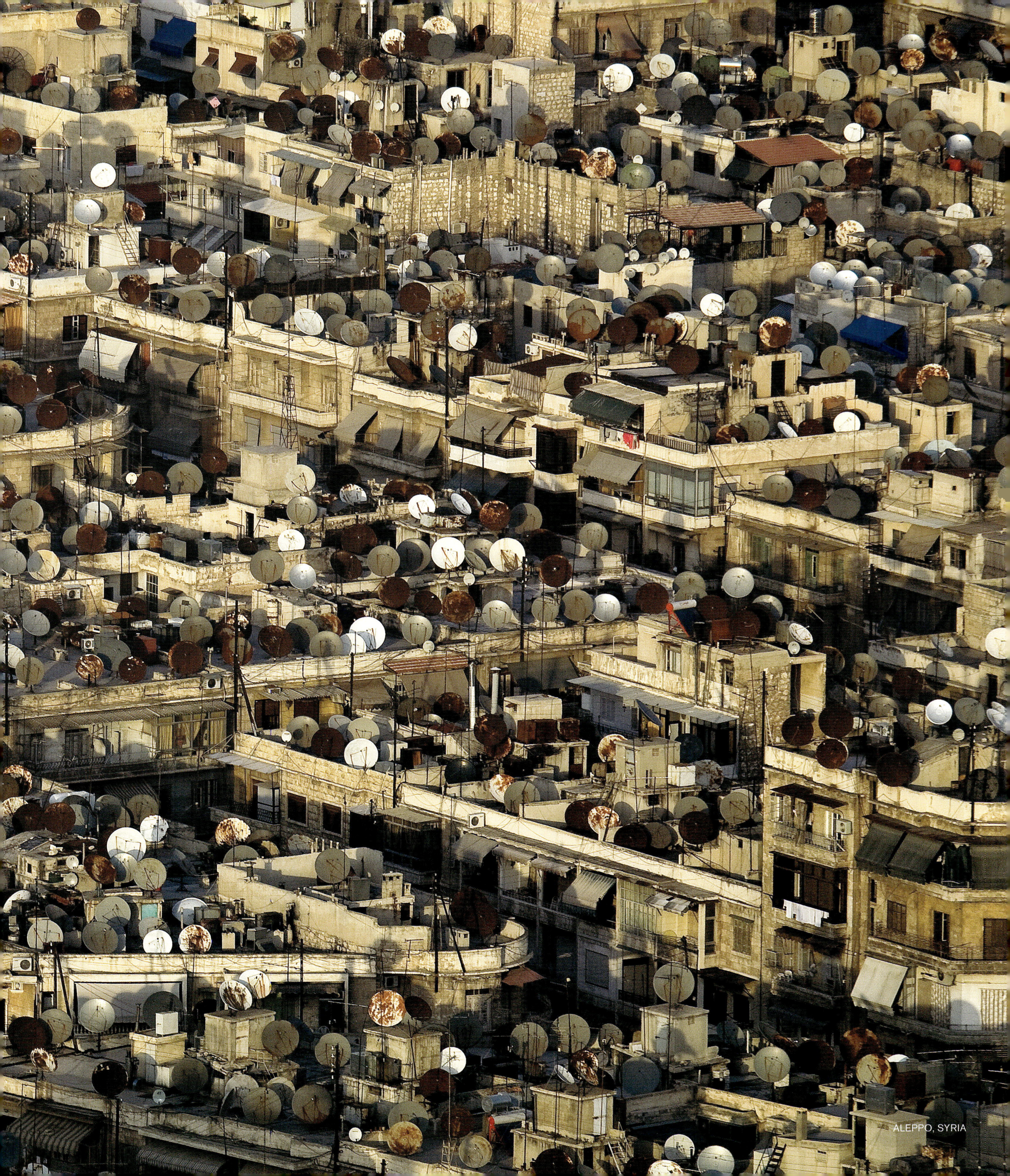

ALEPPO, SYRIA

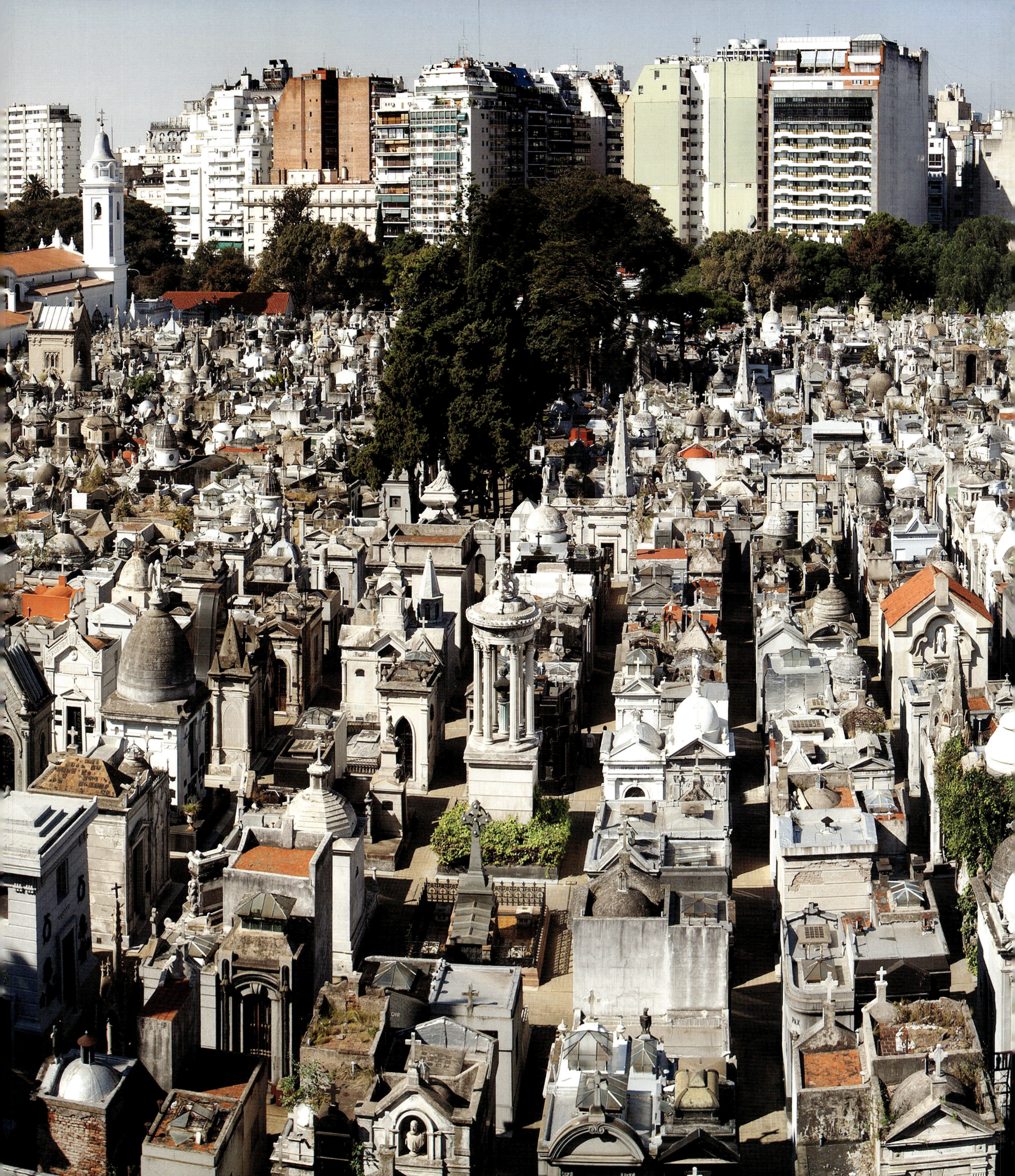

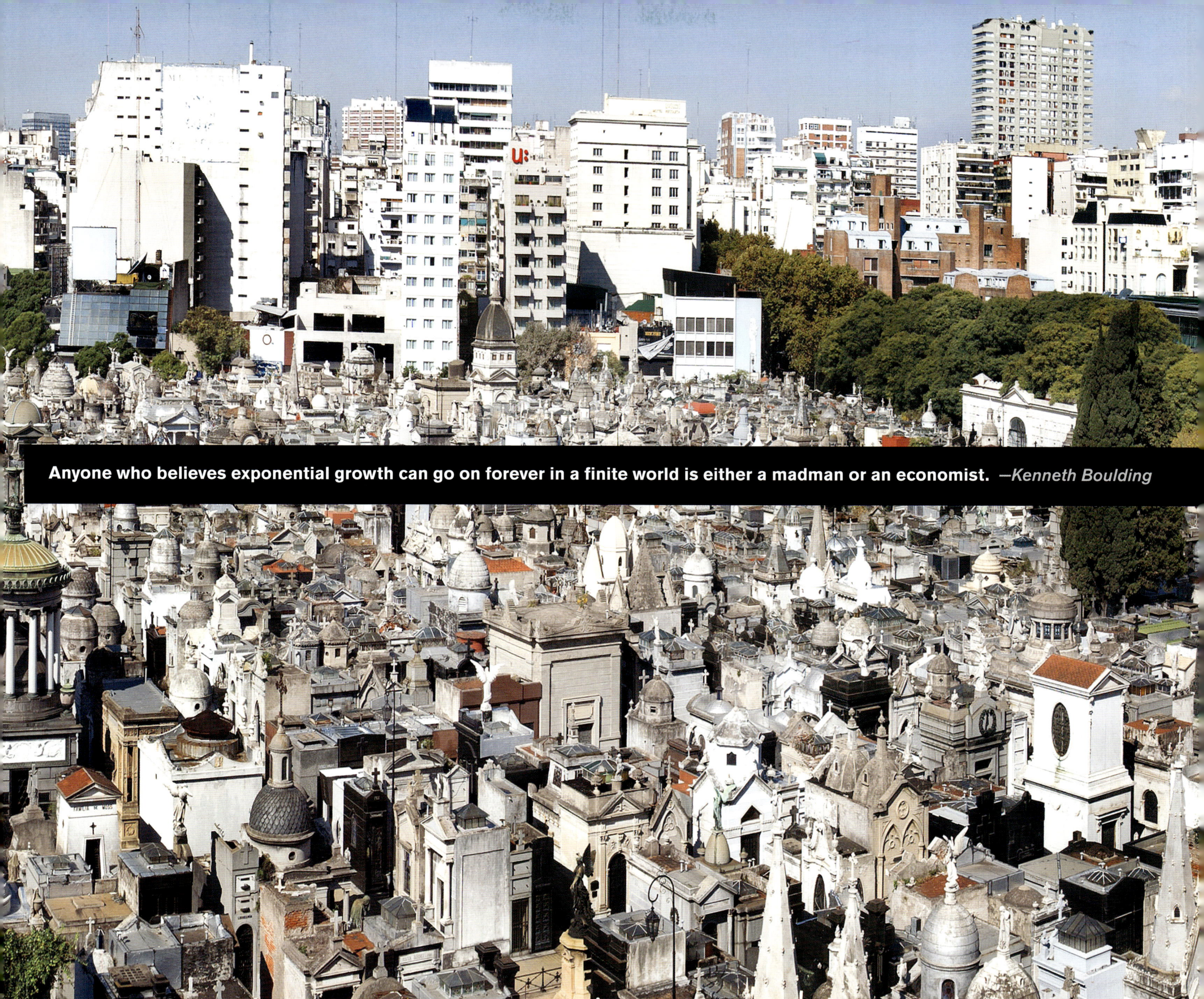

Anyone who believes exponential growth can go on forever in a finite world is either a madman or an economist. —Kenneth Boulding
BUENOS AIRES, ARGENTINA

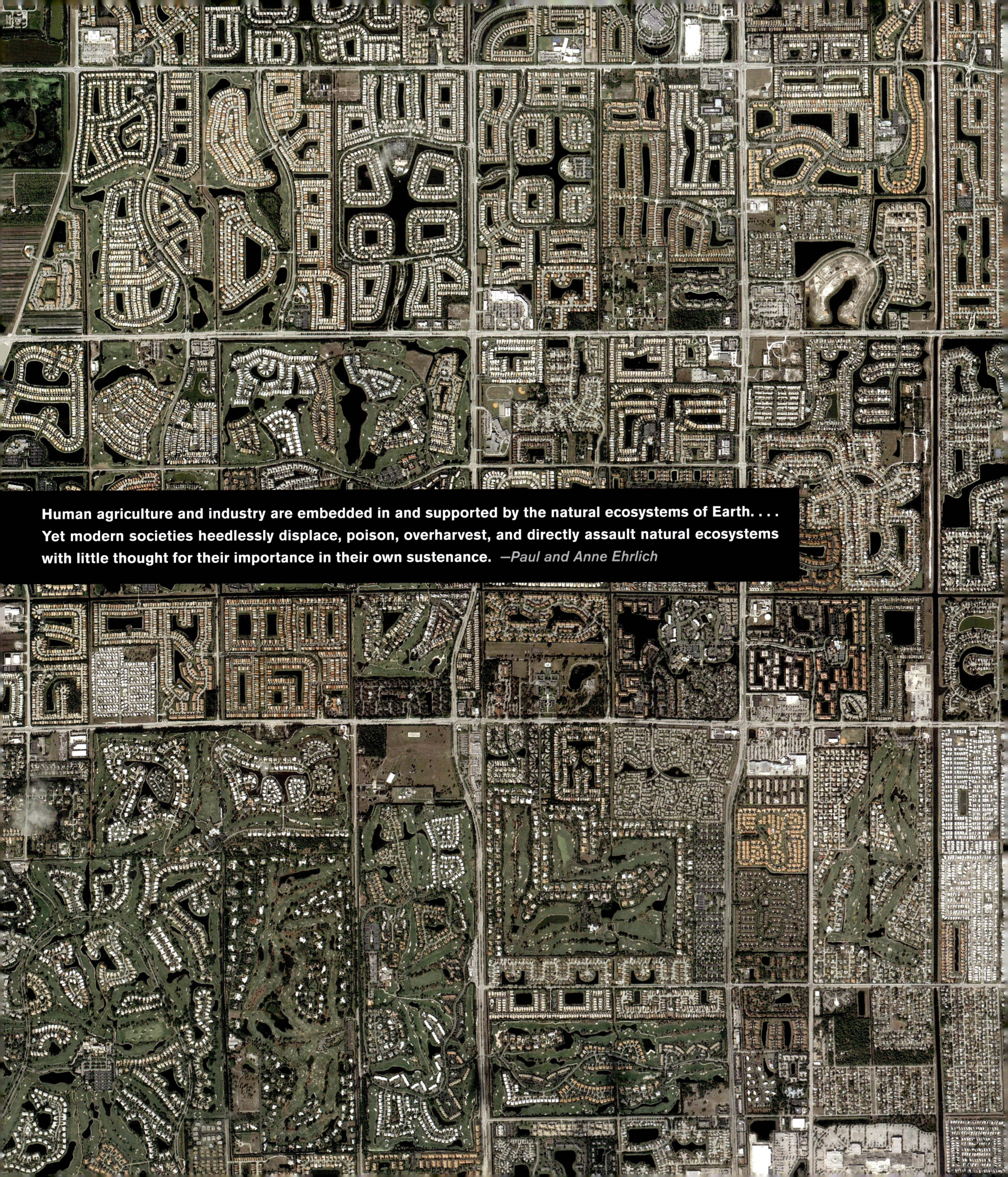
Human agriculture and industry are embedded in and supported by the natural ecosystems of Earth. . . .
Yet modern societies heedlessly displace, poison, overharvest, and directly assault natural ecosystems
with little thought for their importance in their own sustenance. —Paul and Anne Ehrlich

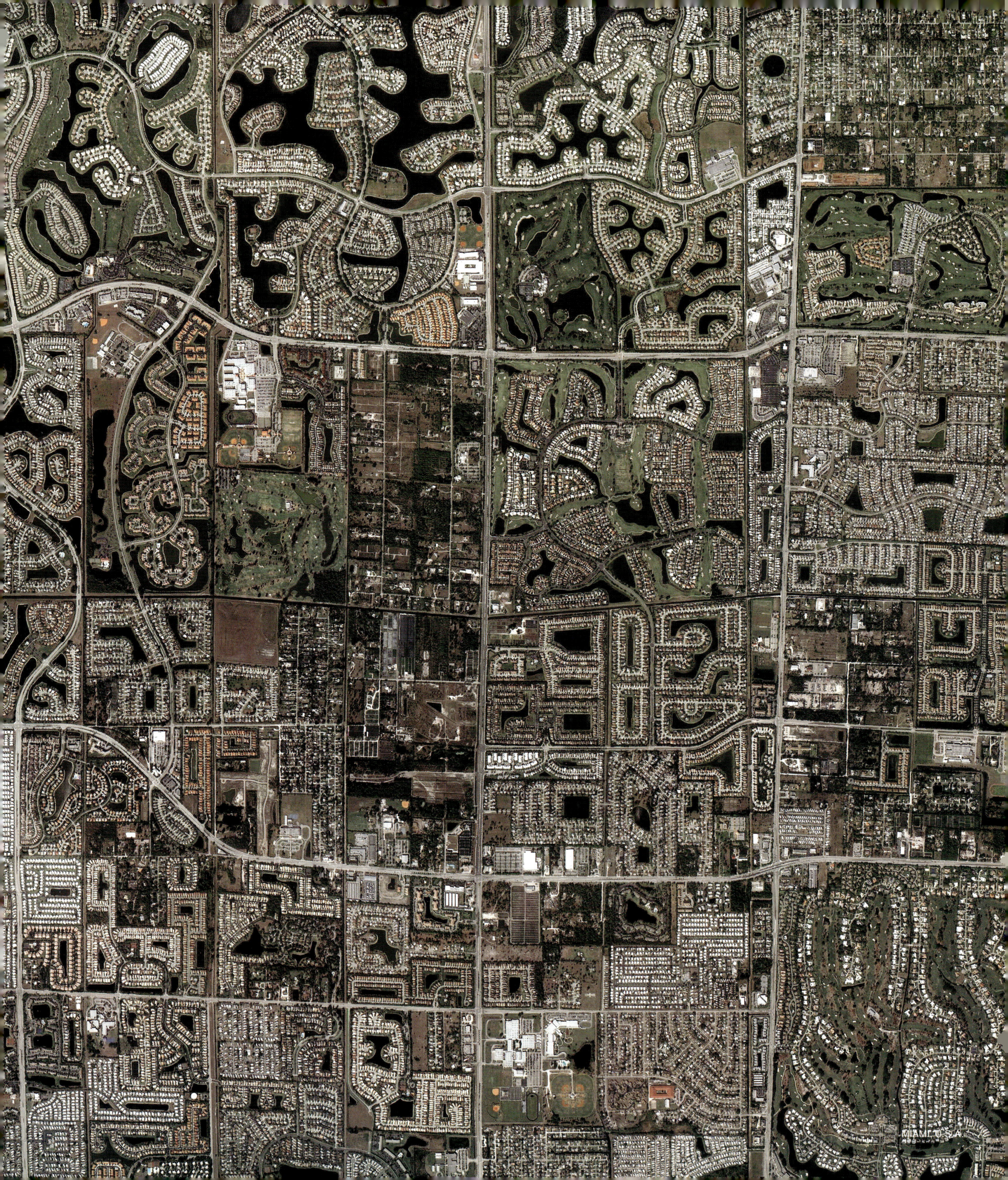
MIAMI, U.S.A.

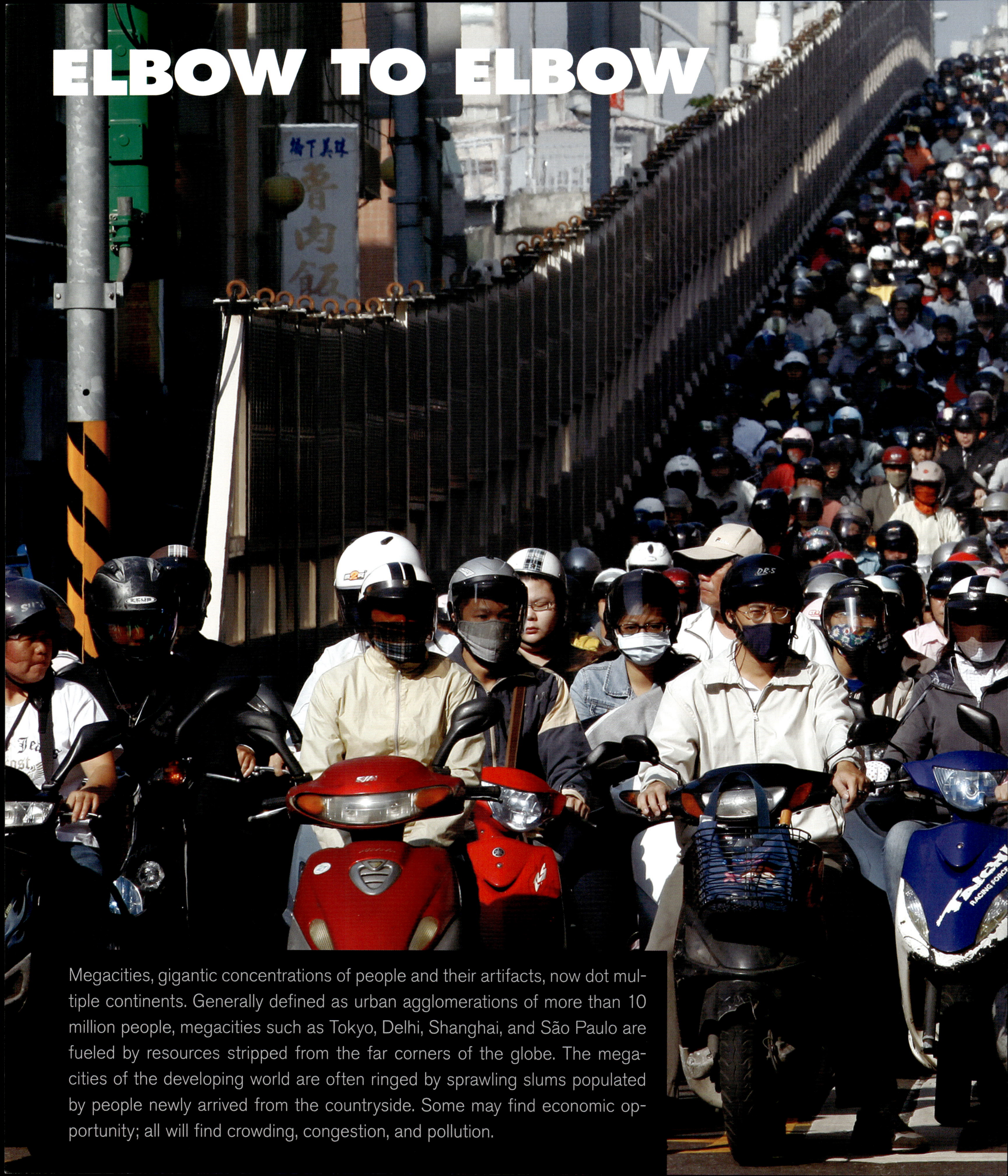

ELBOW TO ELBOW

Megacities, gigantic concentrations of people and their artifacts, now dot multiple continents. Generally defined as urban agglomerations of more than 10 million people, megacities such as Tokyo, Delhi, Shanghai, and São Paulo are fueled by resources stripped from the far corners of the globe. The megacities of the developing world are often ringed by sprawling slums populated by people newly arrived from the countryside. Some may find economic opportunity; all will find crowding, congestion, and pollution.

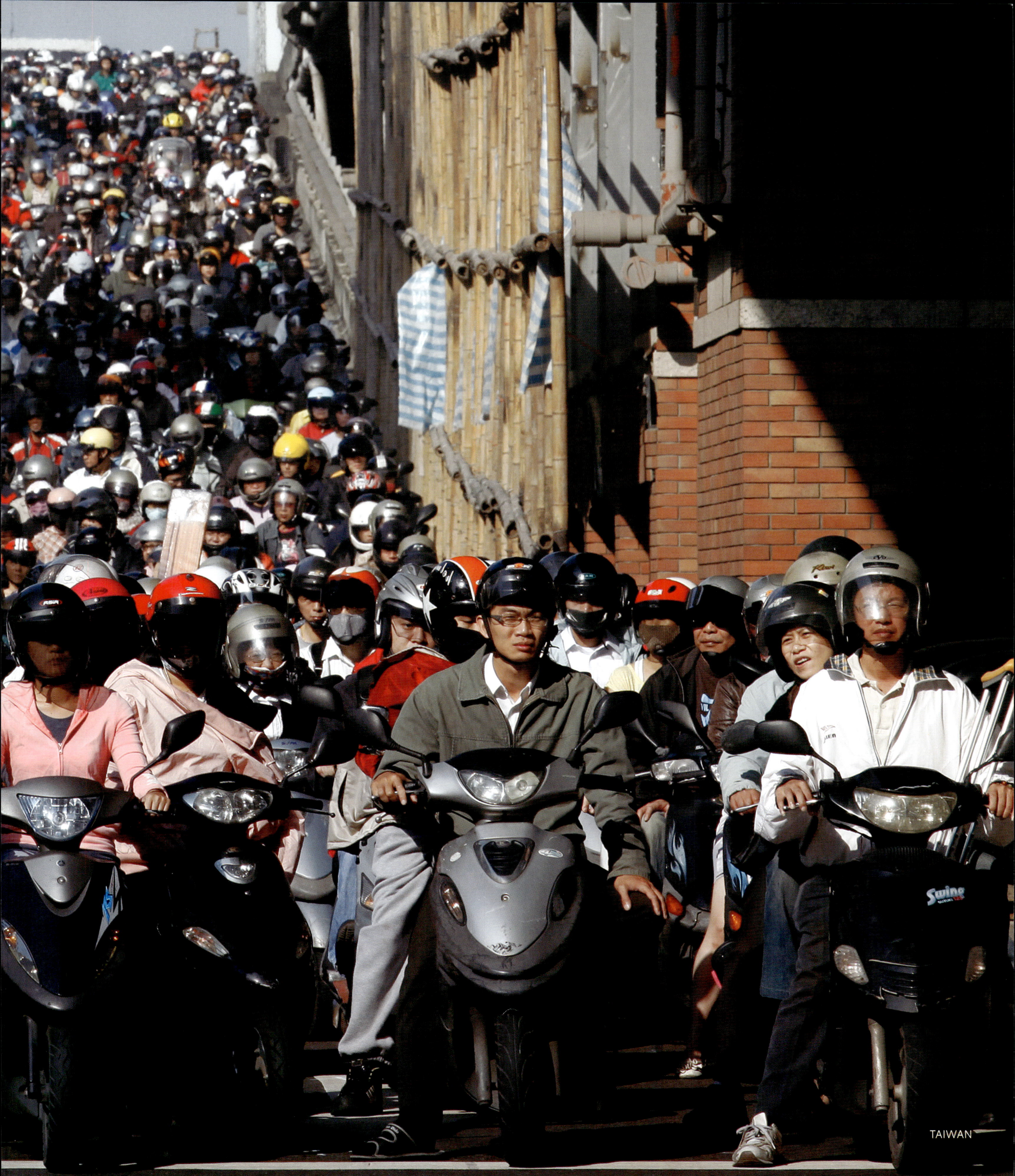

TAIWAN

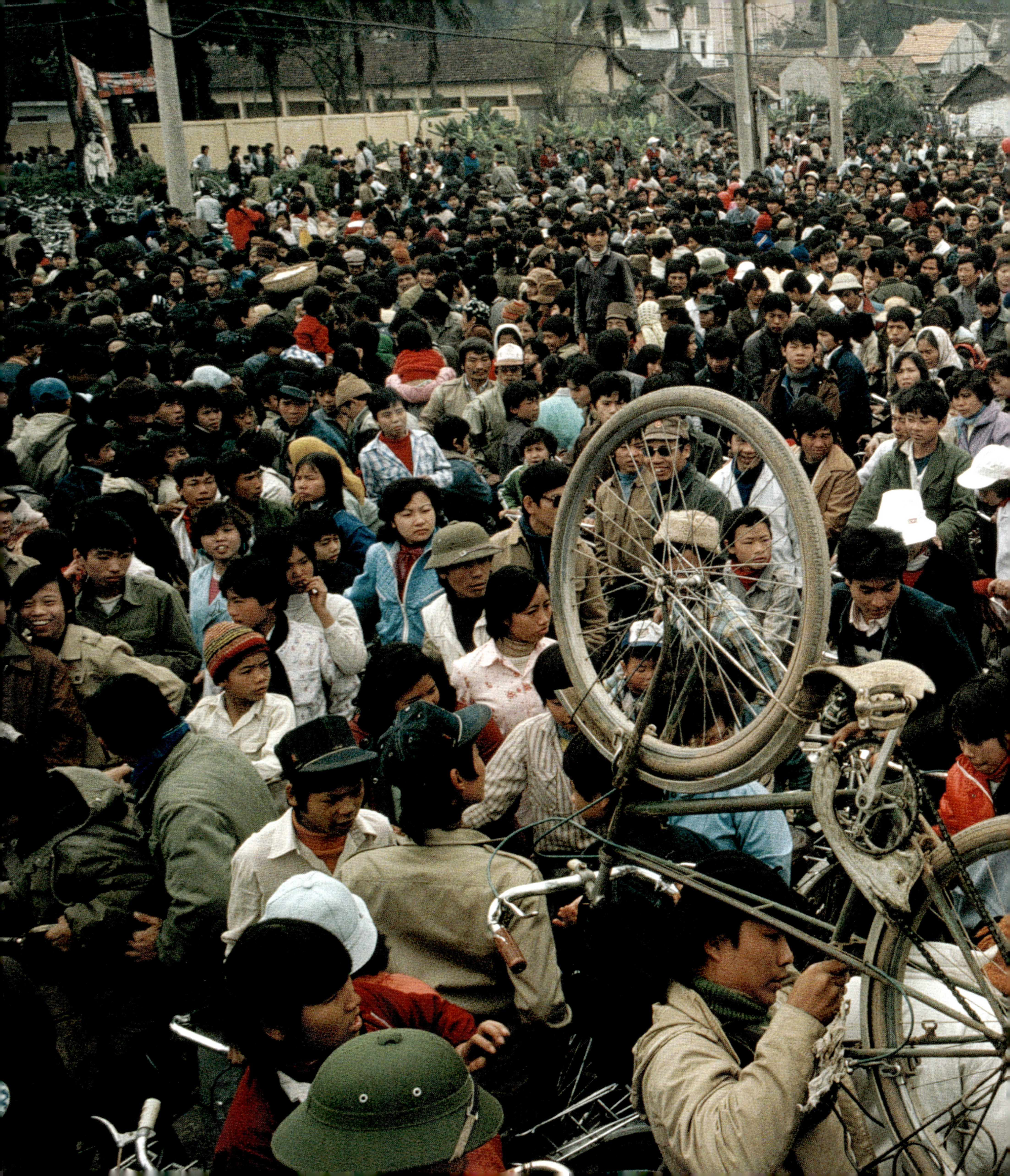

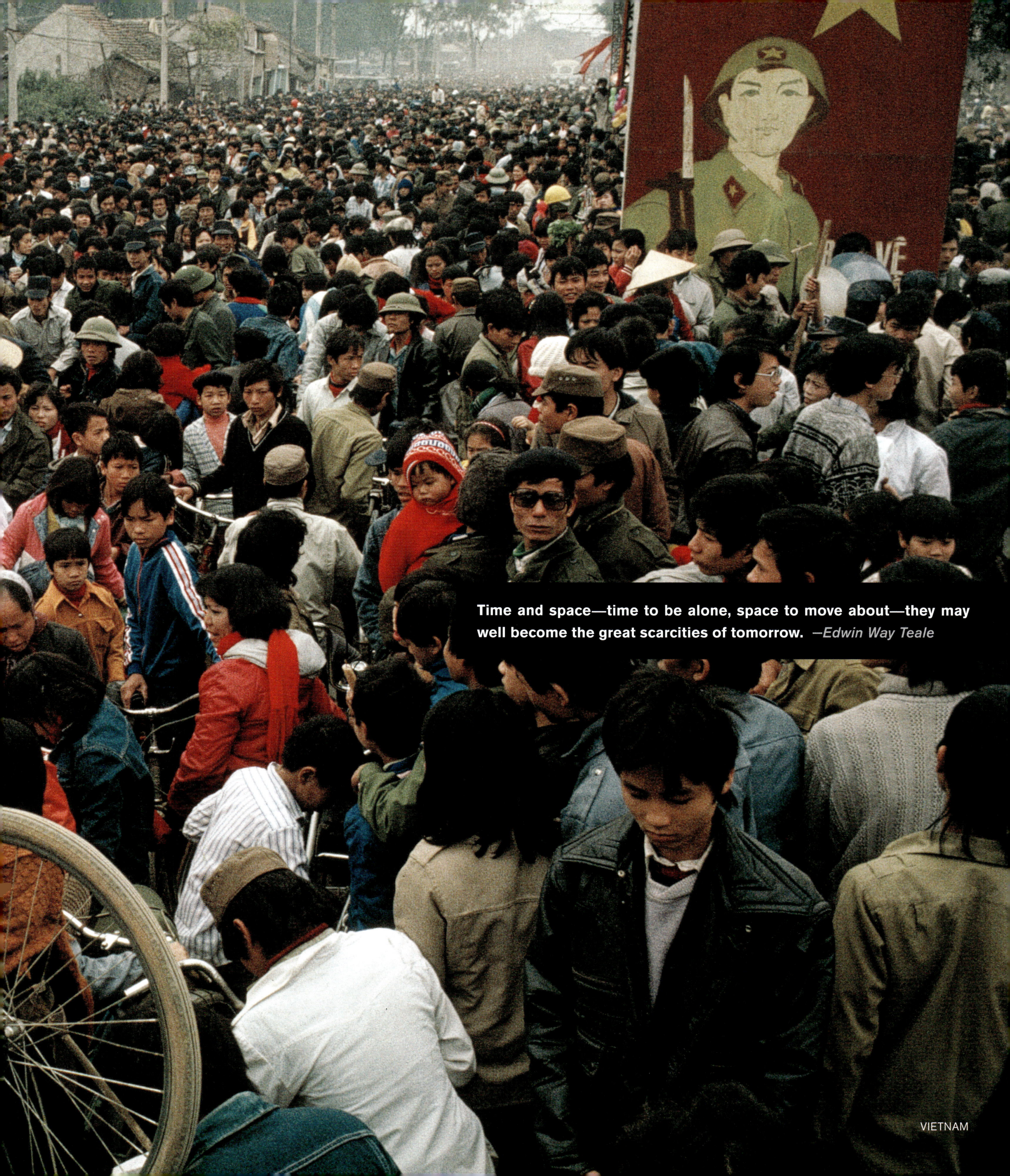
Time and space—time to be alone, space to move about—they may
well become the great scarcities of tomorrow. —Edwin Way Teale

VIETNAM

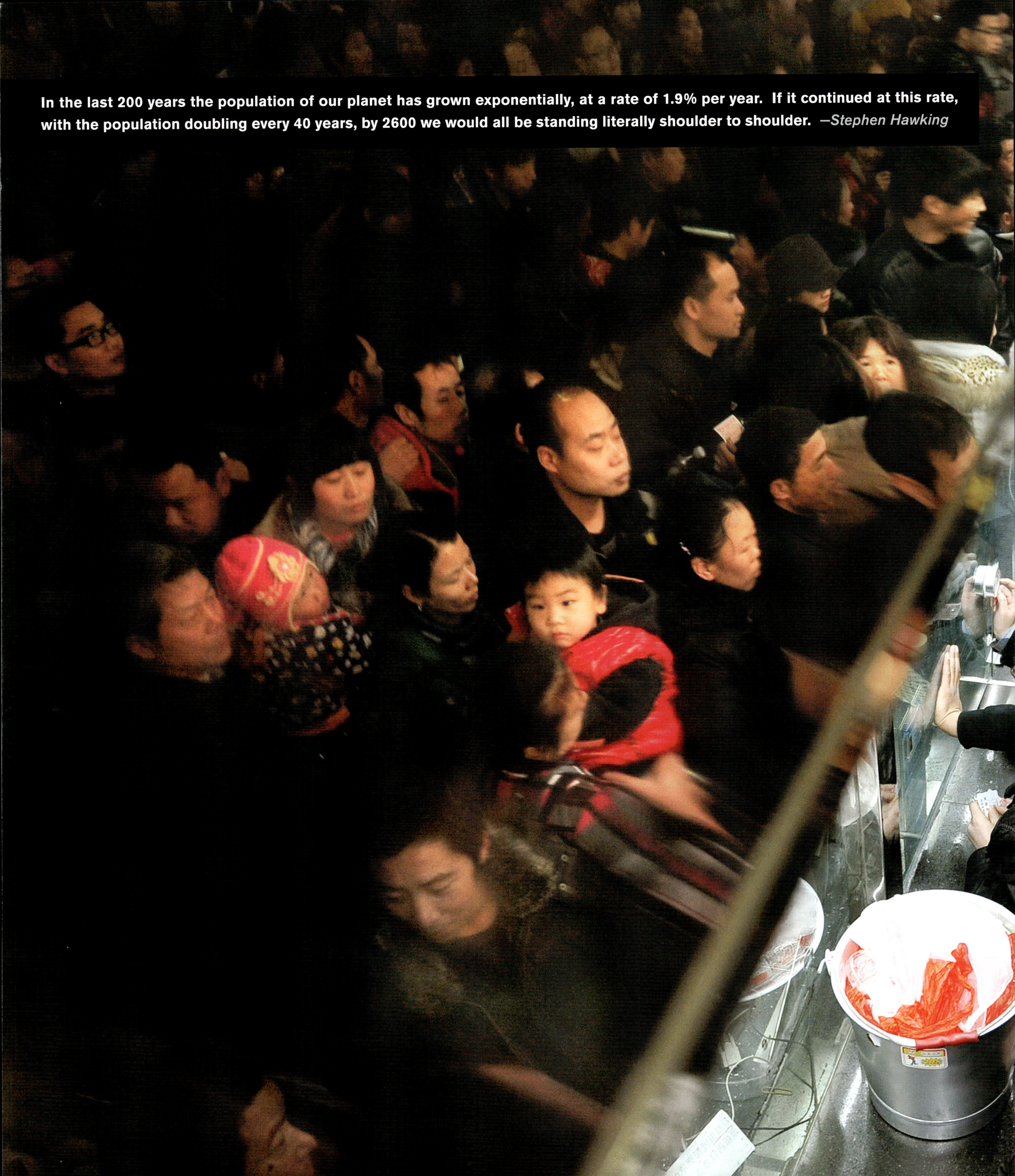

In the last 200 years the population of our planet has grown exponentially, at a rate of 1.9% per year. If it continued at this rate, with the population doubling every 40 years, by 2600 we would all be standing literally shoulder to shoulder. —Stephen Hawking

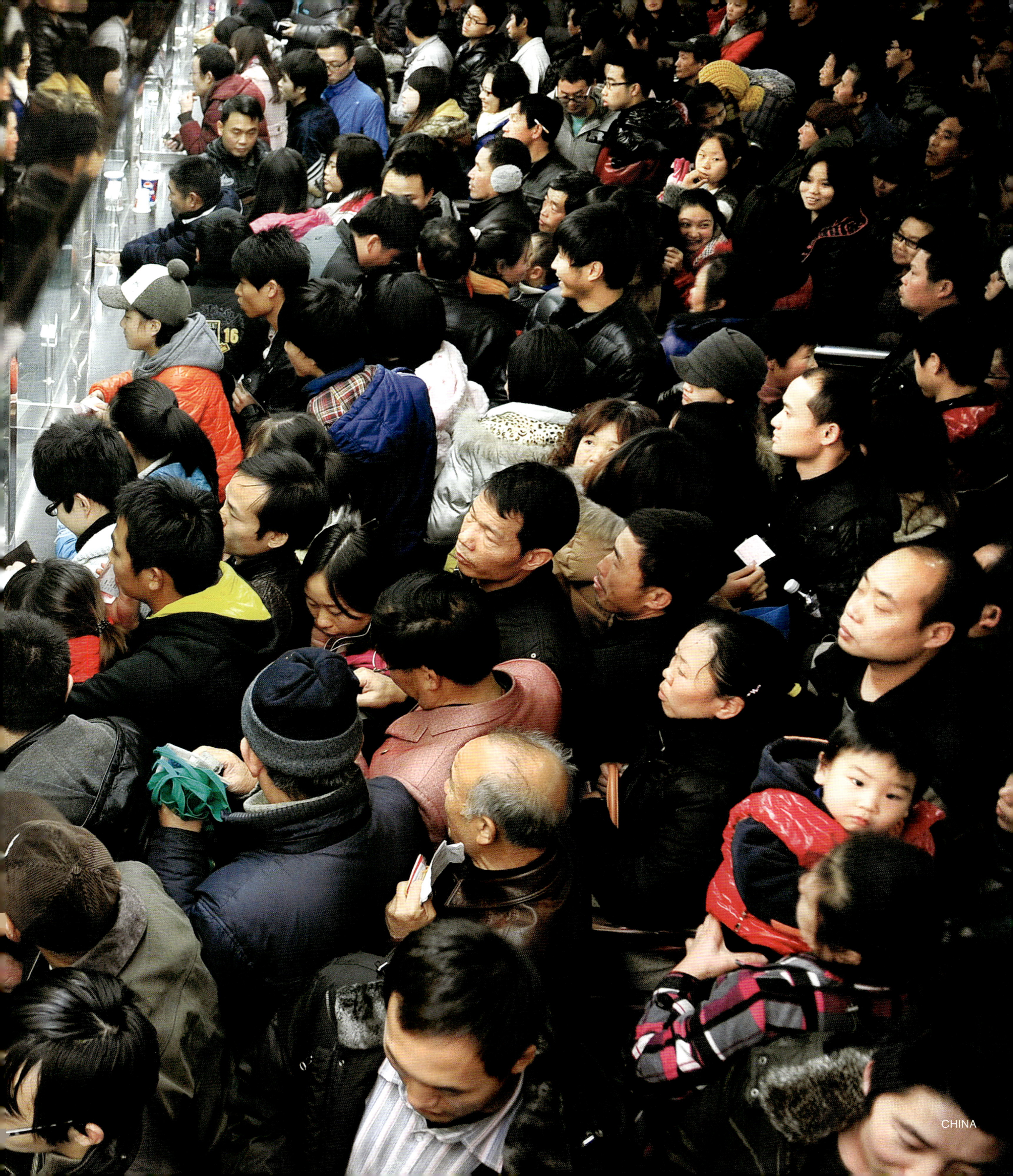
CHINA

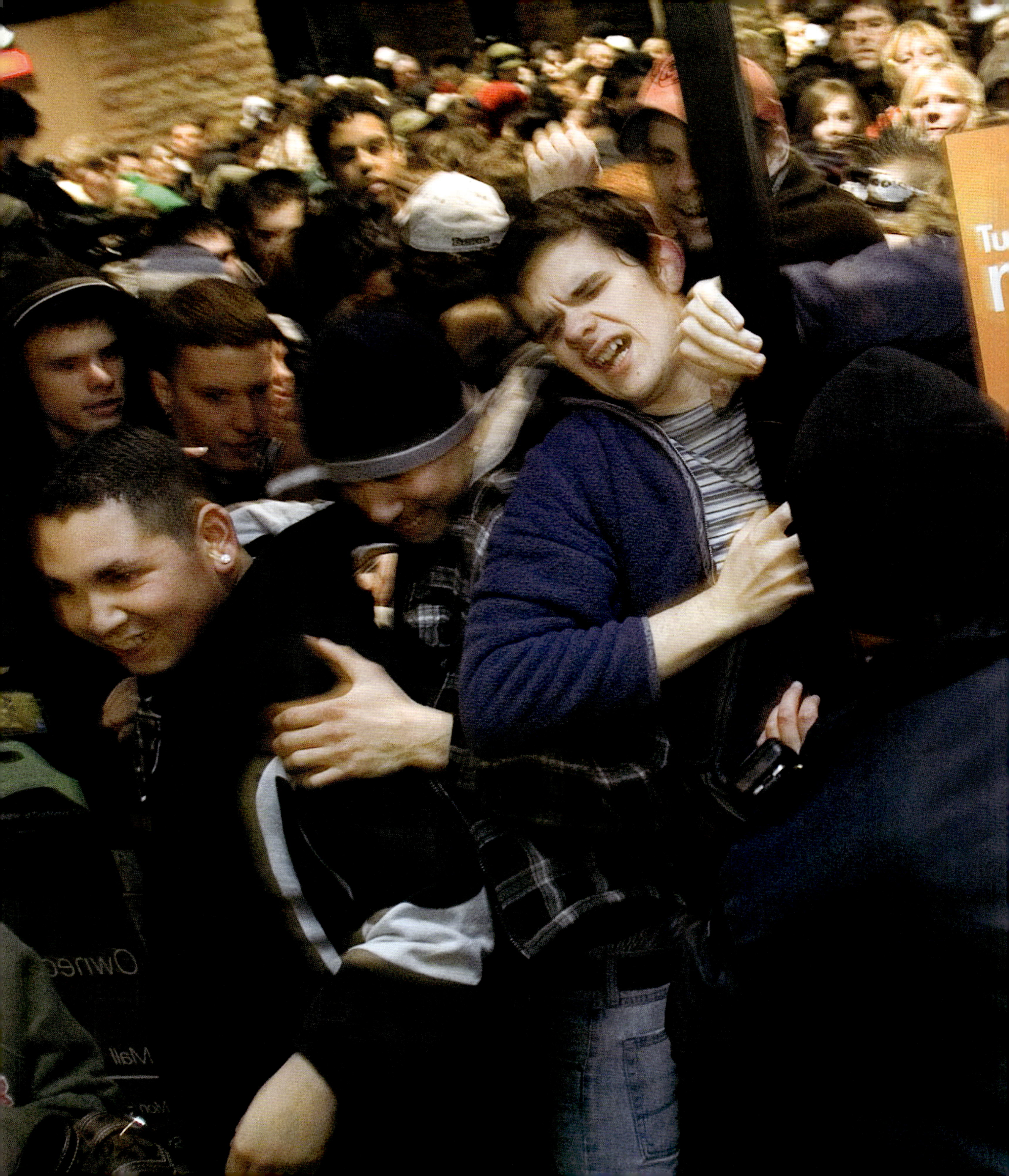

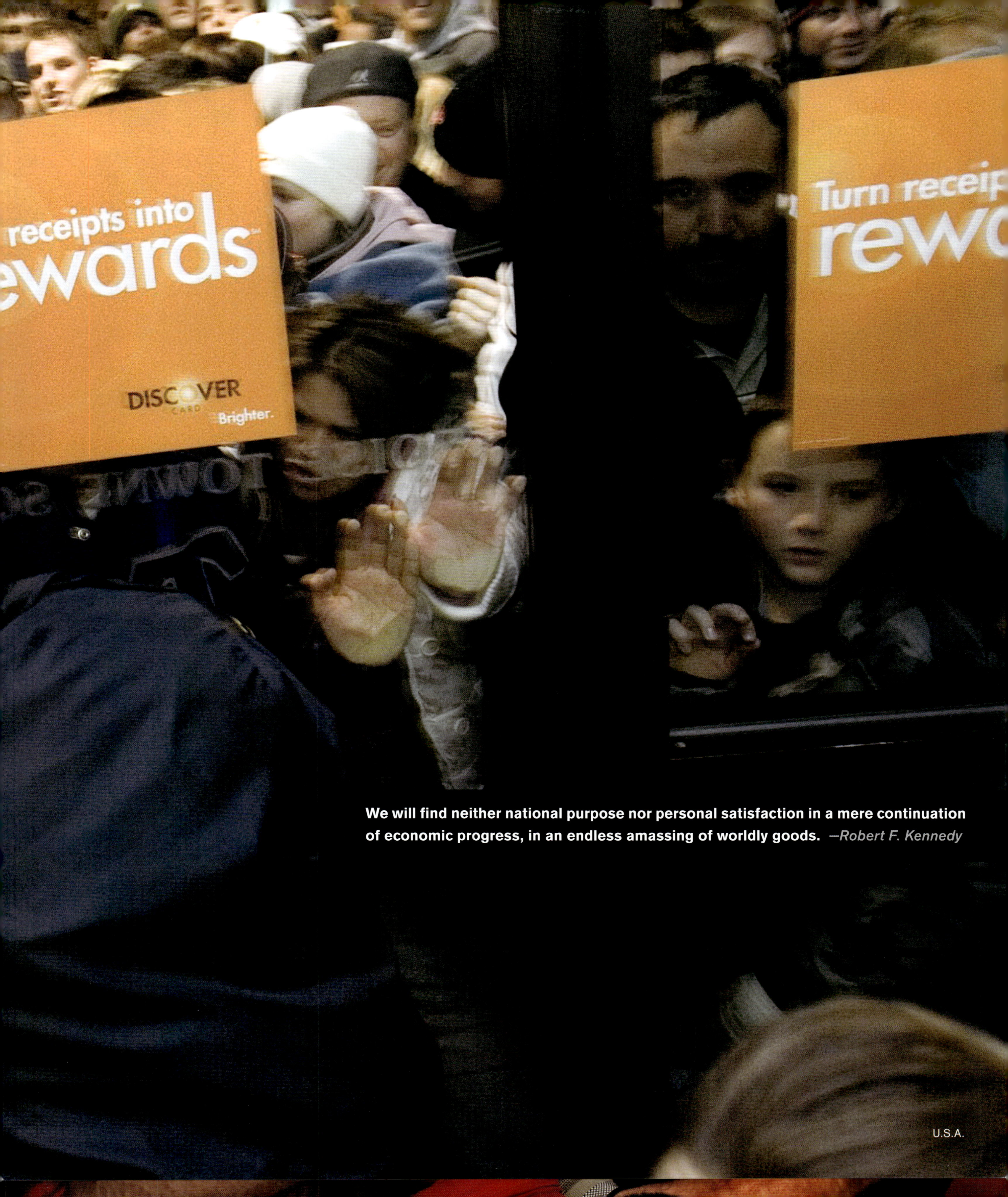

receipts into
ewards
DISCOVER
CARD
Brighter.
Turn receip
rewa
We will find neither national purpose nor personal satisfaction in a mere continuation
of economic progress, in an endless amassing of worldly goods. —Robert F. Kennedy
U.S.A.

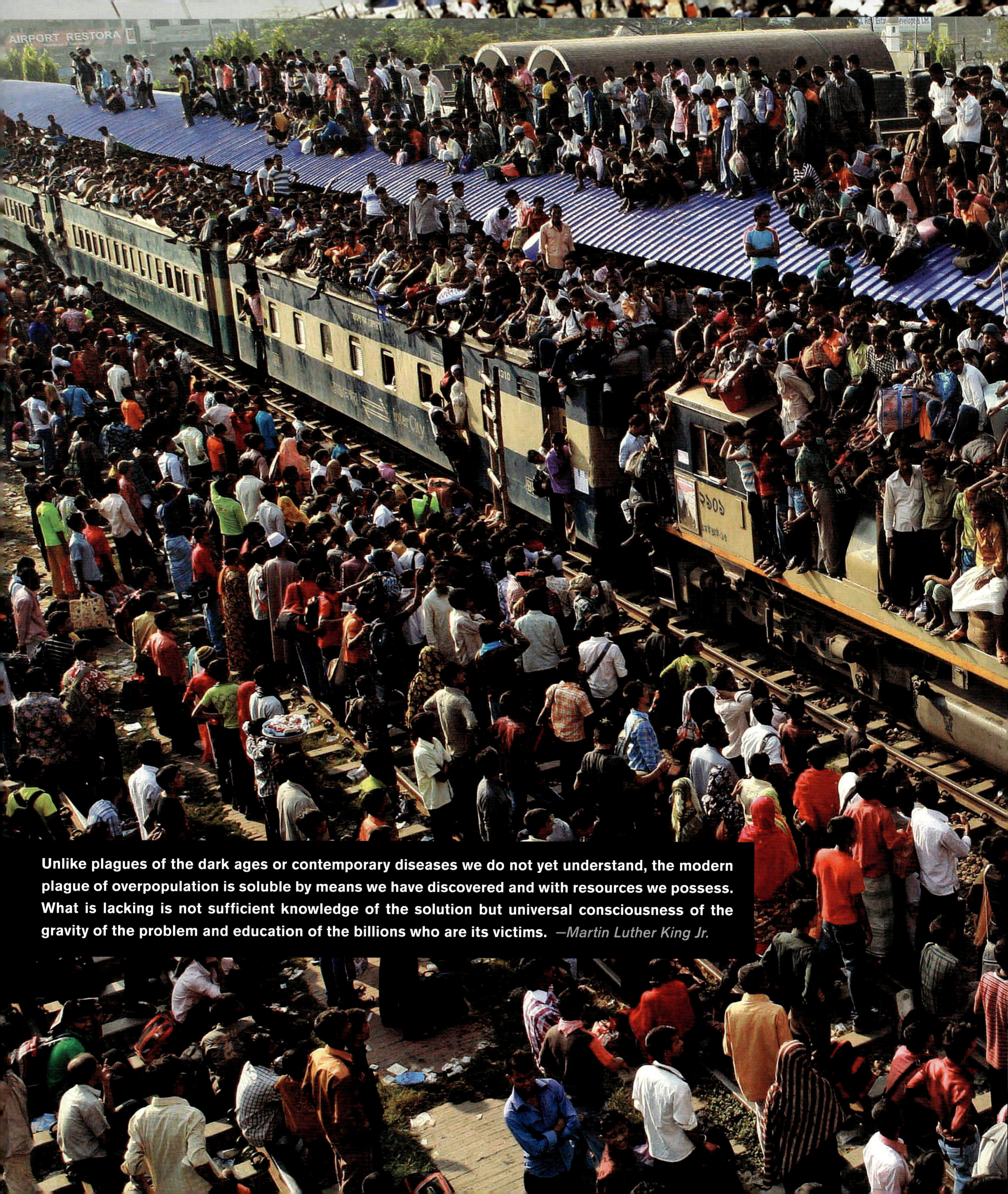

Unlike plagues of the dark ages or contemporary diseases we do not yet understand, the modern plague of overpopulation is soluble by means we have discovered and with resources we possess. What is lacking is not sufficient knowledge of the solution but universal consciousness of the gravity of the problem and education of the billions who are its victims. —Martin Luther King Jr.

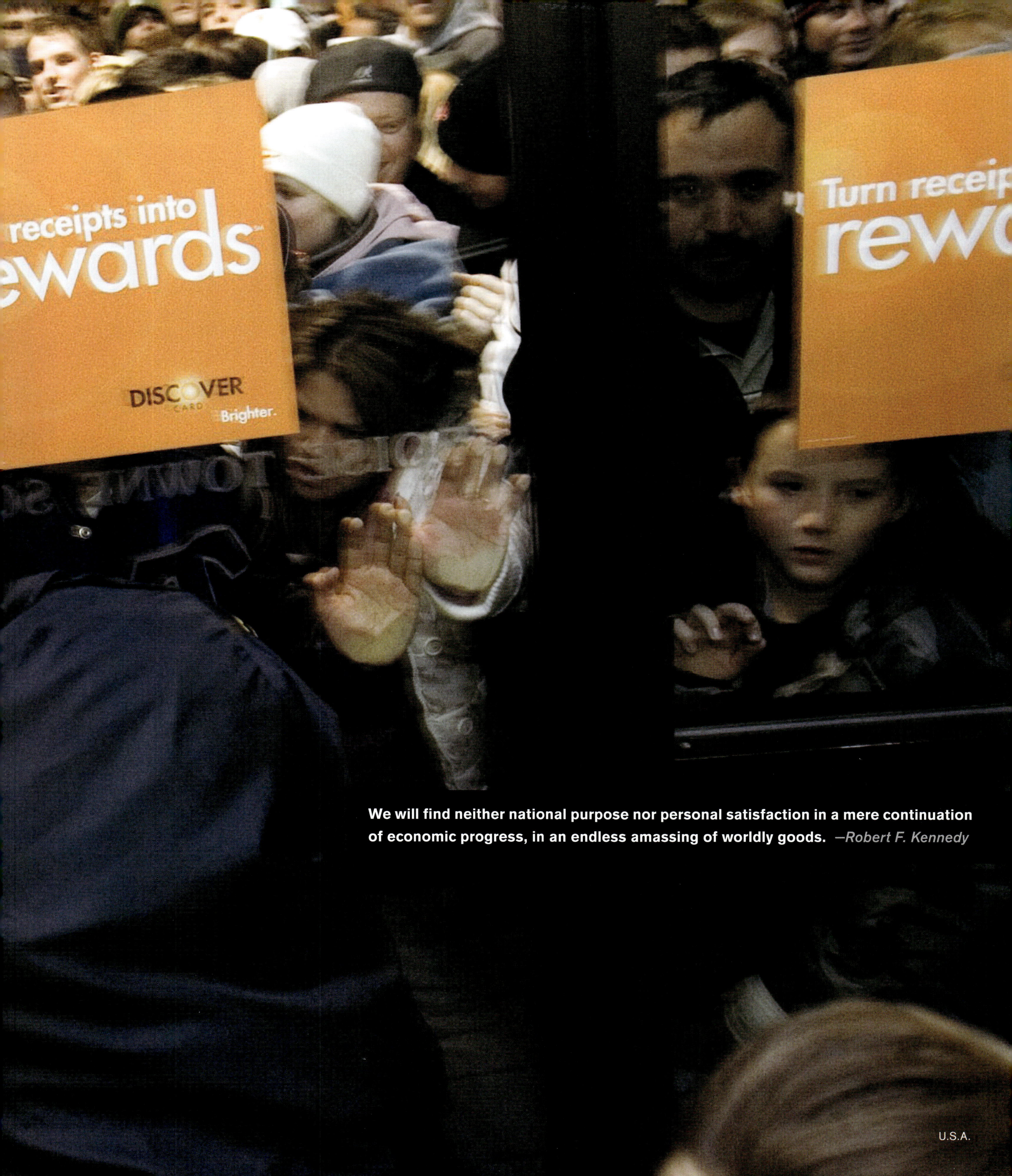
receipts into
ewards
DISCOVER
Brighter.
Turn receip
rewc
We will find neither national purpose nor personal satisfaction in a mere continuation
of economic progress, in an endless amassing of worldly goods. —Robert F. Kennedy
U.S.A.

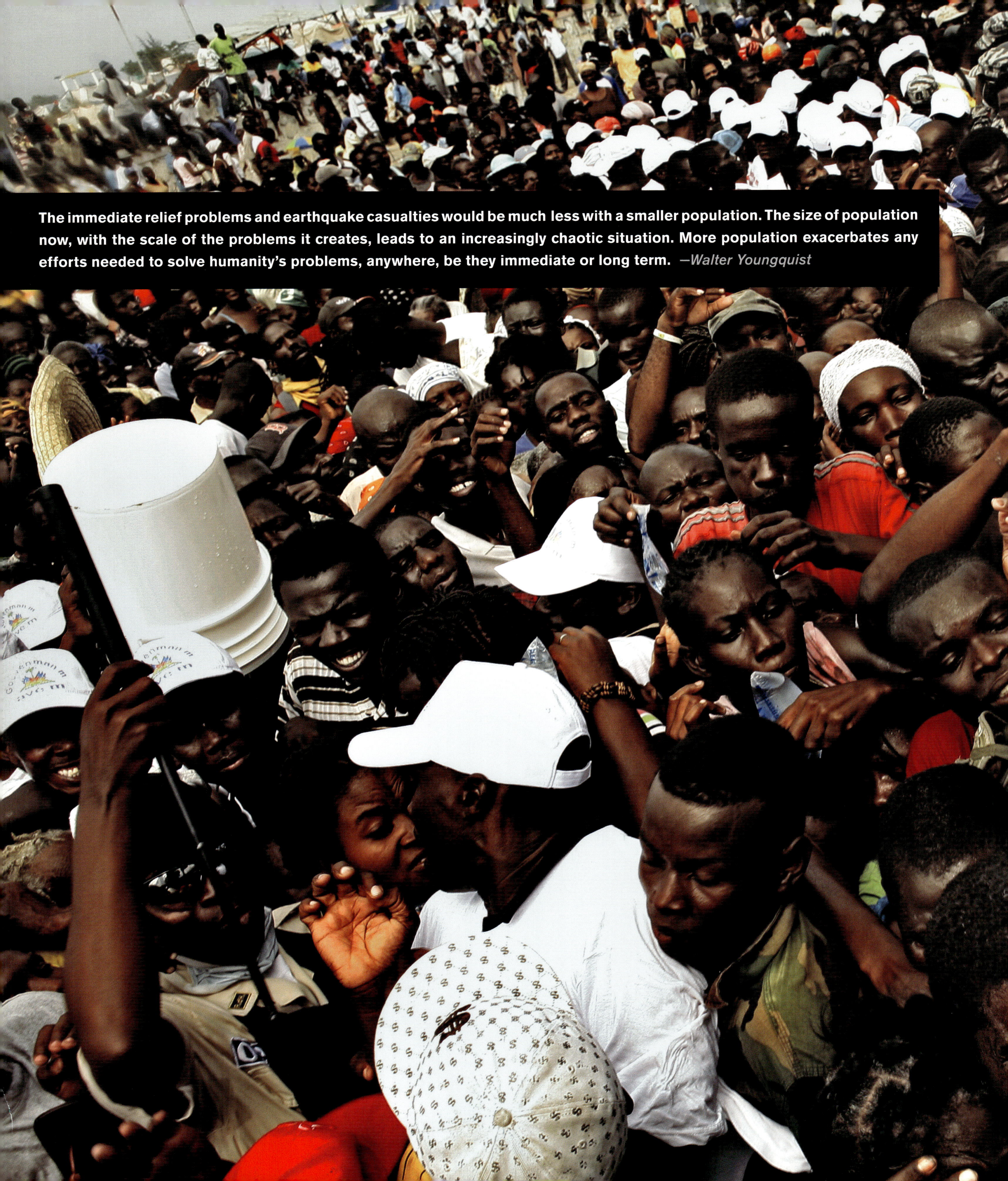
The immediate relief problems and earthquake casualties would be much less with a smaller population. The size of population now, with the scale of the problems it creates, leads to an increasingly chaotic situation. More population exacerbates any efforts needed to solve humanity's problems, anywhere, be they immediate or long term. —Walter Youngquist

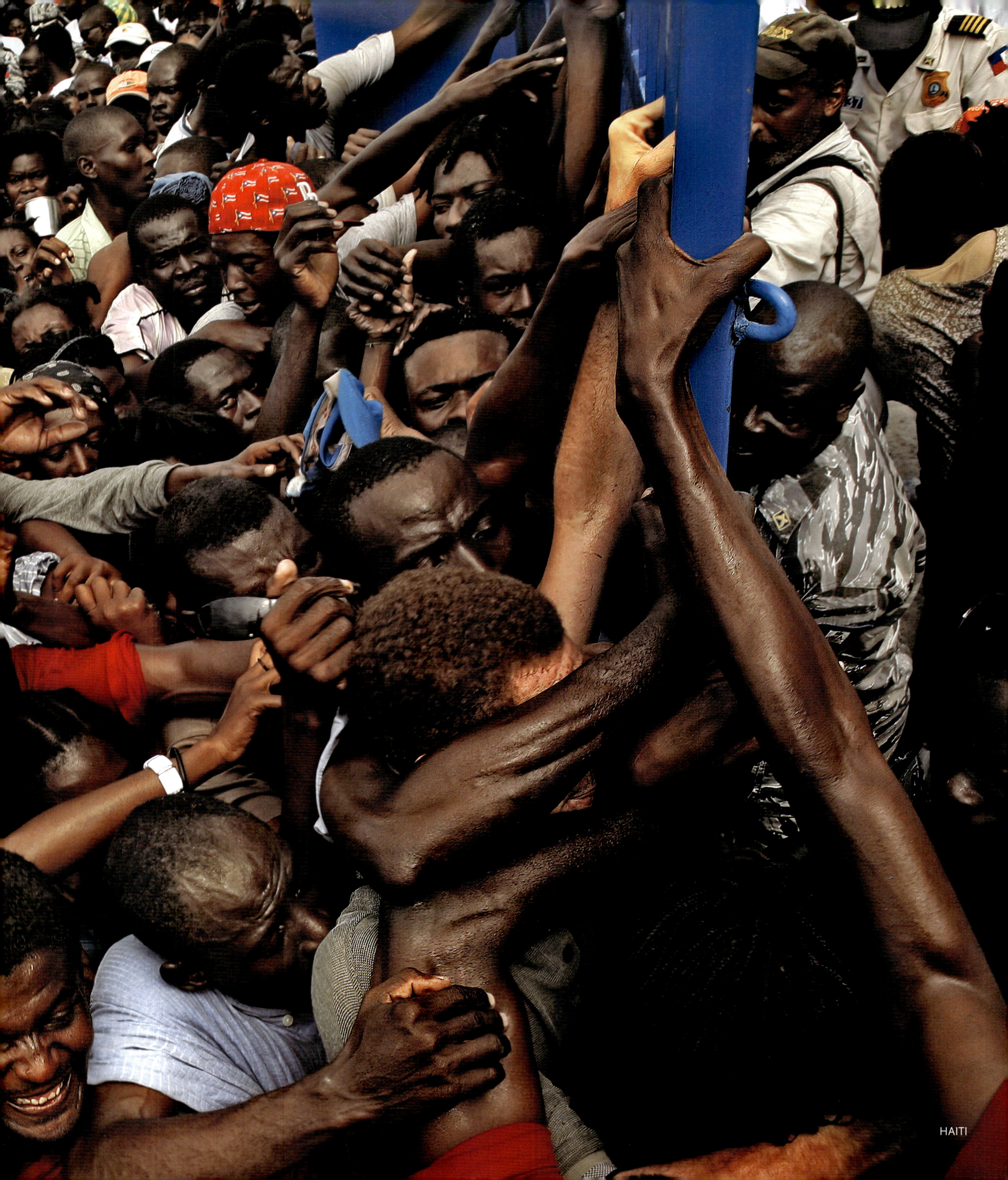
HAITI

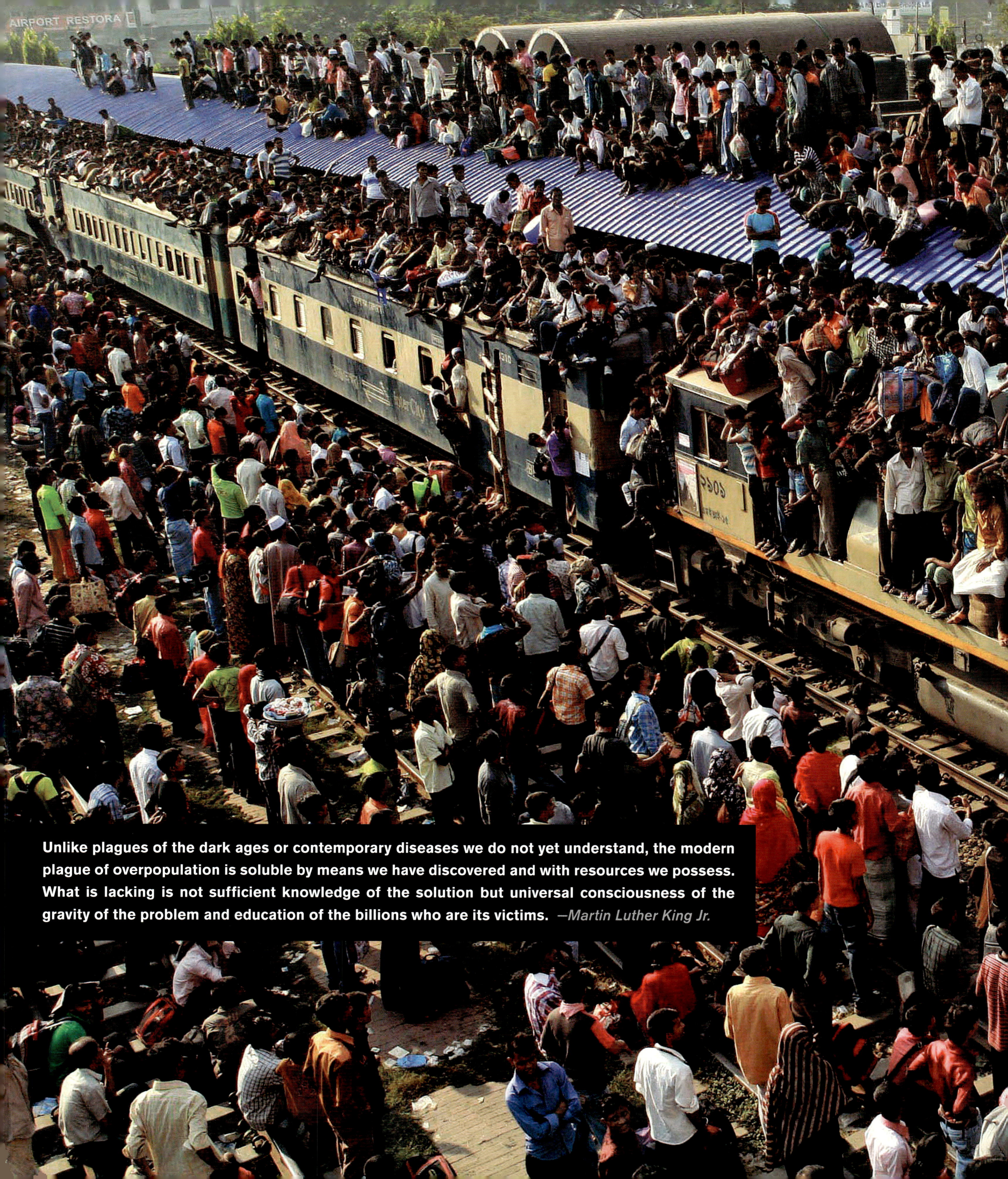

Unlike plagues of the dark ages or contemporary diseases we do not yet understand, the modern plague of overpopulation is soluble by means we have discovered and with resources we possess. What is lacking is not sufficient knowledge of the solution but universal consciousness of the gravity of the problem and education of the billions who are its victims. —*Martin Luther King Jr.*

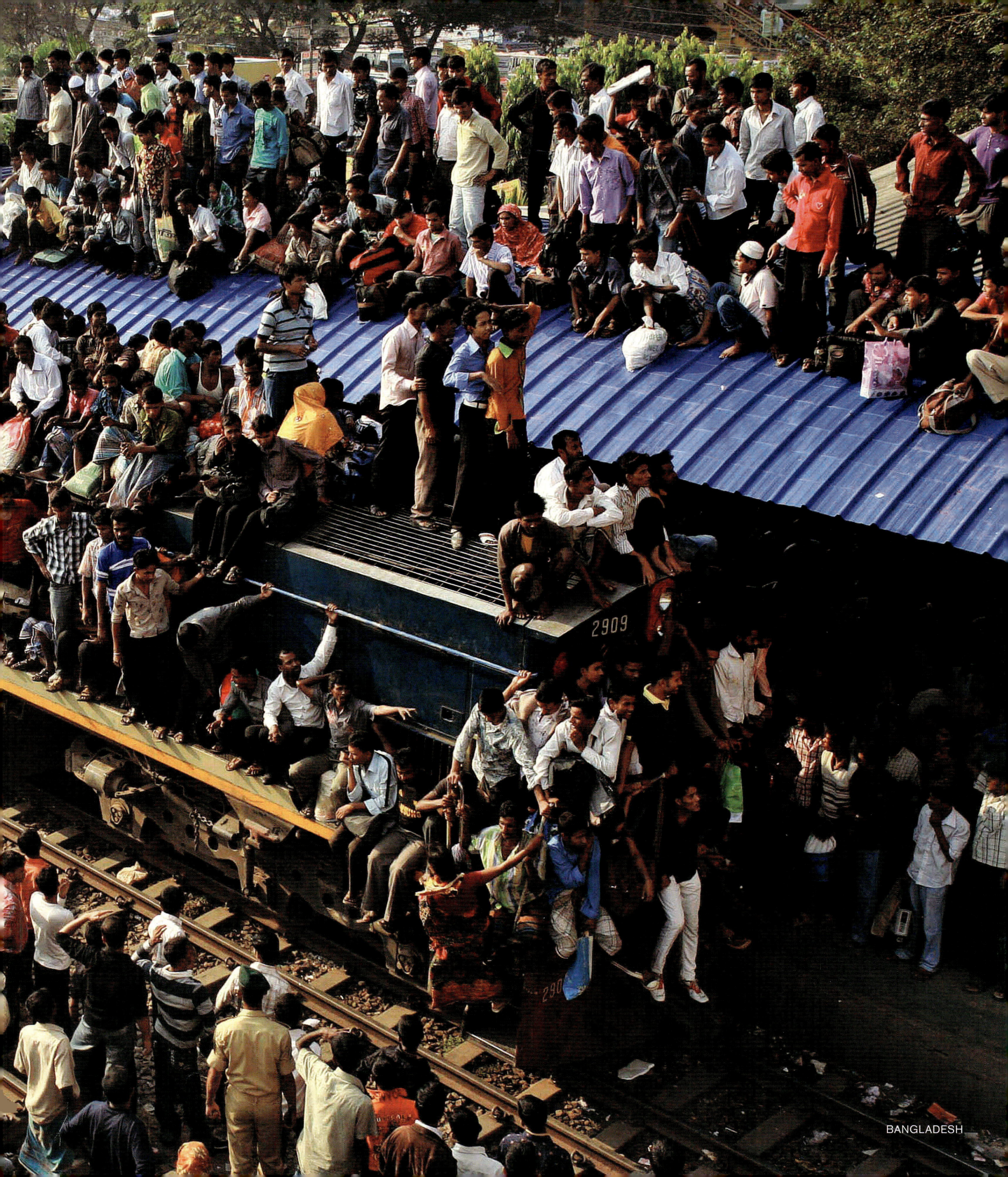
BANGLADESH

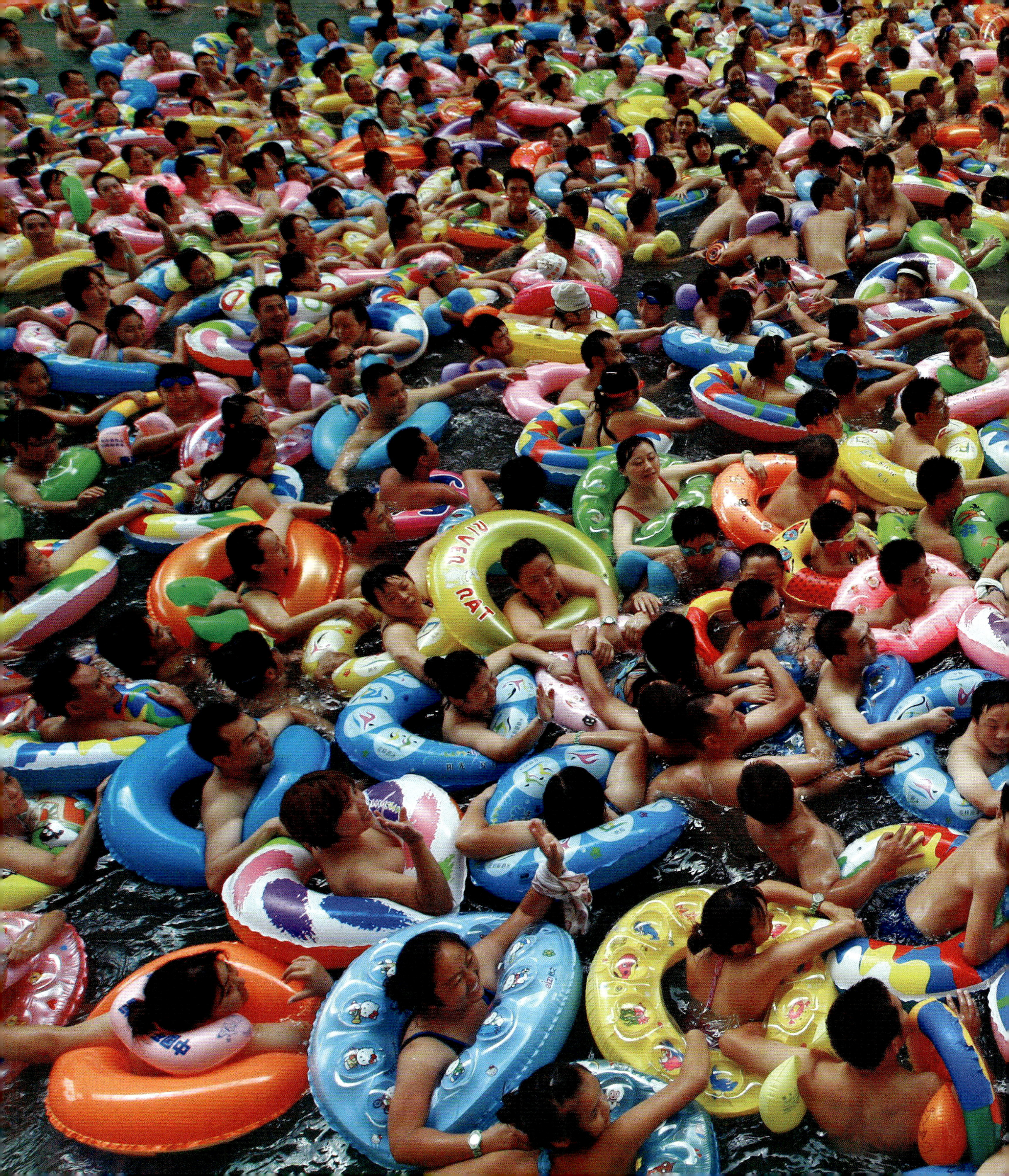

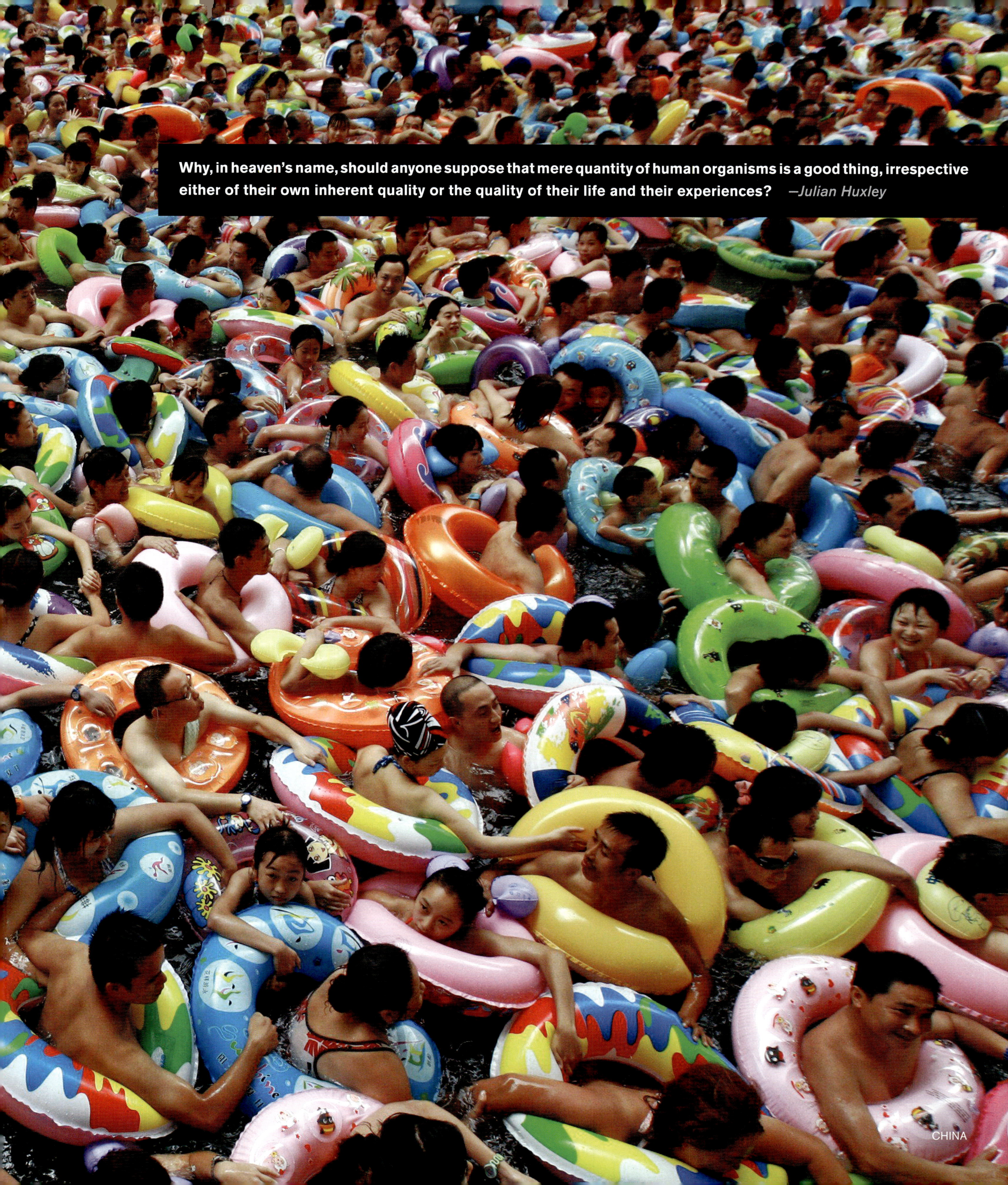
Why, in heaven's name, should anyone suppose that mere quantity of human organisms is a good thing, irrespective either of their own inherent quality or the quality of their life and their experiences?   —Julian Huxley
CHINA

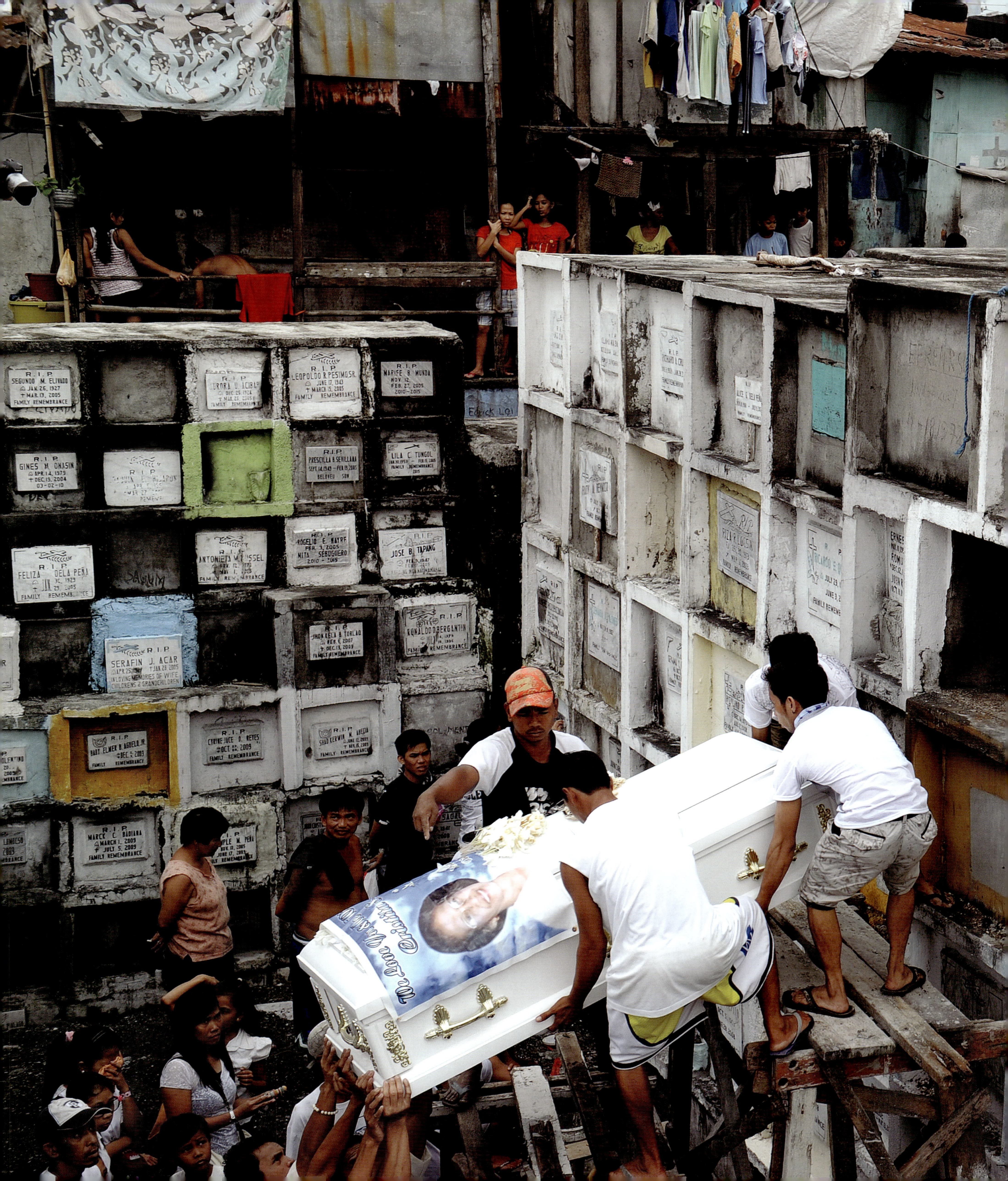

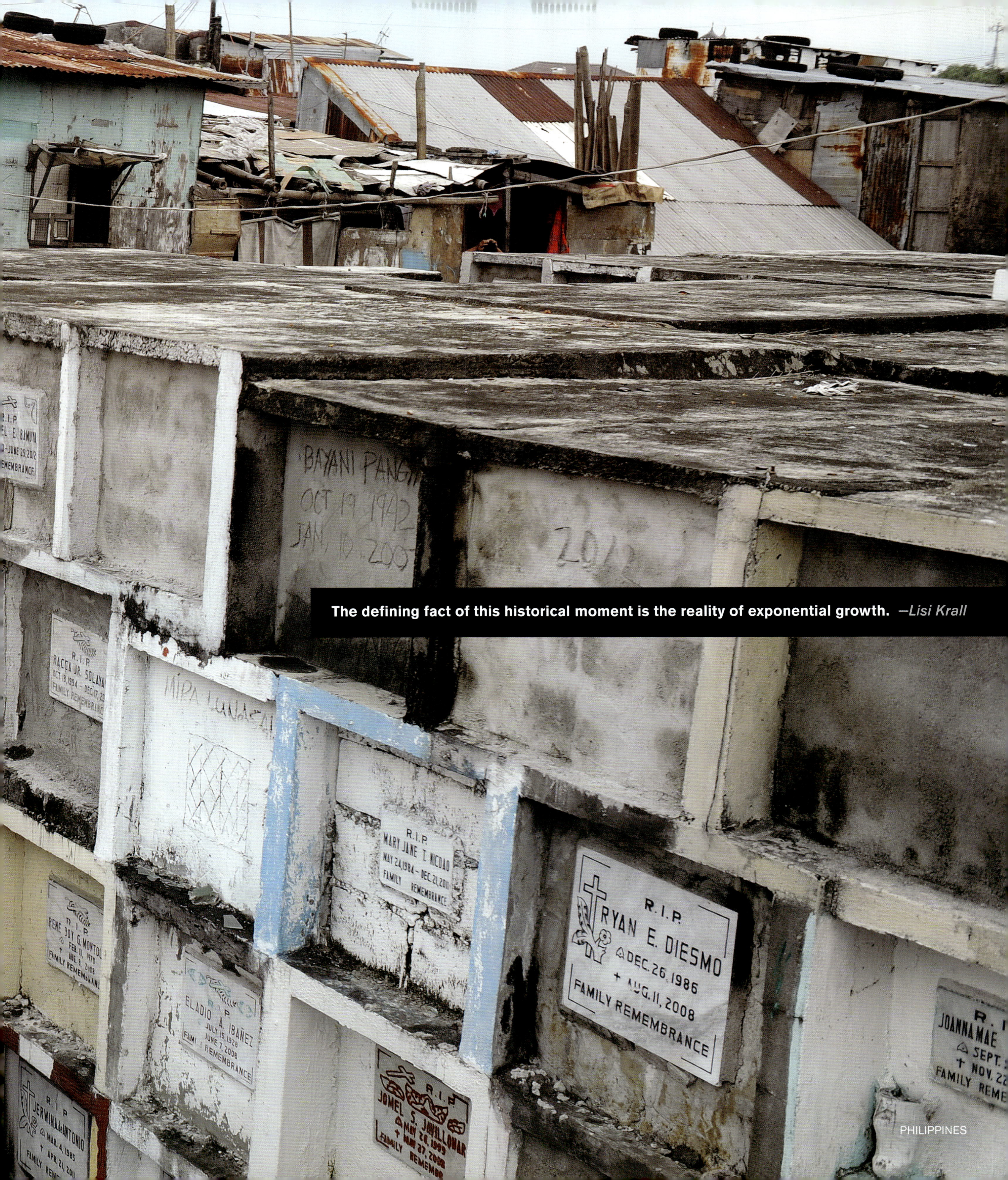

The defining fact of this historical moment is the reality of exponential growth. —Lisi Krall
PHILIPPINES

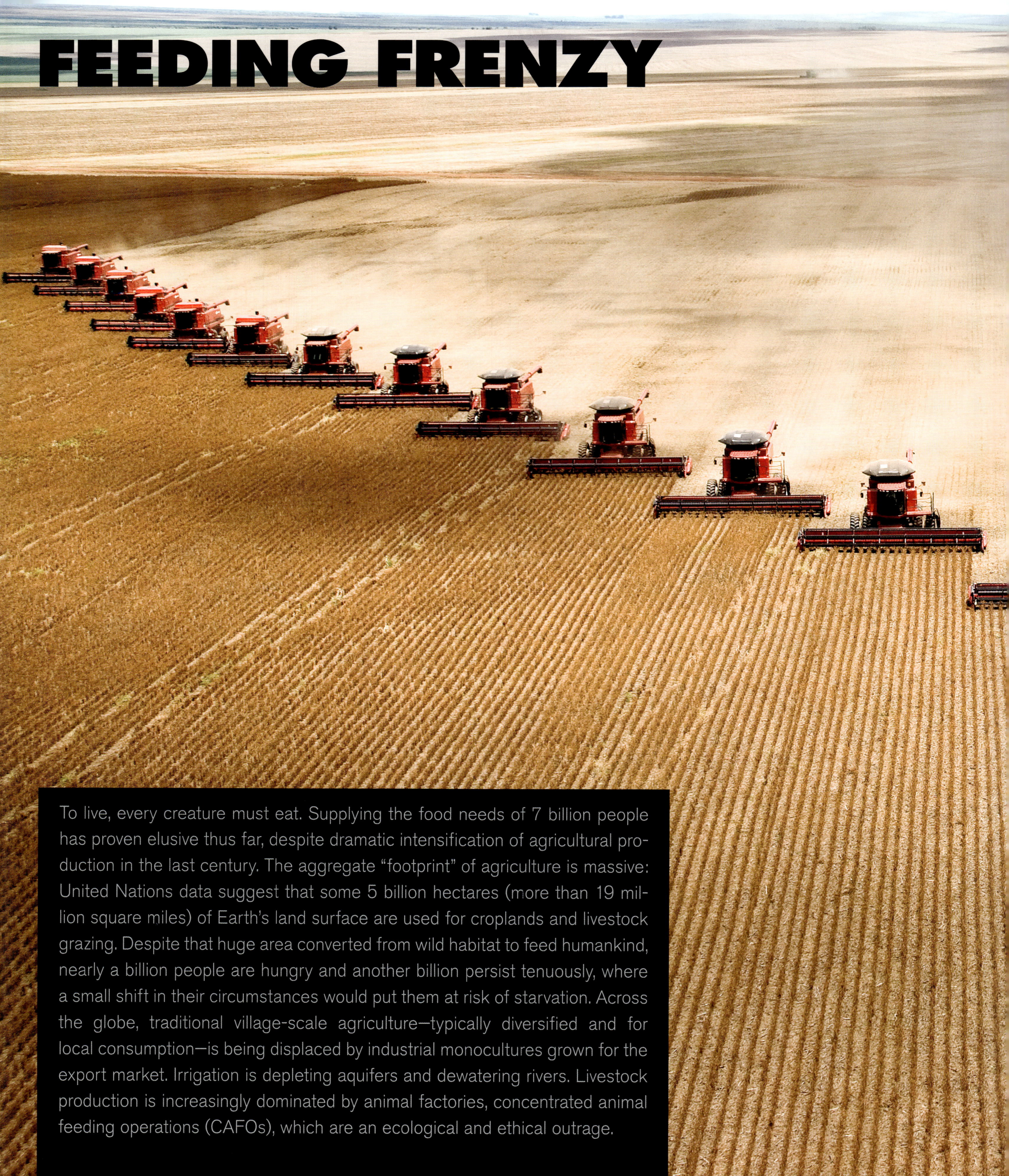

FEEDING FRENZY

To live, every creature must eat. Supplying the food needs of 7 billion people has proven elusive thus far, despite dramatic intensification of agricultural production in the last century. The aggregate "footprint" of agriculture is massive: United Nations data suggest that some 5 billion hectares (more than 19 million square miles) of Earth's land surface are used for croplands and livestock grazing. Despite that huge area converted from wild habitat to feed humankind, nearly a billion people are hungry and another billion persist tenuously, where a small shift in their circumstances would put them at risk of starvation. Across the globe, traditional village-scale agriculture—typically diversified and for local consumption—is being displaced by industrial monocultures grown for the export market. Irrigation is depleting aquifers and dewatering rivers. Livestock production is increasingly dominated by animal factories, concentrated animal feeding operations (CAFOs), which are an ecological and ethical outrage.

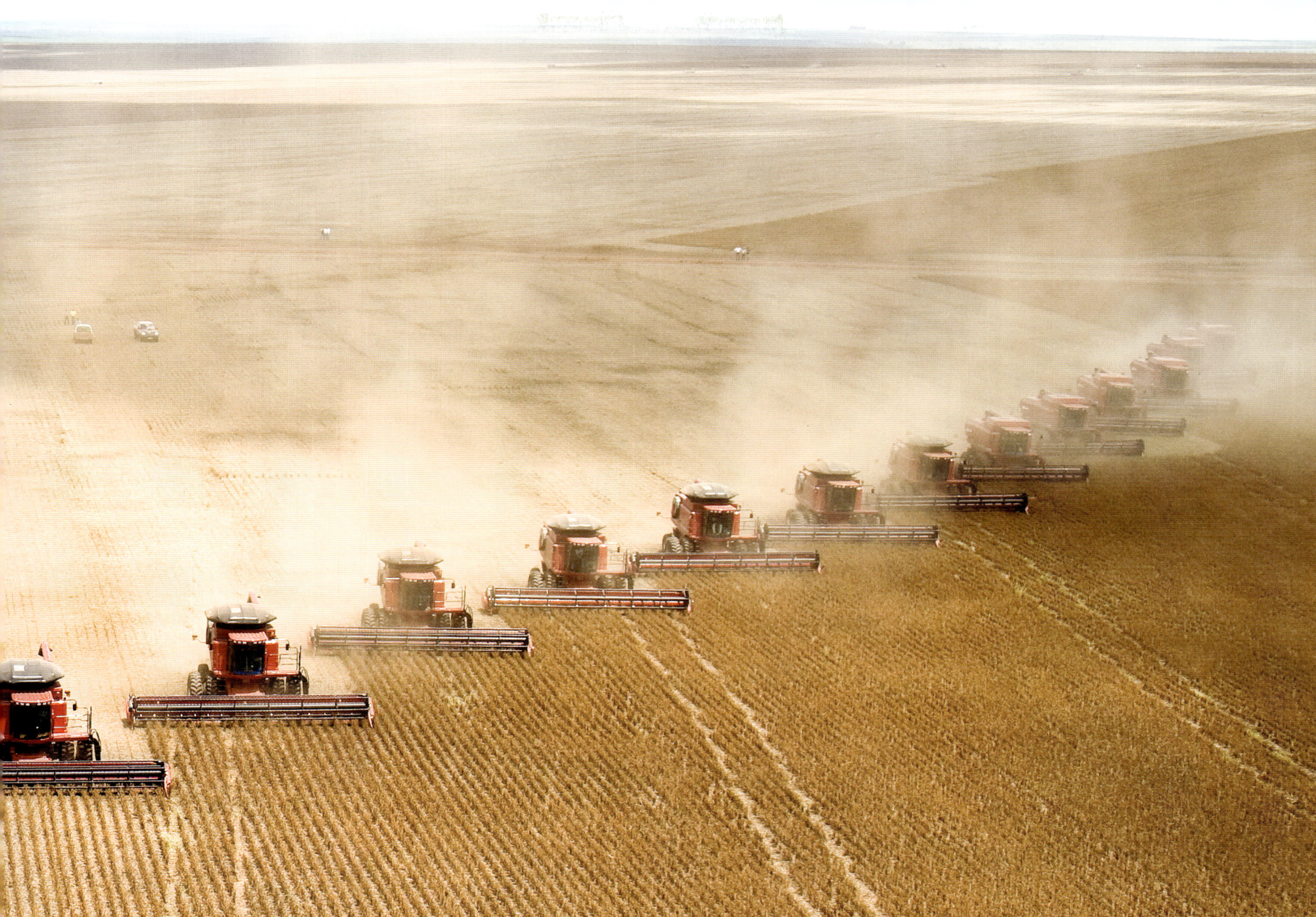

BRAZIL

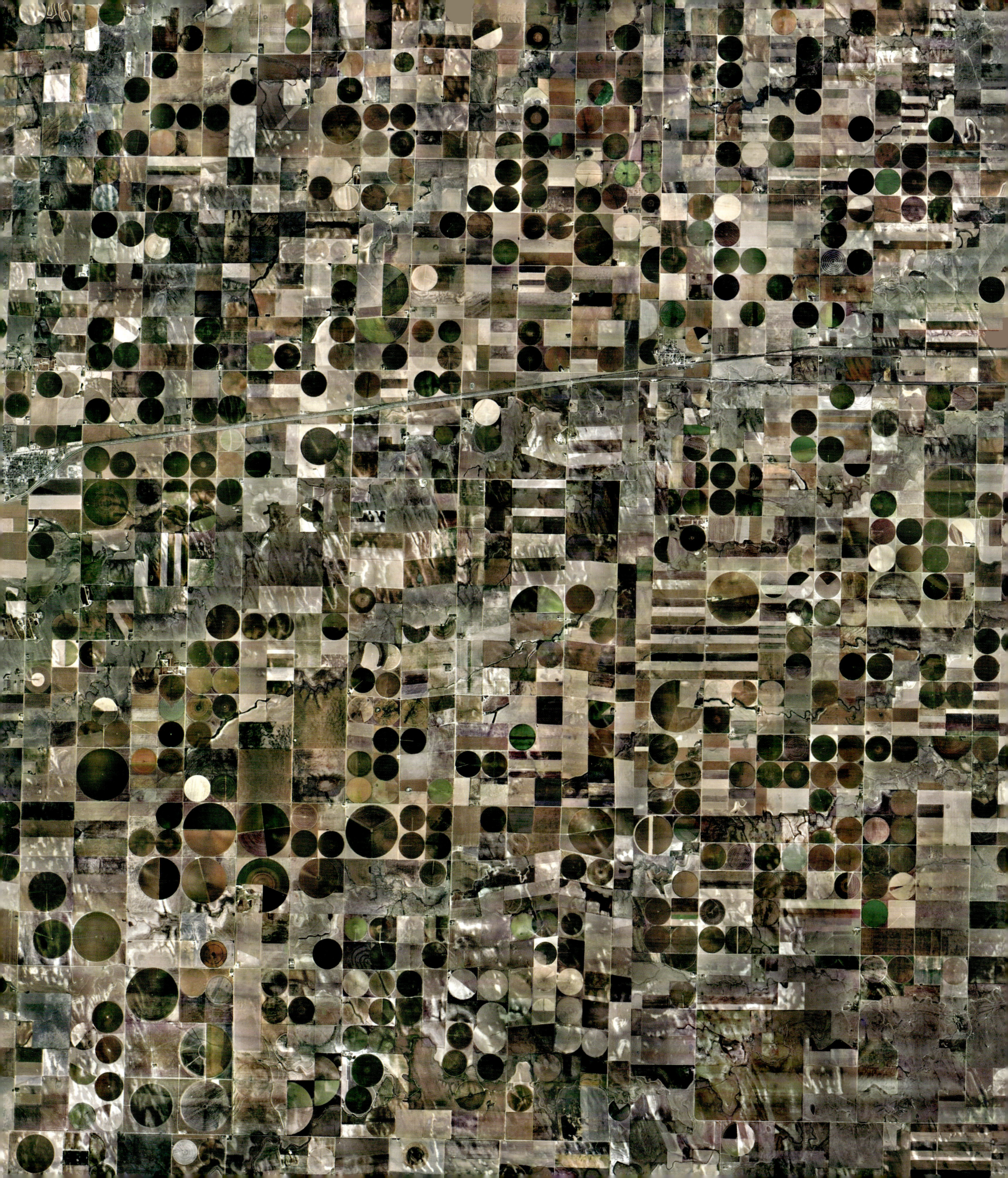

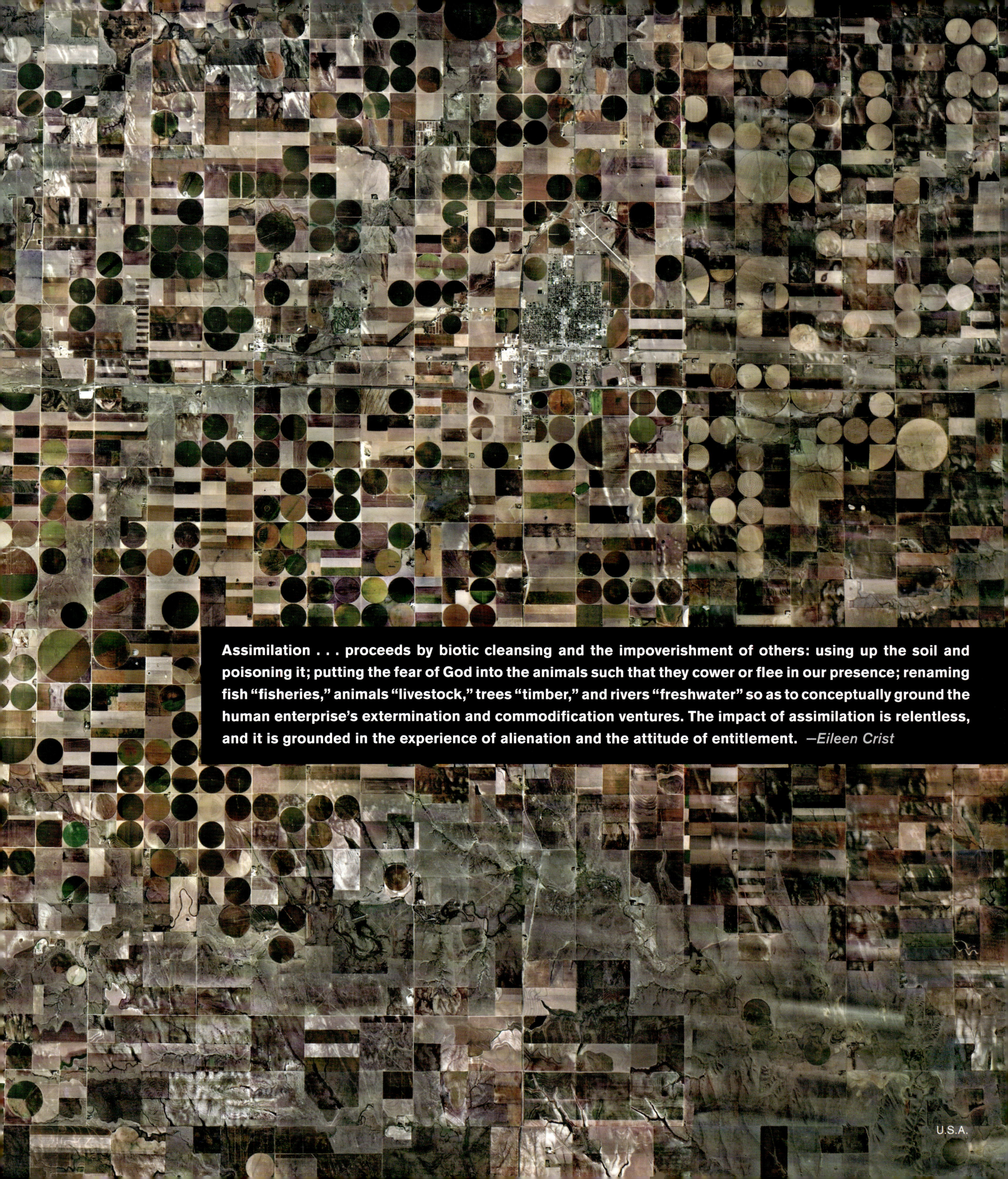
Assimilation . . . proceeds by biotic cleansing and the impoverishment of others: using up the soil and poisoning it; putting the fear of God into the animals such that they cower or flee in our presence; renaming fish "fisheries," animals "livestock," trees "timber," and rivers "freshwater" so as to conceptually ground the human enterprise's extermination and commodification ventures. The impact of assimilation is relentless, and it is grounded in the experience of alienation and the attitude of entitlement. —Eileen Crist
U.S.A.

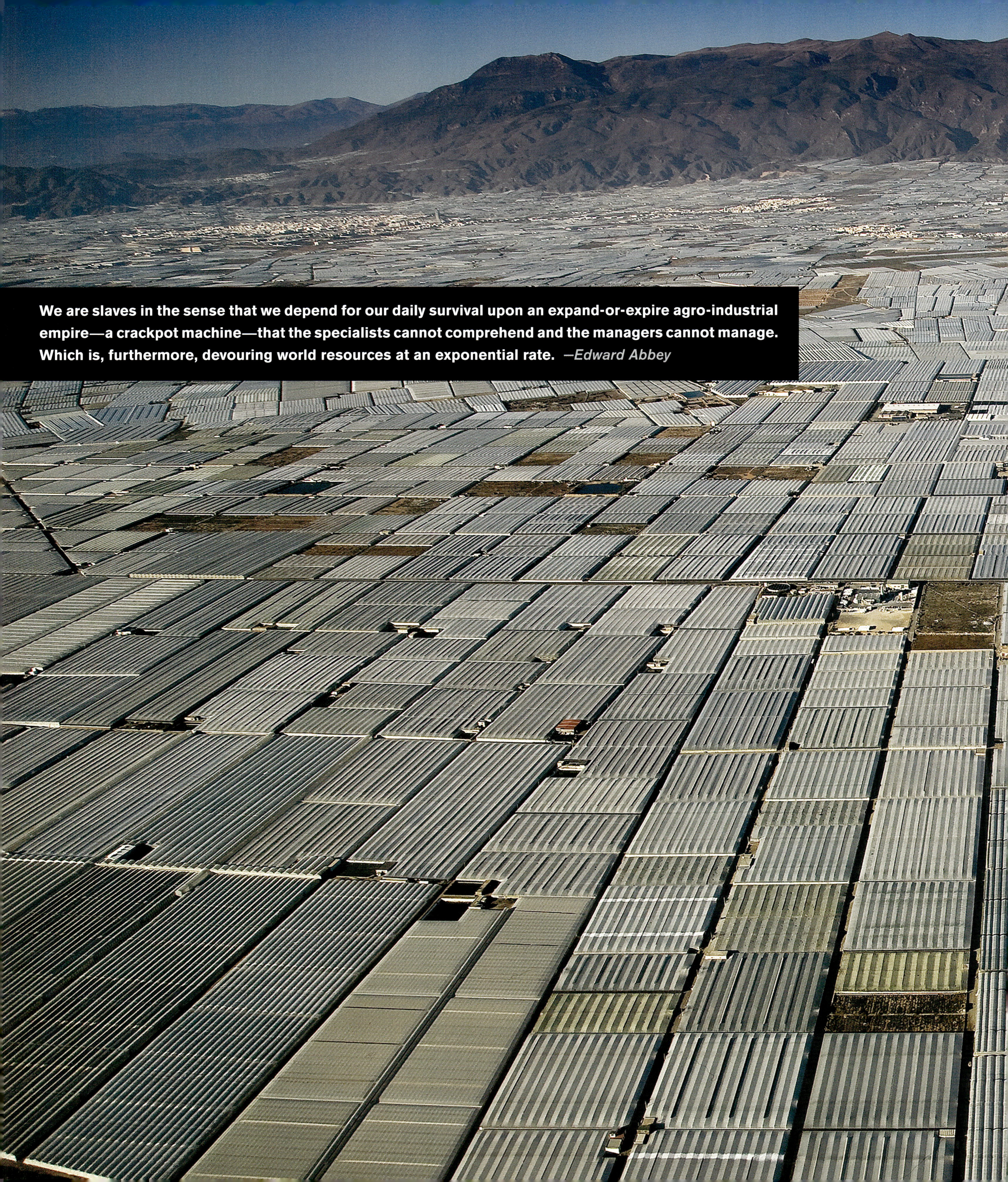

We are slaves in the sense that we depend for our daily survival upon an expand-or-expire agro-industrial empire—a crackpot machine—that the specialists cannot comprehend and the managers cannot manage. Which is, furthermore, devouring world resources at an exponential rate. —Edward Abbey

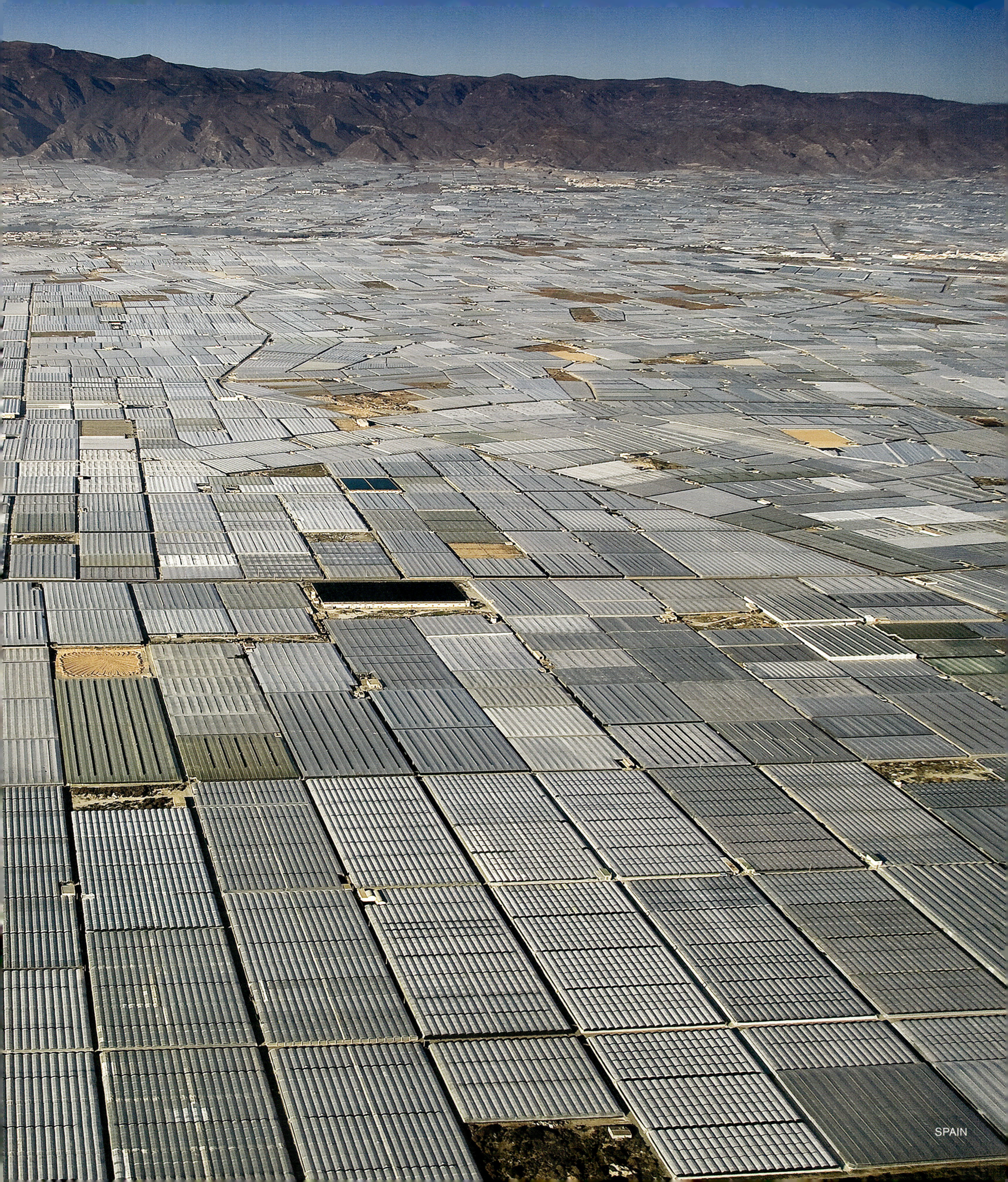
SPAIN

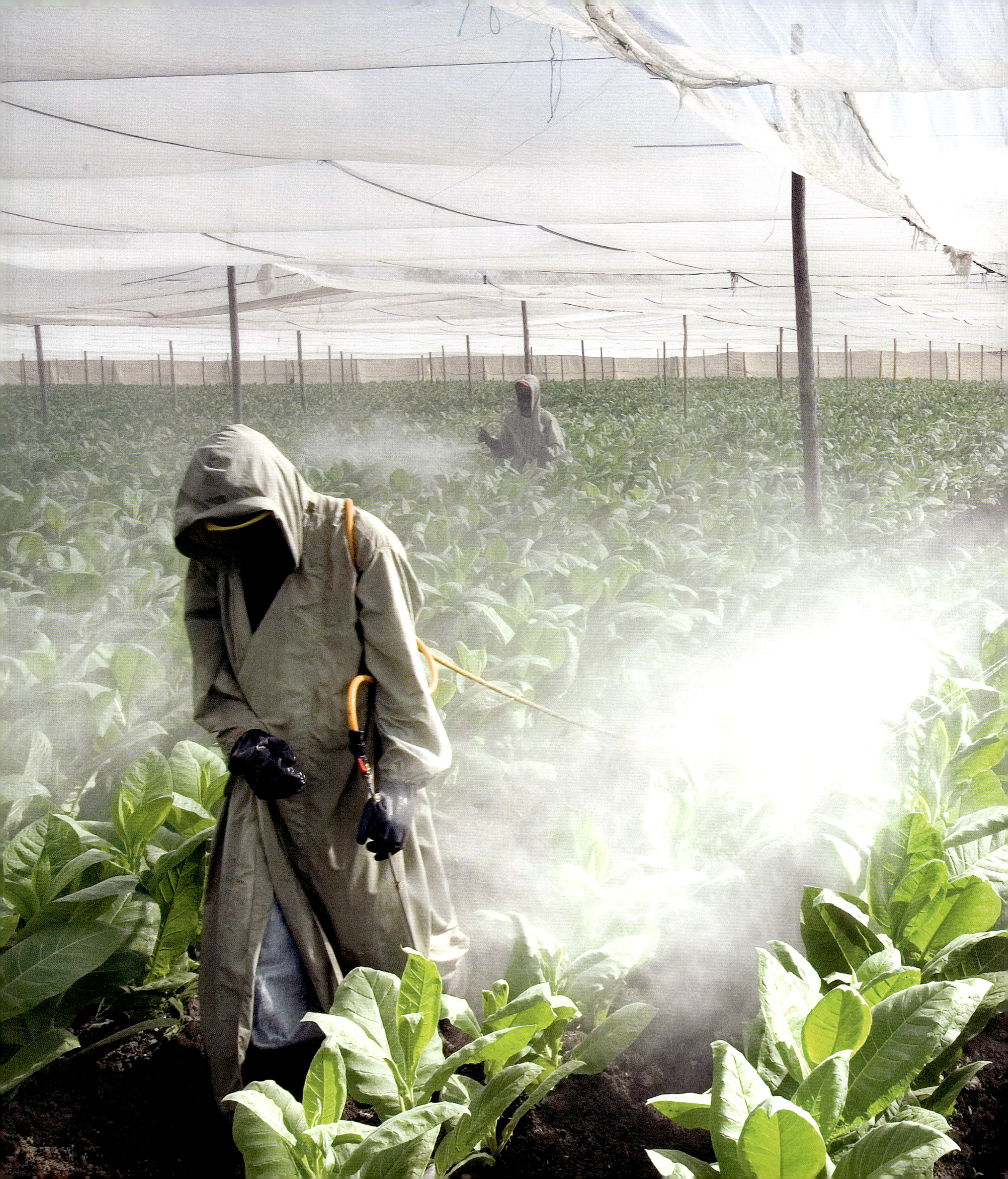

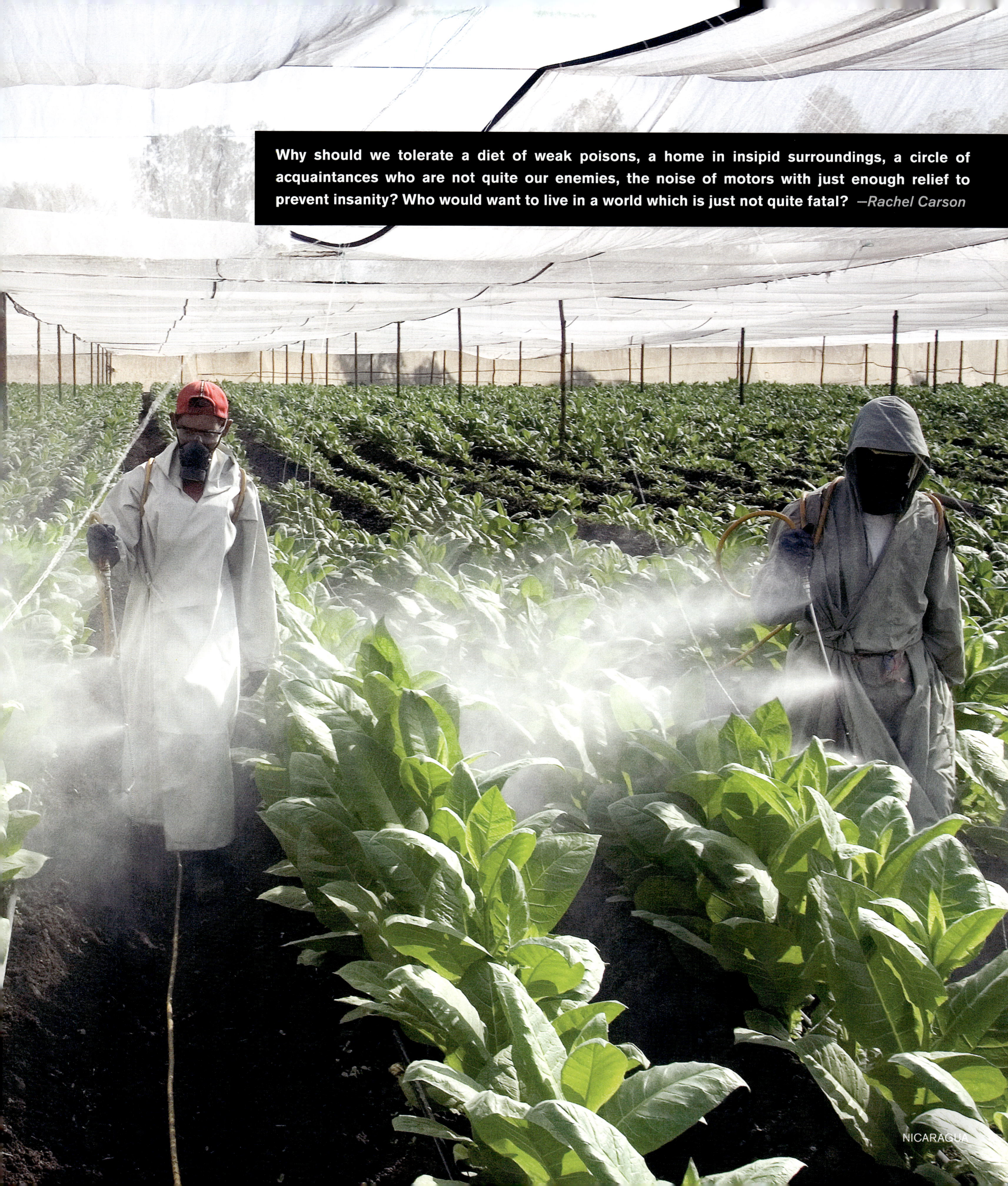

Why should we tolerate a diet of weak poisons, a home in insipid surroundings, a circle of acquaintances who are not quite our enemies, the noise of motors with just enough relief to prevent insanity? Who would want to live in a world which is just not quite fatal? —Rachel Carson
NICARAGUA

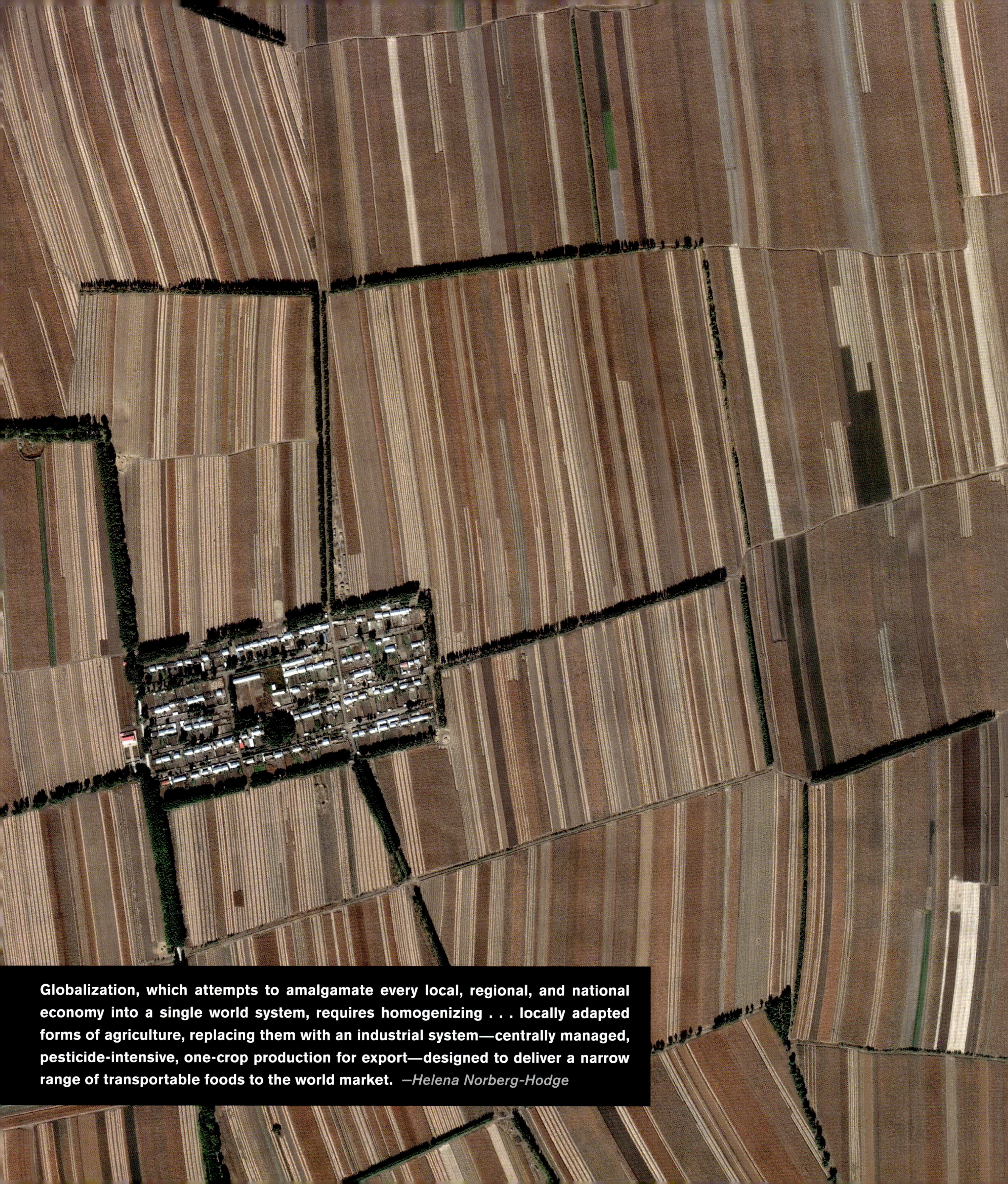
Globalization, which attempts to amalgamate every local, regional, and national economy into a single world system, requires homogenizing . . . locally adapted forms of agriculture, replacing them with an industrial system—centrally managed, pesticide-intensive, one-crop production for export—designed to deliver a narrow range of transportable foods to the world market. —Helena Norberg-Hodge

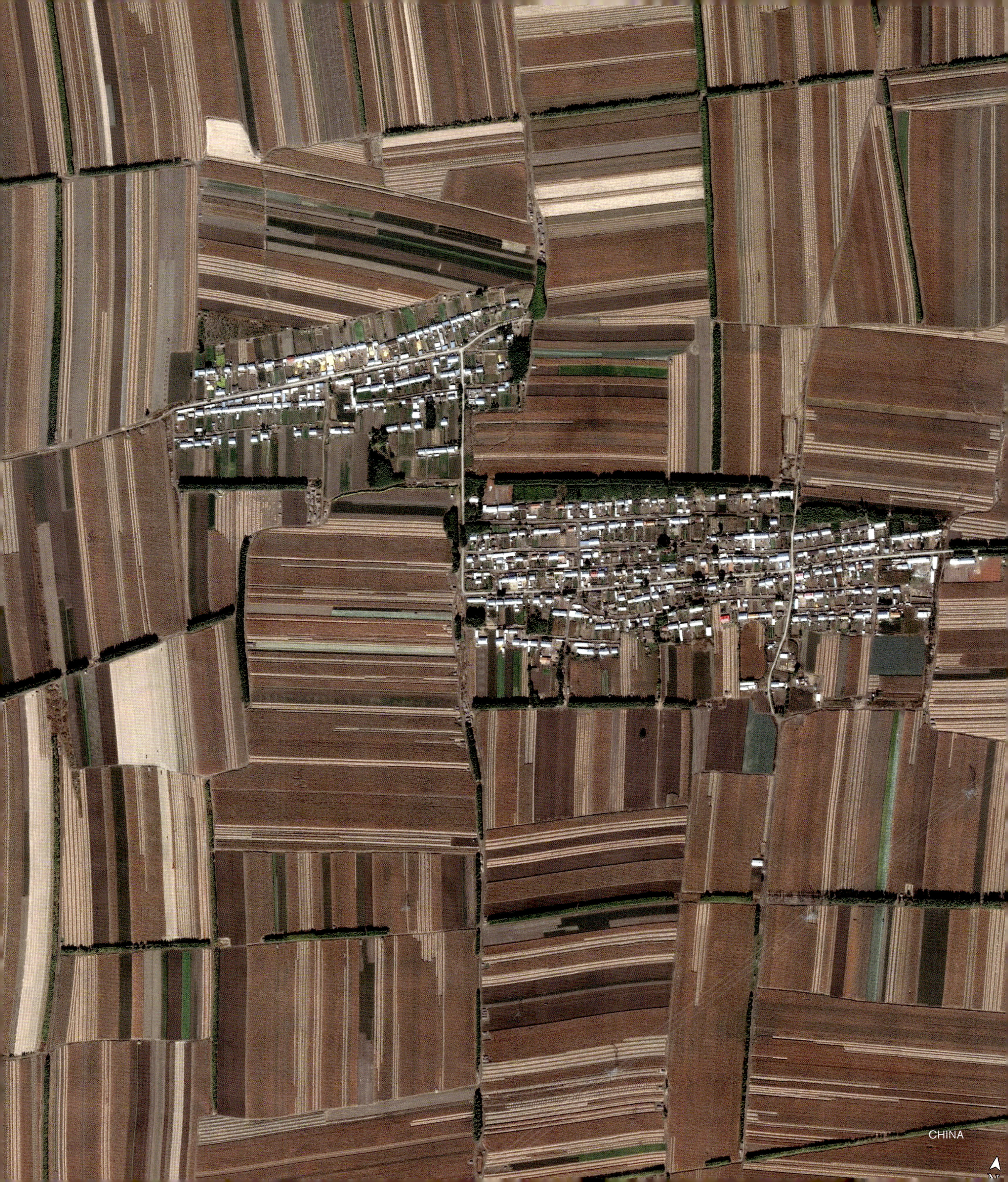

CHINA

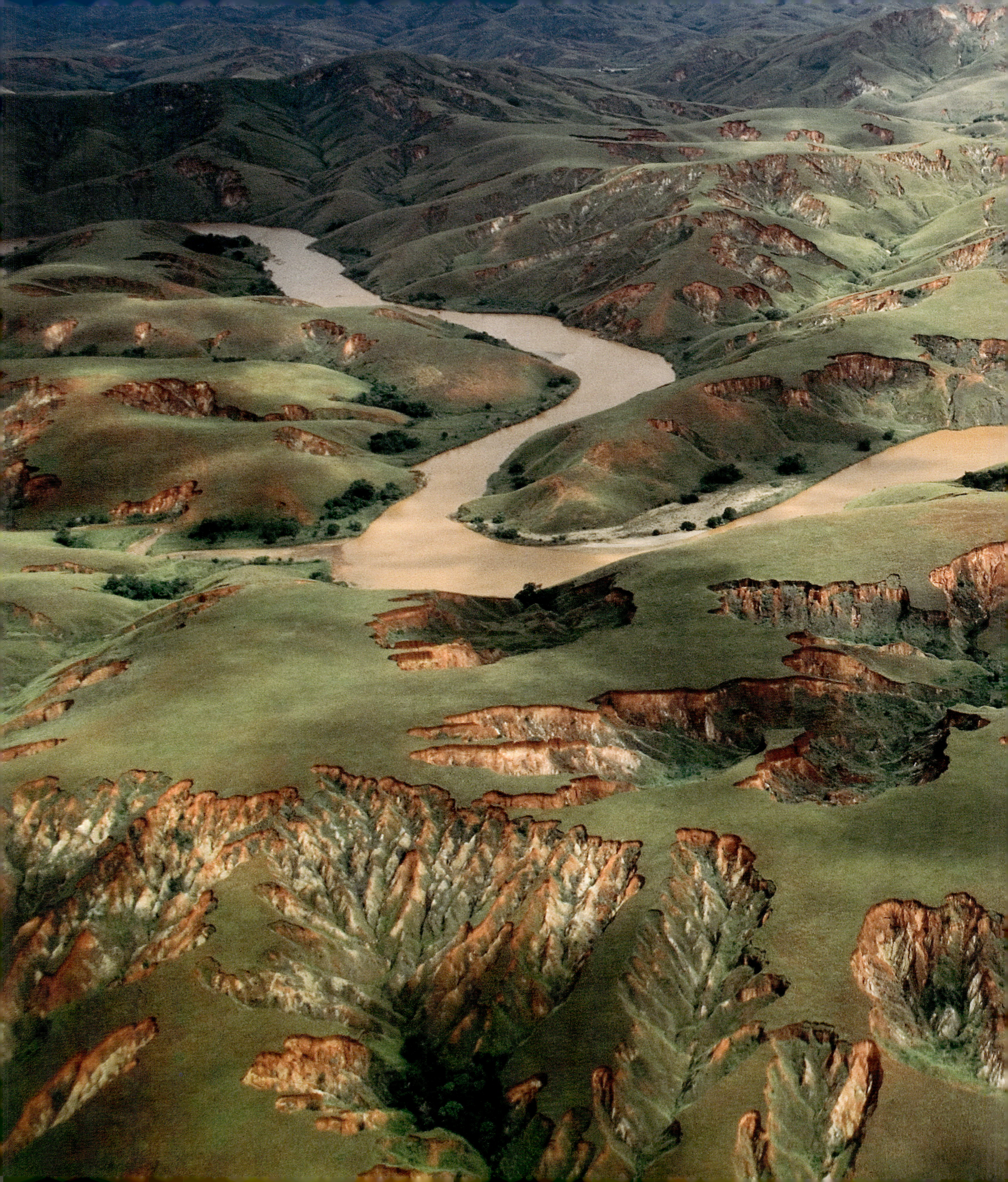

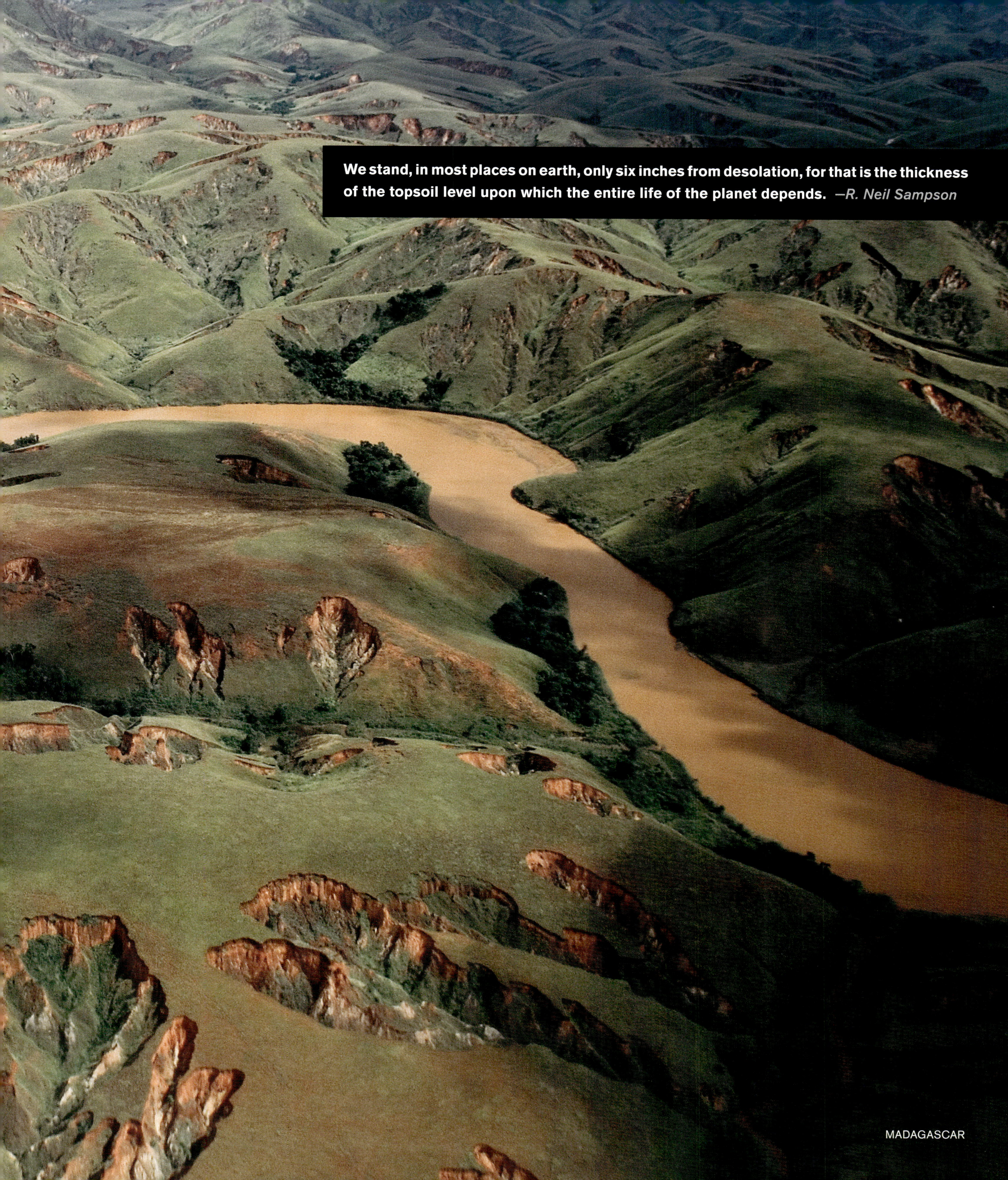
We stand, in most places on earth, only six inches from desolation, for that is the thickness
of the topsoil level upon which the entire life of the planet depends. —R. Neil Sampson
MADAGASCAR

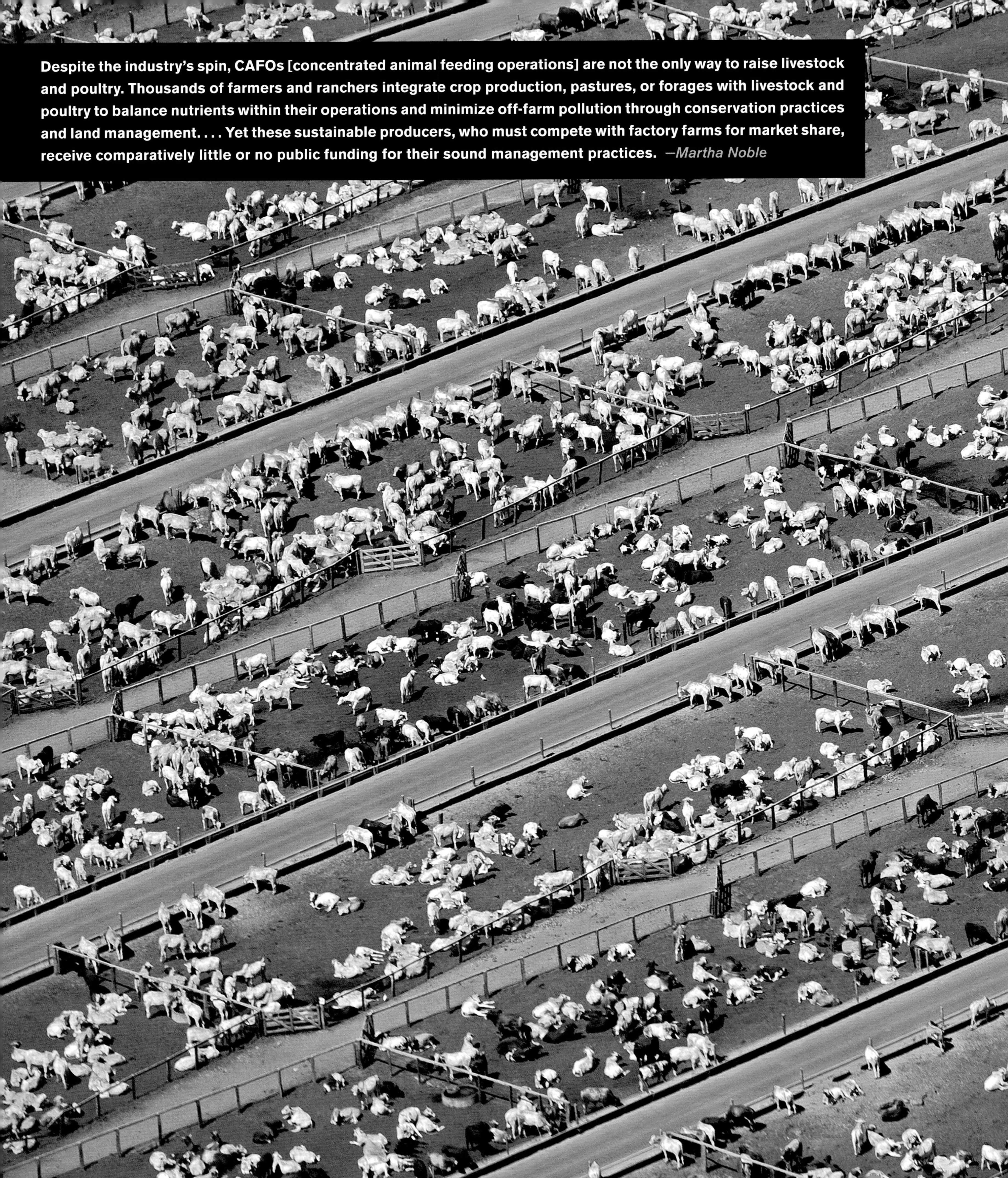

Despite the industry's spin, CAFOs [concentrated animal feeding operations] are not the only way to raise livestock and poultry. Thousands of farmers and ranchers integrate crop production, pastures, or forages with livestock and poultry to balance nutrients within their operations and minimize off-farm pollution through conservation practices and land management. . . . Yet these sustainable producers, who must compete with factory farms for market share, receive comparatively little or no public funding for their sound management practices. —Martha Noble

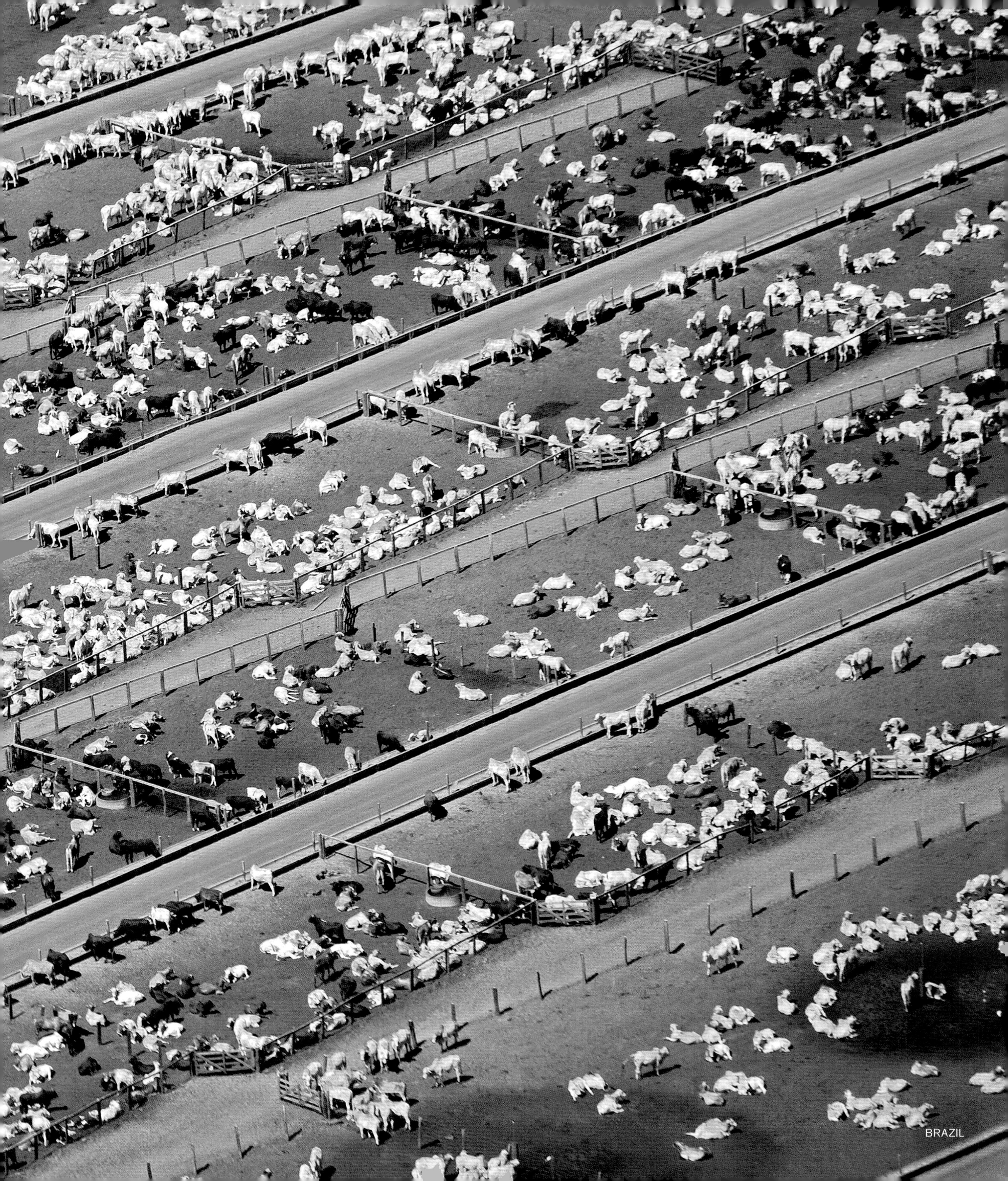

BRAZIL

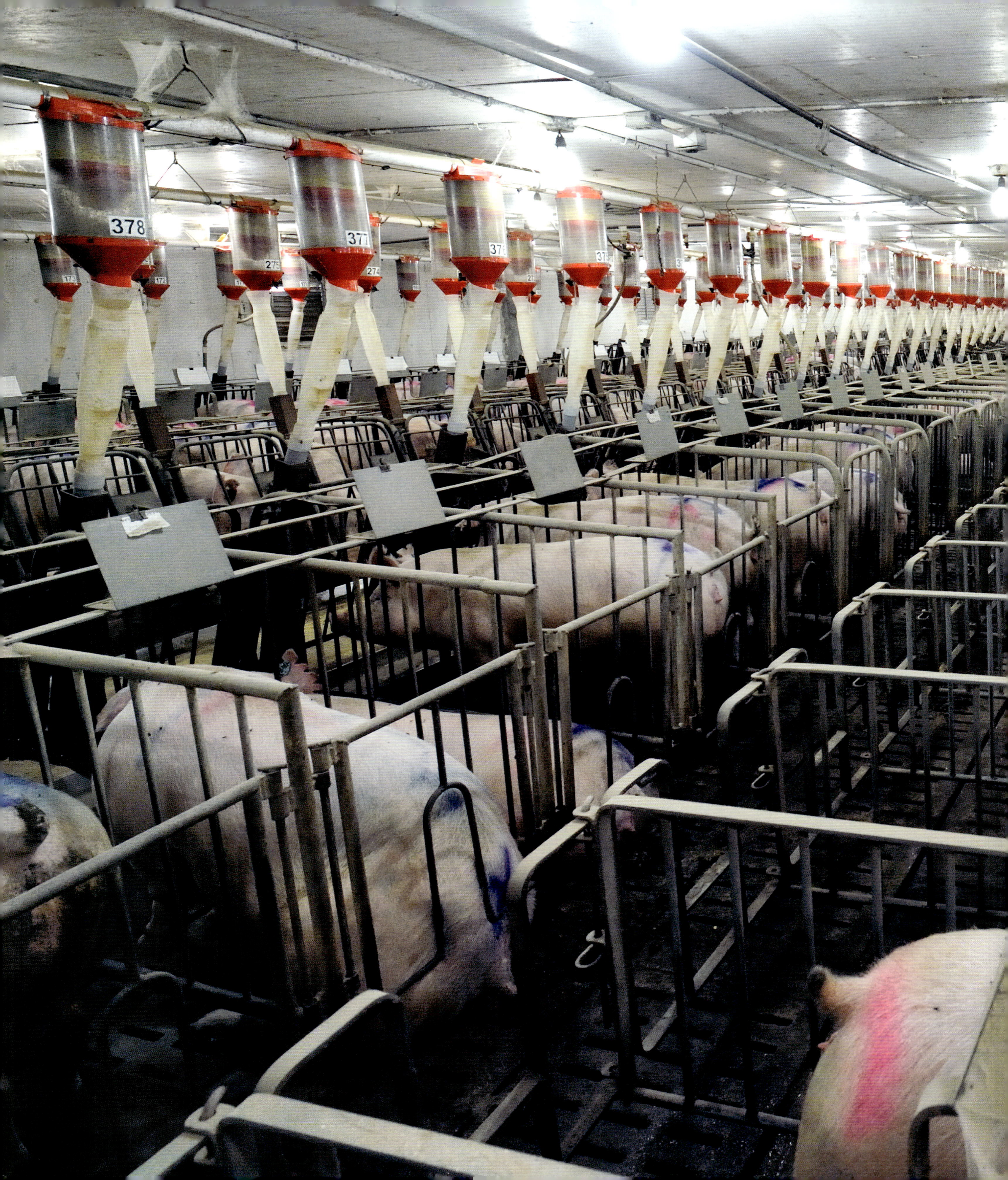

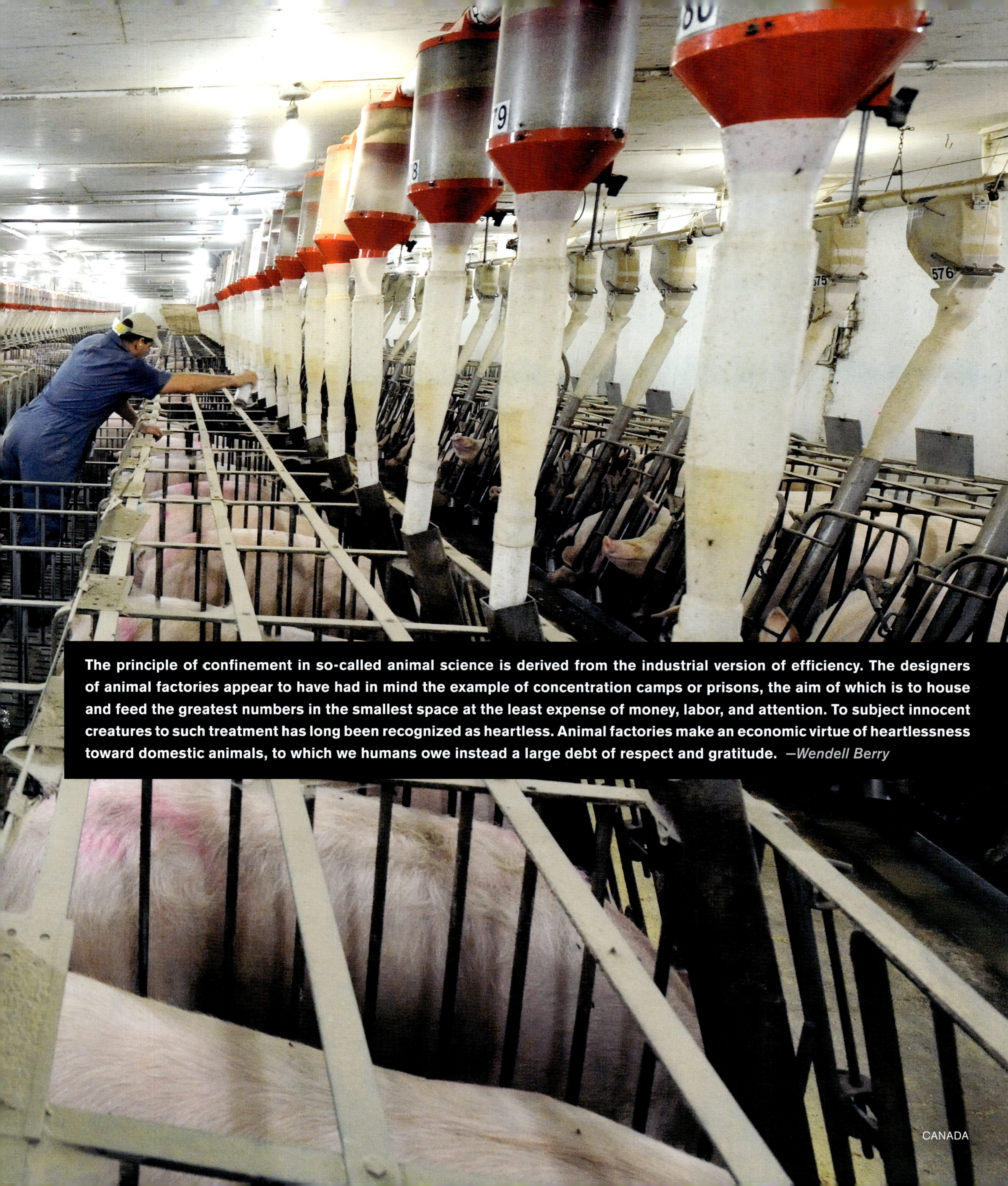

The principle of confinement in so-called animal science is derived from the industrial version of efficiency. The designers of animal factories appear to have had in mind the example of concentration camps or prisons, the aim of which is to house and feed the greatest numbers in the smallest space at the least expense of money, labor, and attention. To subject innocent creatures to such treatment has long been recognized as heartless. Animal factories make an economic virtue of heartlessness toward domestic animals, to which we humans owe instead a large debt of respect and gratitude. —Wendell Berry
CANADA

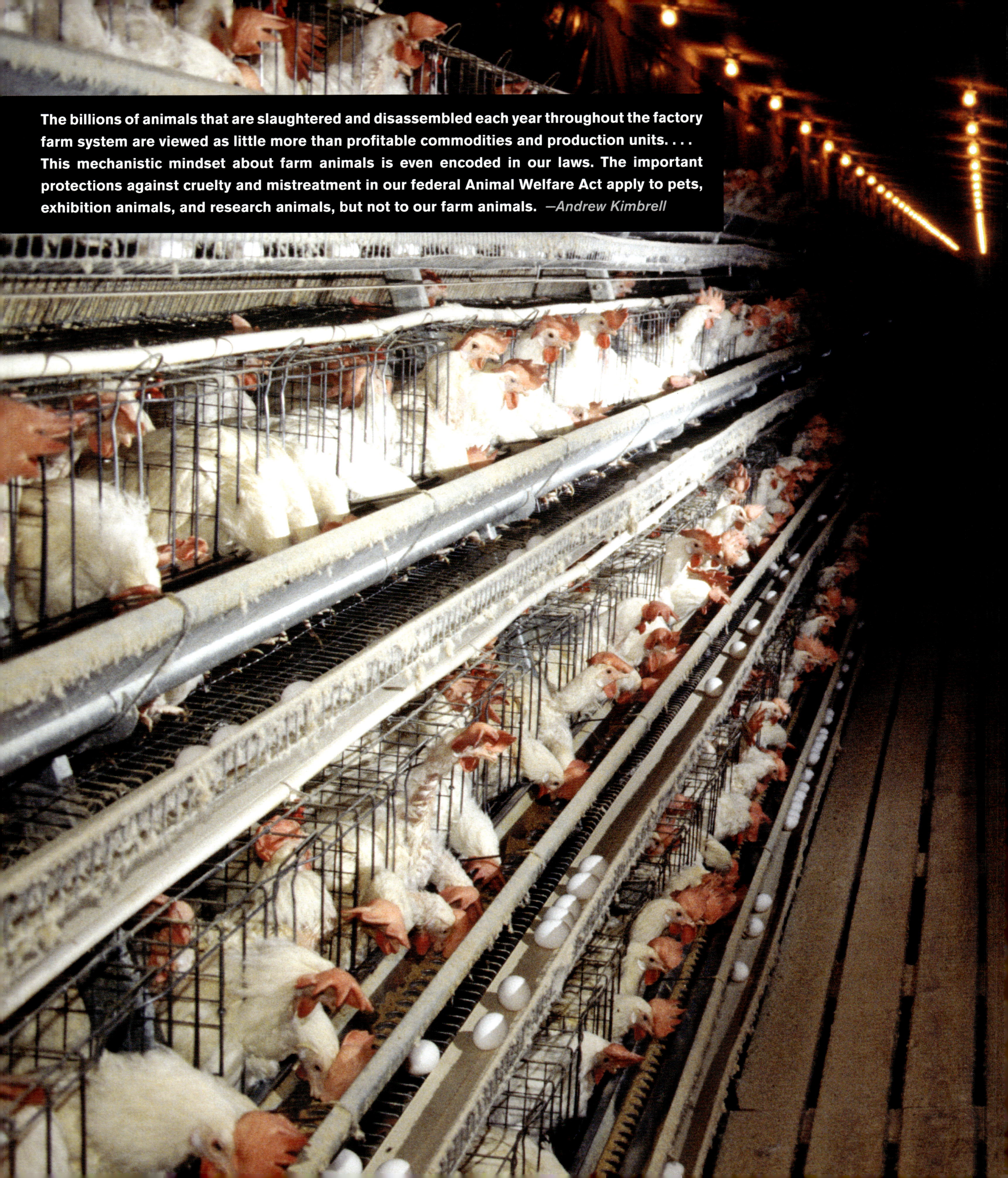

The billions of animals that are slaughtered and disassembled each year throughout the factory farm system are viewed as little more than profitable commodities and production units. . . . This mechanistic mindset about farm animals is even encoded in our laws. The important protections against cruelty and mistreatment in our federal Animal Welfare Act apply to pets, exhibition animals, and research animals, but not to our farm animals. —Andrew Kimbrell

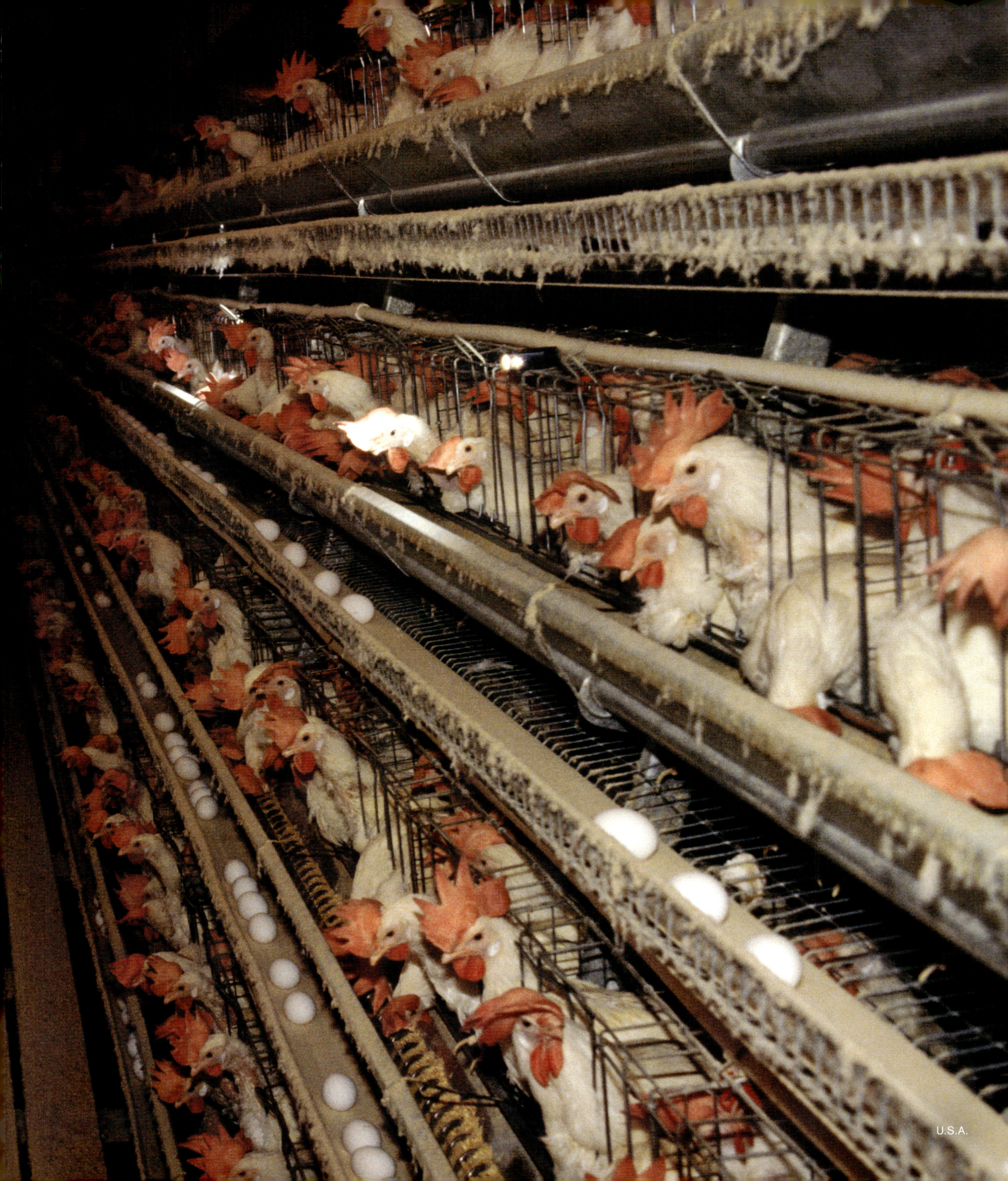
U.S.A.

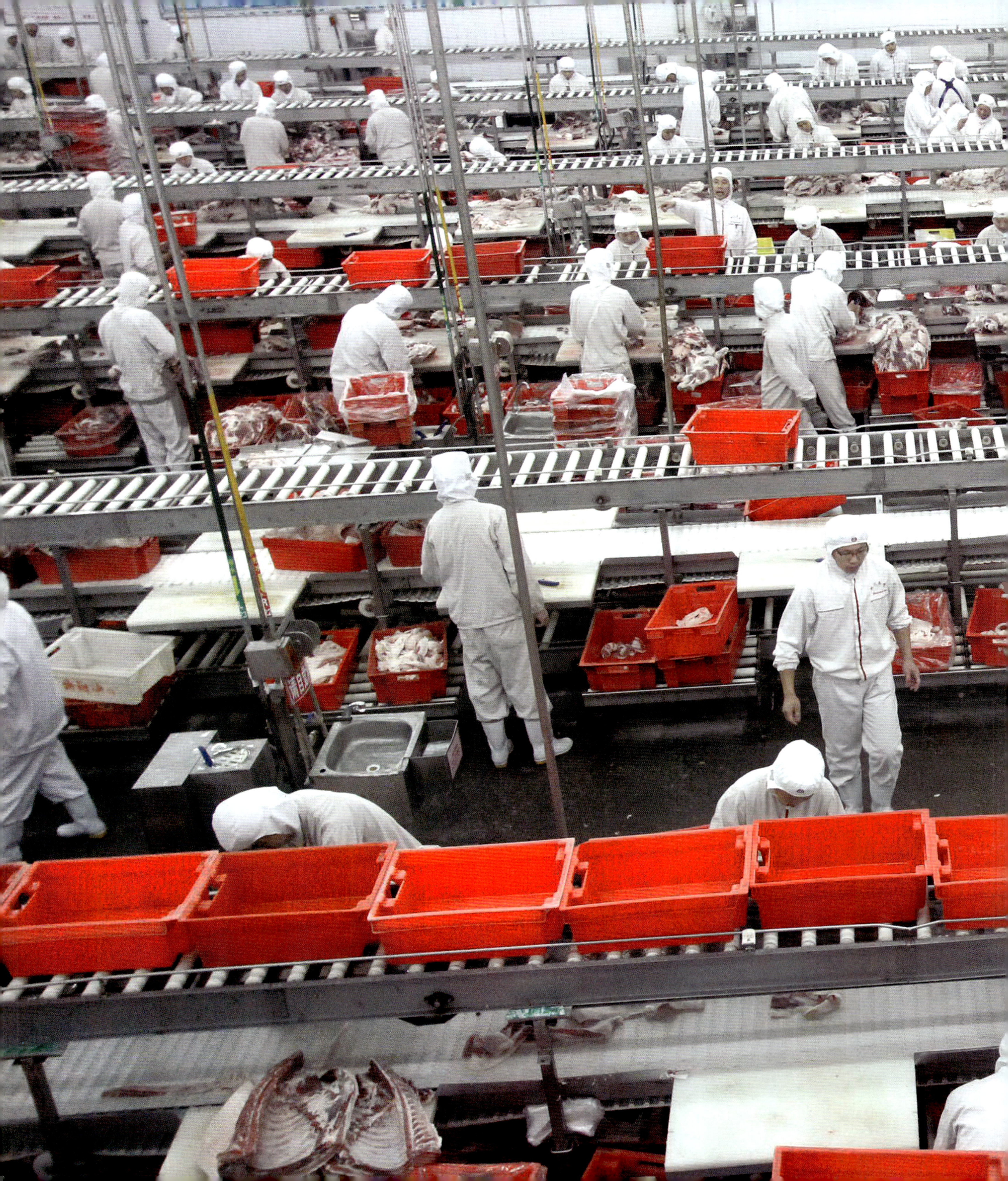

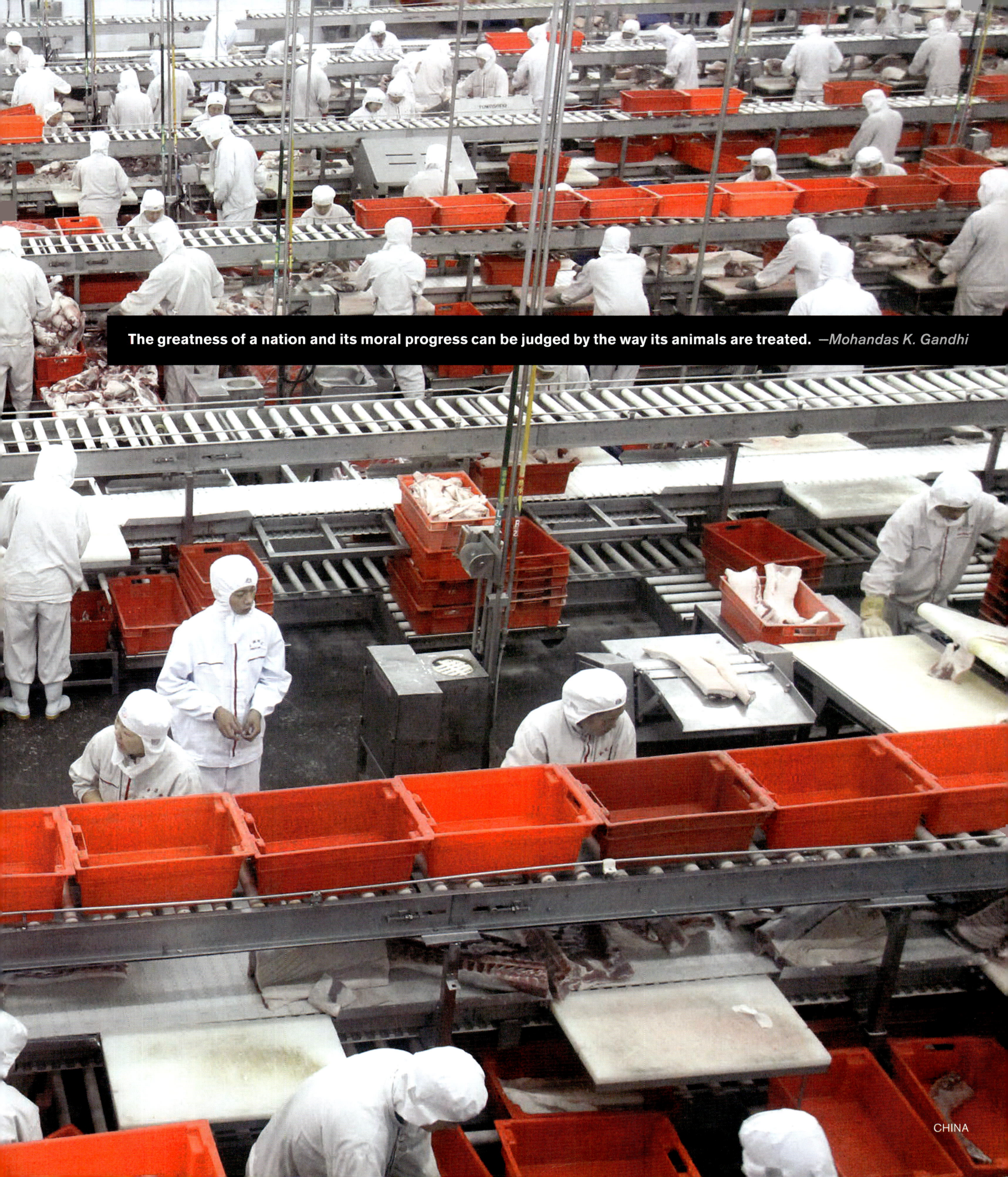

The greatness of a nation and its moral progress can be judged by the way its animals are treated. —Mohandas K. Gandhi
CHINA

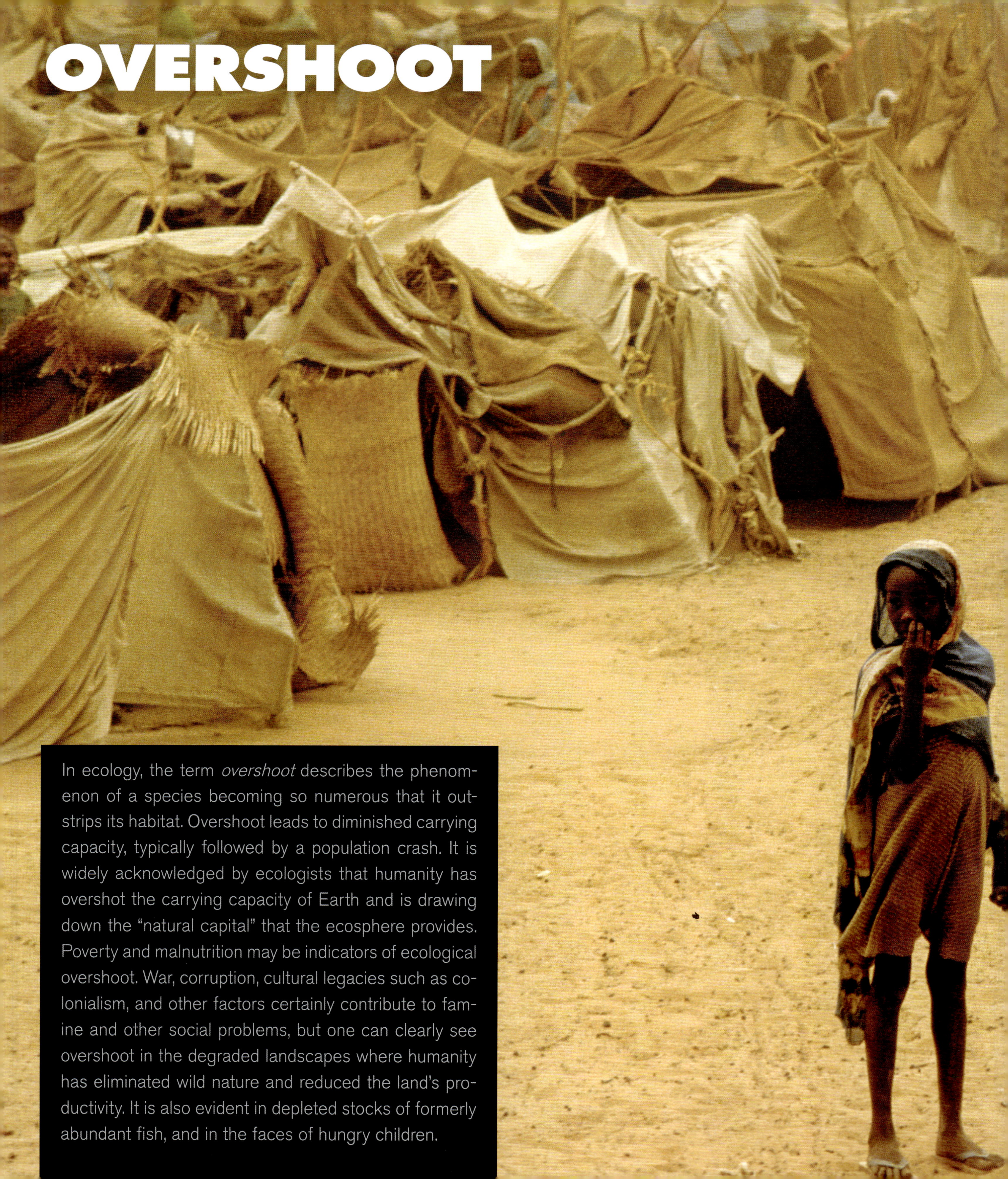

In ecology, the term *overshoot* describes the phenomenon of a species becoming so numerous that it outstrips its habitat. Overshoot leads to diminished carrying capacity, typically followed by a population crash. It is widely acknowledged by ecologists that humanity has overshot the carrying capacity of Earth and is drawing down the "natural capital" that the ecosphere provides. Poverty and malnutrition may be indicators of ecological overshoot. War, corruption, cultural legacies such as colonialism, and other factors certainly contribute to famine and other social problems, but one can clearly see overshoot in the degraded landscapes where humanity has eliminated wild nature and reduced the land's productivity. It is also evident in depleted stocks of formerly abundant fish, and in the faces of hungry children.

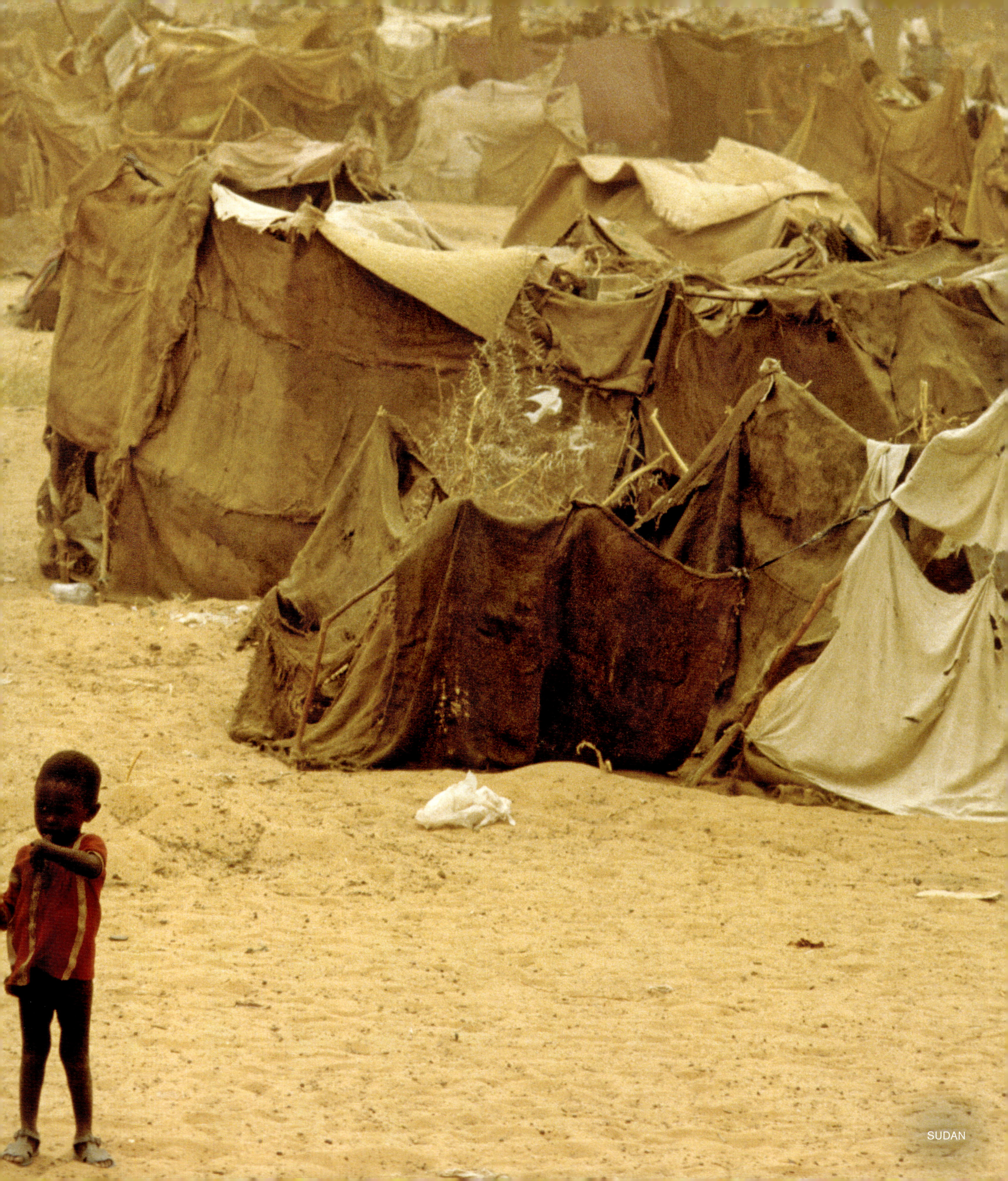

SUDAN

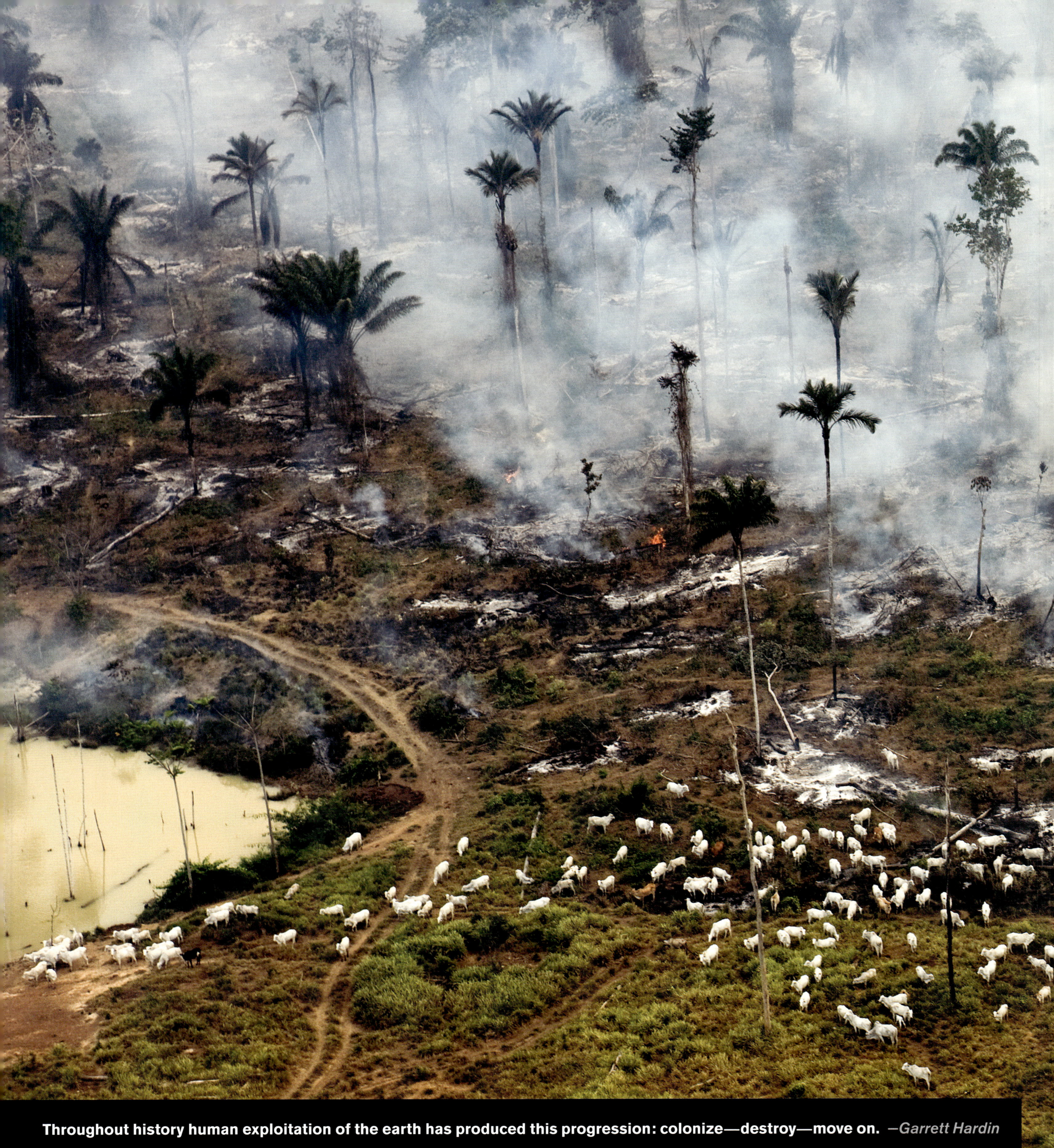

Throughout history human exploitation of the earth has produced this progression: colonize—destroy—move on. —Garrett Hardin

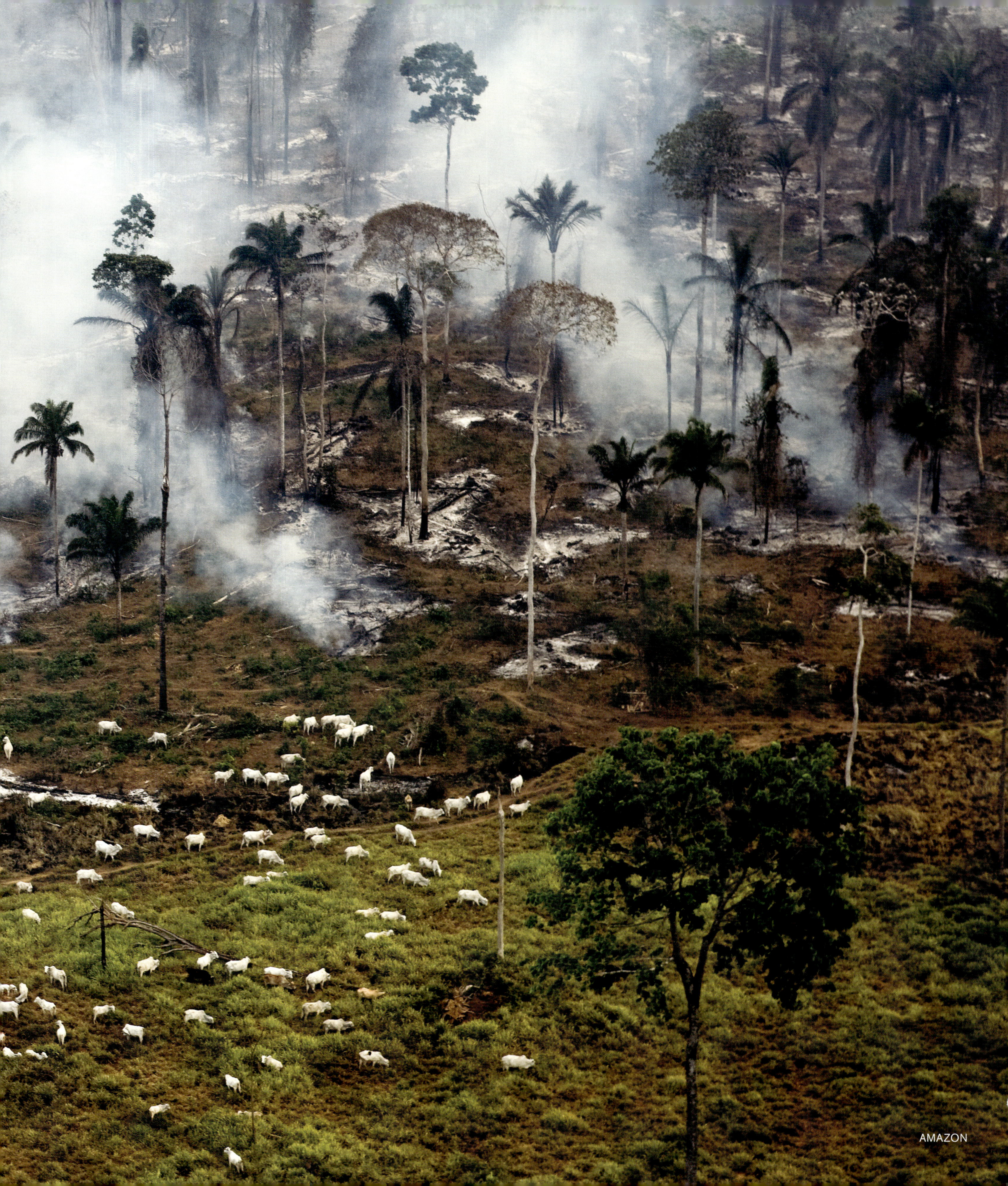
AMAZON

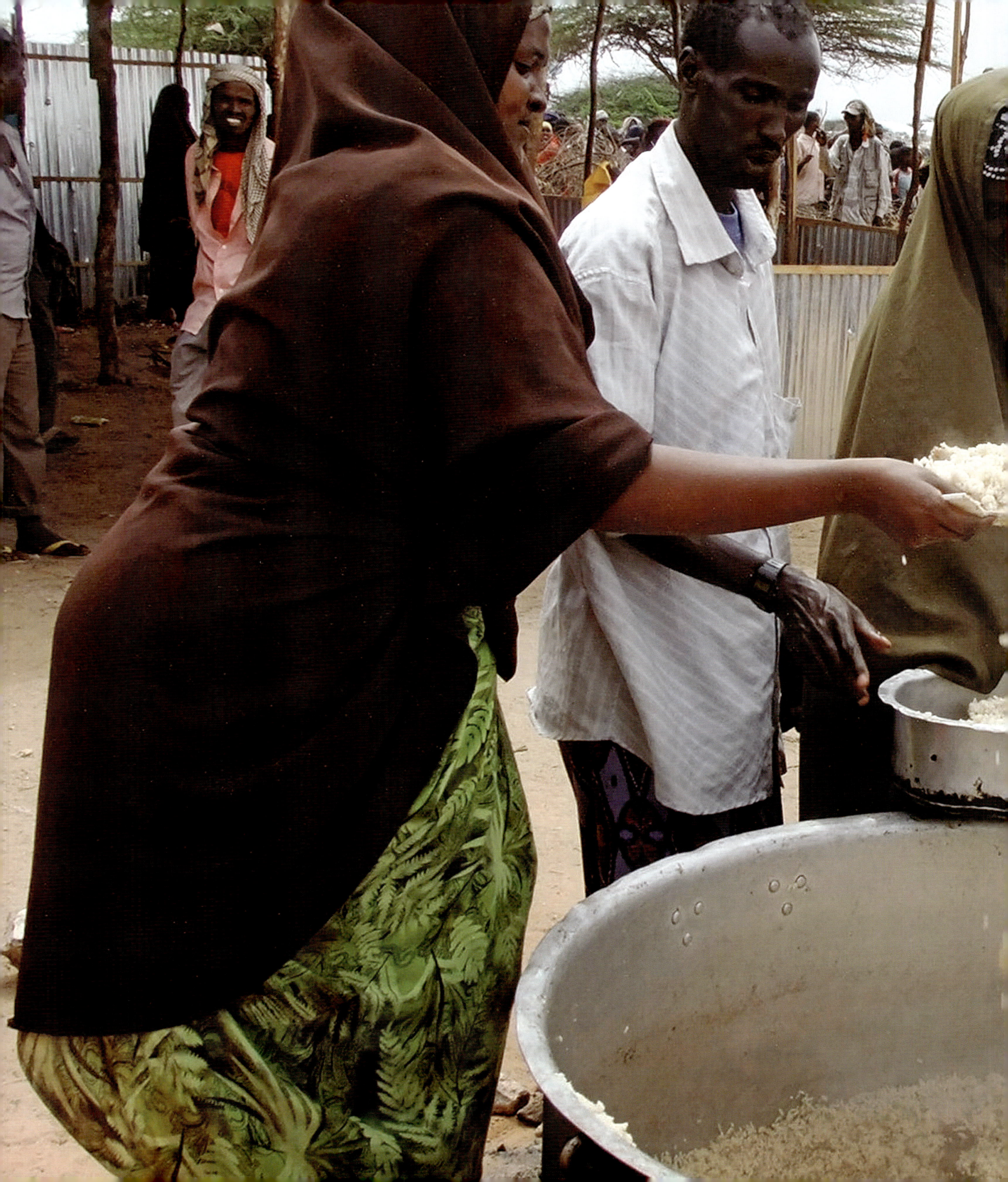

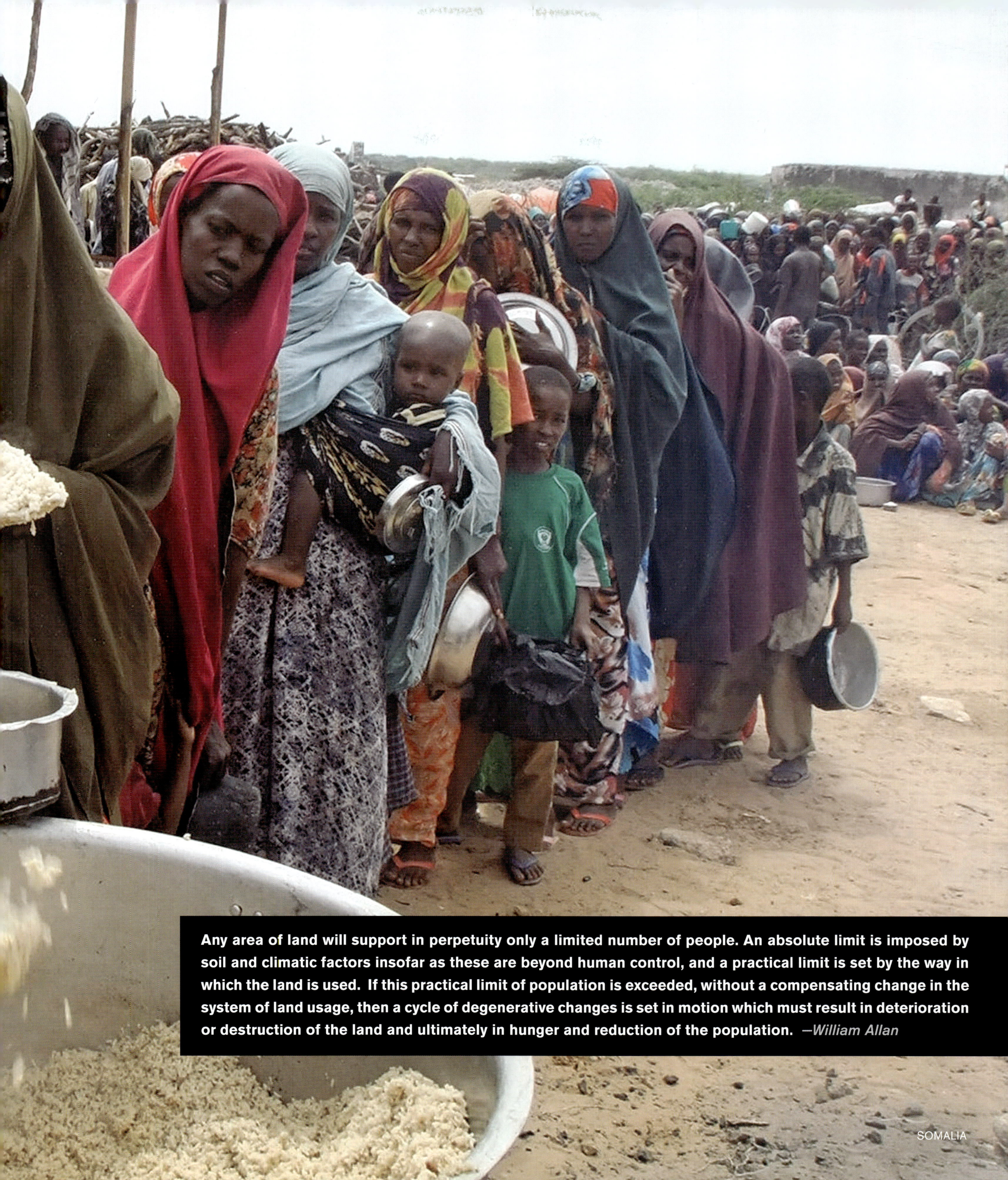

Any area of land will support in perpetuity only a limited number of people. An absolute limit is imposed by soil and climatic factors insofar as these are beyond human control, and a practical limit is set by the way in which the land is used. If this practical limit of population is exceeded, without a compensating change in the system of land usage, then a cycle of degenerative changes is set in motion which must result in deterioration or destruction of the land and ultimately in hunger and reduction of the population. —*William Allan*

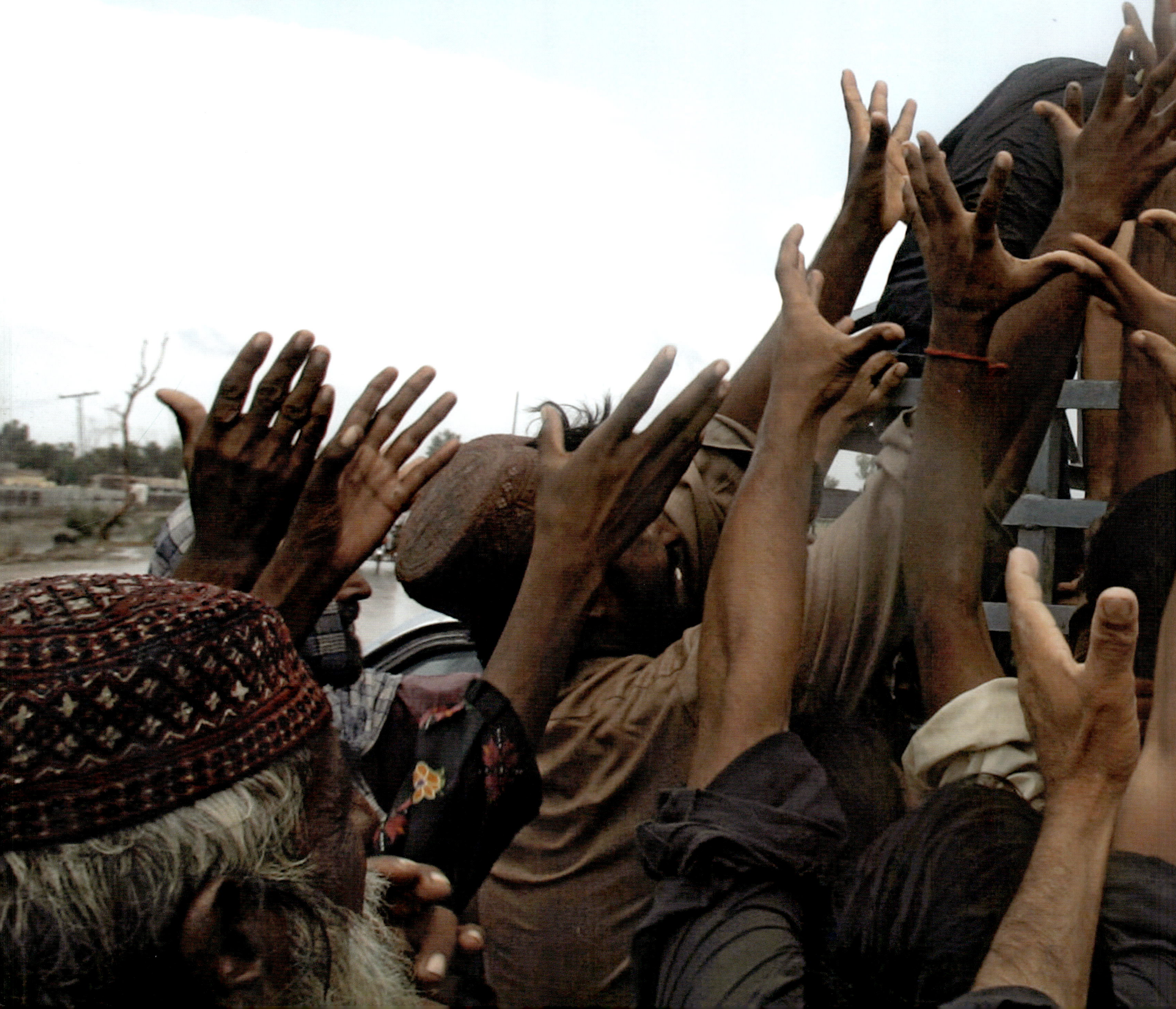

A population and economy are in overshoot mode when they are drawing resources or emitting pollutants at an unsustainable rate, but the stresses on the support system are not yet strong enough to reduce the rates of withdrawal or emission. Overshoot comes from delays in feedback—from the fact that decision makers in the system do not get, or believe, or act upon the information that limits have been exceeded until long after they have been exceeded. —Donella Meadows, Dennis Meadows, and Jørgen Randers

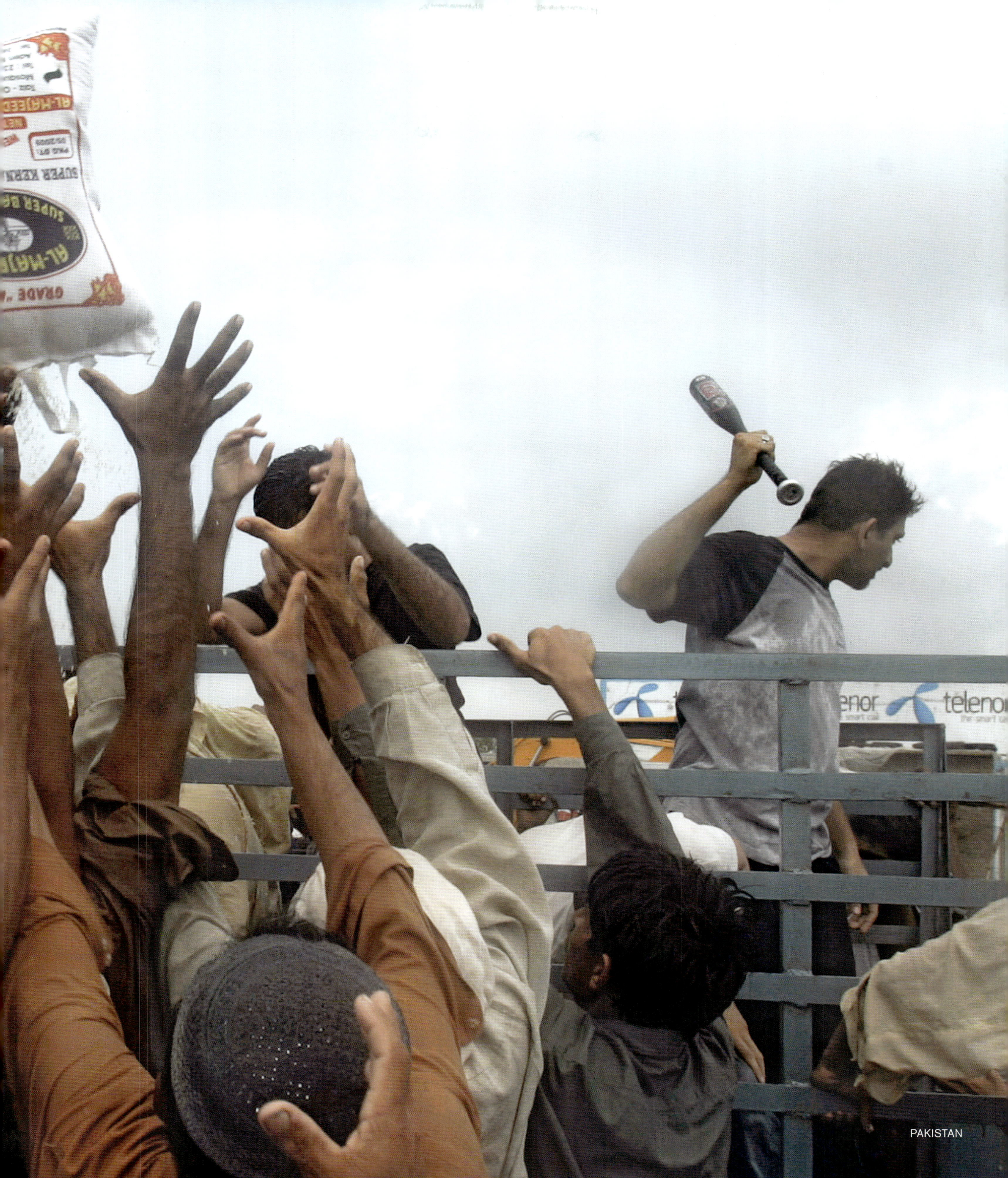

PAKISTAN

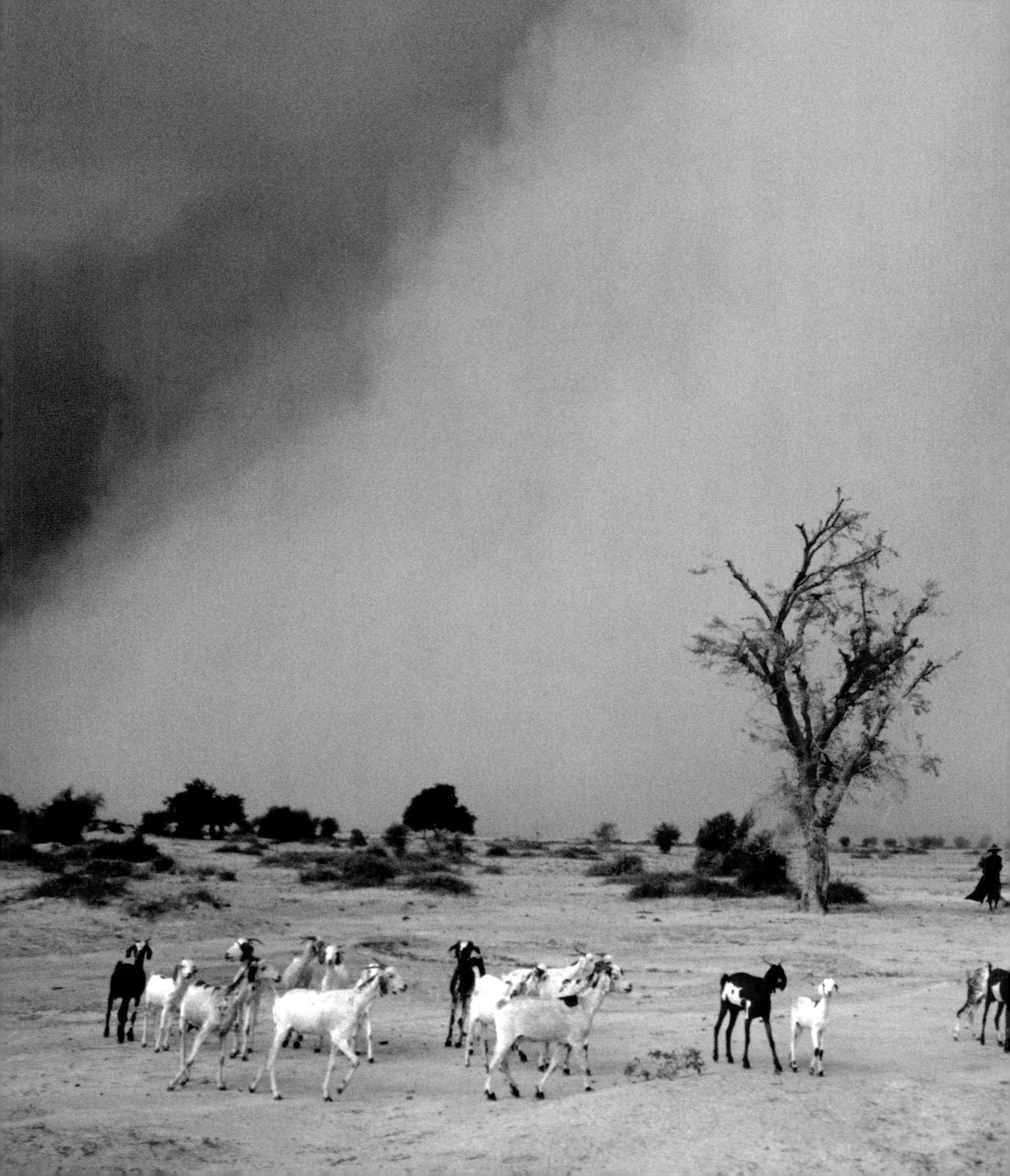

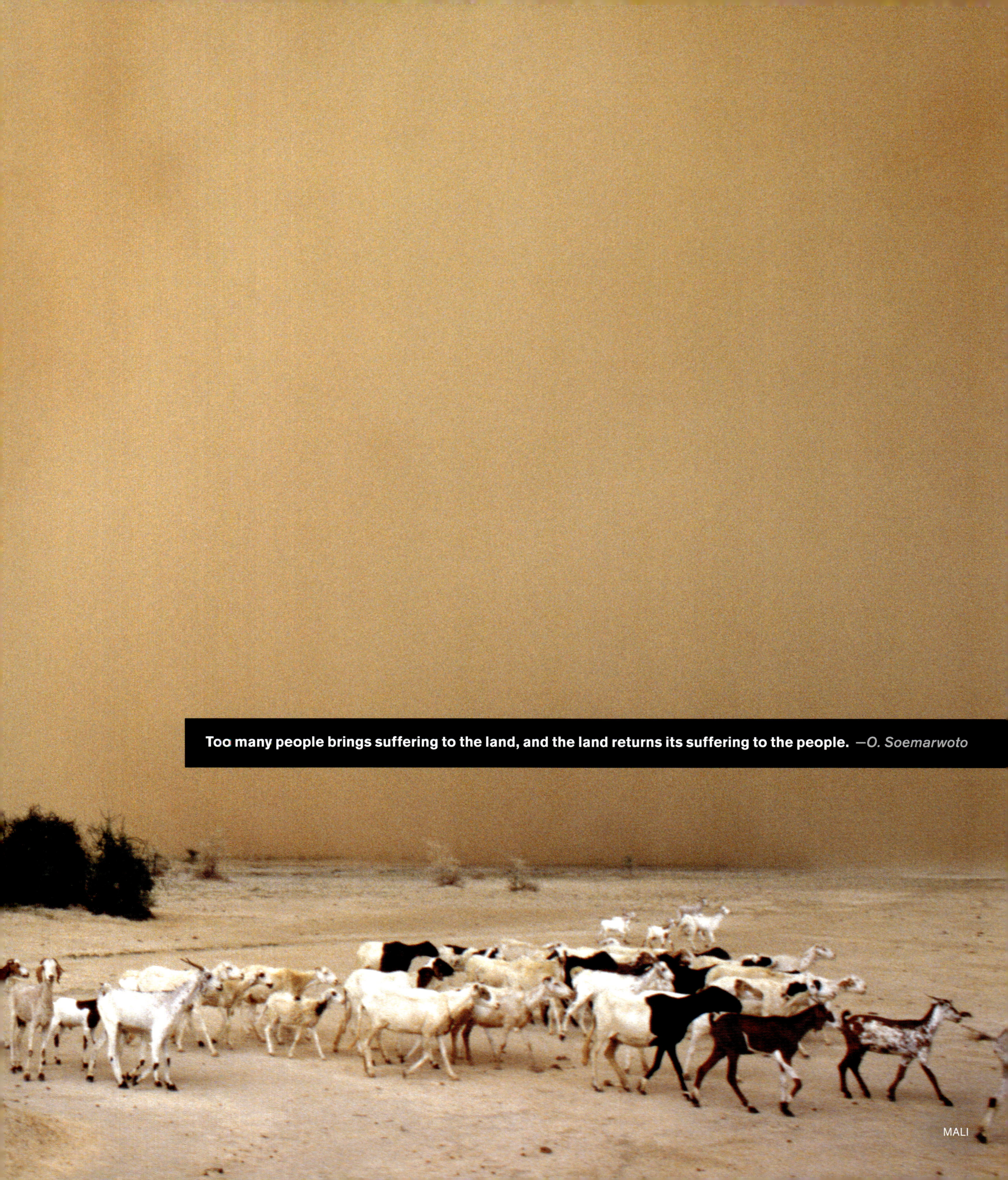

Too many people brings suffering to the land, and the land returns its suffering to the people. —O. Soemarwoto
MALI

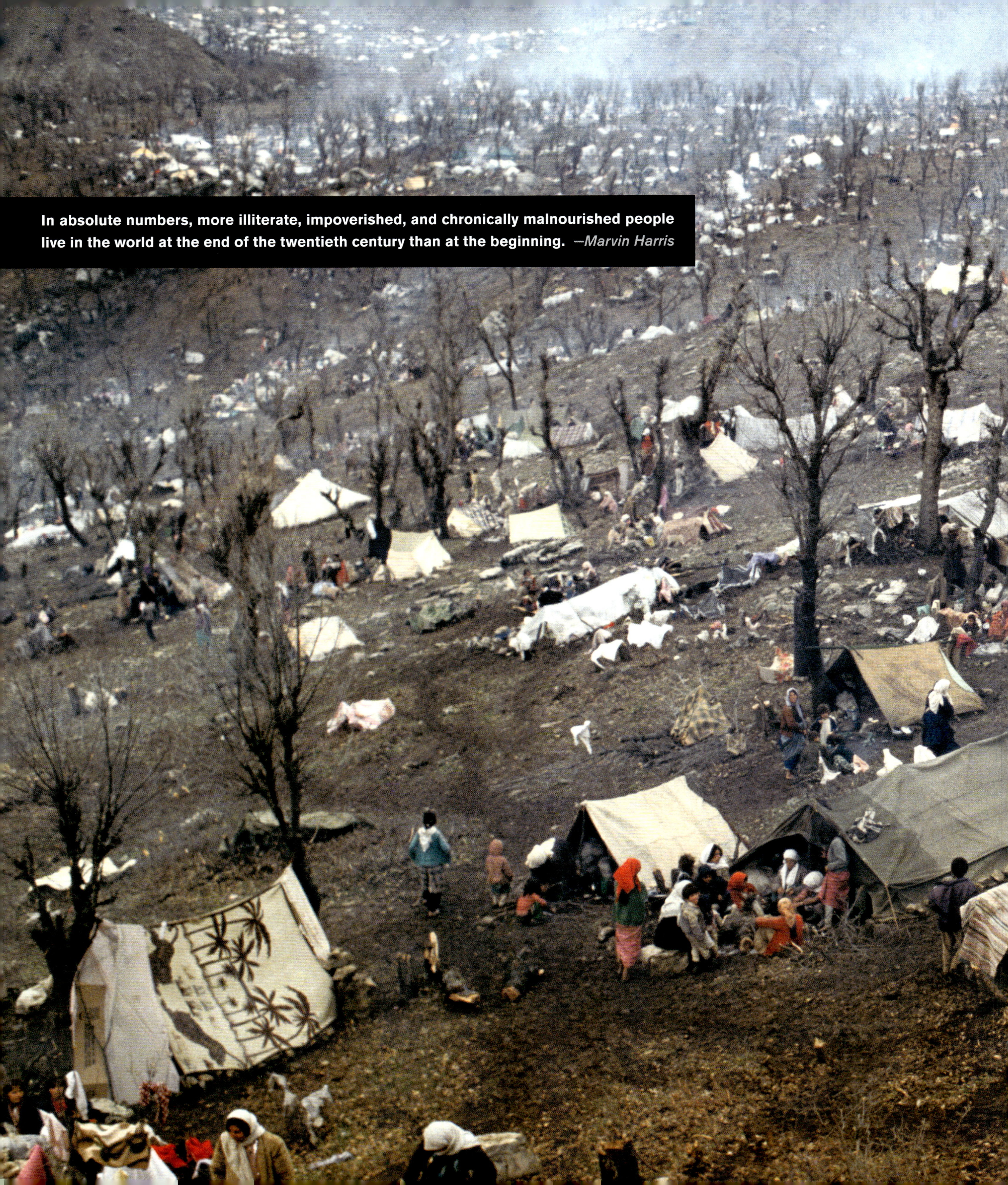

In absolute numbers, more illiterate, impoverished, and chronically malnourished people live in the world at the end of the twentieth century than at the beginning. —Marvin Harris

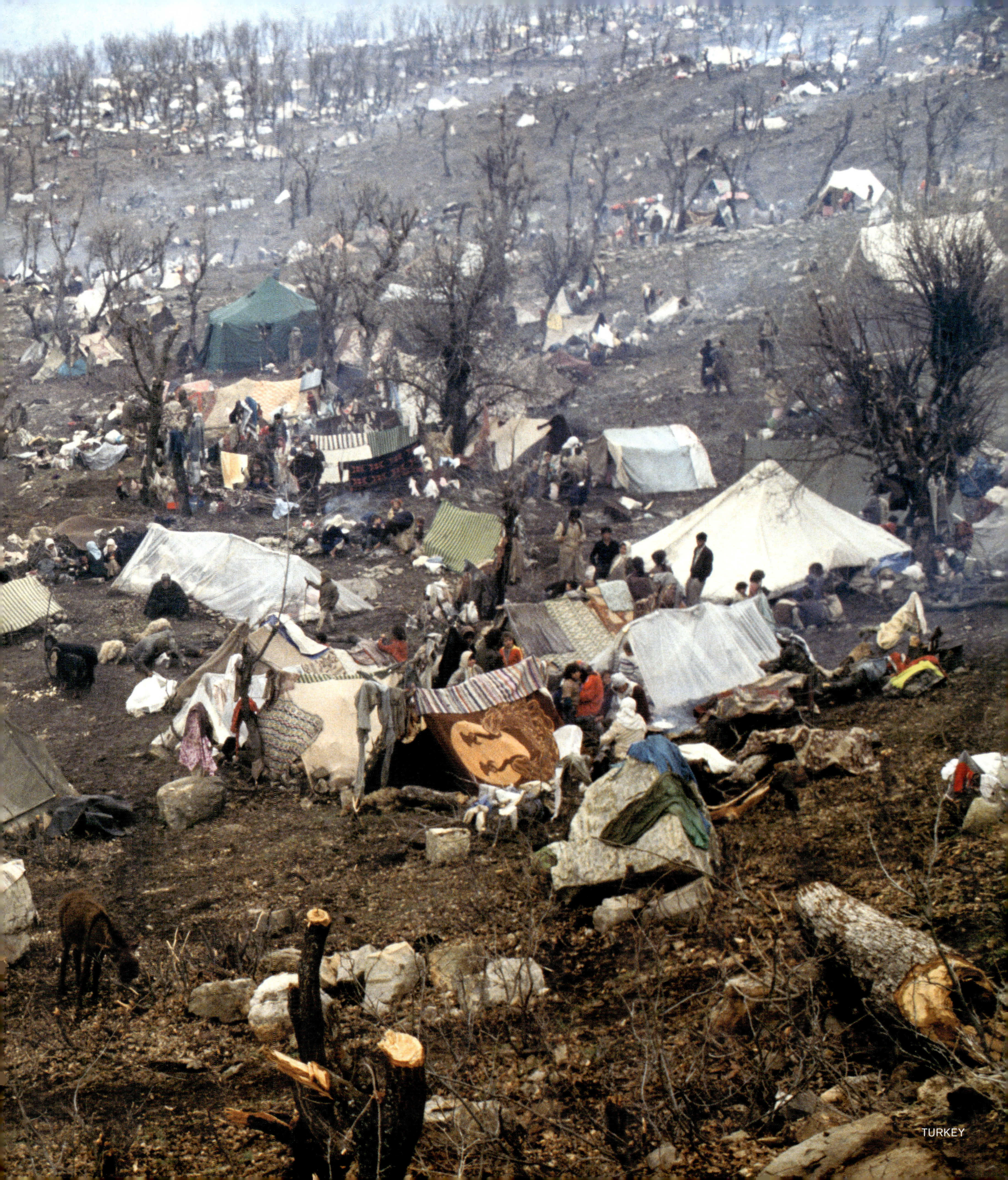

TURKEY

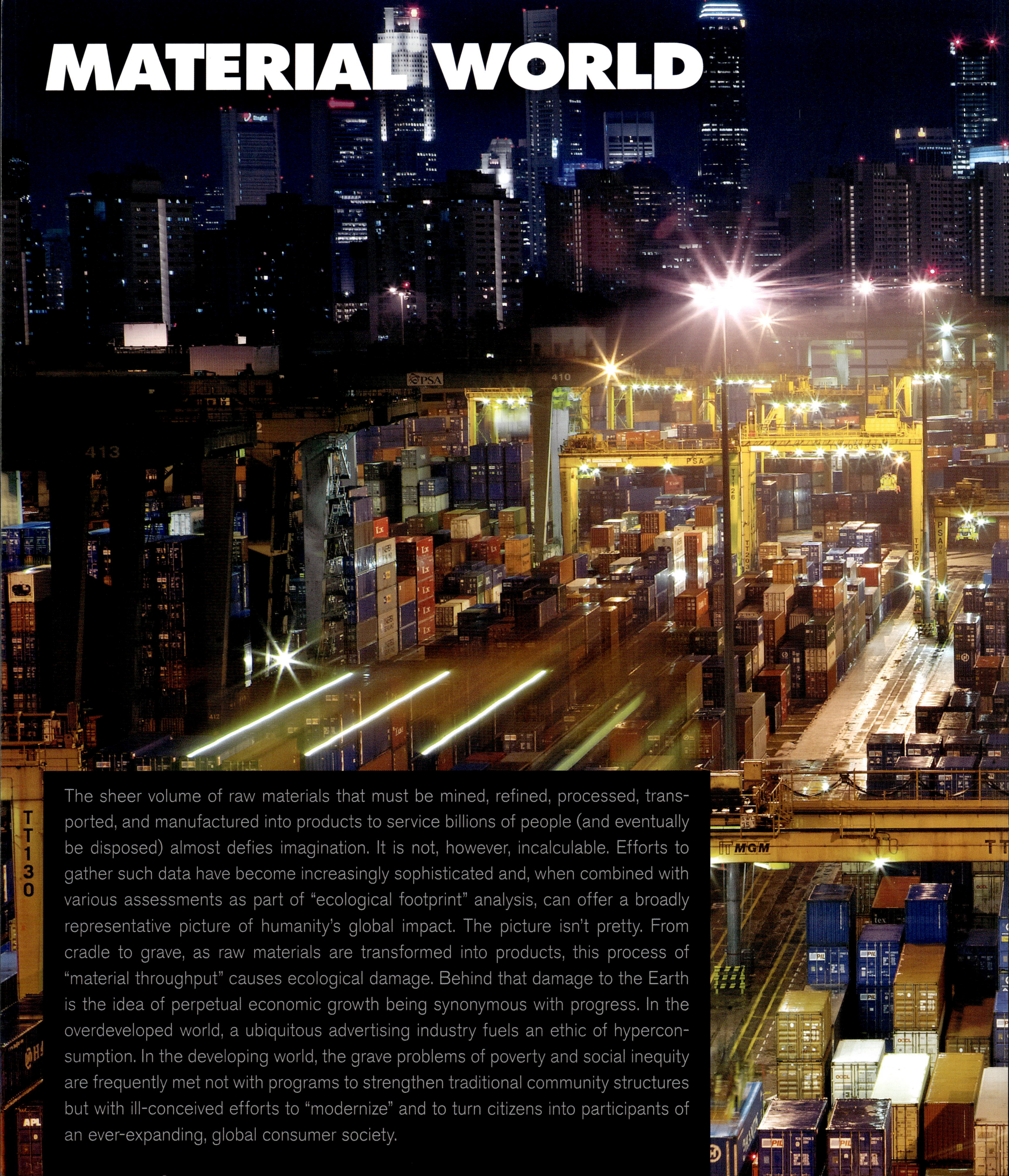

MATERIAL WORLD

The sheer volume of raw materials that must be mined, refined, processed, transported, and manufactured into products to service billions of people (and eventually be disposed) almost defies imagination. It is not, however, incalculable. Efforts to gather such data have become increasingly sophisticated and, when combined with various assessments as part of "ecological footprint" analysis, can offer a broadly representative picture of humanity's global impact. The picture isn't pretty. From cradle to grave, as raw materials are transformed into products, this process of "material throughput" causes ecological damage. Behind that damage to the Earth is the idea of perpetual economic growth being synonymous with progress. In the overdeveloped world, a ubiquitous advertising industry fuels an ethic of hyperconsumption. In the developing world, the grave problems of poverty and social inequity are frequently met not with programs to strengthen traditional community structures but with ill-conceived efforts to "modernize" and to turn citizens into participants of an ever-expanding, global consumer society.

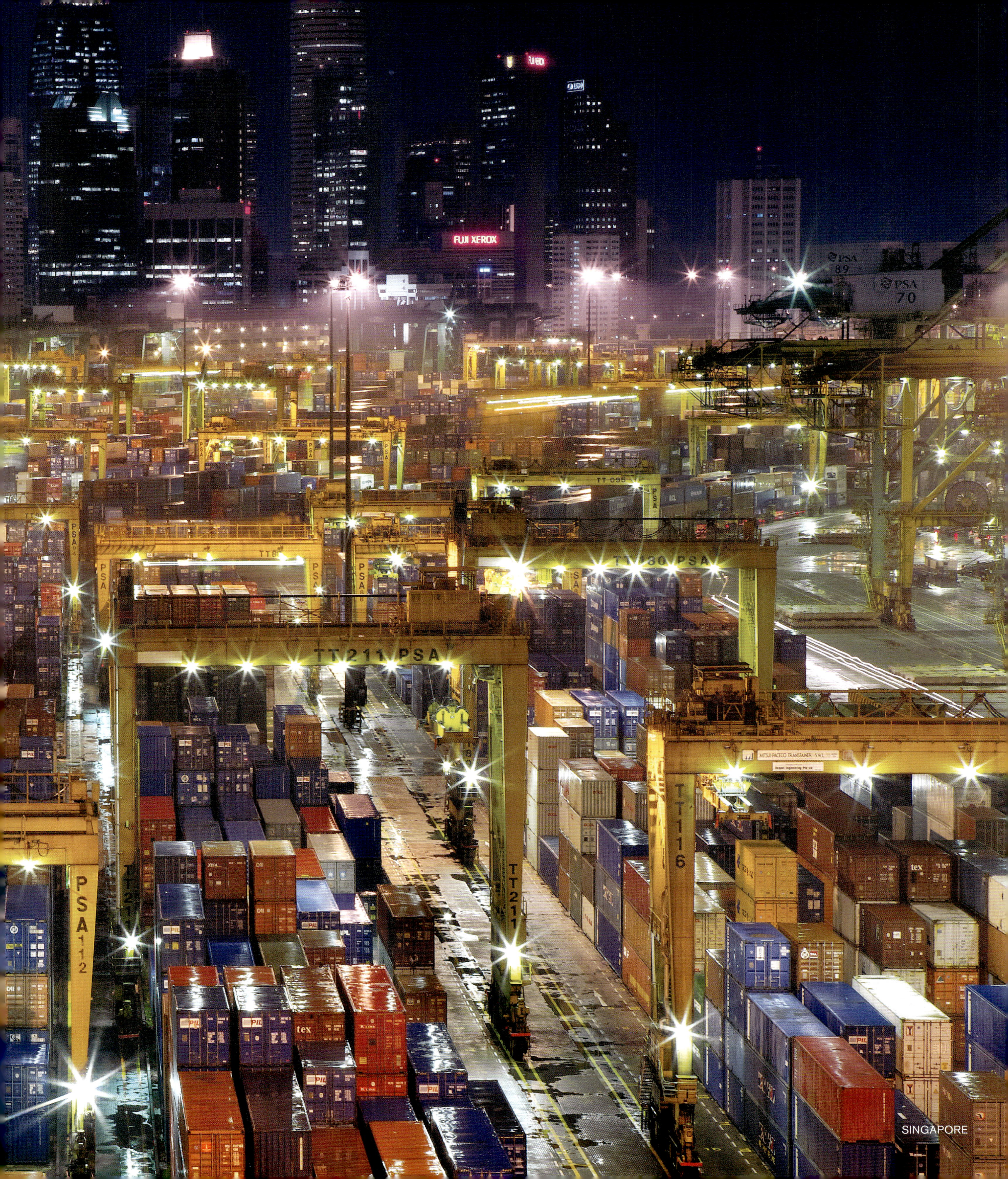

FUJI XEROX
PSA 89
PSA 70
TT 095
TT 030 PSA
TT 211 PSA
TT 116
PSA 112
PIL
tex
SINGAPORE

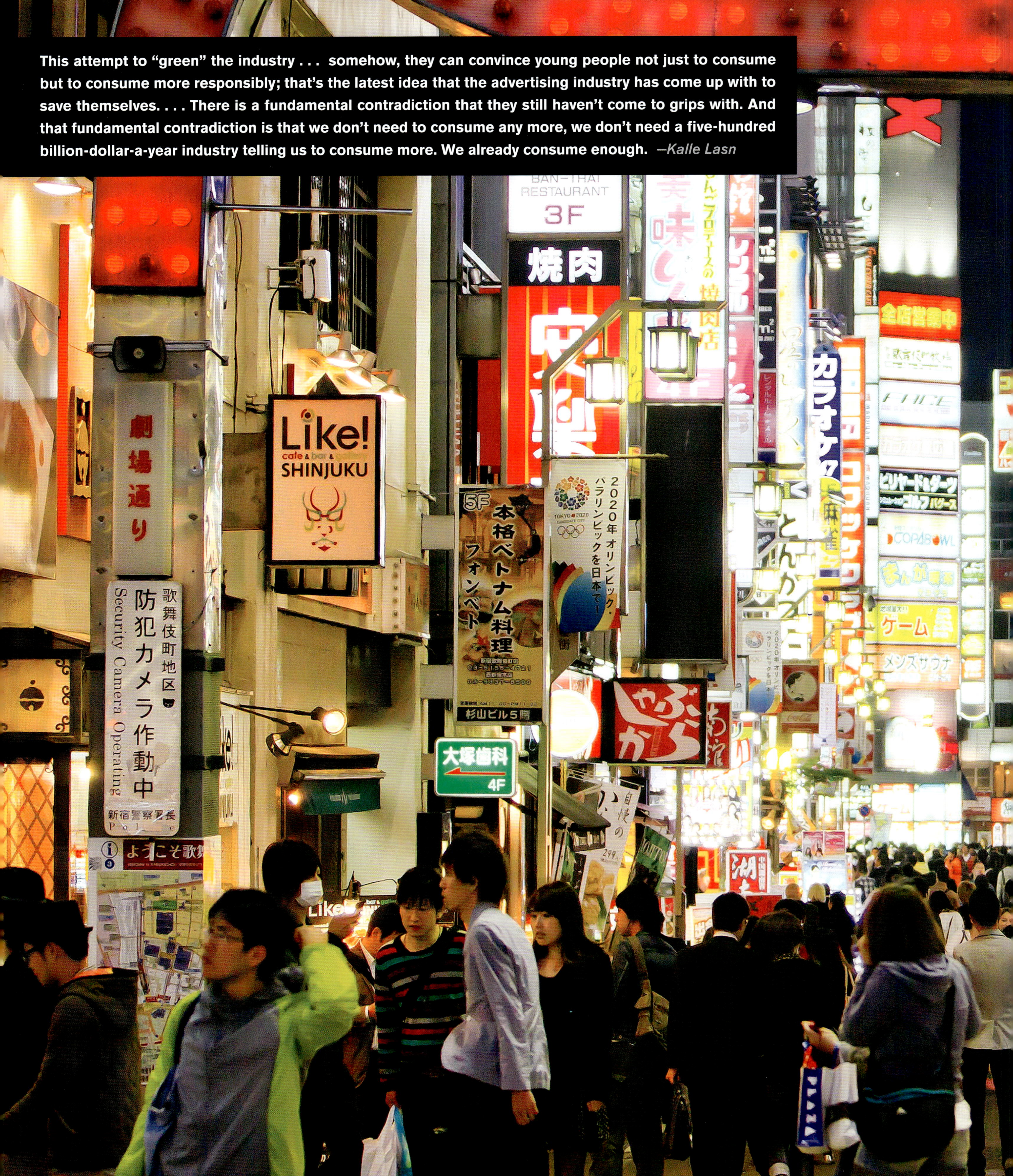

This attempt to "green" the industry . . . somehow, they can convince young people not just to consume but to consume more responsibly; that's the latest idea that the advertising industry has come up with to save themselves. . . . There is a fundamental contradiction that they still haven't come to grips with. And that fundamental contradiction is that we don't need to consume any more, we don't need a five-hundred billion-dollar-a-year industry telling us to consume more. We already consume enough. —Kalle Lasn

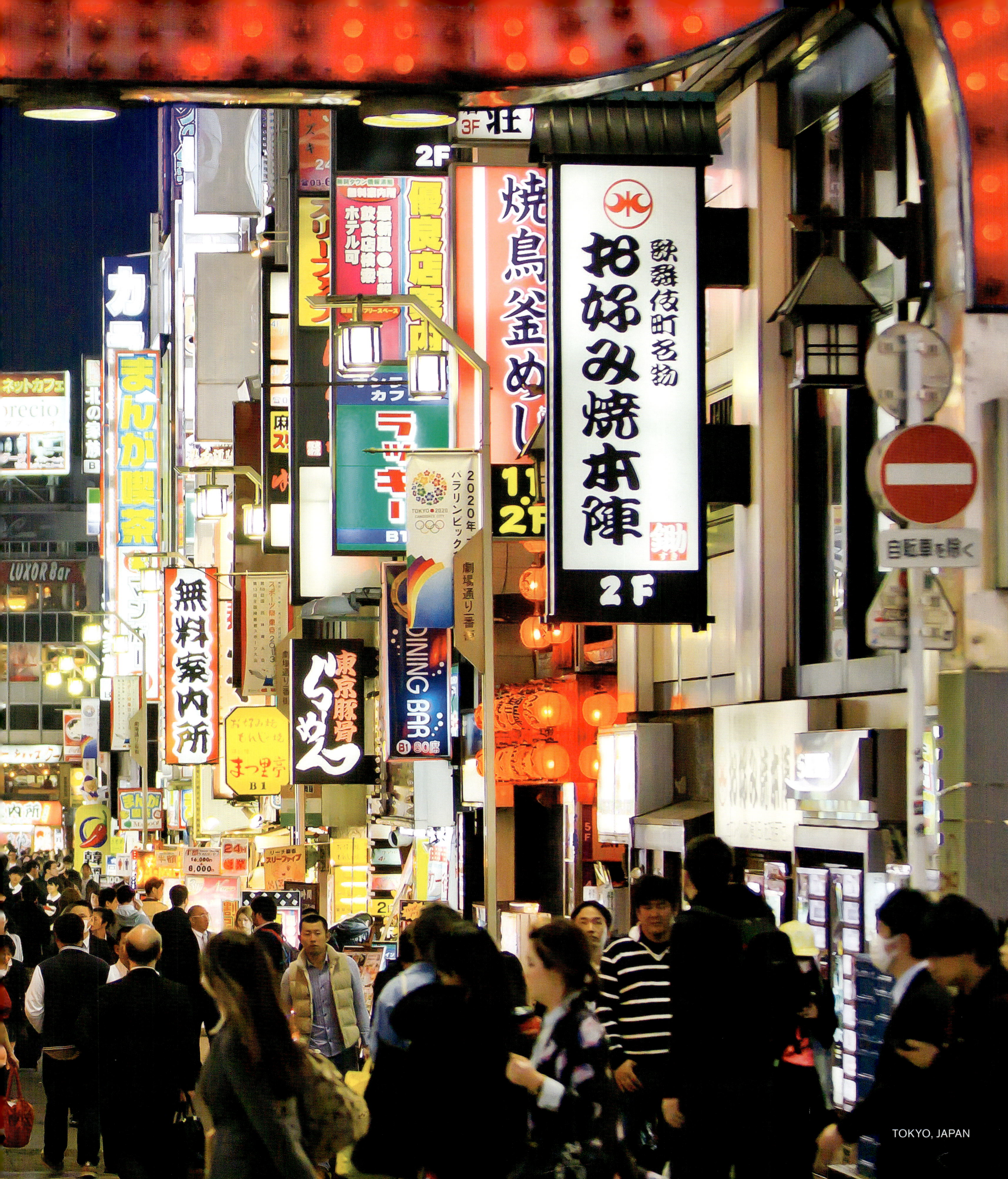
TOKYO, JAPAN

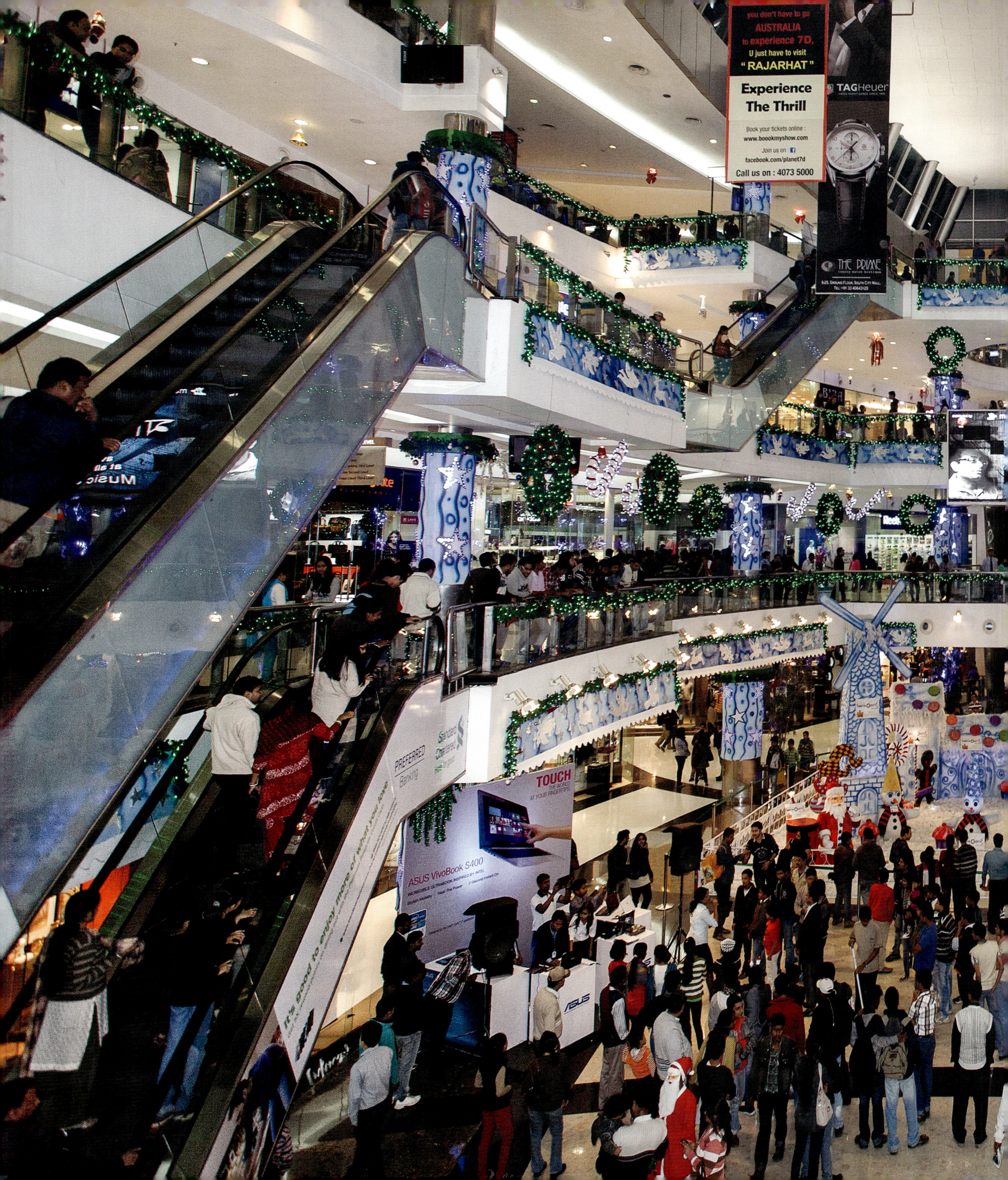

you don't have to go
AUSTRALIA
to experience 7D,
U just have to visit
"RAJARHAT"
Experience
The Thrill
Book your tickets online :
www.booookmyshow.com
Join us on
facebook.com/planet7d
Call us on : 4073 5000
TAGHeuer
THE PRIME
TOUCH
AT YOUR FINGERTIPS
ASUS VivoBook S400
It's good to enjoy more of what you see
PREFERRED
Banking

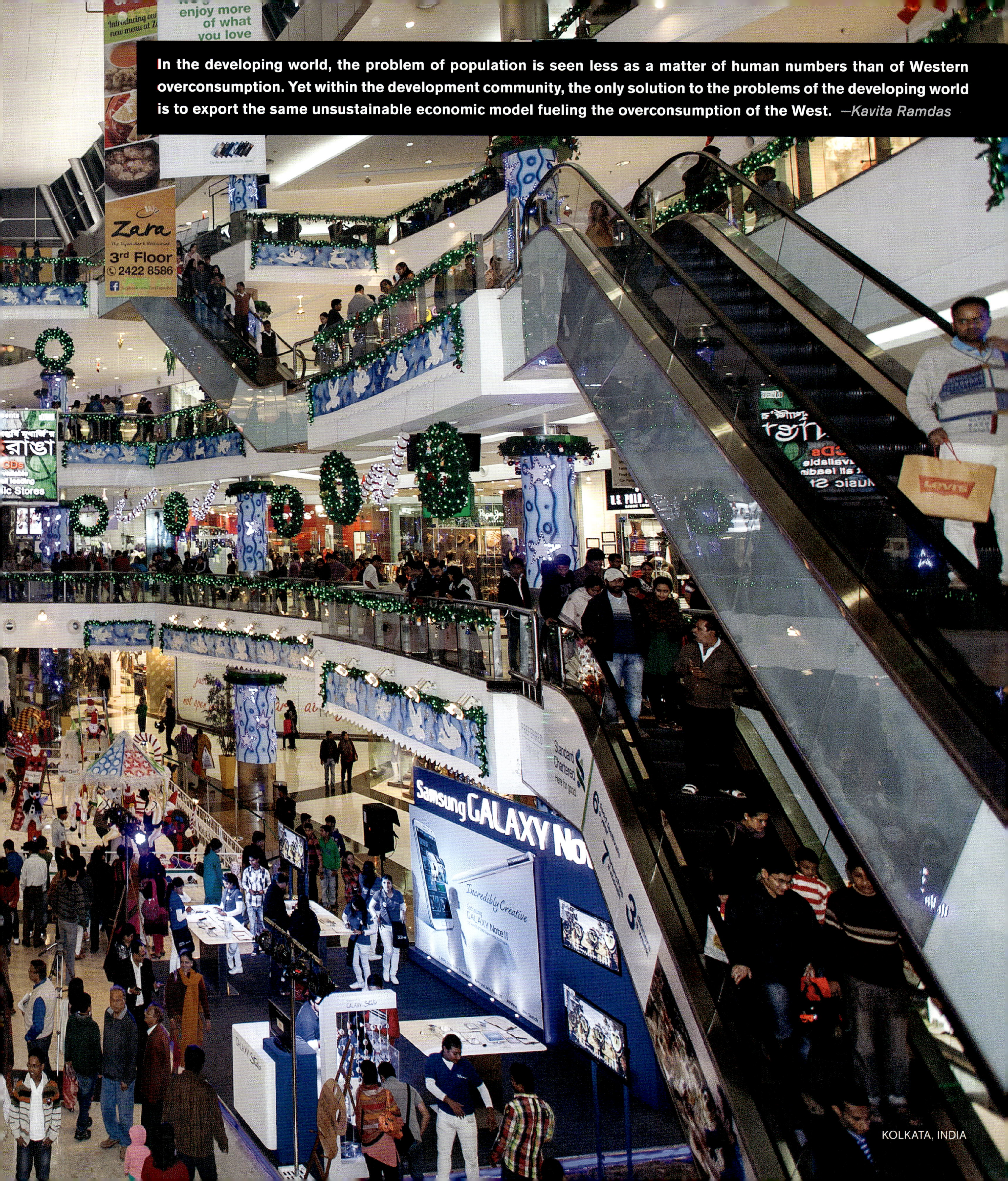

KOLKATA, INDIA

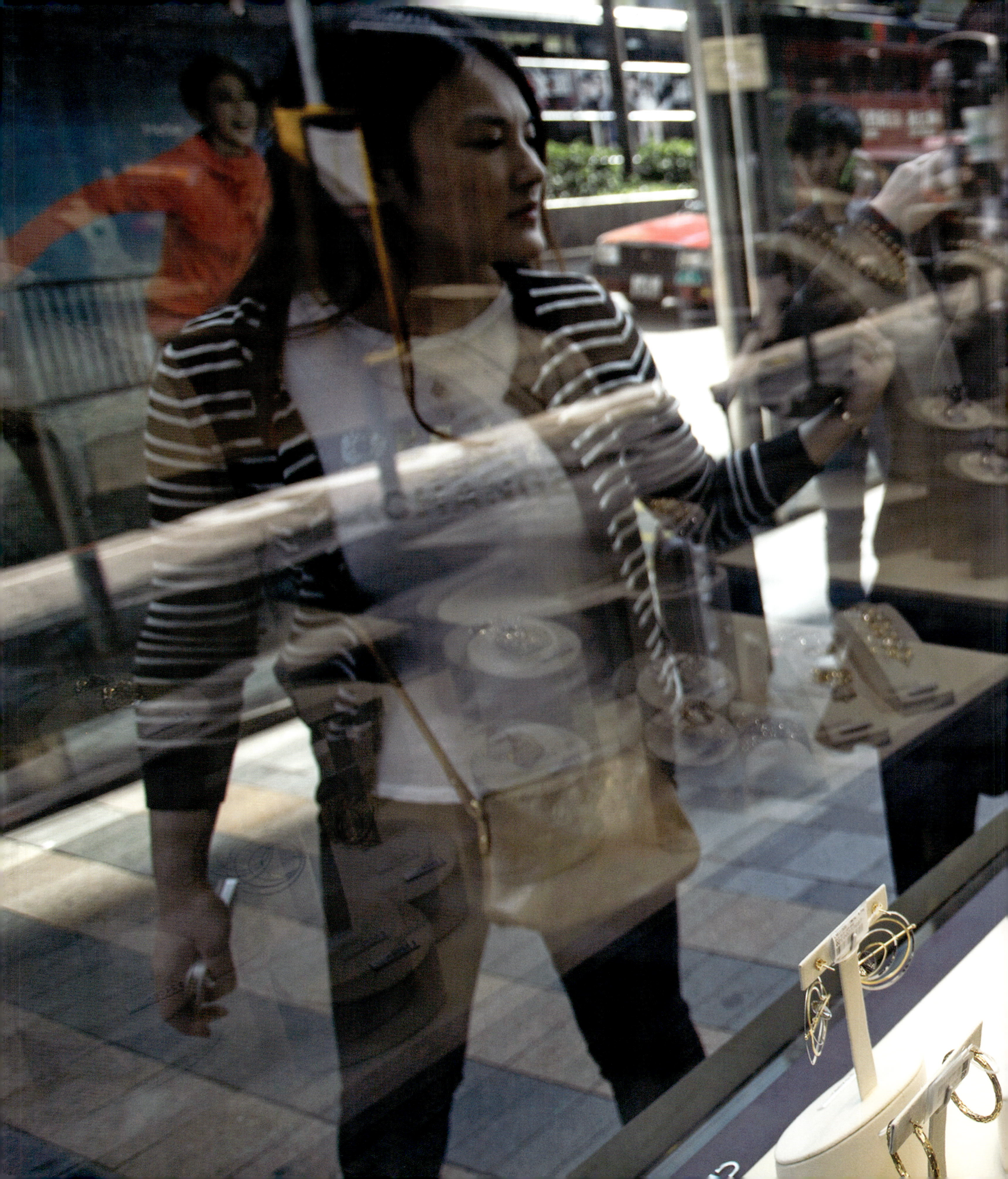

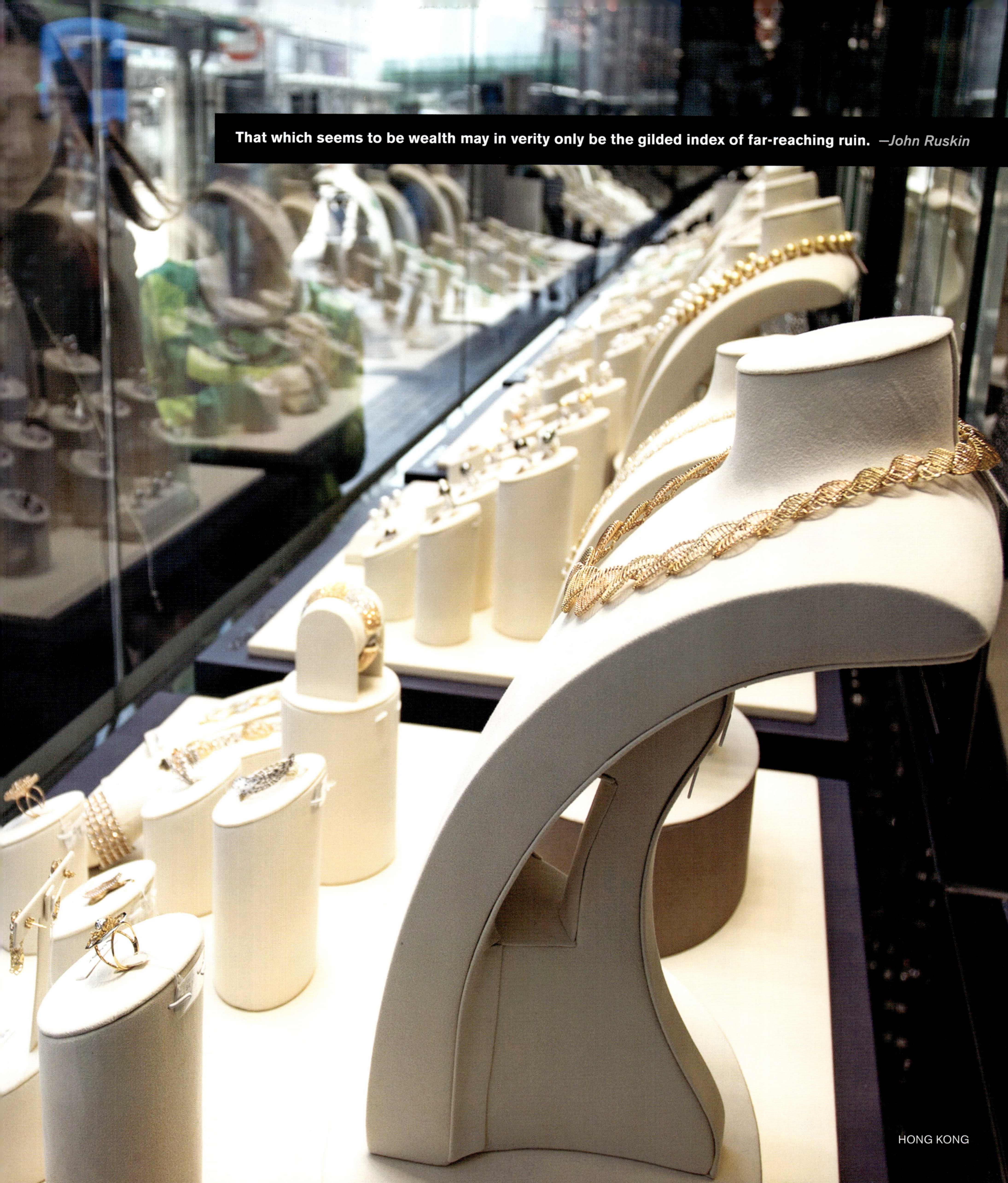
That which seems to be wealth may in verity only be the gilded index of far-reaching ruin. —John Ruskin
HONG KONG

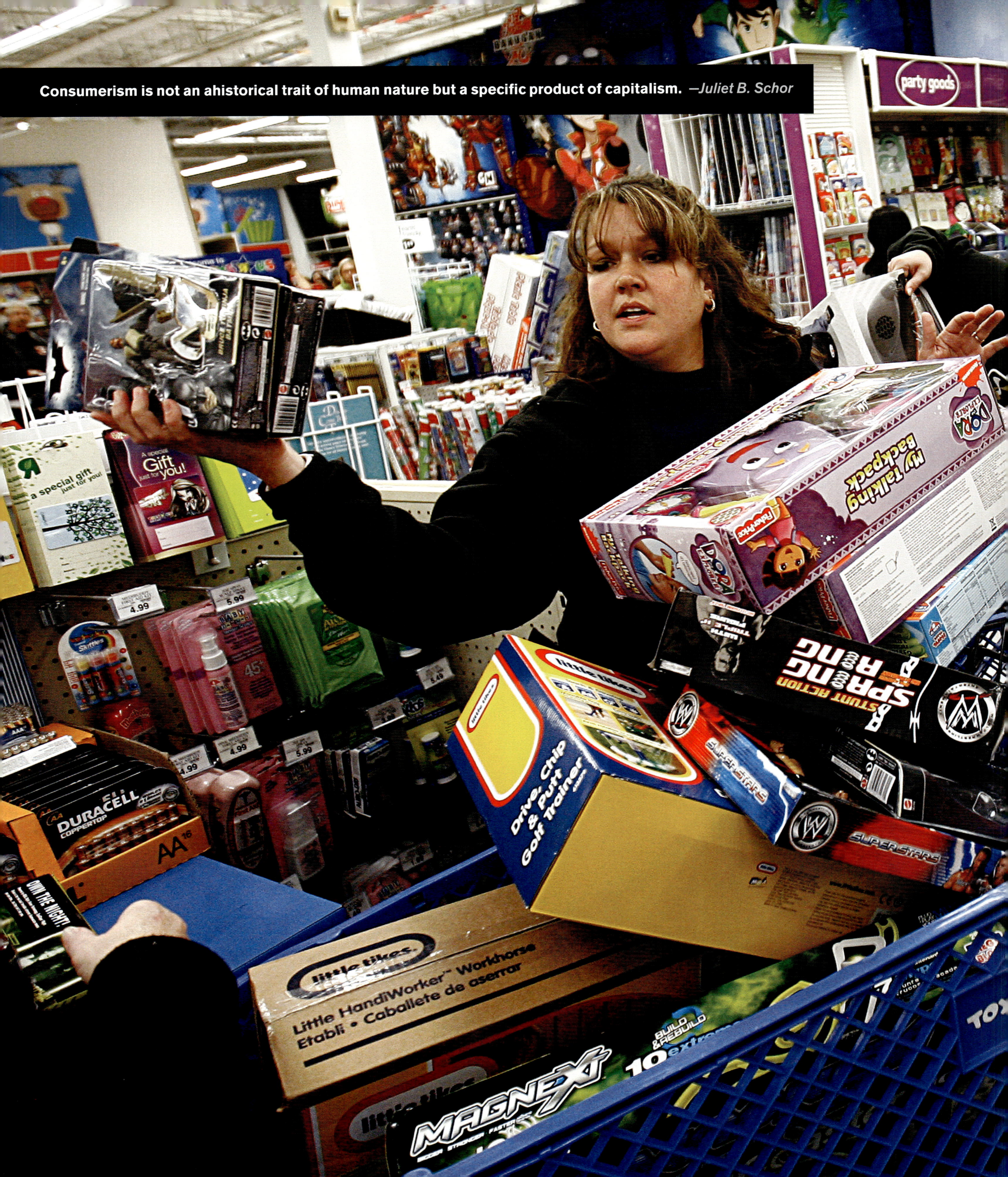

Consumerism is not an ahistorical trait of human nature but a specific product of capitalism. —Juliet B. Schor

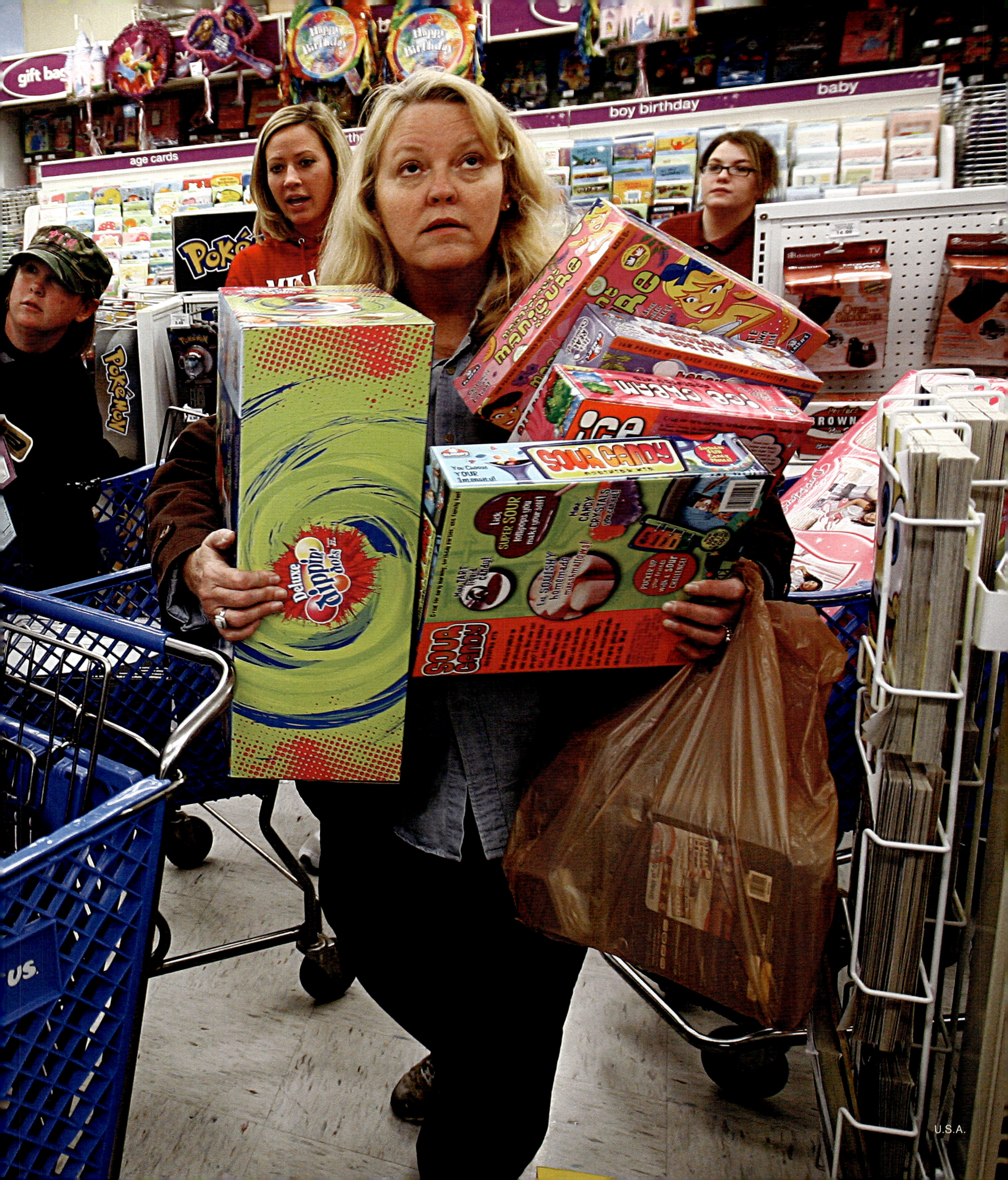

U.S.A.

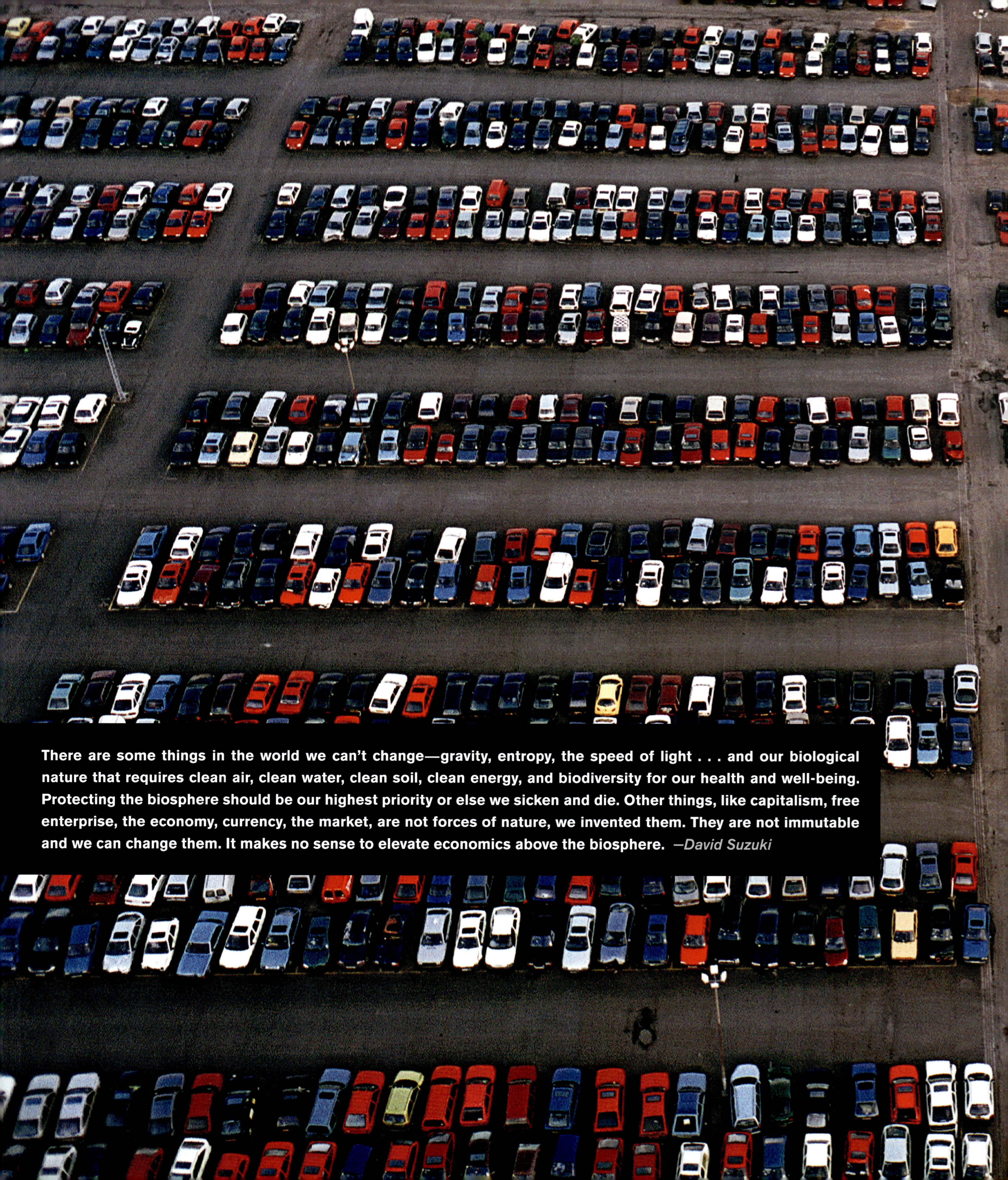

There are some things in the world we can't change—gravity, entropy, the speed of light . . . and our biological nature that requires clean air, clean water, clean soil, clean energy, and biodiversity for our health and well-being. Protecting the biosphere should be our highest priority or else we sicken and die. Other things, like capitalism, free enterprise, the economy, currency, the market, are not forces of nature, we invented them. They are not immutable and we can change them. It makes no sense to elevate economics above the biosphere. —David Suzuki

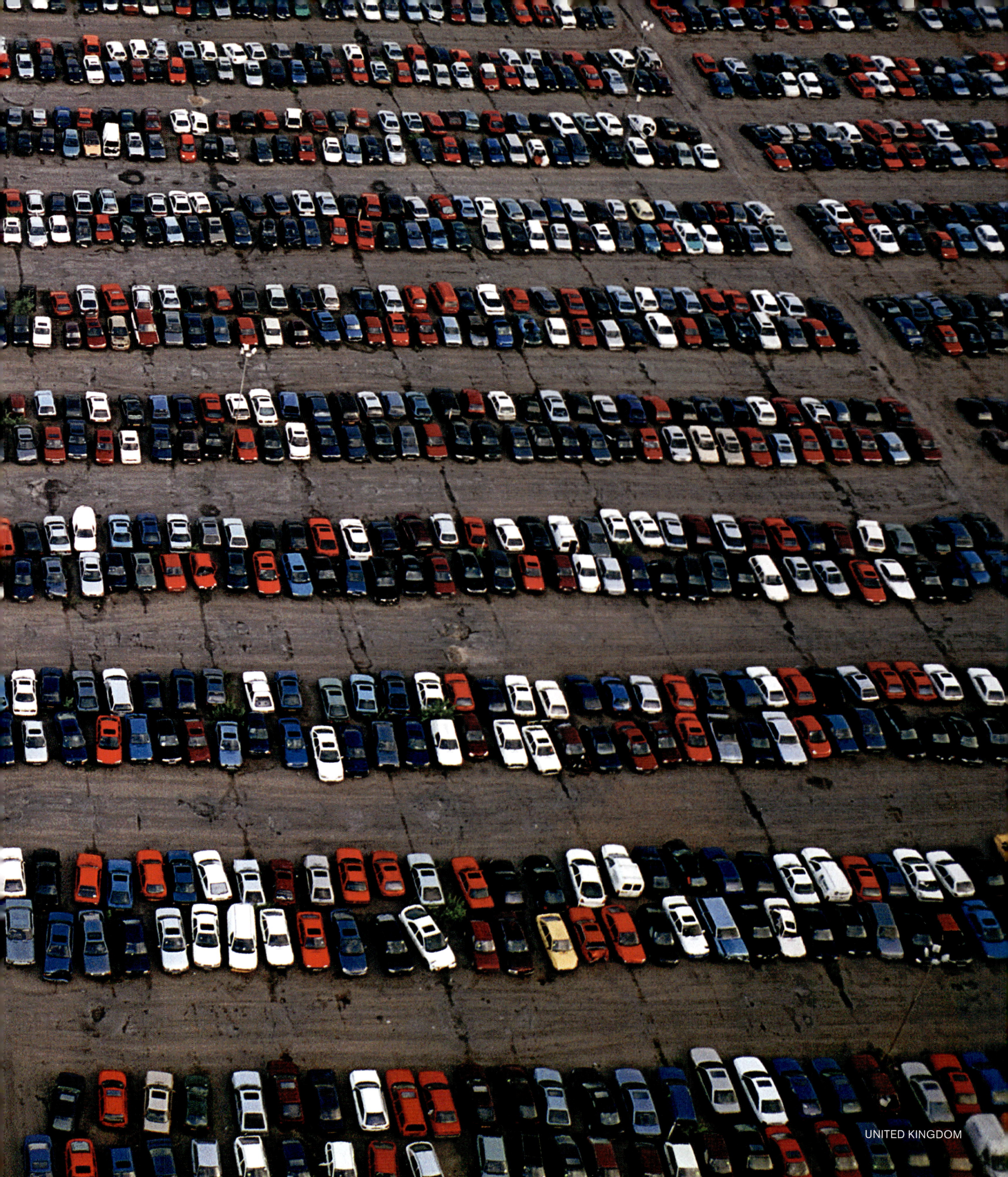

UNITED KINGDOM

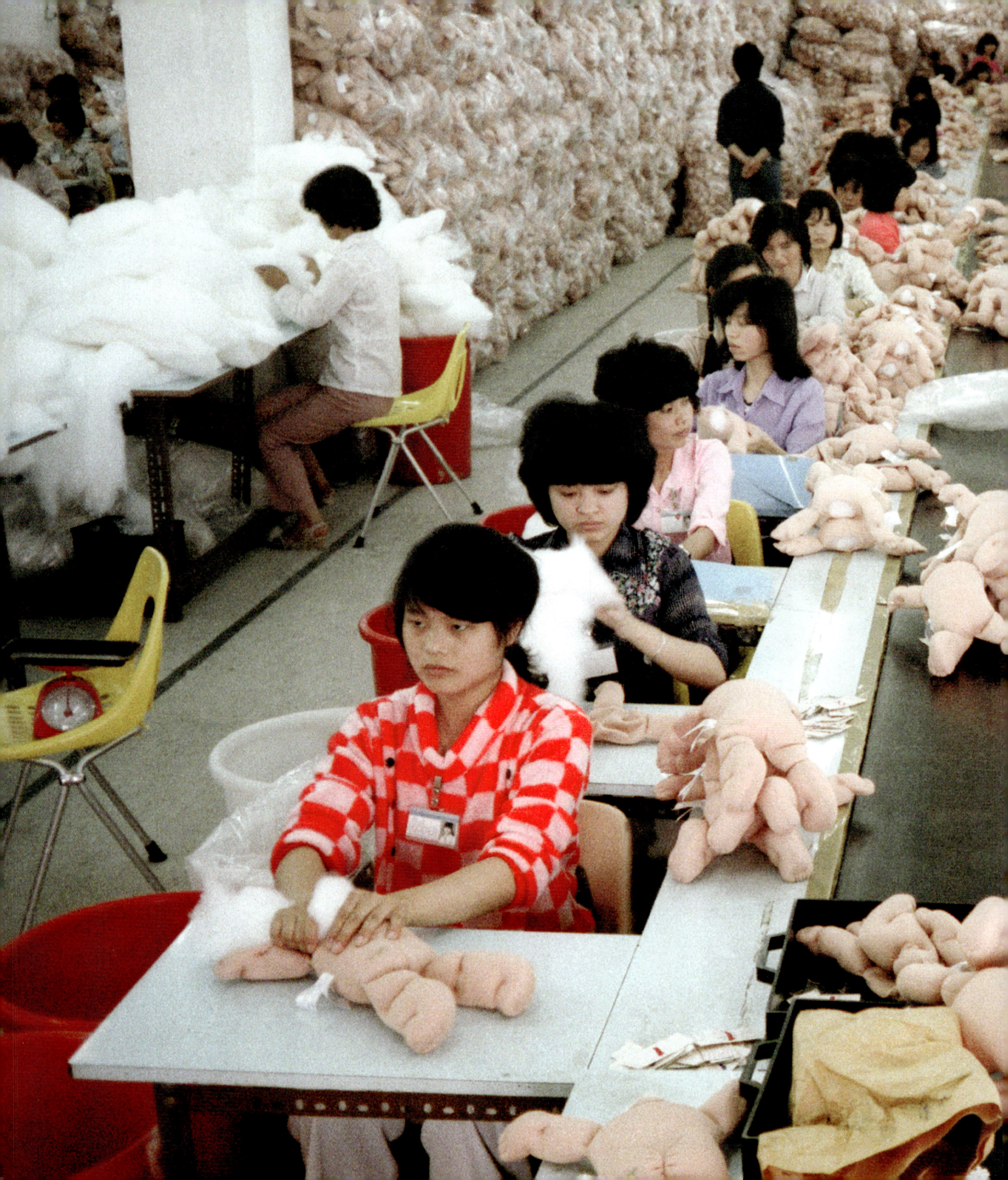

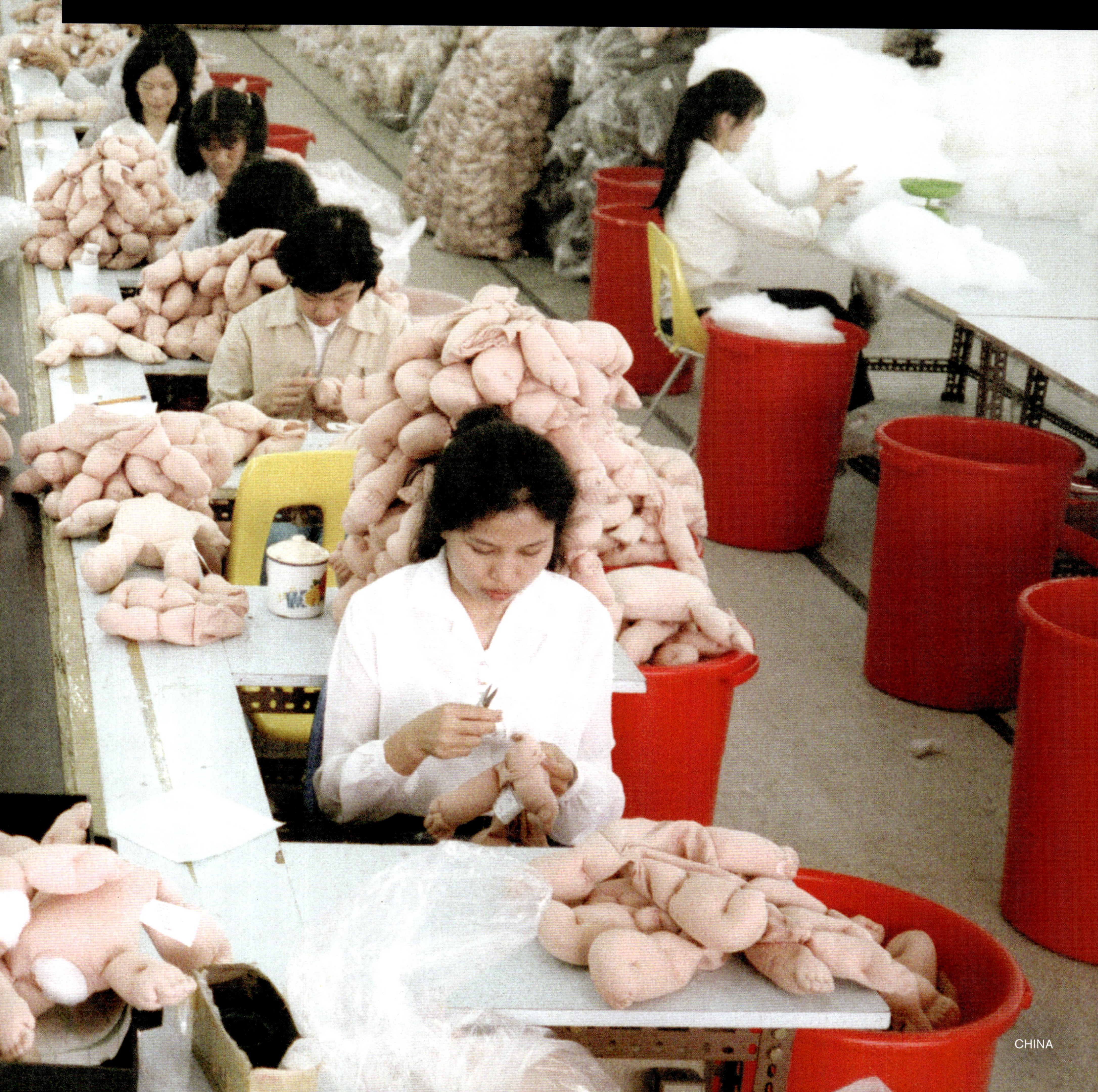

Some people continue to defend trickle-down theories which assume that economic growth, encouraged by a free market, will inevitably succeed in bringing about greater justice and inclusiveness in the world. This opinion, which has never been confirmed by the facts, expresses a crude and naive trust in the goodness of those wielding economic power and in the sacralized workings of the prevailing economic system. Meanwhile, the excluded are still waiting. To sustain a lifestyle which excludes others, or to sustain enthusiasm for that selfish ideal, a globalization of indifference has developed. —Pope Francis
CHINA

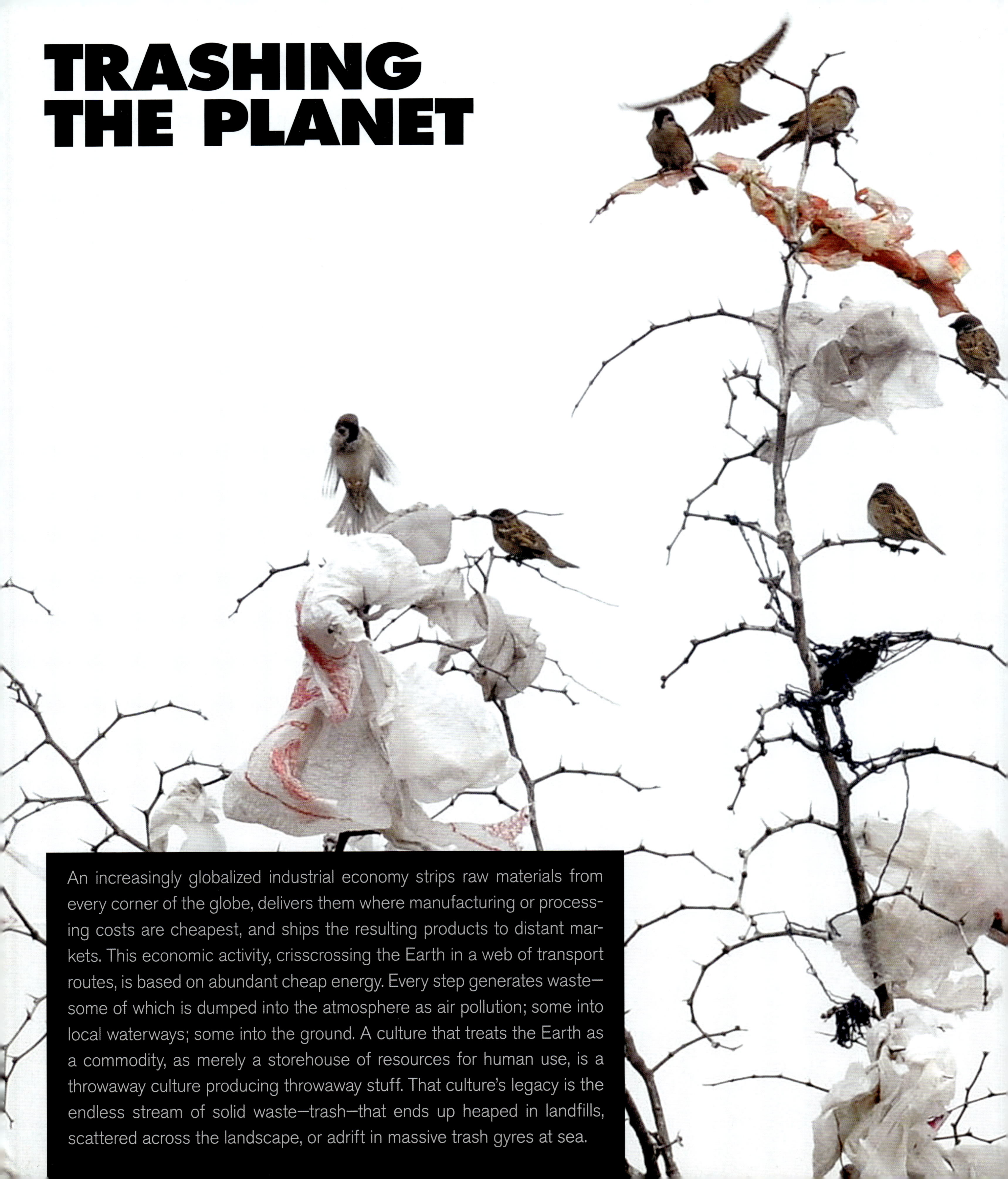

TRASHING THE PLANET

An increasingly globalized industrial economy strips raw materials from every corner of the globe, delivers them where manufacturing or processing costs are cheapest, and ships the resulting products to distant markets. This economic activity, crisscrossing the Earth in a web of transport routes, is based on abundant cheap energy. Every step generates waste—some of which is dumped into the atmosphere as air pollution; some into local waterways; some into the ground. A culture that treats the Earth as a commodity, as merely a storehouse of resources for human use, is a throwaway culture producing throwaway stuff. That culture's legacy is the endless stream of solid waste—trash—that ends up heaped in landfills, scattered across the landscape, or adrift in massive trash gyres at sea.

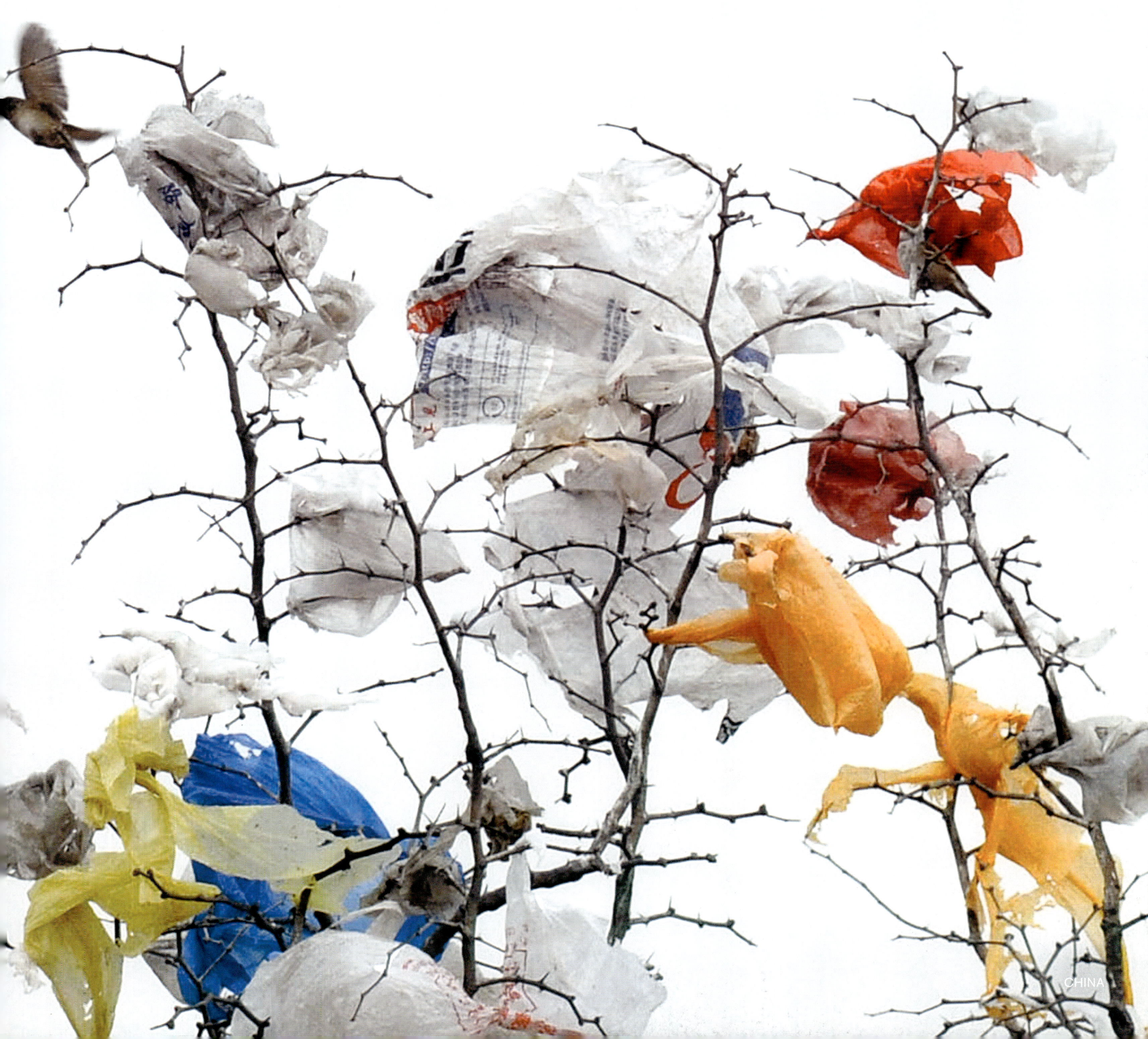

CHINA

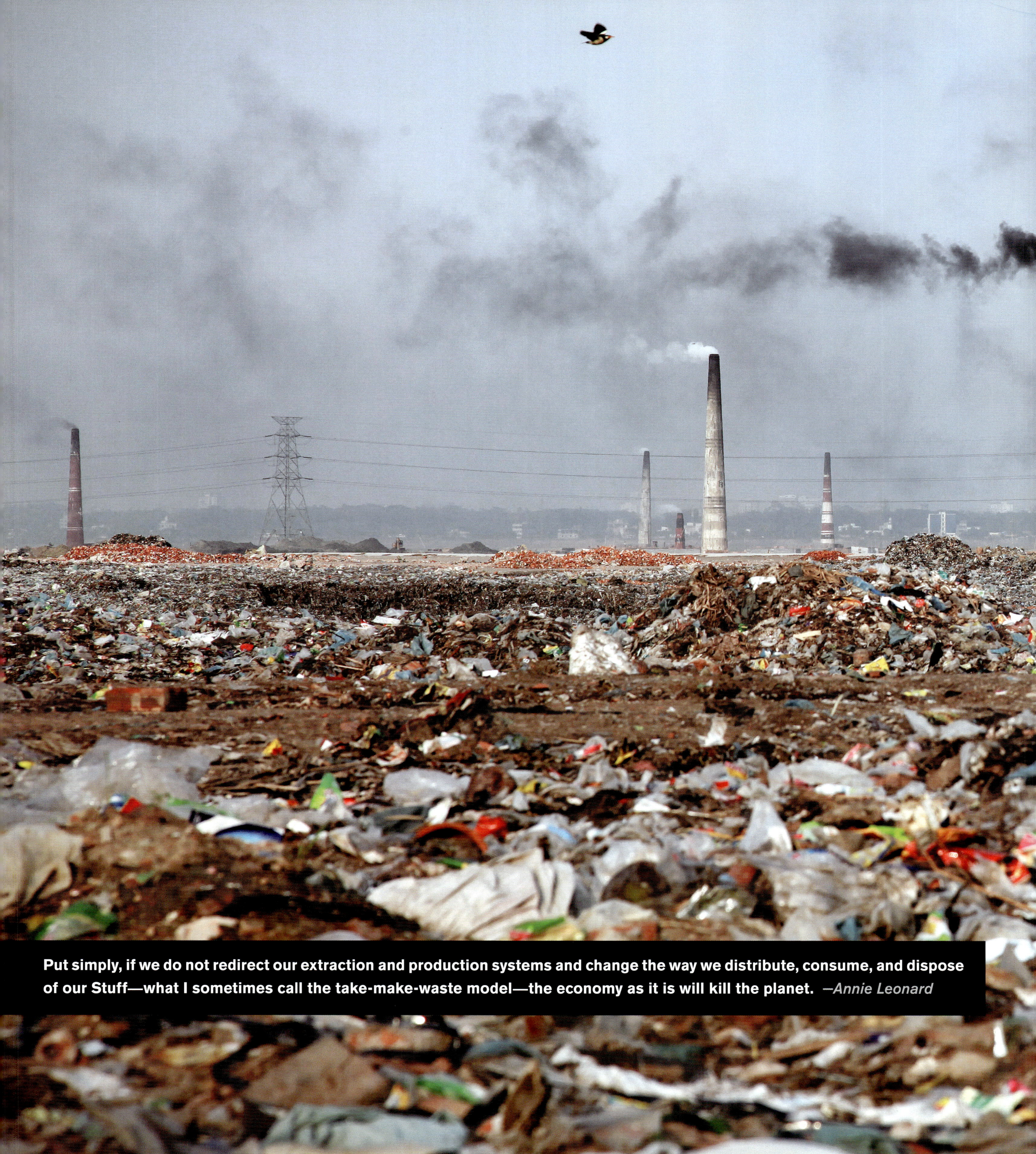

Put simply, if we do not redirect our extraction and production systems and change the way we distribute, consume, and dispose of our Stuff—what I sometimes call the take-make-waste model—the economy as it is will kill the planet. —Annie Leonard

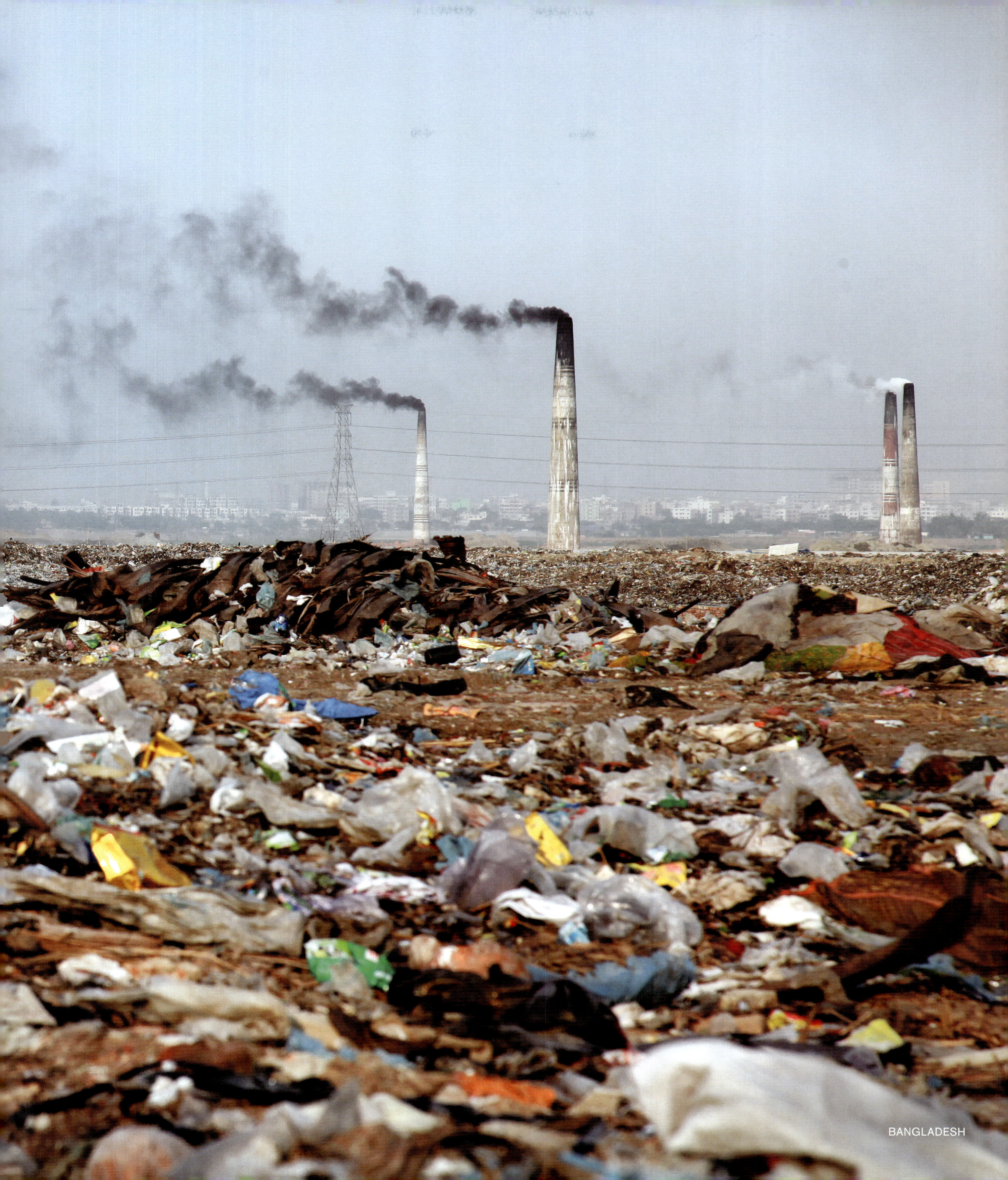
BANGLADESH

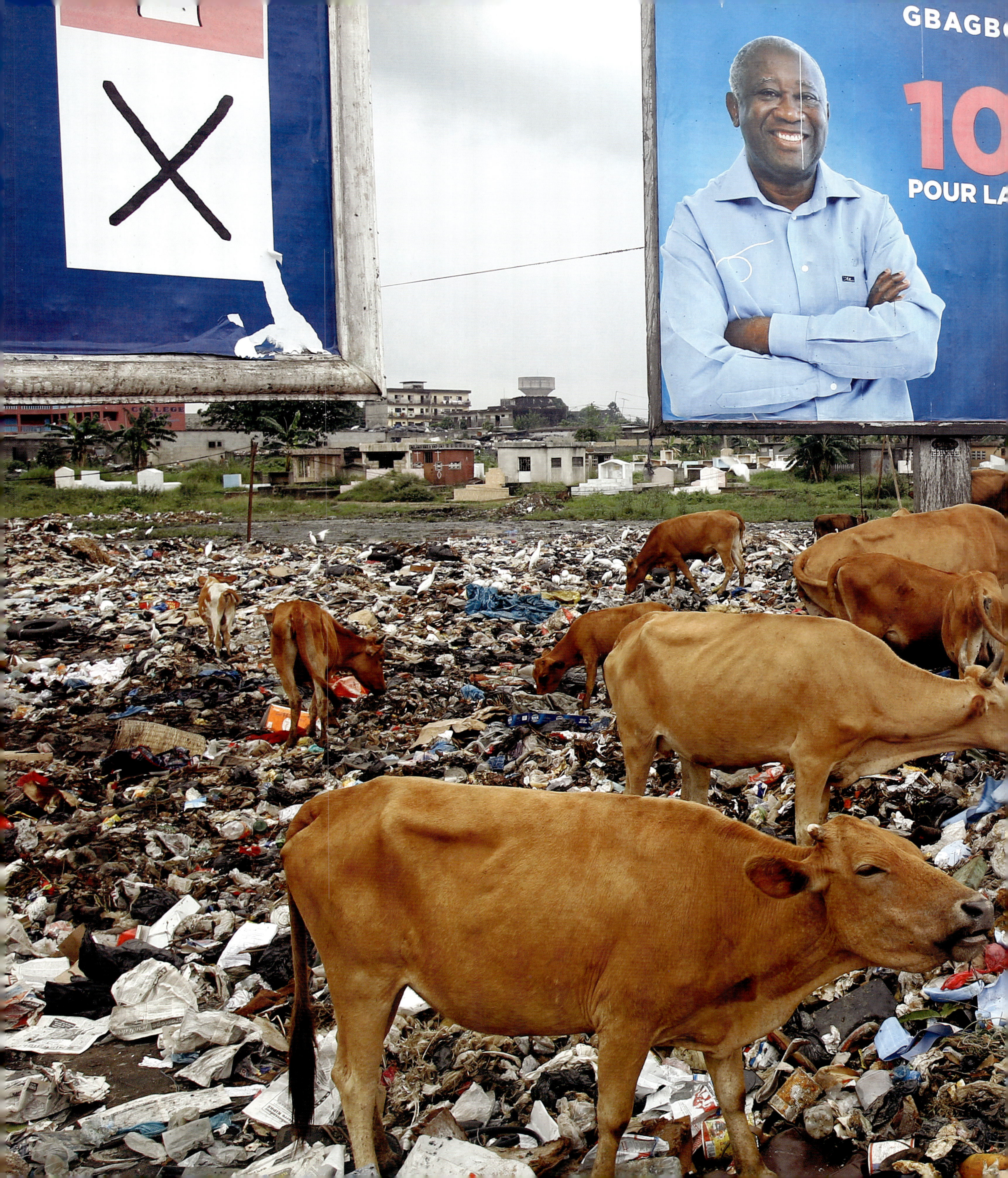

X
GBAGBO
10
POUR LA

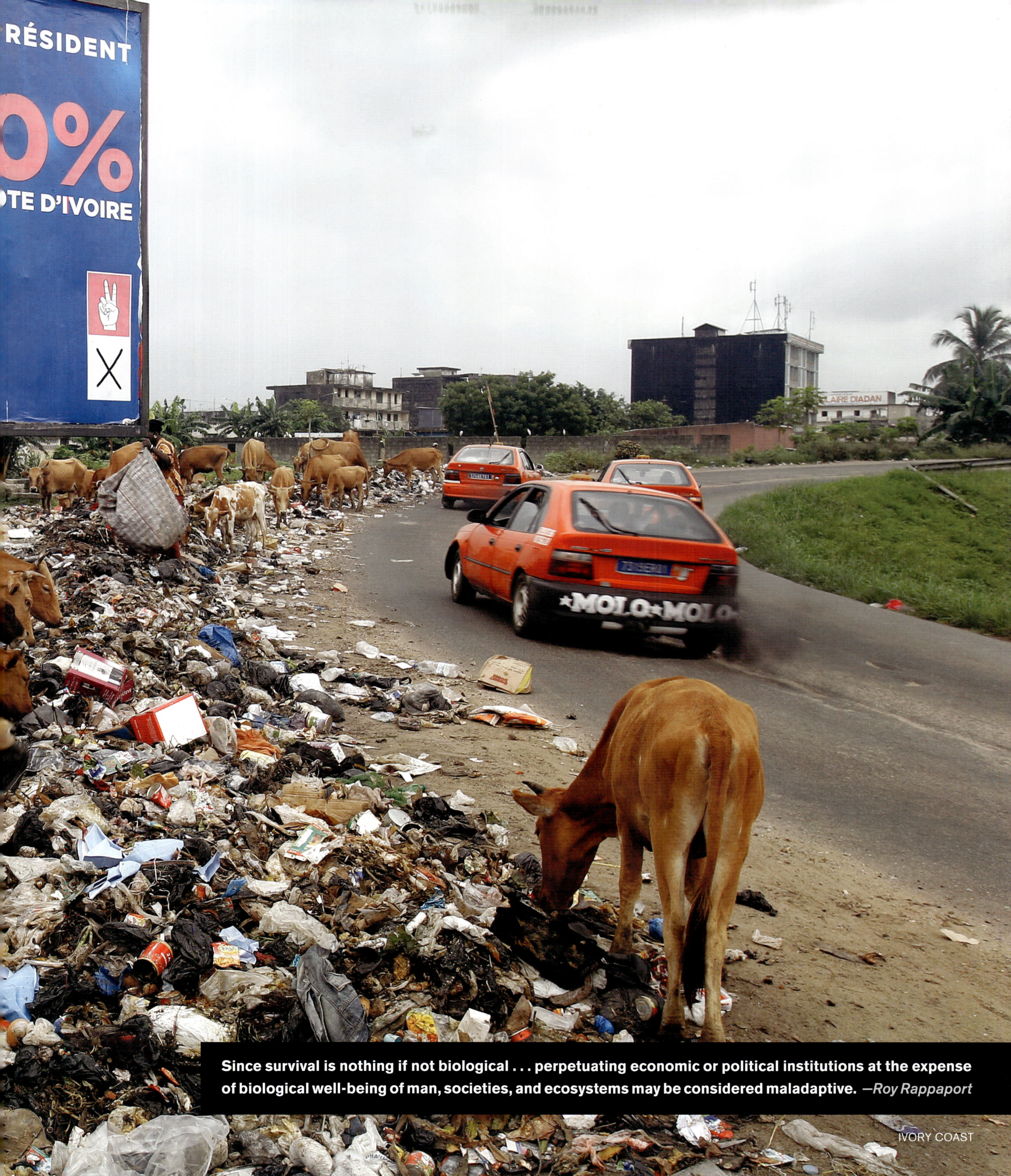

Since survival is nothing if not biological . . . perpetuating economic or political institutions at the expense of biological well-being of man, societies, and ecosystems may be considered maladaptive. —*Roy Rappaport*

IVORY COAST

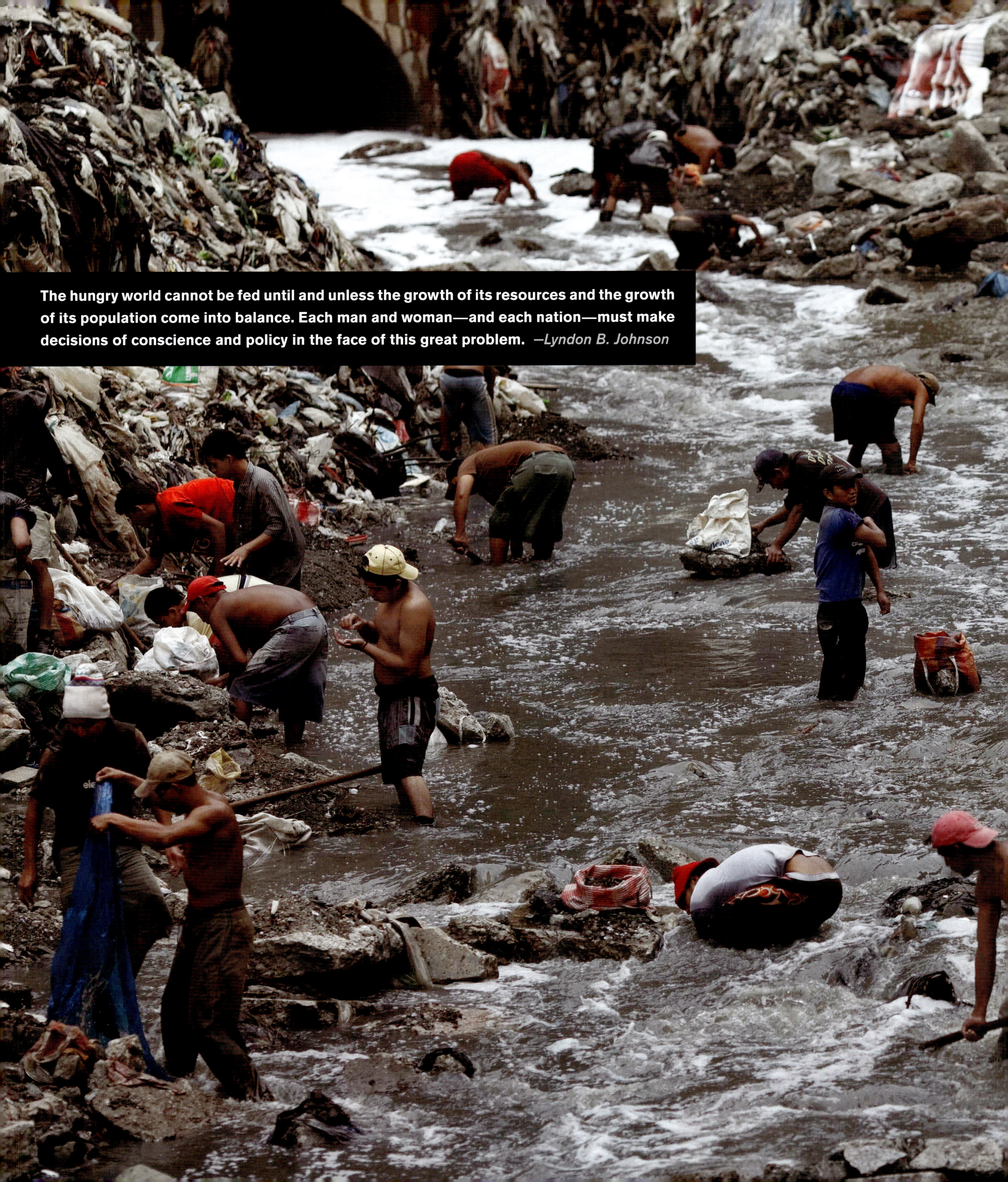

The hungry world cannot be fed until and unless the growth of its resources and the growth of its population come into balance. Each man and woman—and each nation—must make decisions of conscience and policy in the face of this great problem. —Lyndon B. Johnson

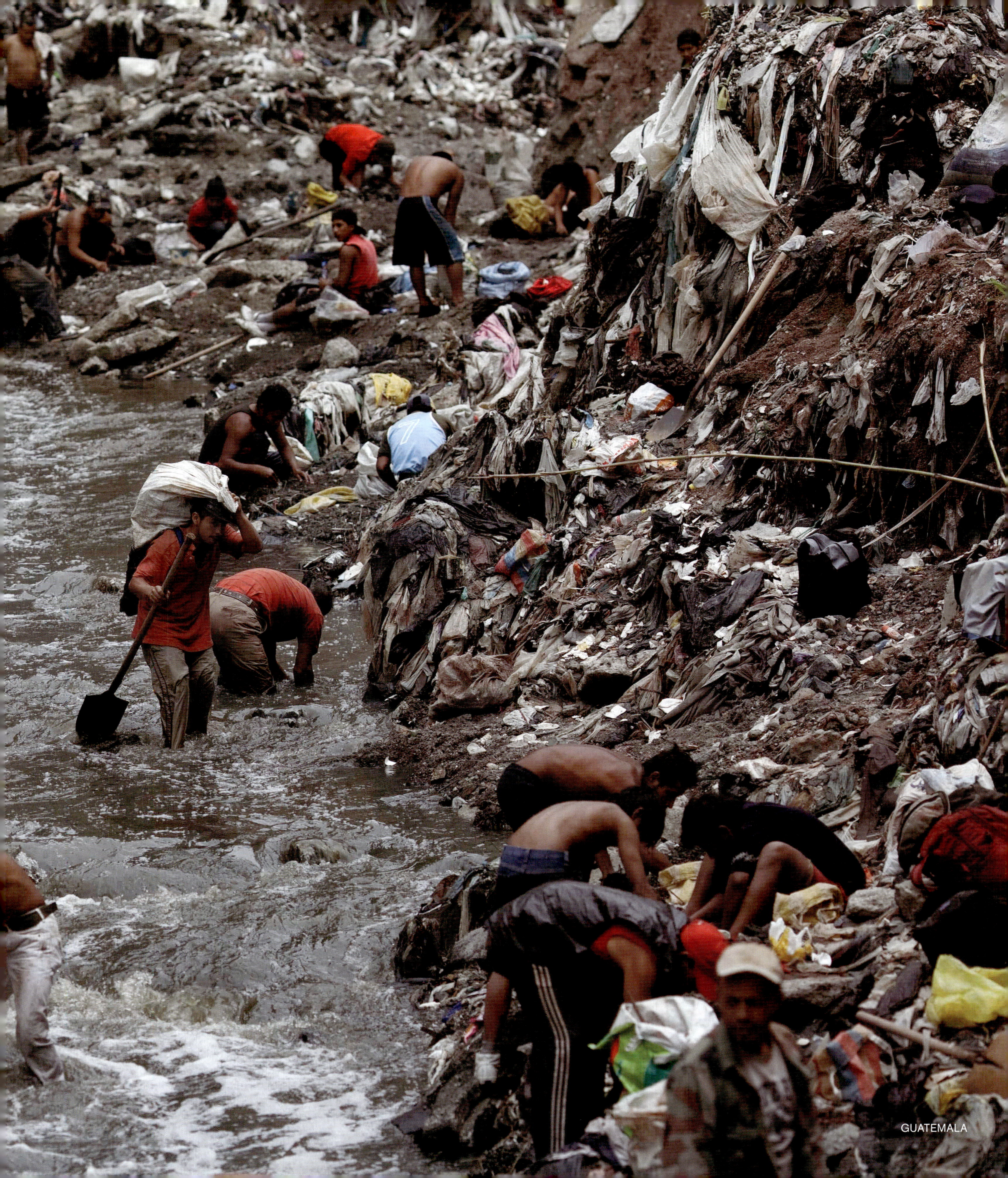
GUATEMALA

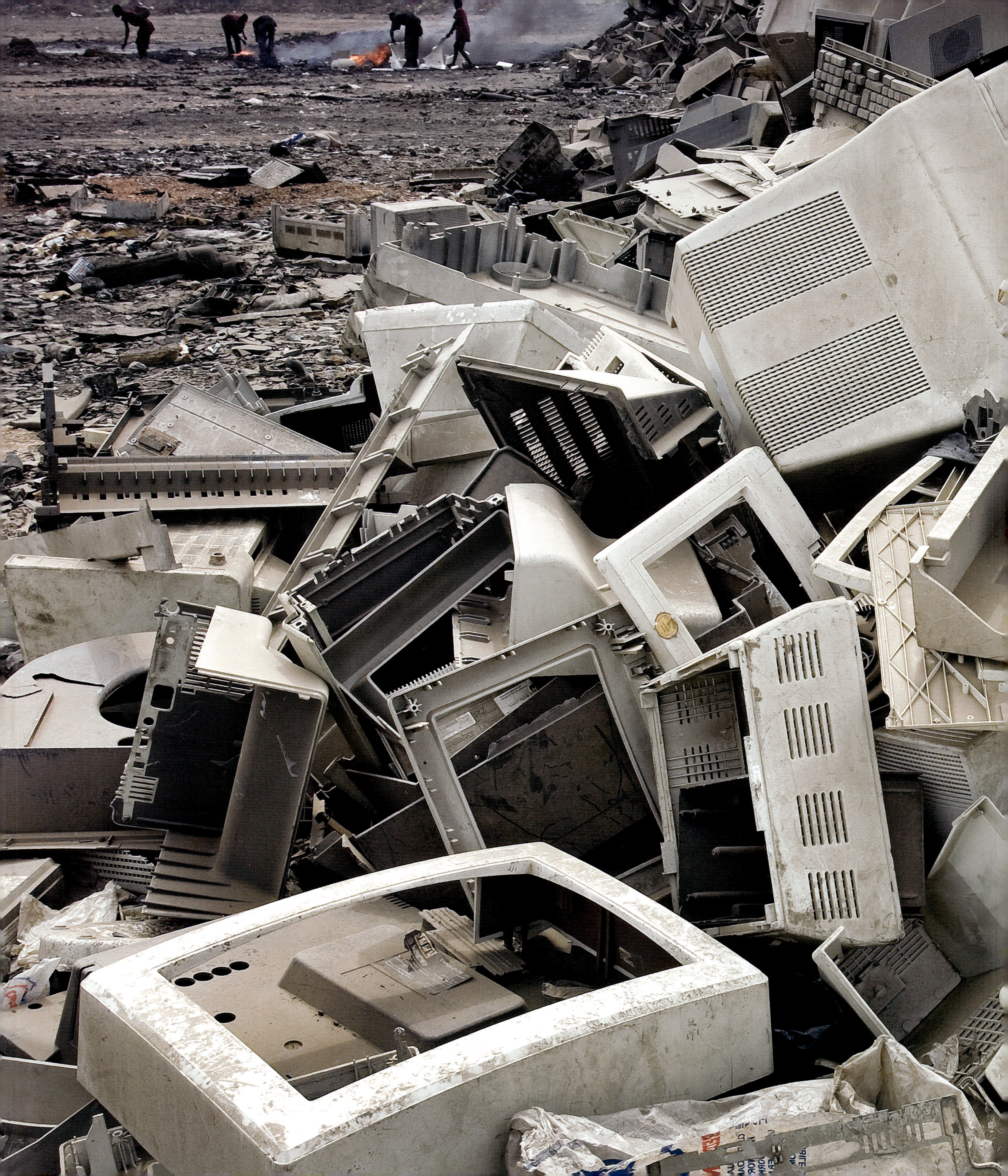

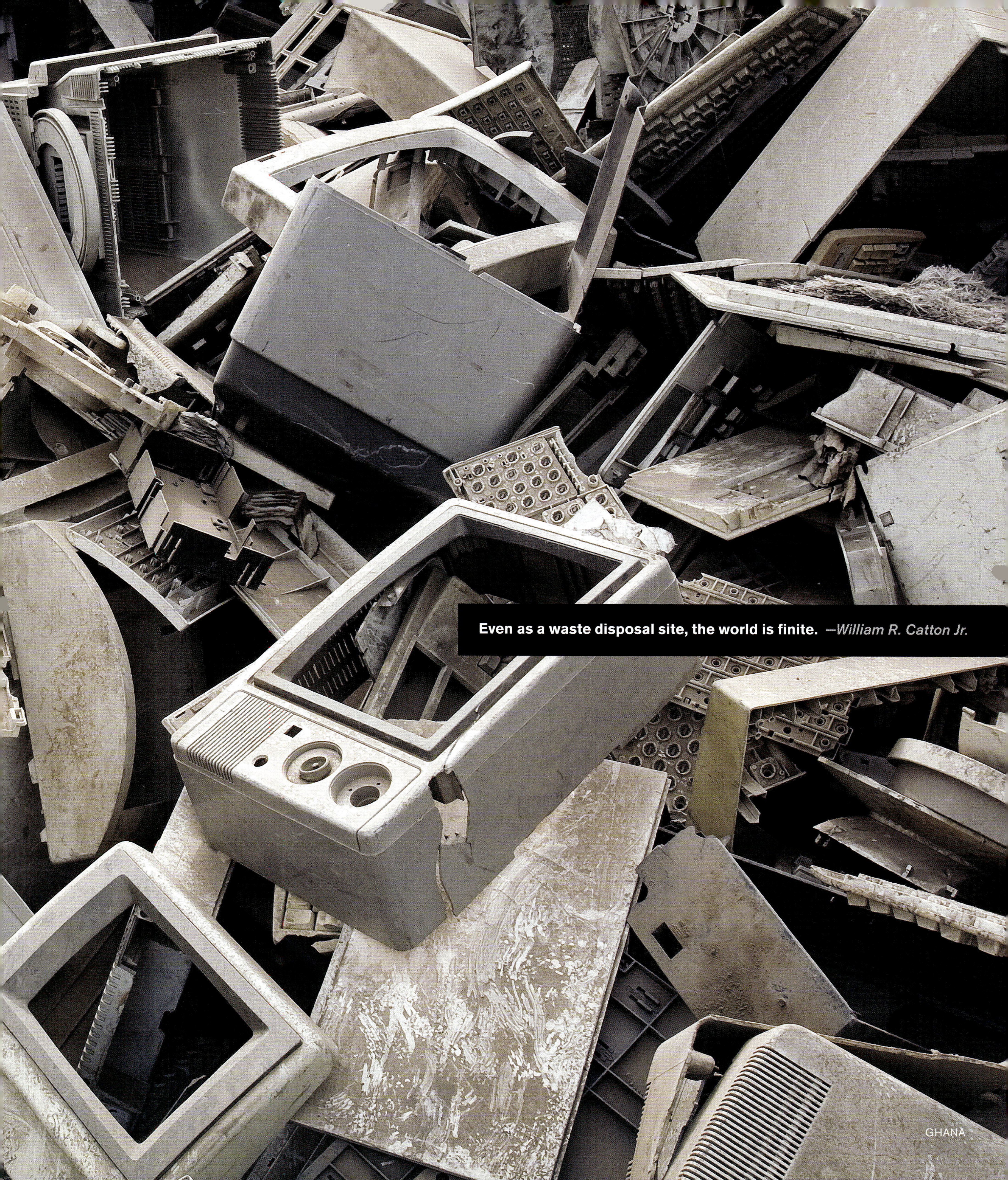
Even as a waste disposal site, the world is finite. —William R. Catton Jr.
GHANA

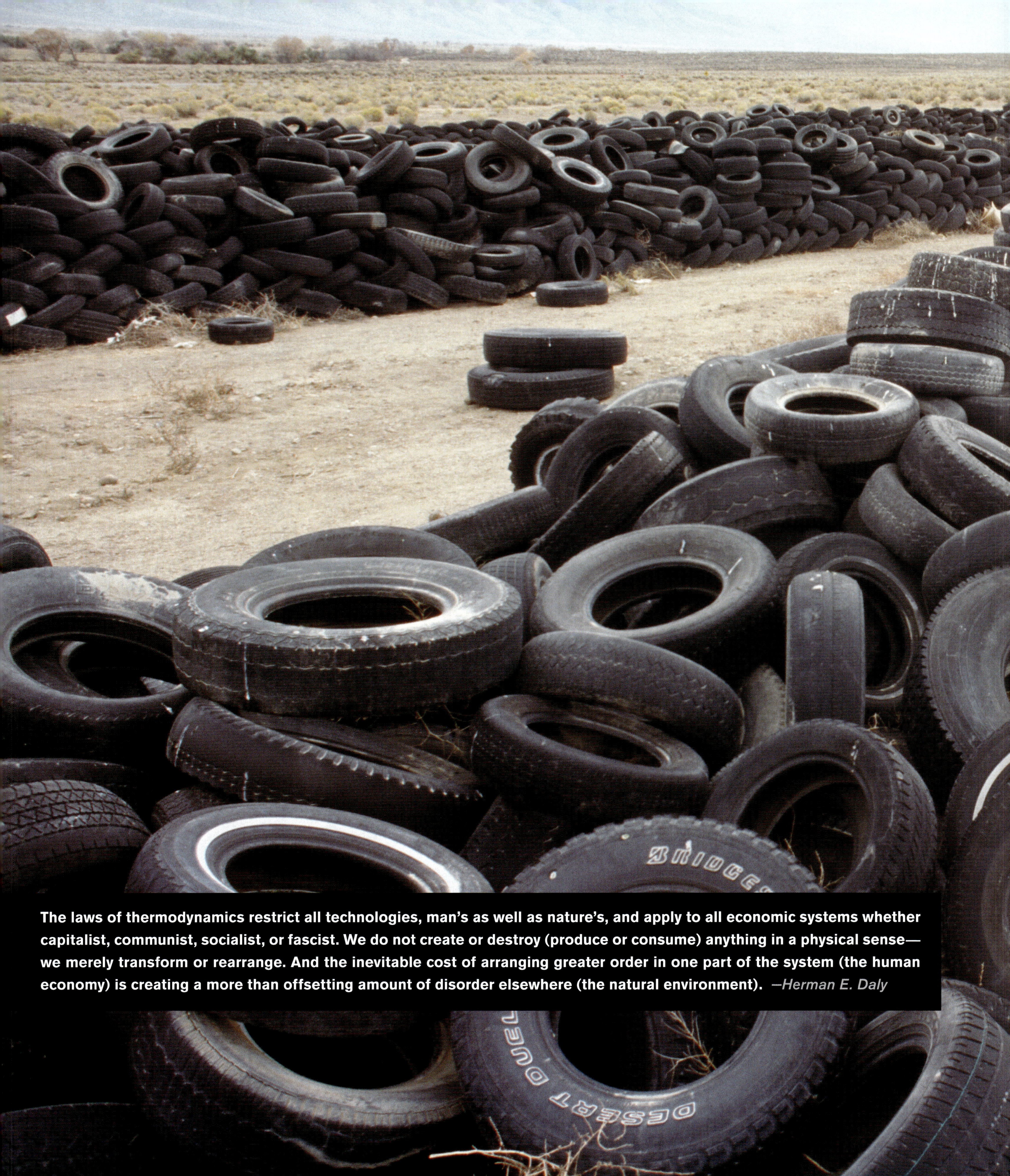
The laws of thermodynamics restrict all technologies, man's as well as nature's, and apply to all economic systems whether capitalist, communist, socialist, or fascist. We do not create or destroy (produce or consume) anything in a physical sense—we merely transform or rearrange. And the inevitable cost of arranging greater order in one part of the system (the human economy) is creating a more than offsetting amount of disorder elsewhere (the natural environment). —Herman E. Daly

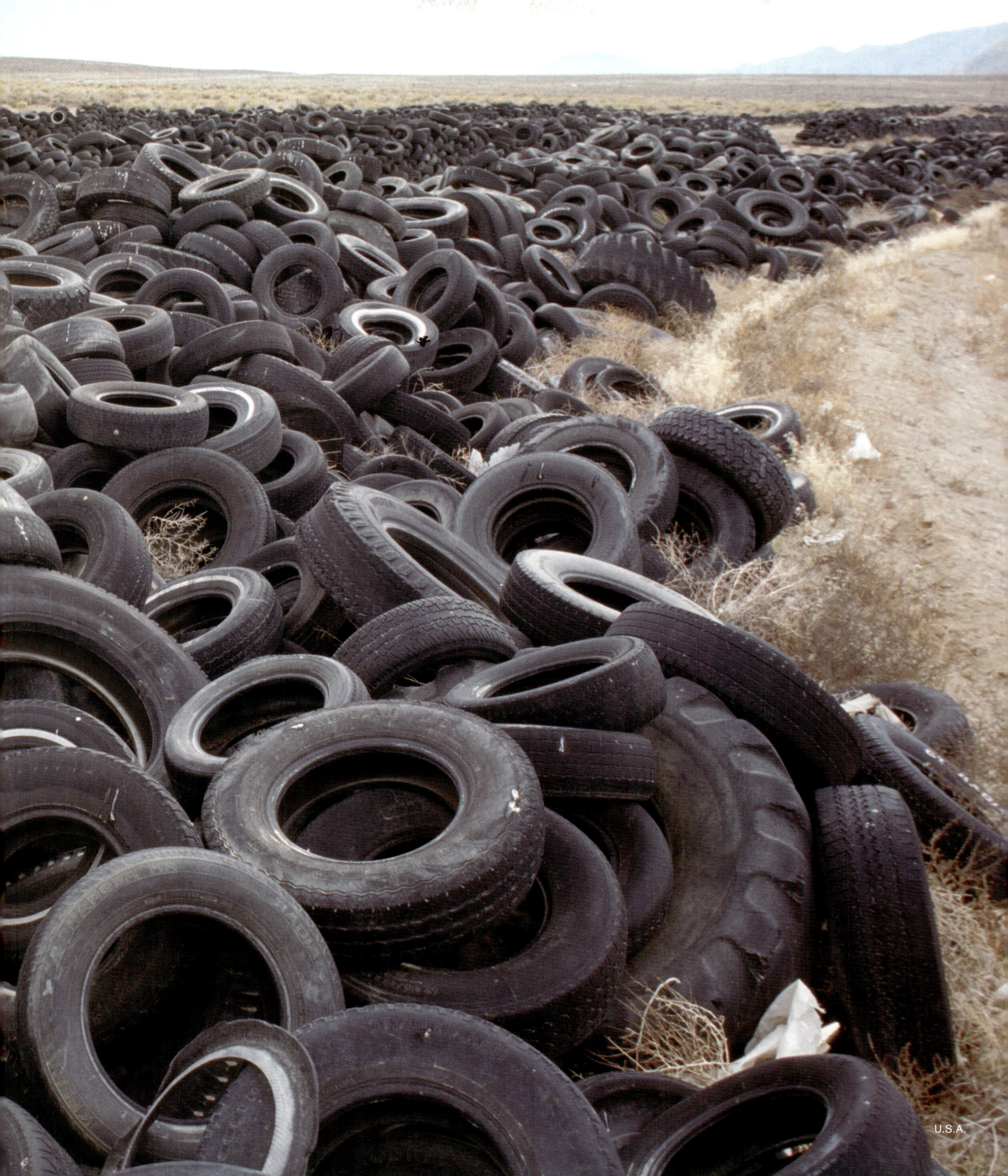
U.S.A.

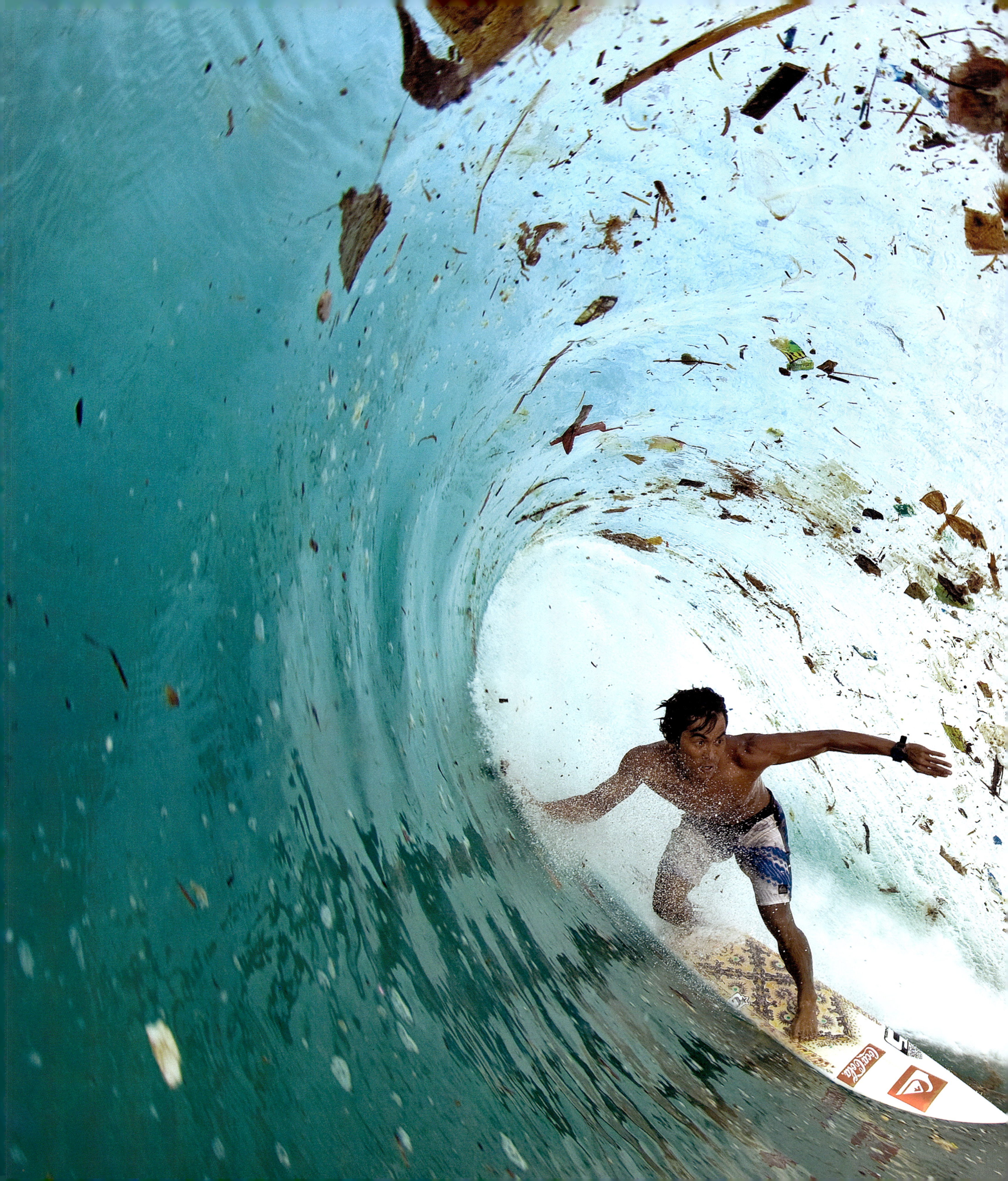

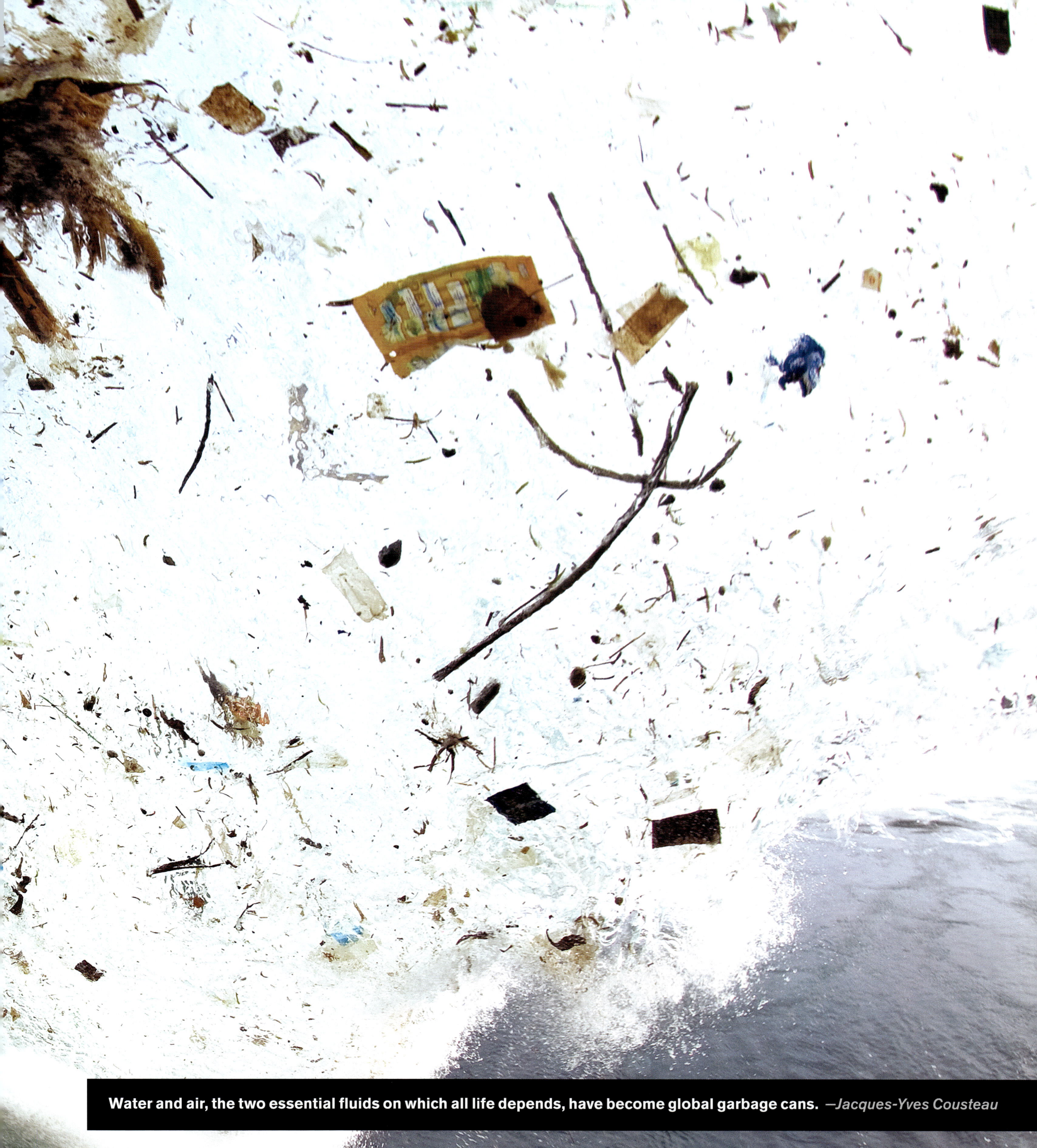

Water and air, the two essential fluids on which all life depends, have become global garbage cans. —Jacques-Yves Cousteau
JAVA, INDONESIA

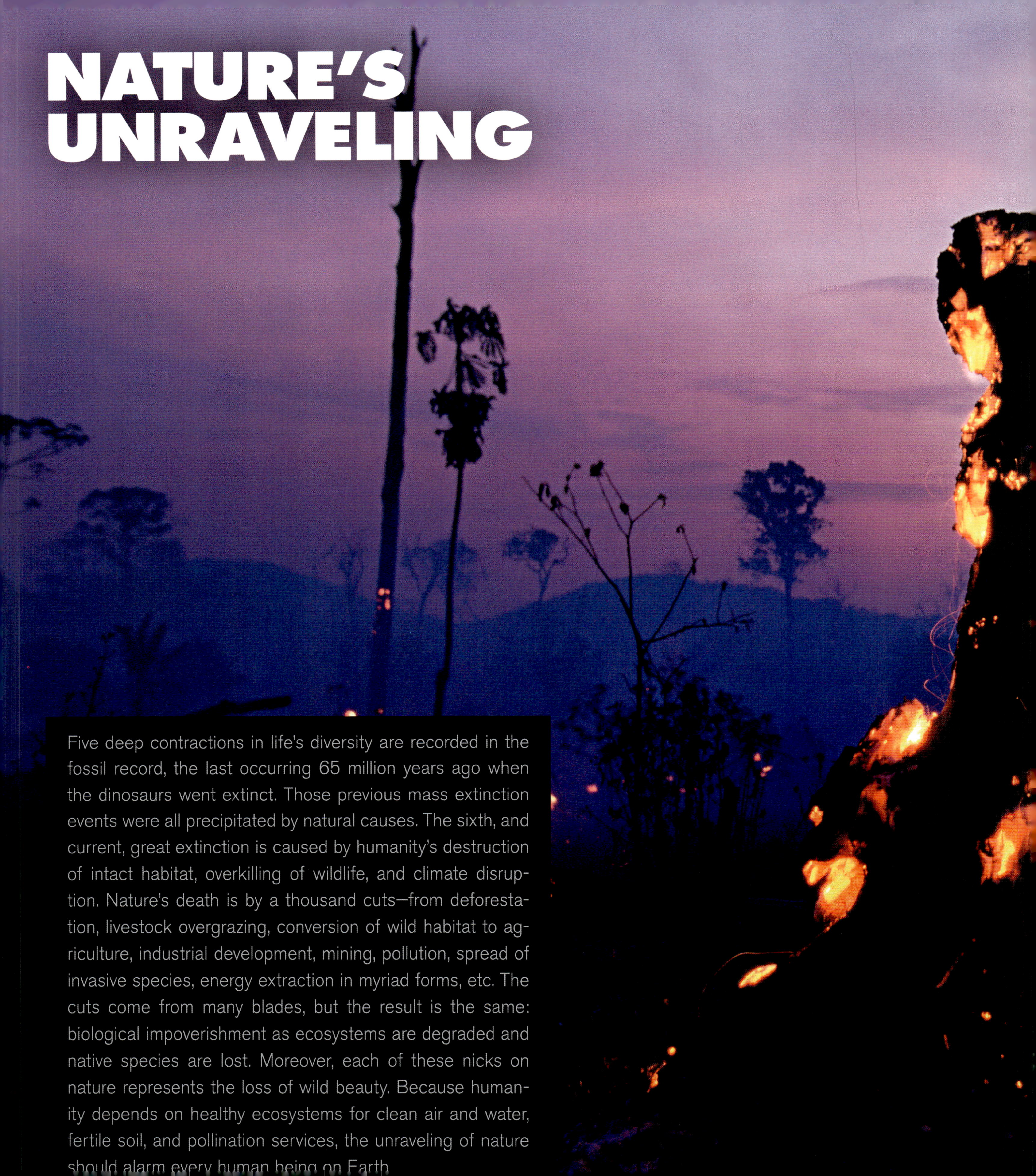

NATURE'S UNRAVELING

Five deep contractions in life's diversity are recorded in the fossil record, the last occurring 65 million years ago when the dinosaurs went extinct. Those previous mass extinction events were all precipitated by natural causes. The sixth, and current, great extinction is caused by humanity's destruction of intact habitat, overkilling of wildlife, and climate disruption. Nature's death is by a thousand cuts—from deforestation, livestock overgrazing, conversion of wild habitat to agriculture, industrial development, mining, pollution, spread of invasive species, energy extraction in myriad forms, etc. The cuts come from many blades, but the result is the same: biological impoverishment as ecosystems are degraded and native species are lost. Moreover, each of these nicks on nature represents the loss of wild beauty. Because humanity depends on healthy ecosystems for clean air and water, fertile soil, and pollination services, the unraveling of nature should alarm every human being on Earth.

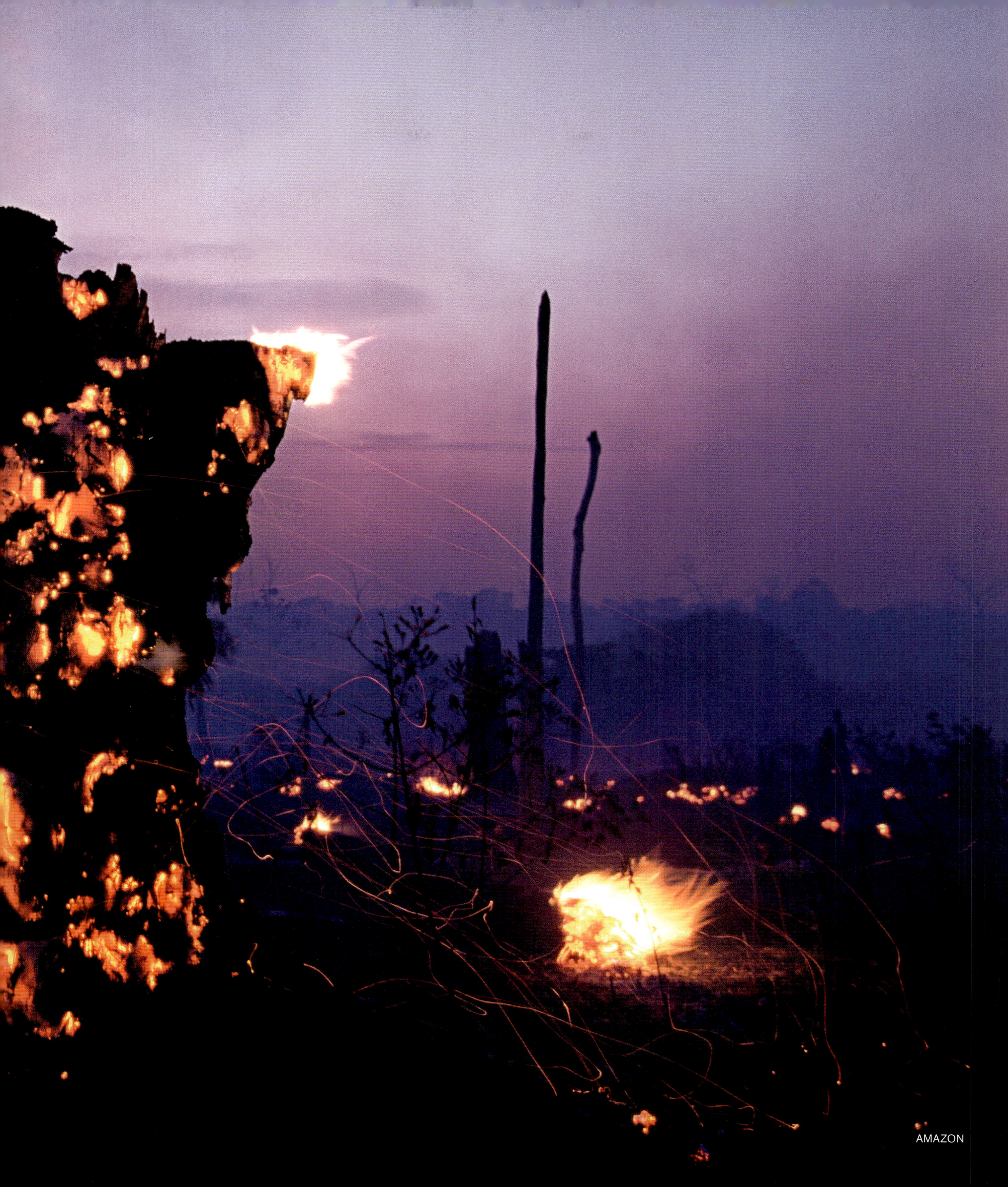
AMAZON

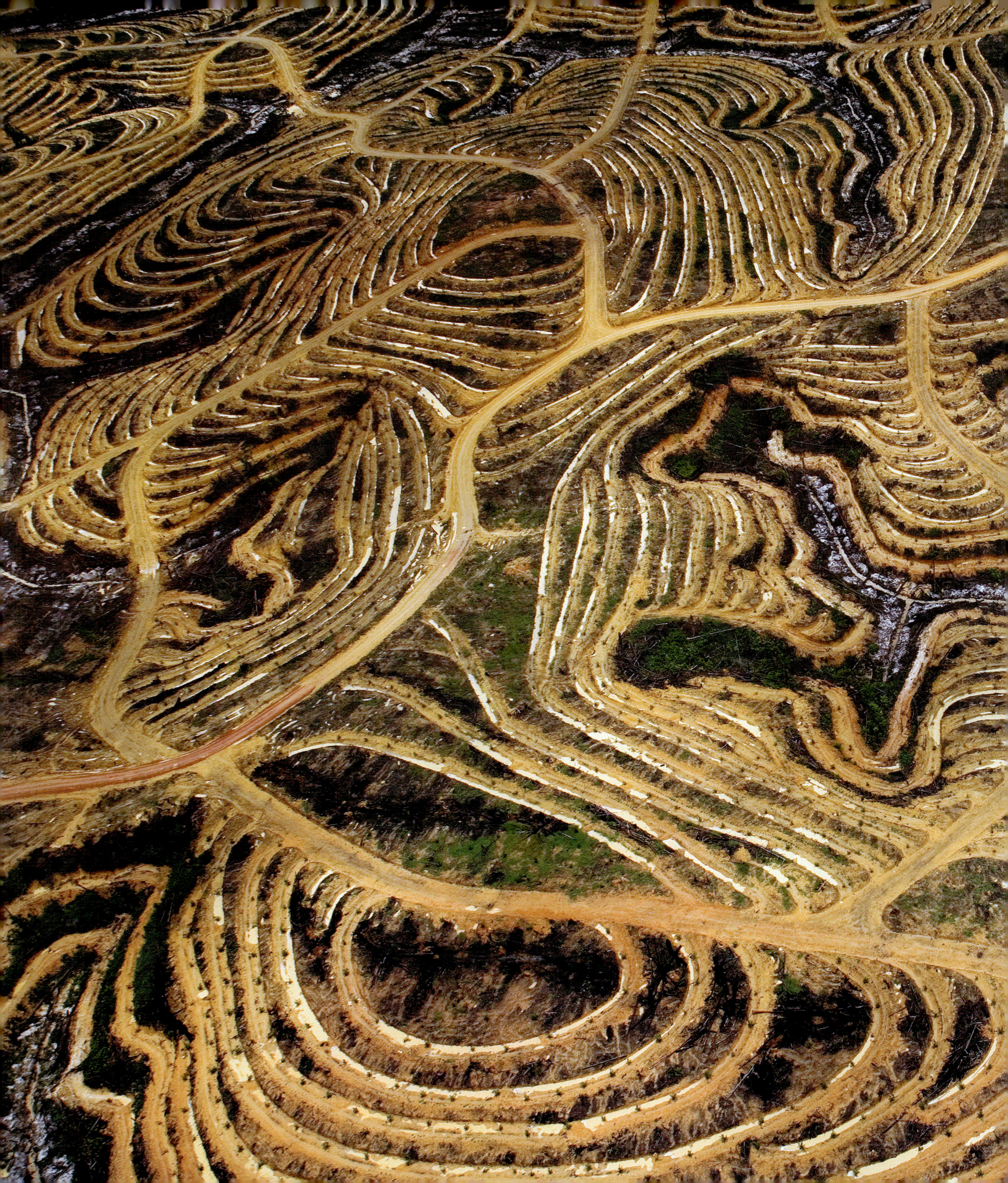

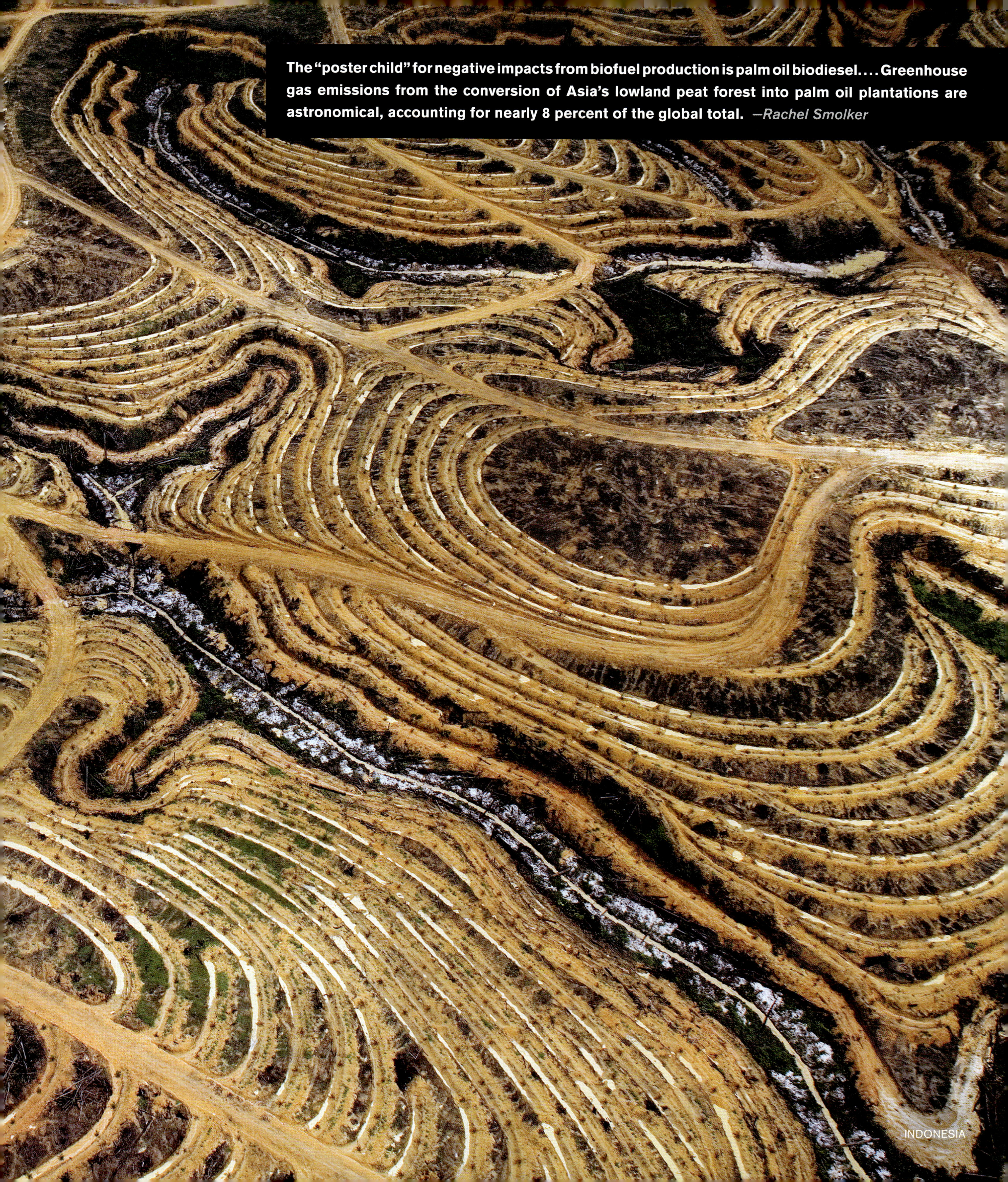
The "poster child" for negative impacts from biofuel production is palm oil biodiesel.... Greenhouse gas emissions from the conversion of Asia's lowland peat forest into palm oil plantations are astronomical, accounting for nearly 8 percent of the global total. —Rachel Smolker
INDONESIA

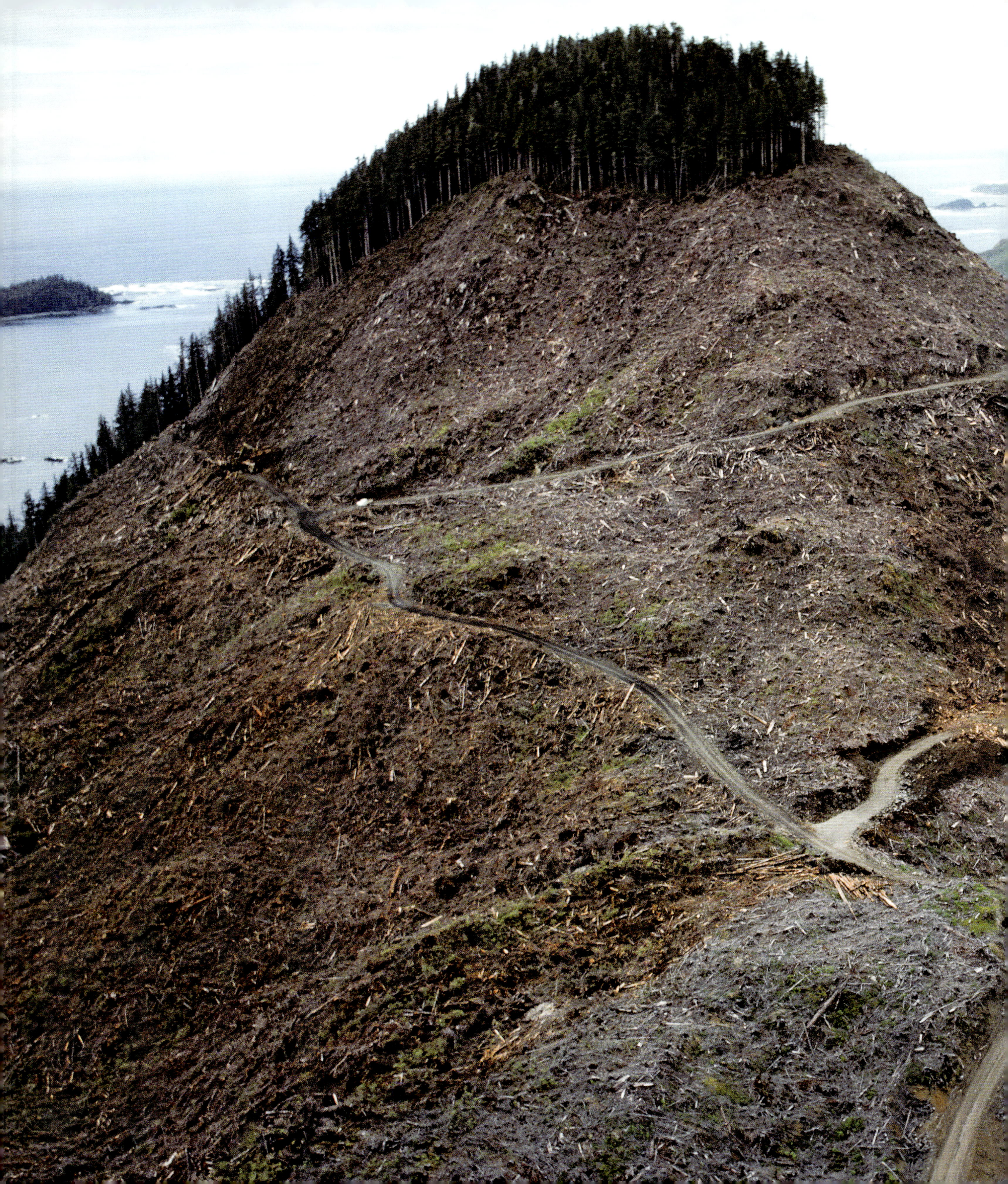

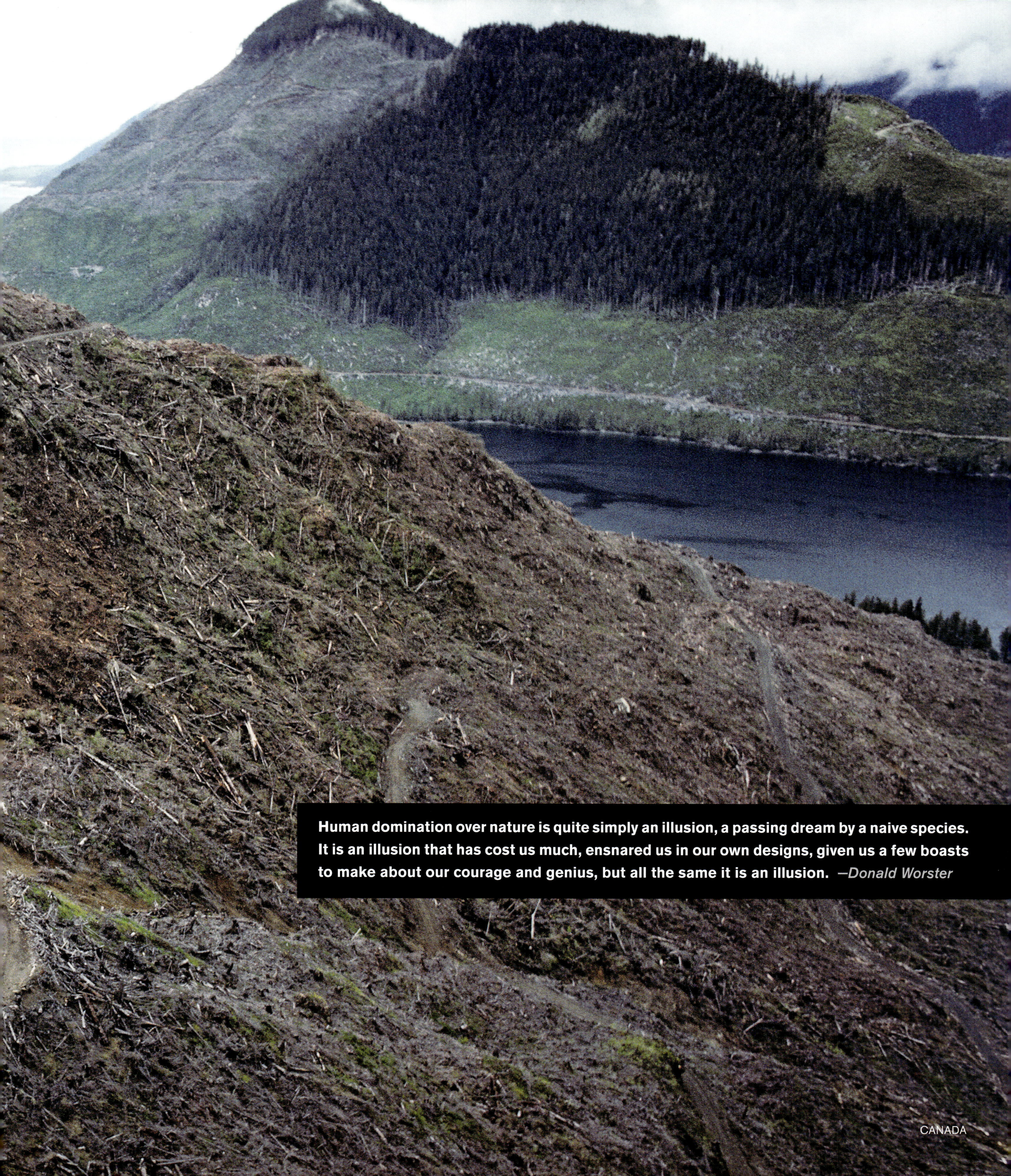
Human domination over nature is quite simply an illusion, a passing dream by a naive species.
It is an illusion that has cost us much, ensnared us in our own designs, given us a few boasts
to make about our courage and genius, but all the same it is an illusion. —Donald Worster
CANADA

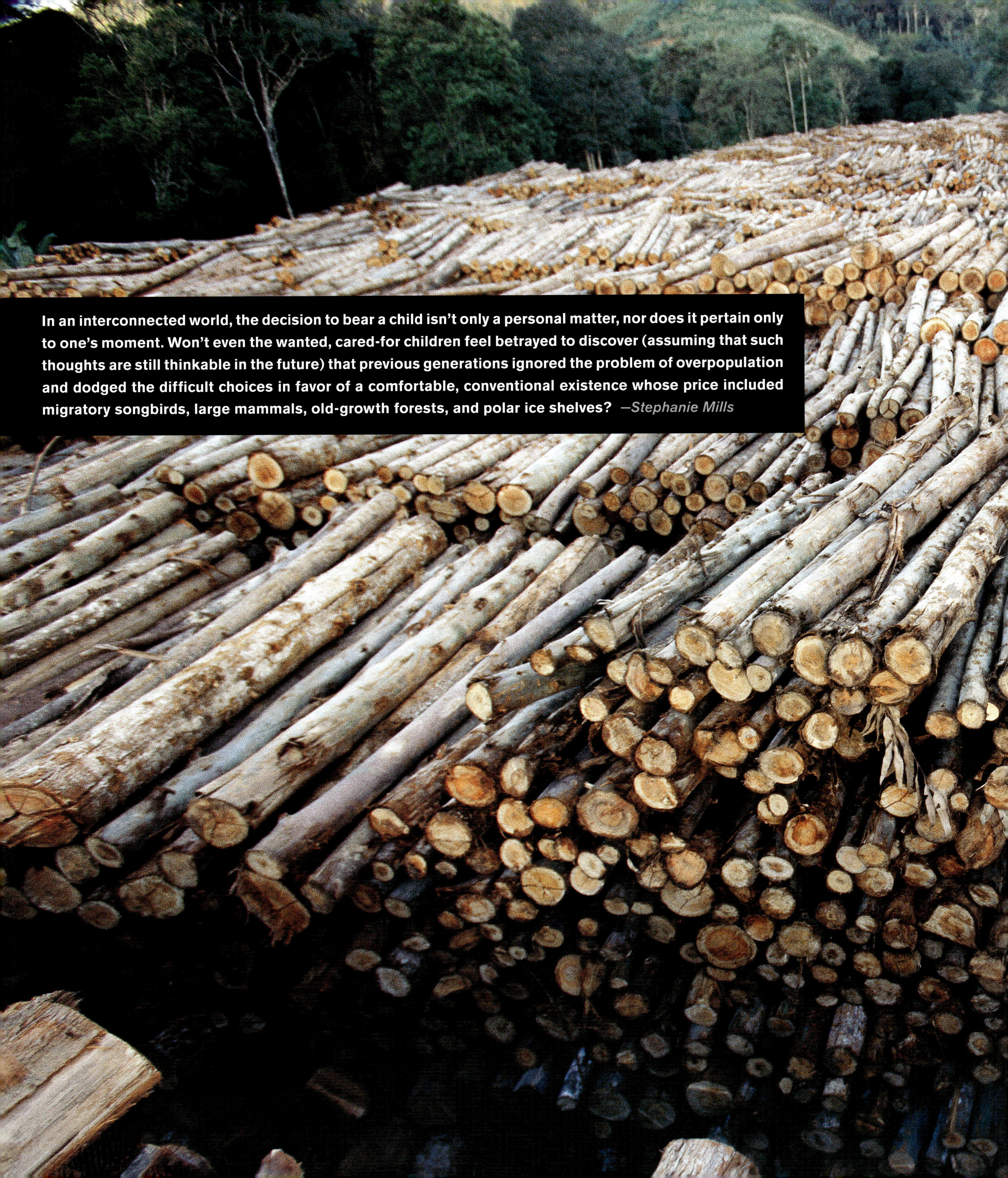

In an interconnected world, the decision to bear a child isn't only a personal matter, nor does it pertain only to one's moment. Won't even the wanted, cared-for children feel betrayed to discover (assuming that such thoughts are still thinkable in the future) that previous generations ignored the problem of overpopulation and dodged the difficult choices in favor of a comfortable, conventional existence whose price included migratory songbirds, large mammals, old-growth forests, and polar ice shelves? —Stephanie Mills

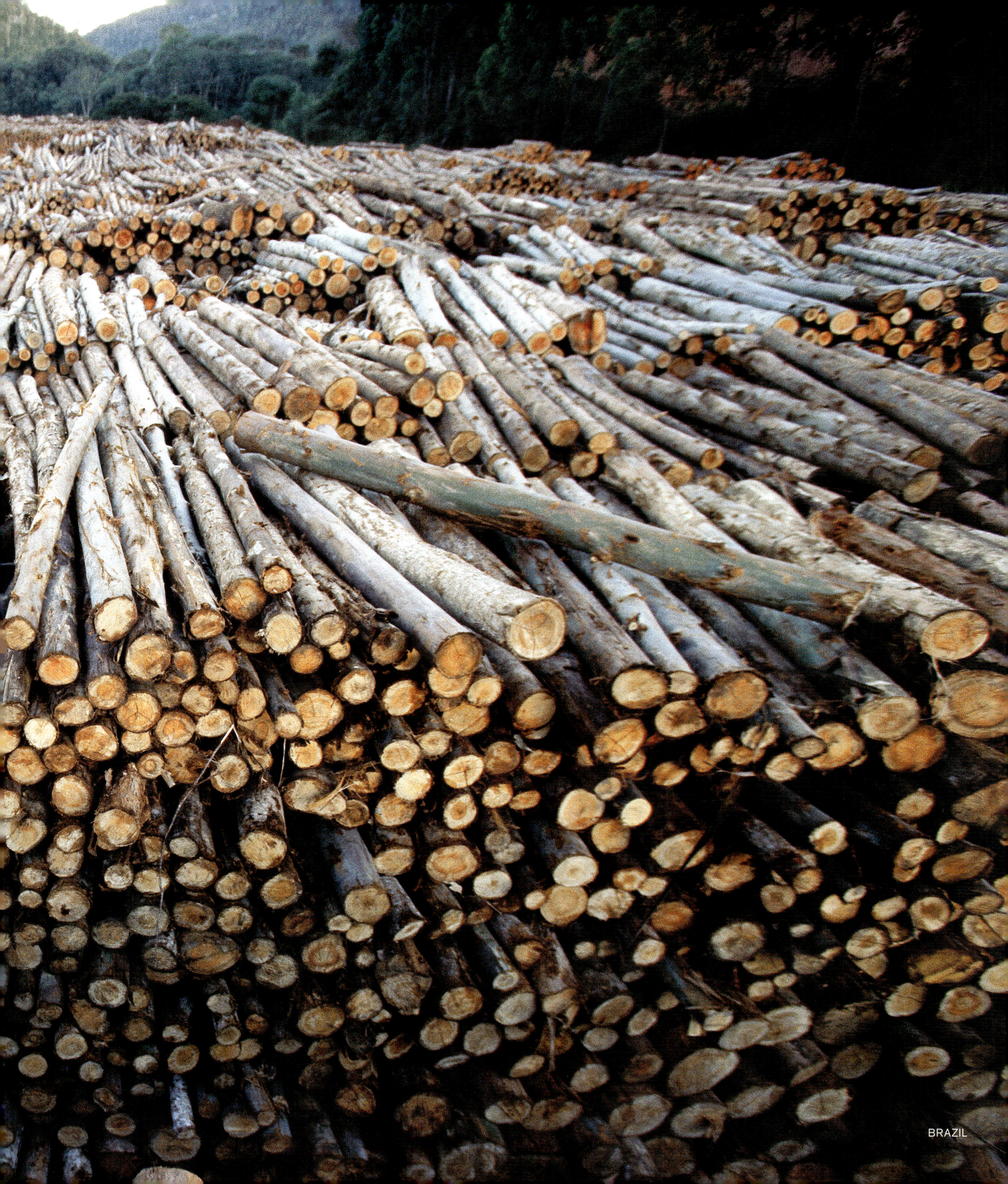
BRAZIL

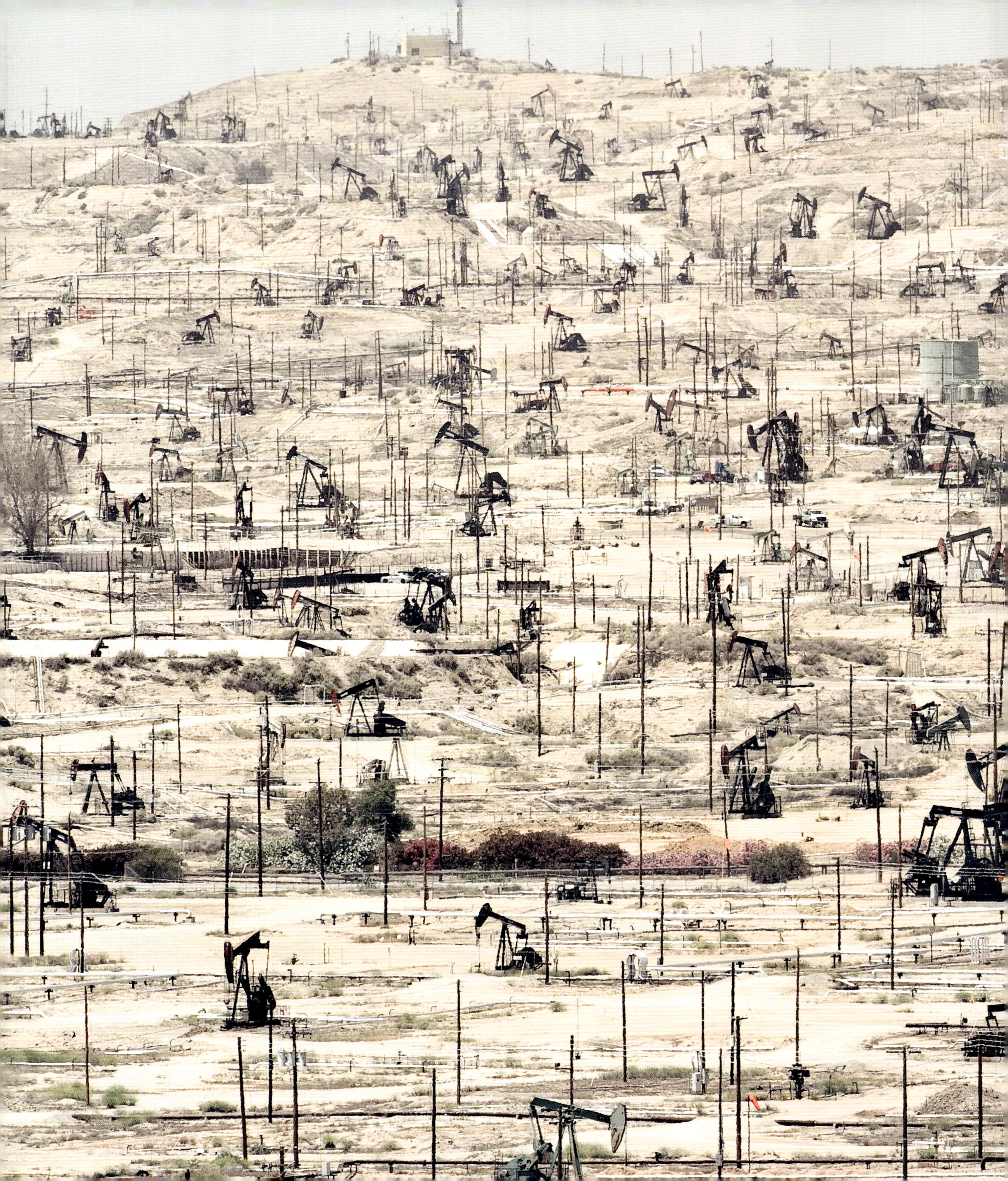

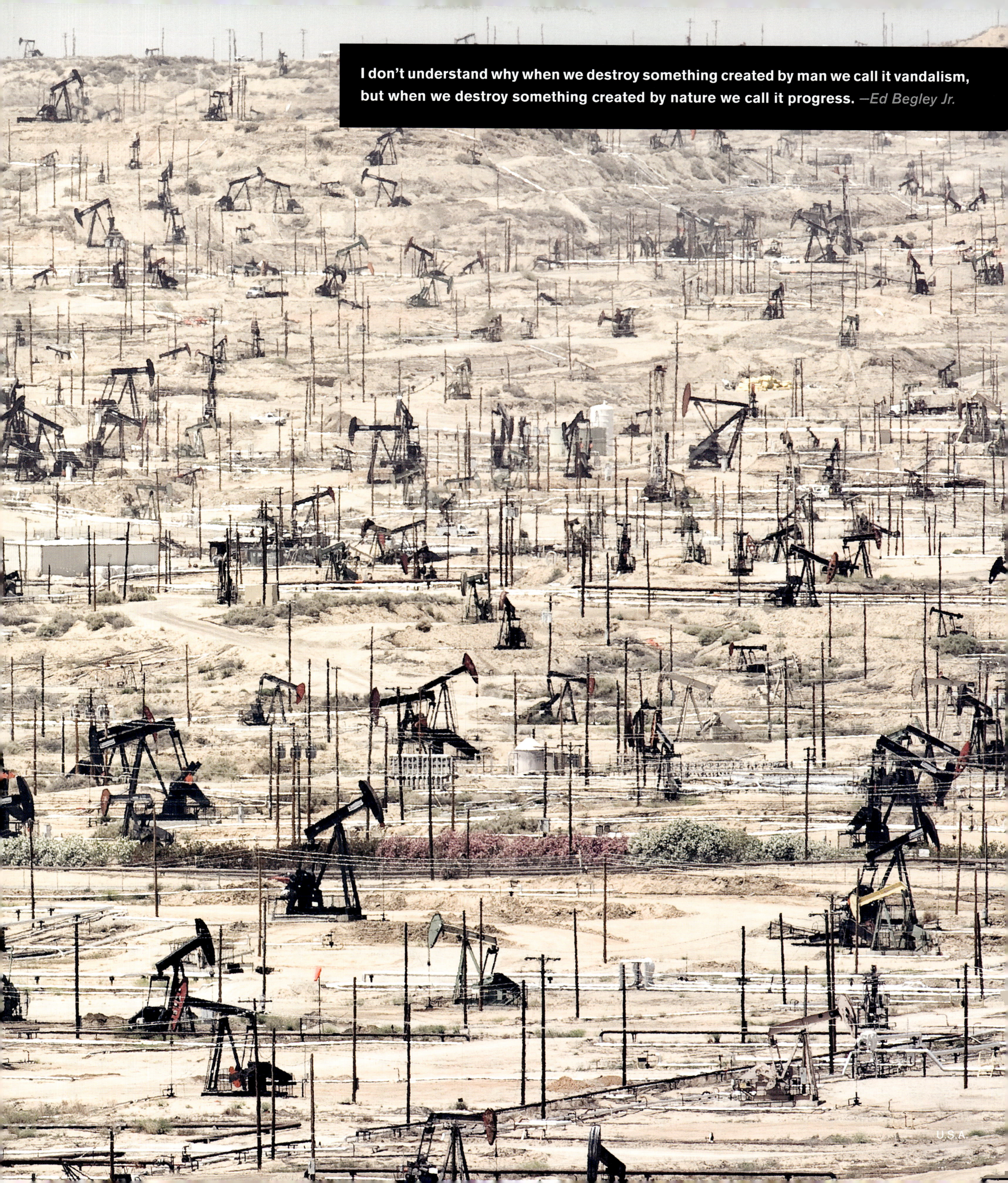

I don't understand why when we destroy something created by man we call it vandalism, but when we destroy something created by nature we call it progress. —Ed Begley Jr.
U.S.A

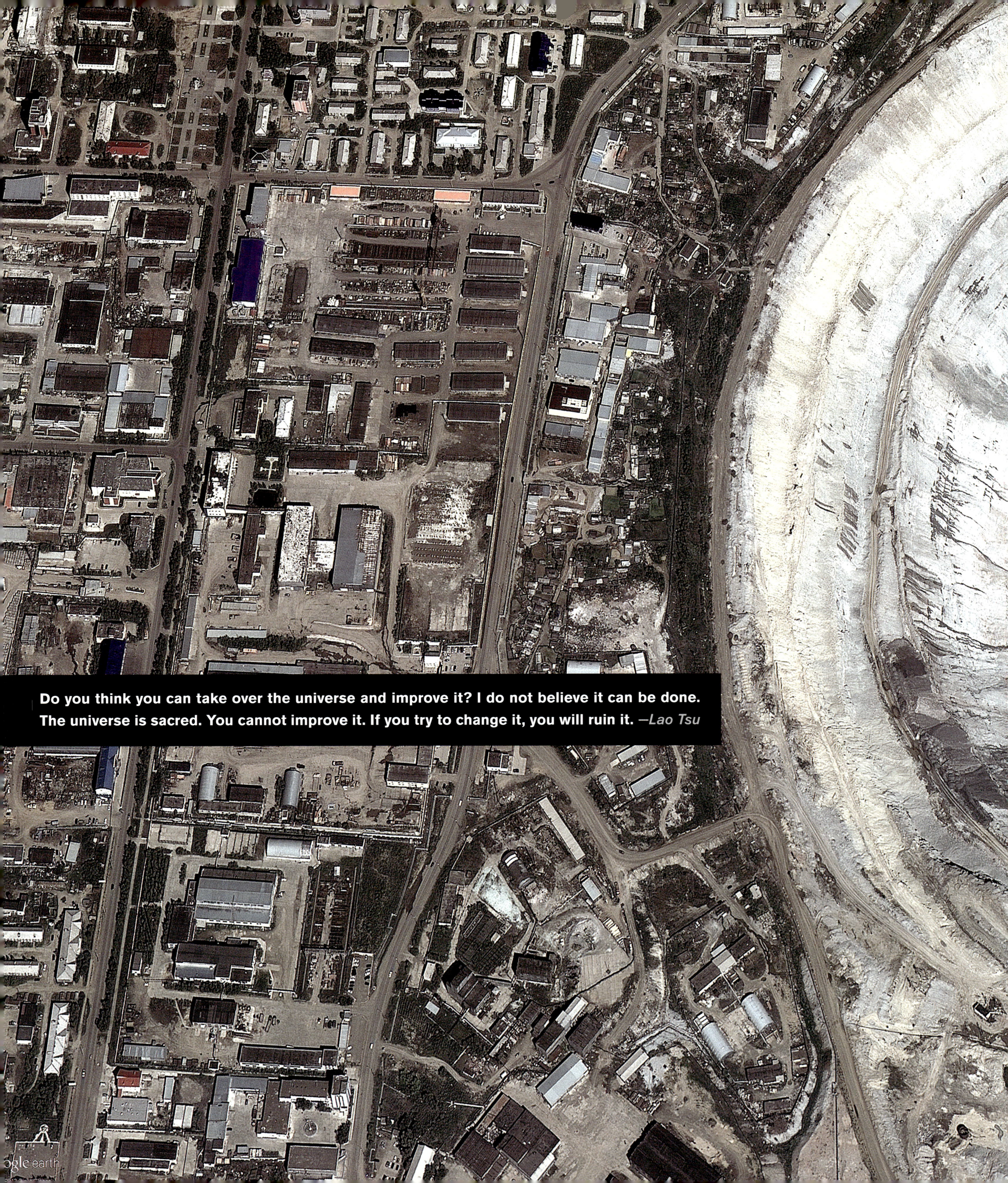
Do you think you can take over the universe and improve it? I do not believe it can be done.
The universe is sacred. You cannot improve it. If you try to change it, you will ruin it. —Lao Tsu

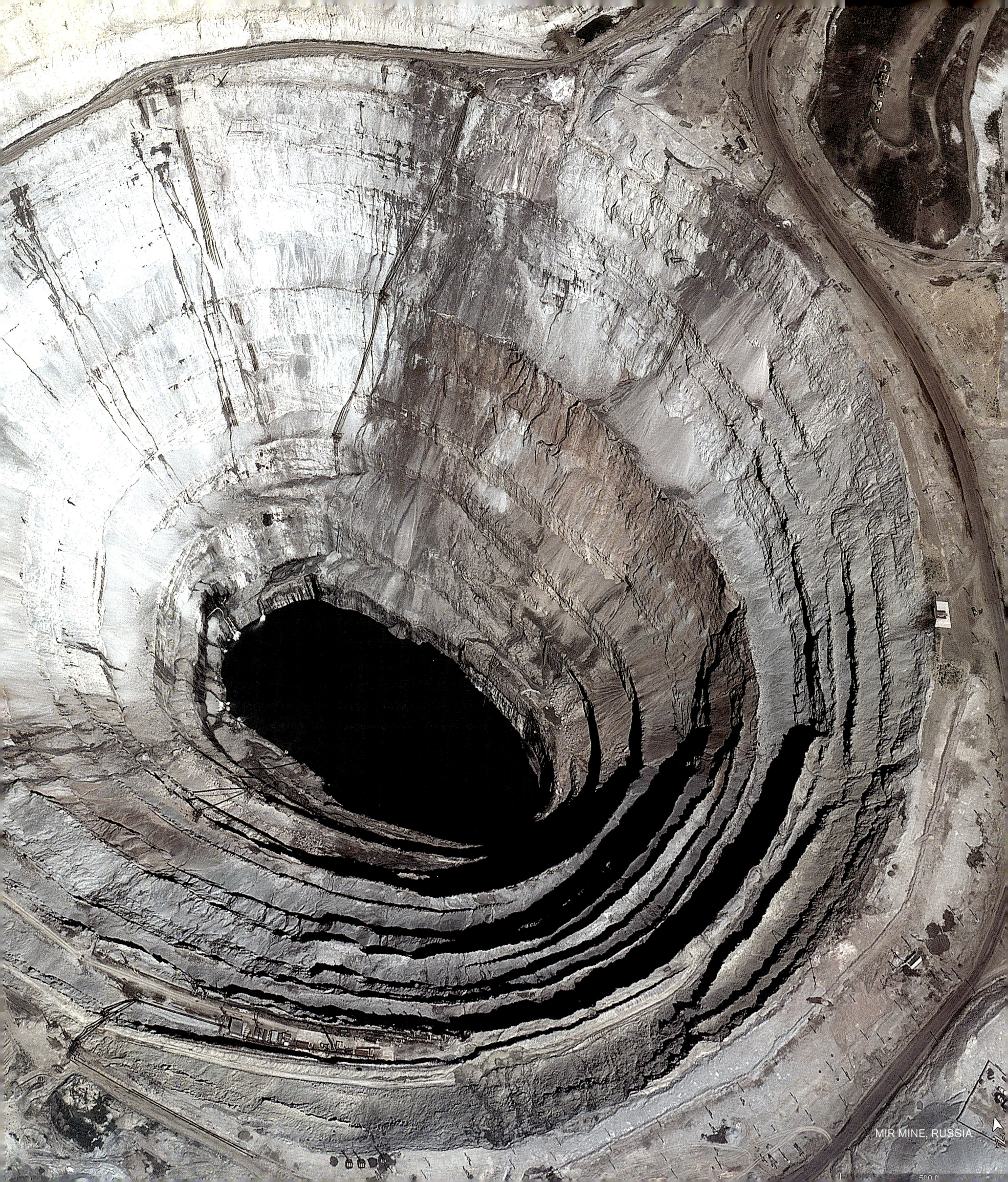

MIR MINE, RUSSIA

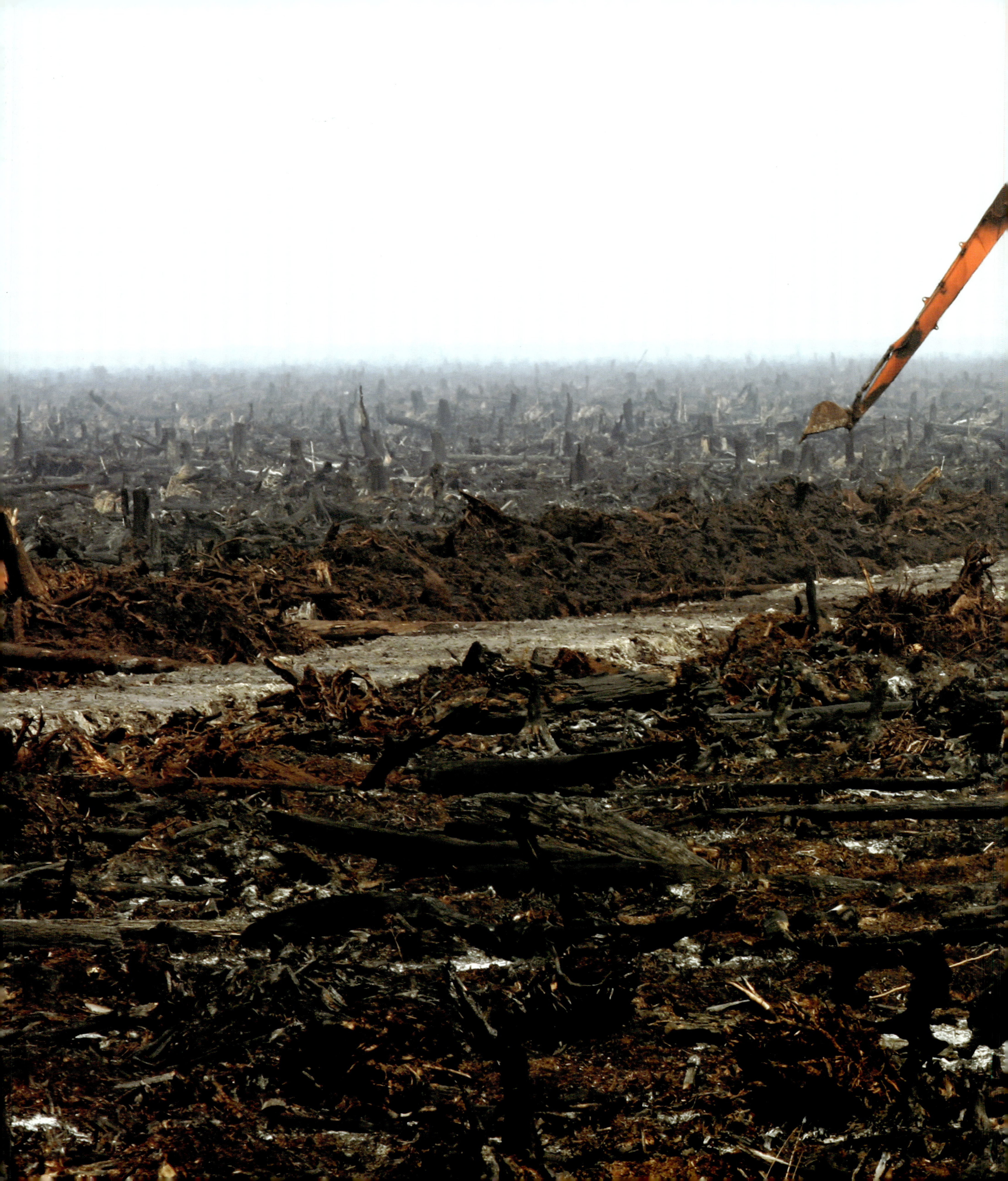

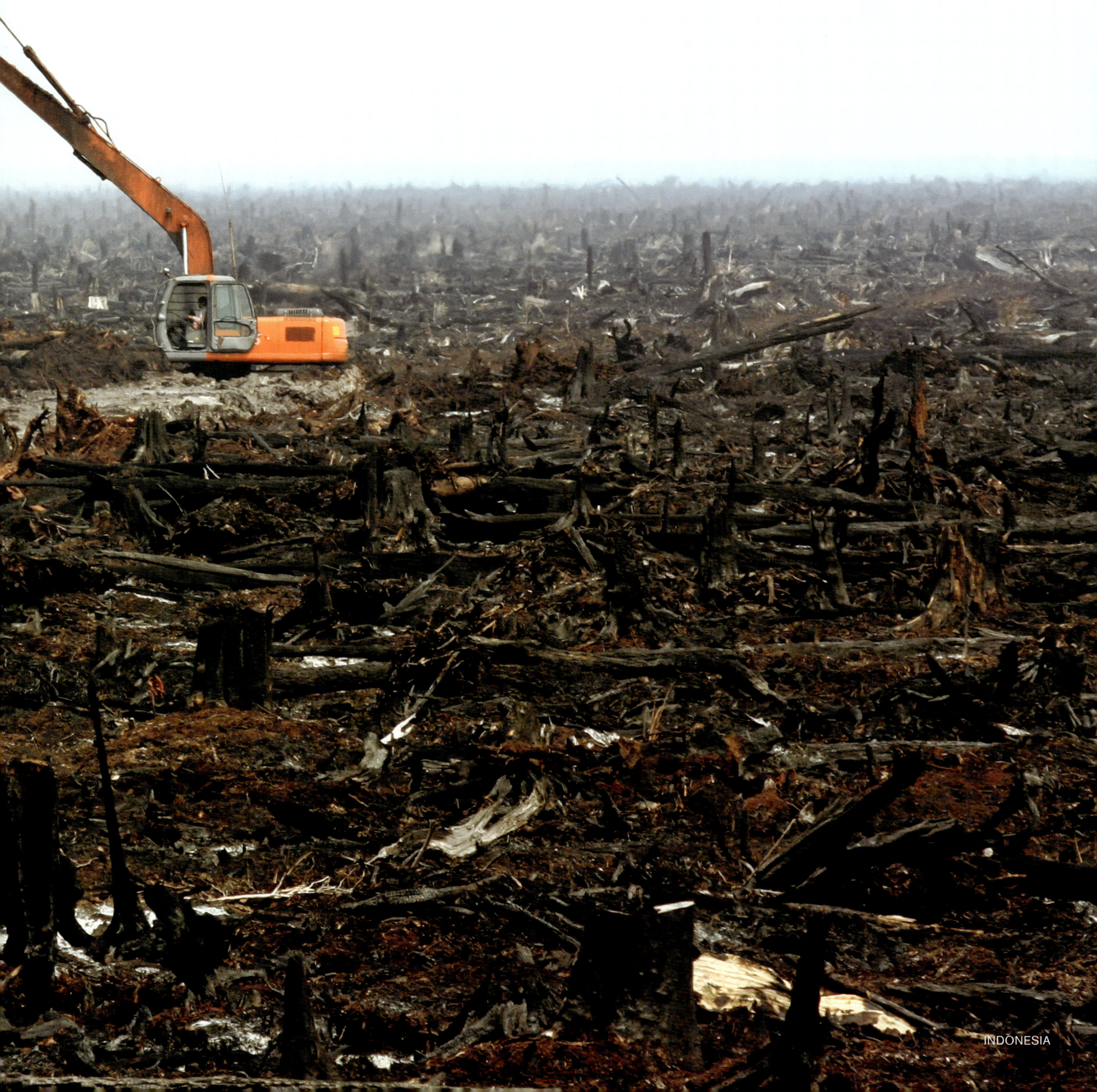

Our primordial spontaneities, which give us a delight in existence and enable us to interact creatively with natural phenomena, are being stifled. Somehow we have become autistic. We don't hear the voices. We are not entranced with the universe, with the natural world. We are entranced instead with domination over the natural world, with bringing about violent transformation. —Thomas Berry

INDONESIA

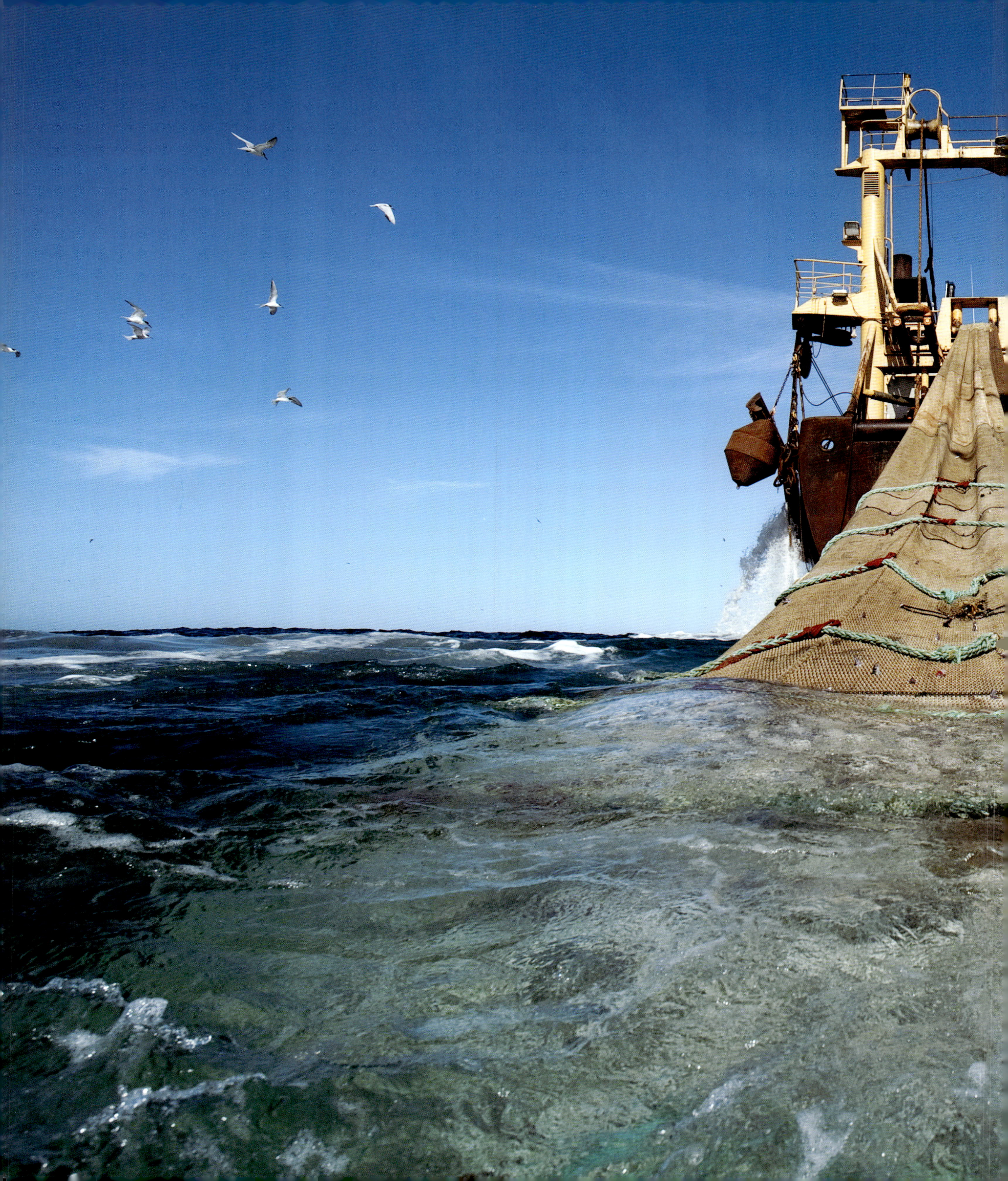

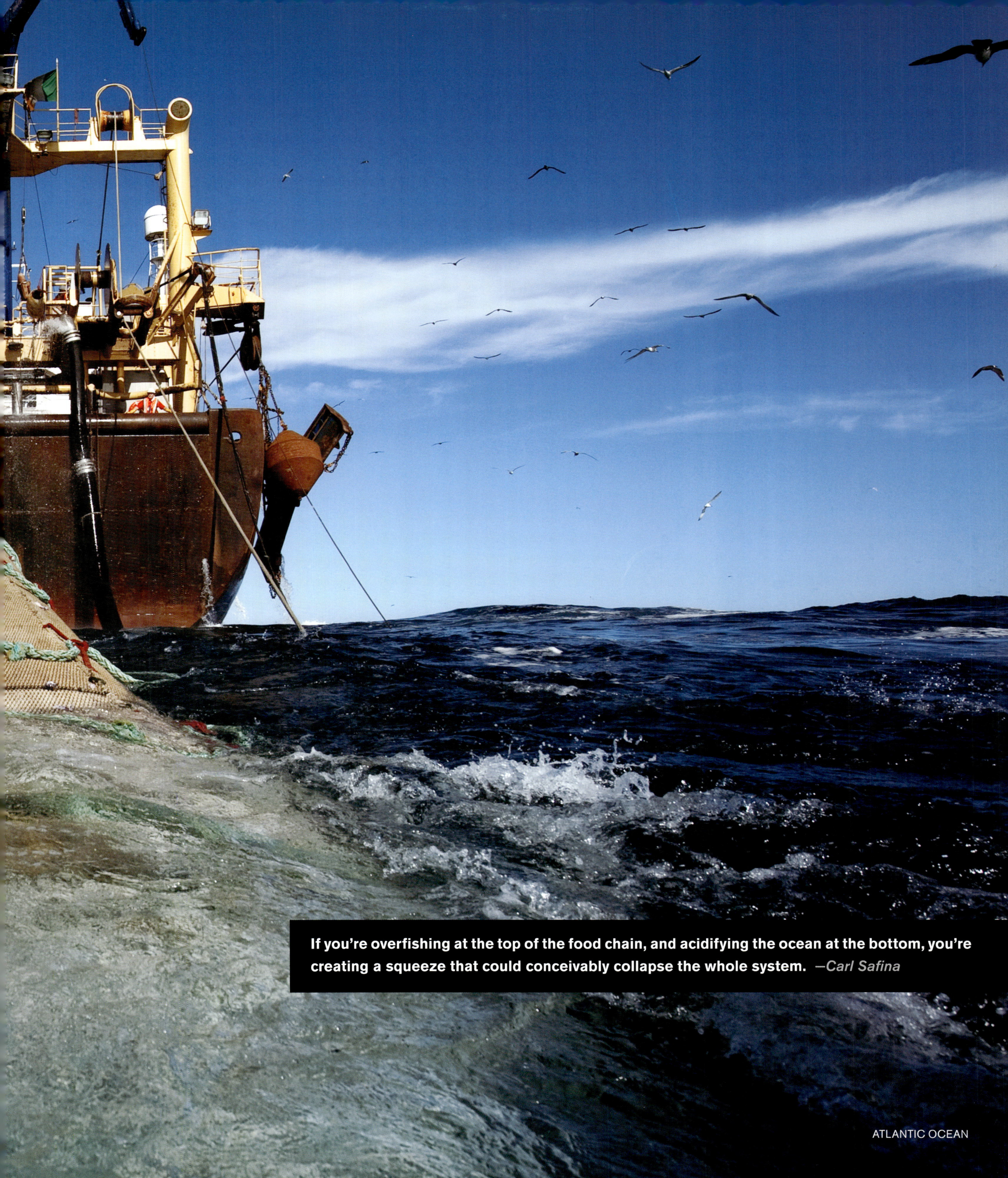
If you're overfishing at the top of the food chain, and acidifying the ocean at the bottom, you're creating a squeeze that could conceivably collapse the whole system. —Carl Safina

ATLANTIC OCEAN

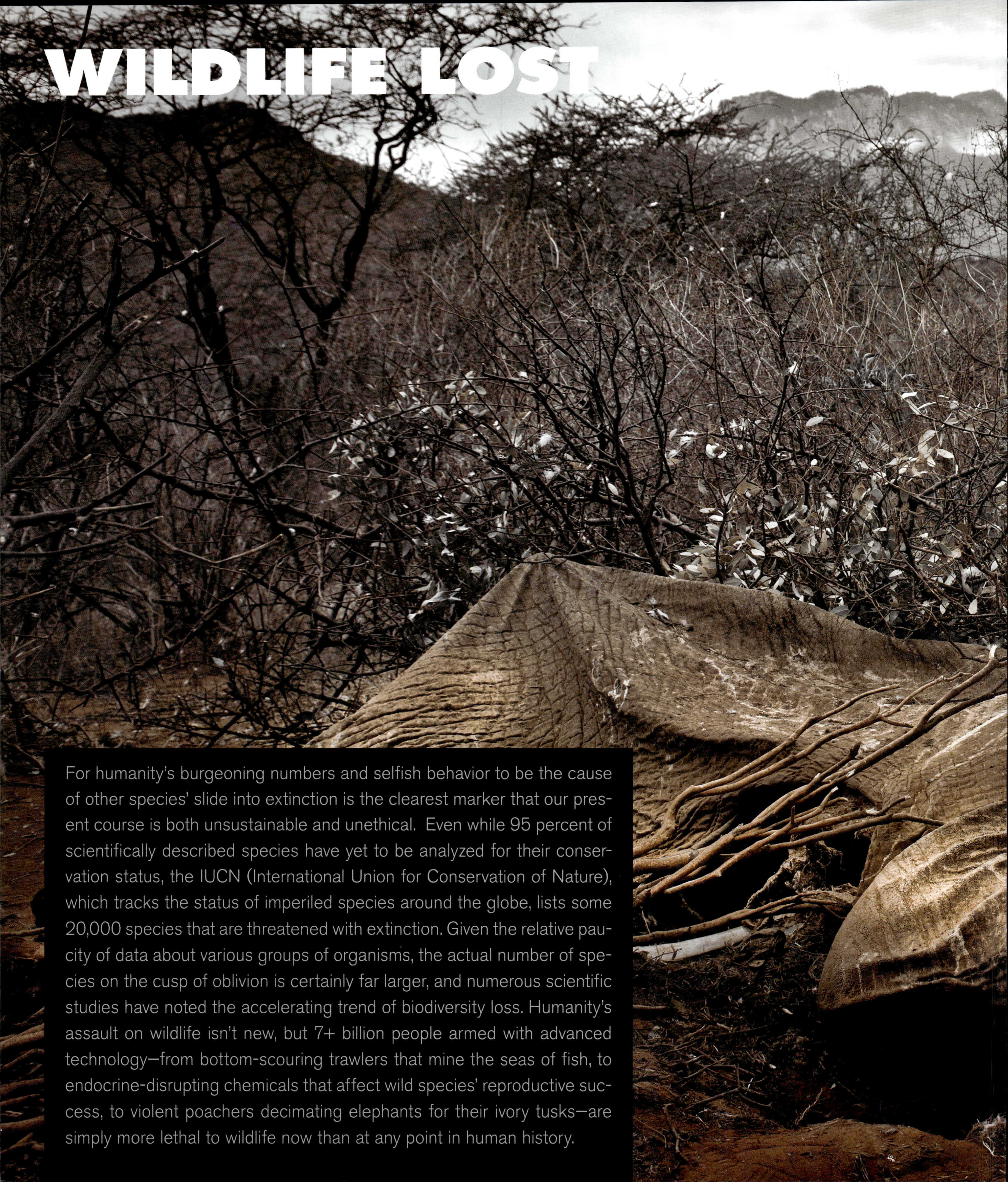

For humanity's burgeoning numbers and selfish behavior to be the cause of other species' slide into extinction is the clearest marker that our present course is both unsustainable and unethical. Even while 95 percent of scientifically described species have yet to be analyzed for their conservation status, the IUCN (International Union for Conservation of Nature), which tracks the status of imperiled species around the globe, lists some 20,000 species that are threatened with extinction. Given the relative paucity of data about various groups of organisms, the actual number of species on the cusp of oblivion is certainly far larger, and numerous scientific studies have noted the accelerating trend of biodiversity loss. Humanity's assault on wildlife isn't new, but 7+ billion people armed with advanced technology—from bottom-scouring trawlers that mine the seas of fish, to endocrine-disrupting chemicals that affect wild species' reproductive success, to violent poachers decimating elephants for their ivory tusks—are simply more lethal to wildlife now than at any point in human history.

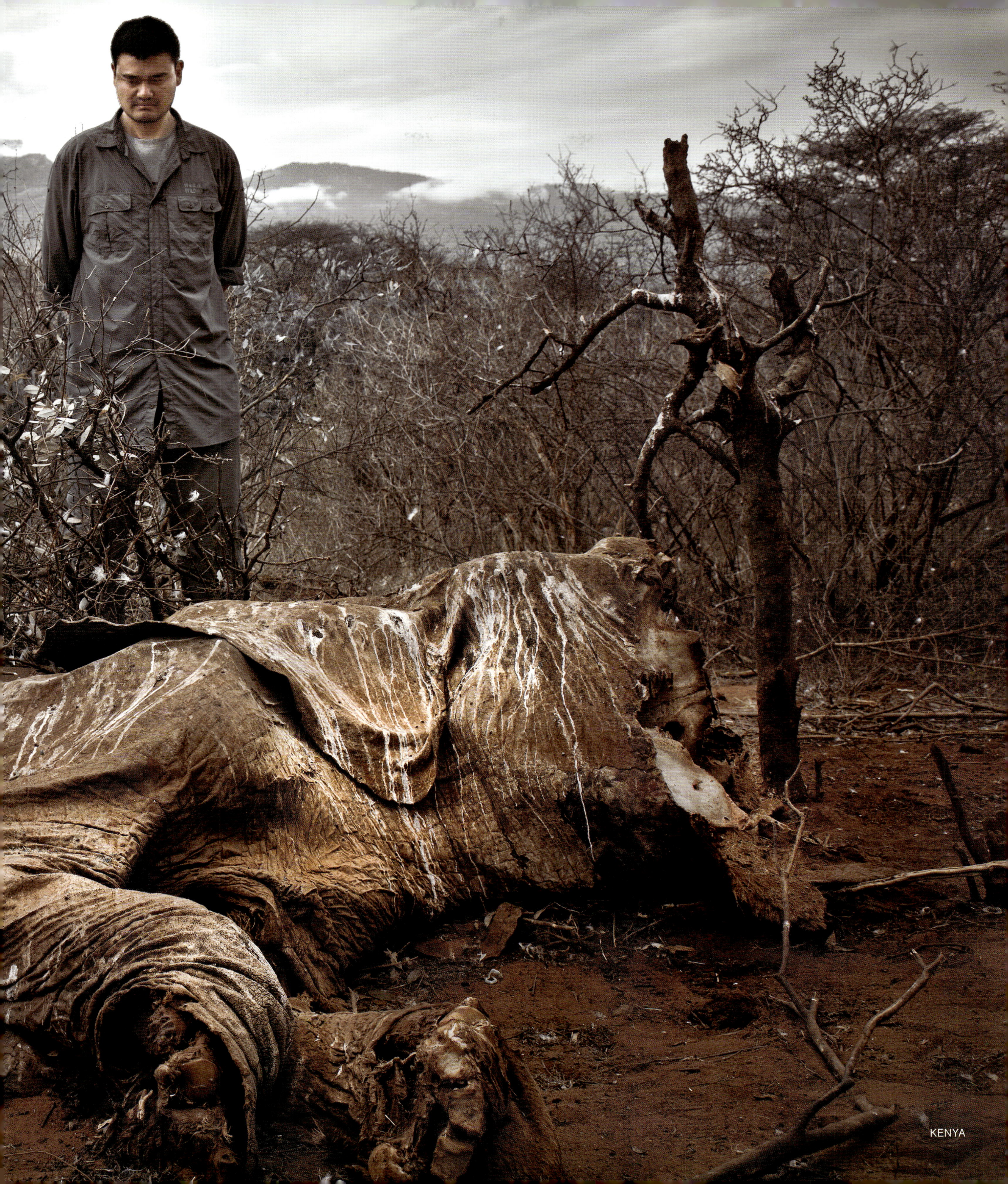

KENYA

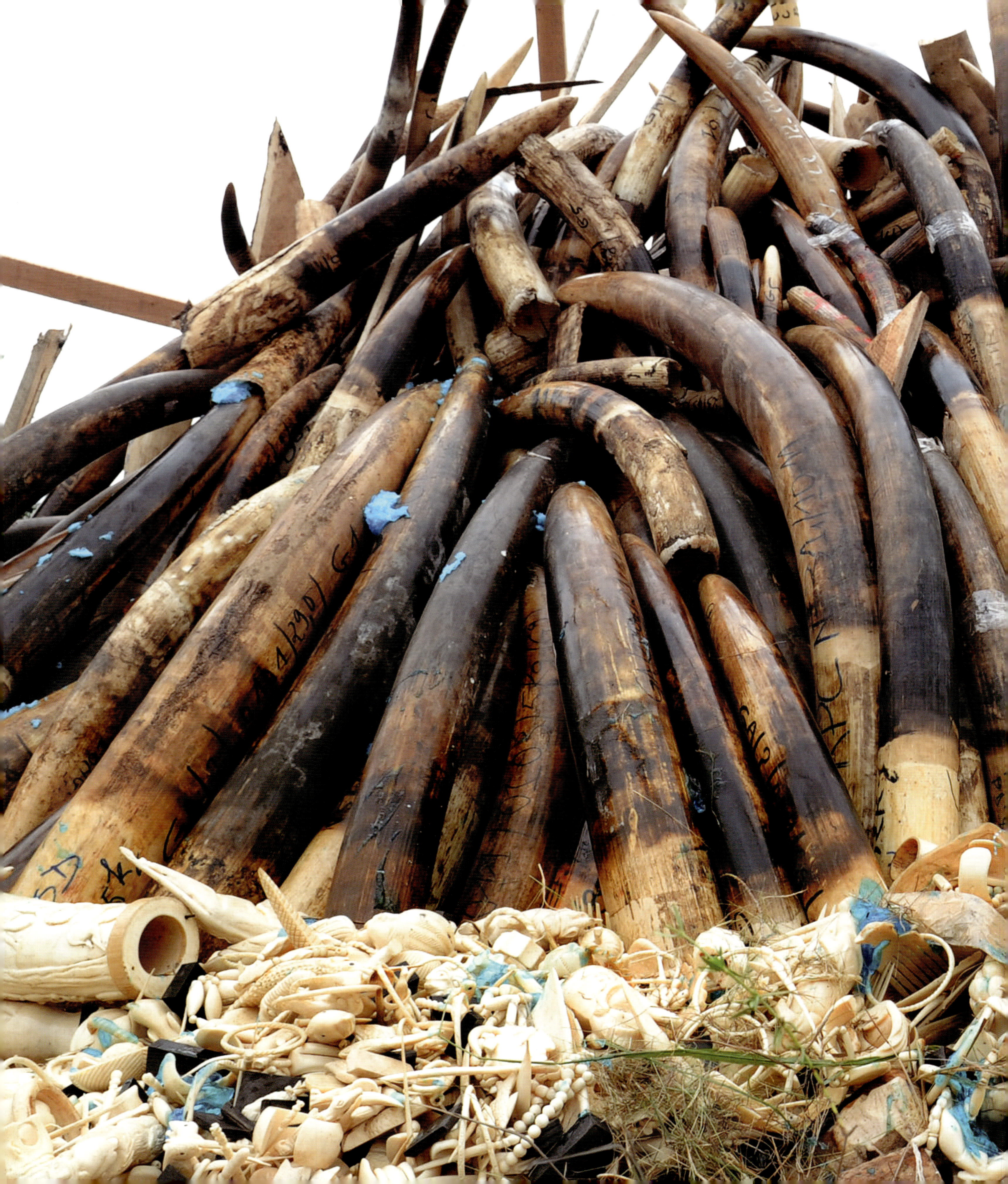

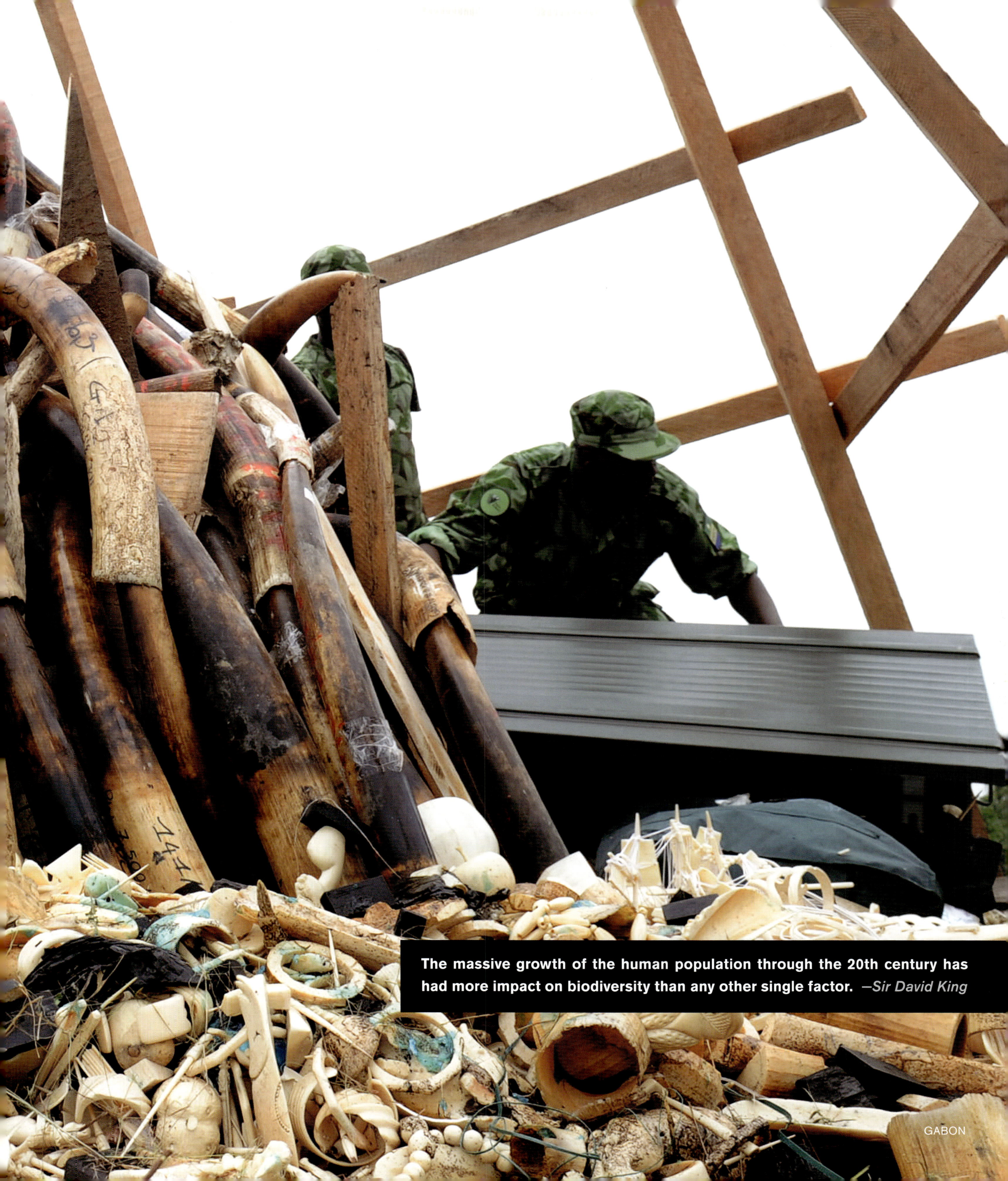

The massive growth of the human population through the 20th century has had more impact on biodiversity than any other single factor. —Sir David King

GABON

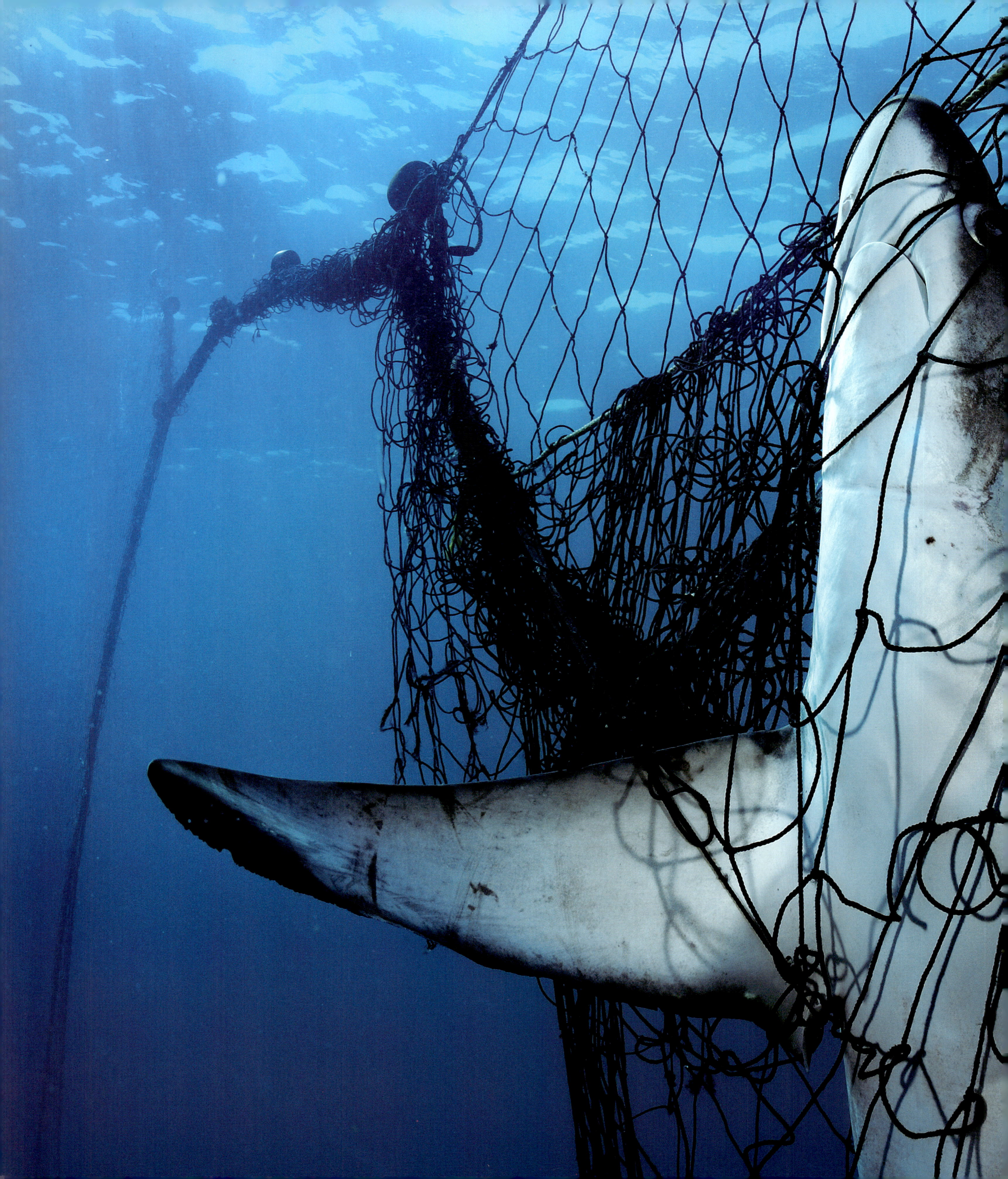

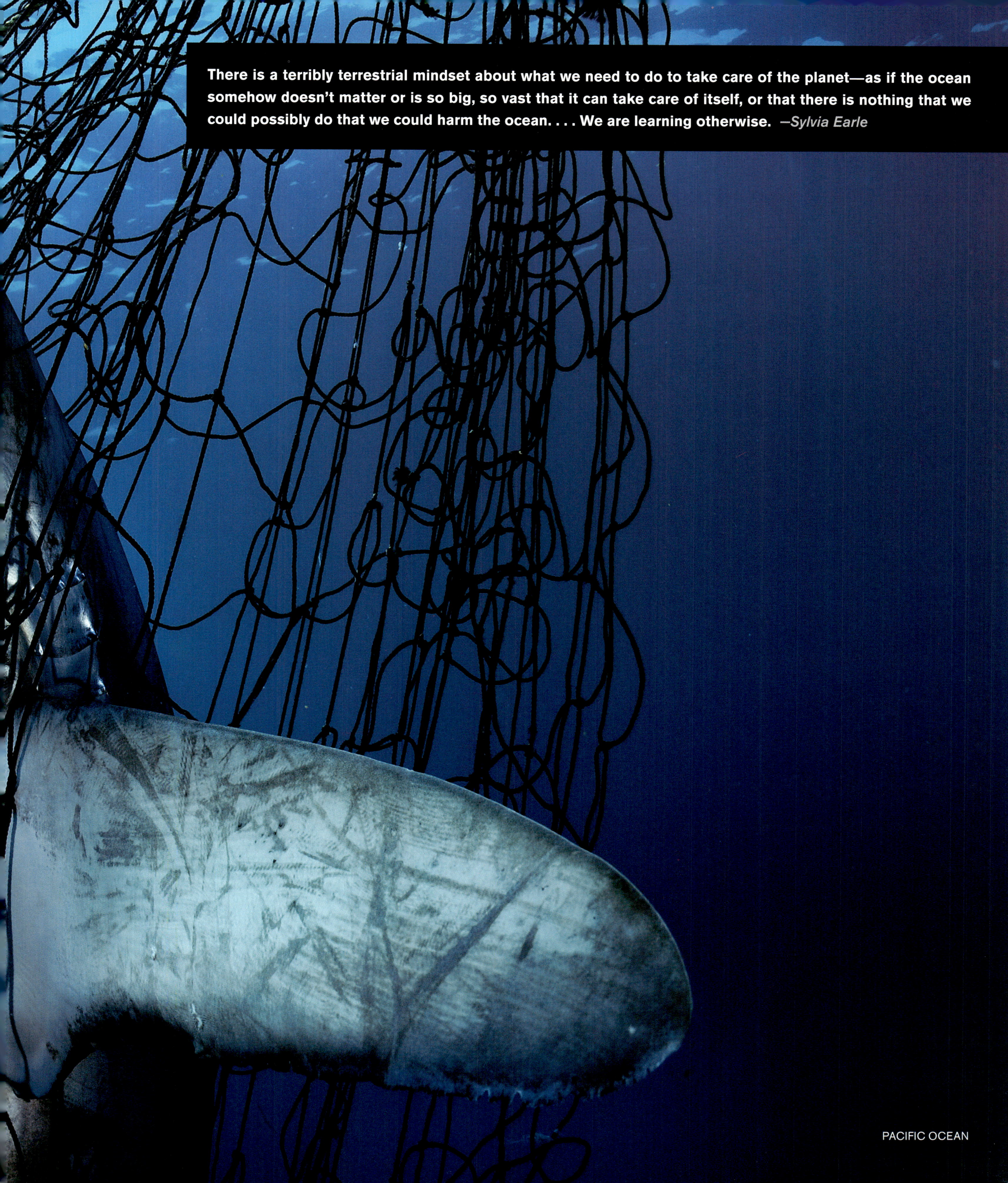

There is a terribly terrestrial mindset about what we need to do to take care of the planet—as if the ocean somehow doesn't matter or is so big, so vast that it can take care of itself, or that there is nothing that we could possibly do that we could harm the ocean. . . . We are learning otherwise. —Sylvia Earle
PACIFIC OCEAN

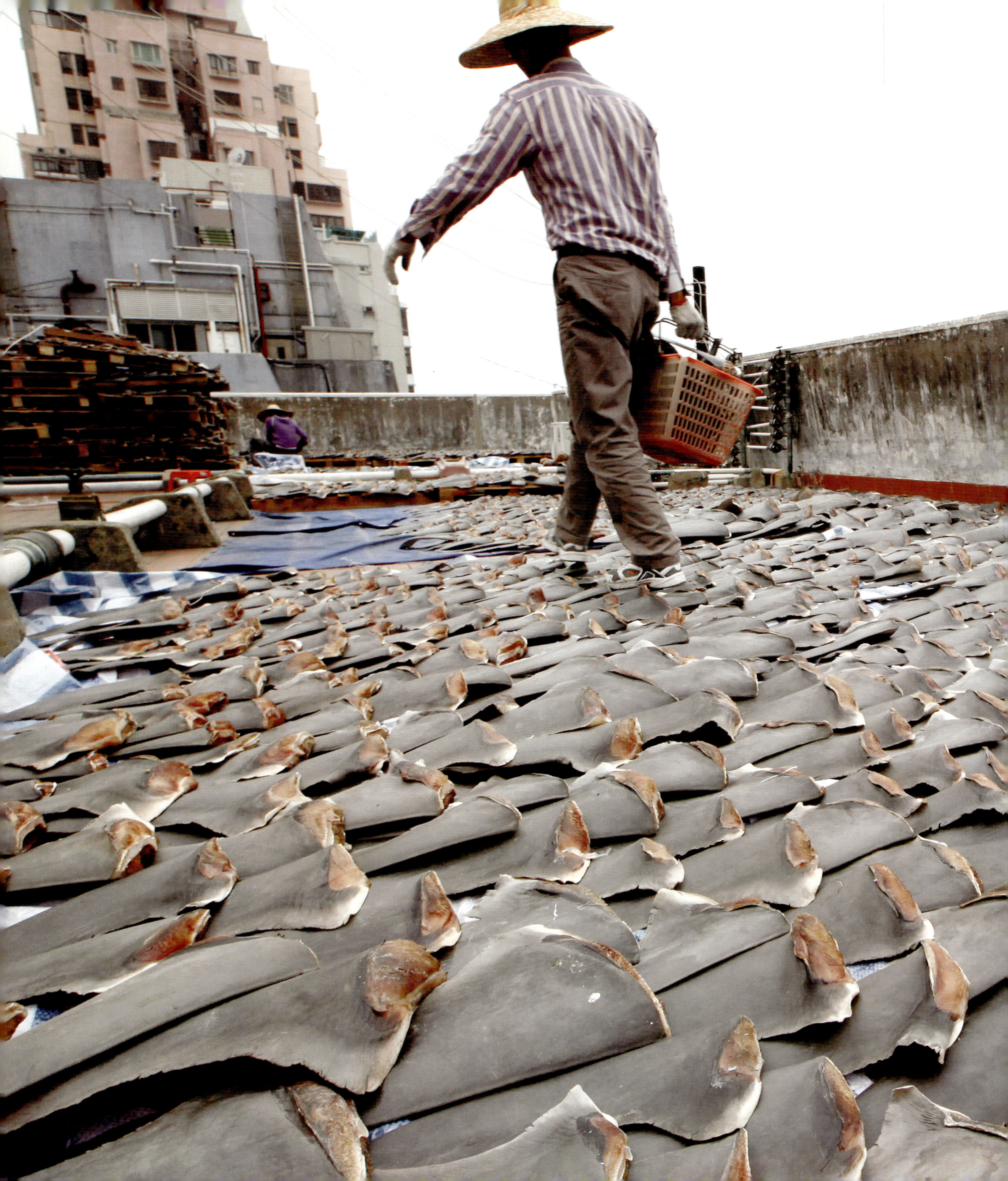

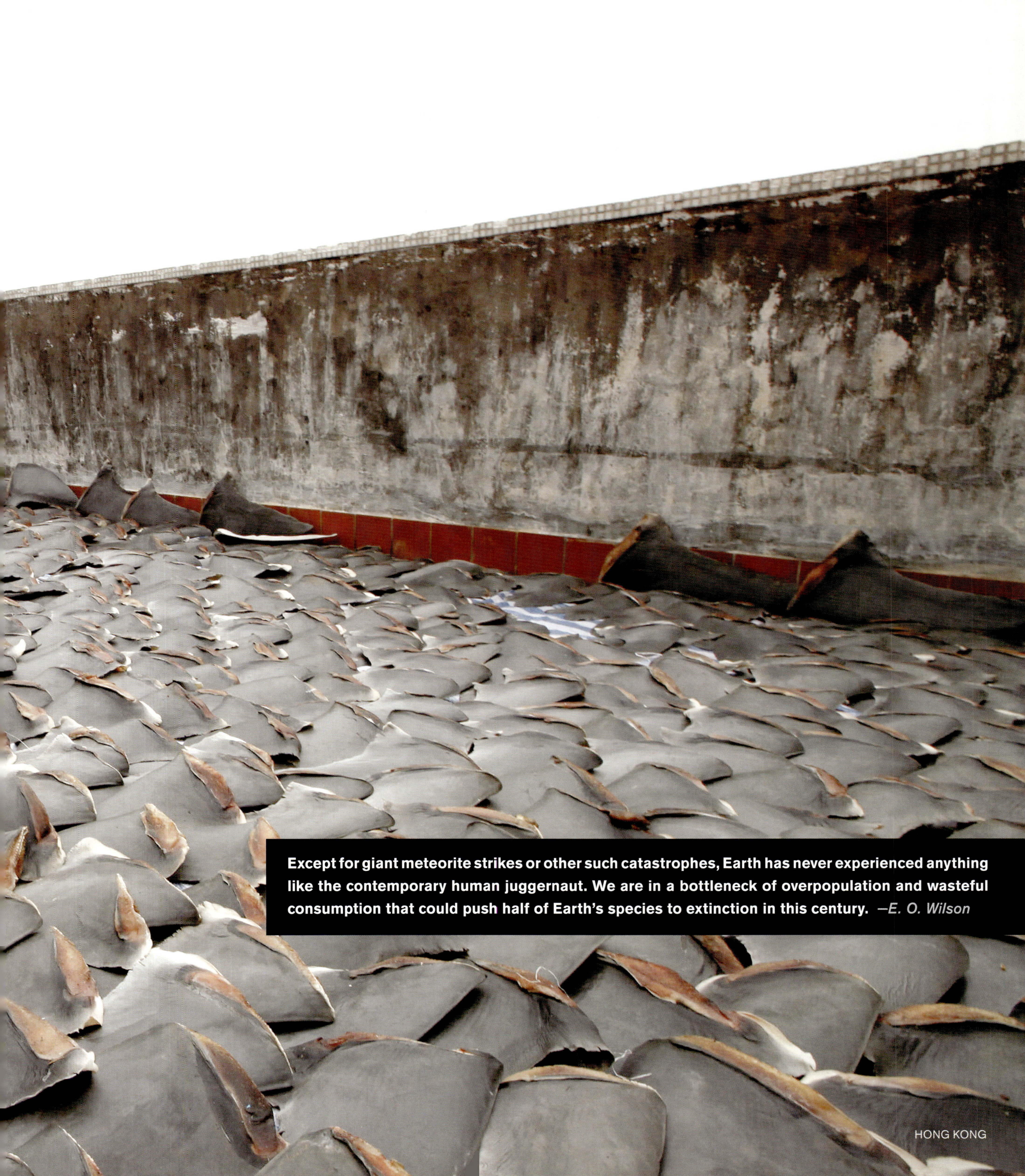

Except for giant meteorite strikes or other such catastrophes, Earth has never experienced anything like the contemporary human juggernaut. We are in a bottleneck of overpopulation and wasteful consumption that could push half of Earth's species to extinction in this century. —E. O. Wilson

HONG KONG

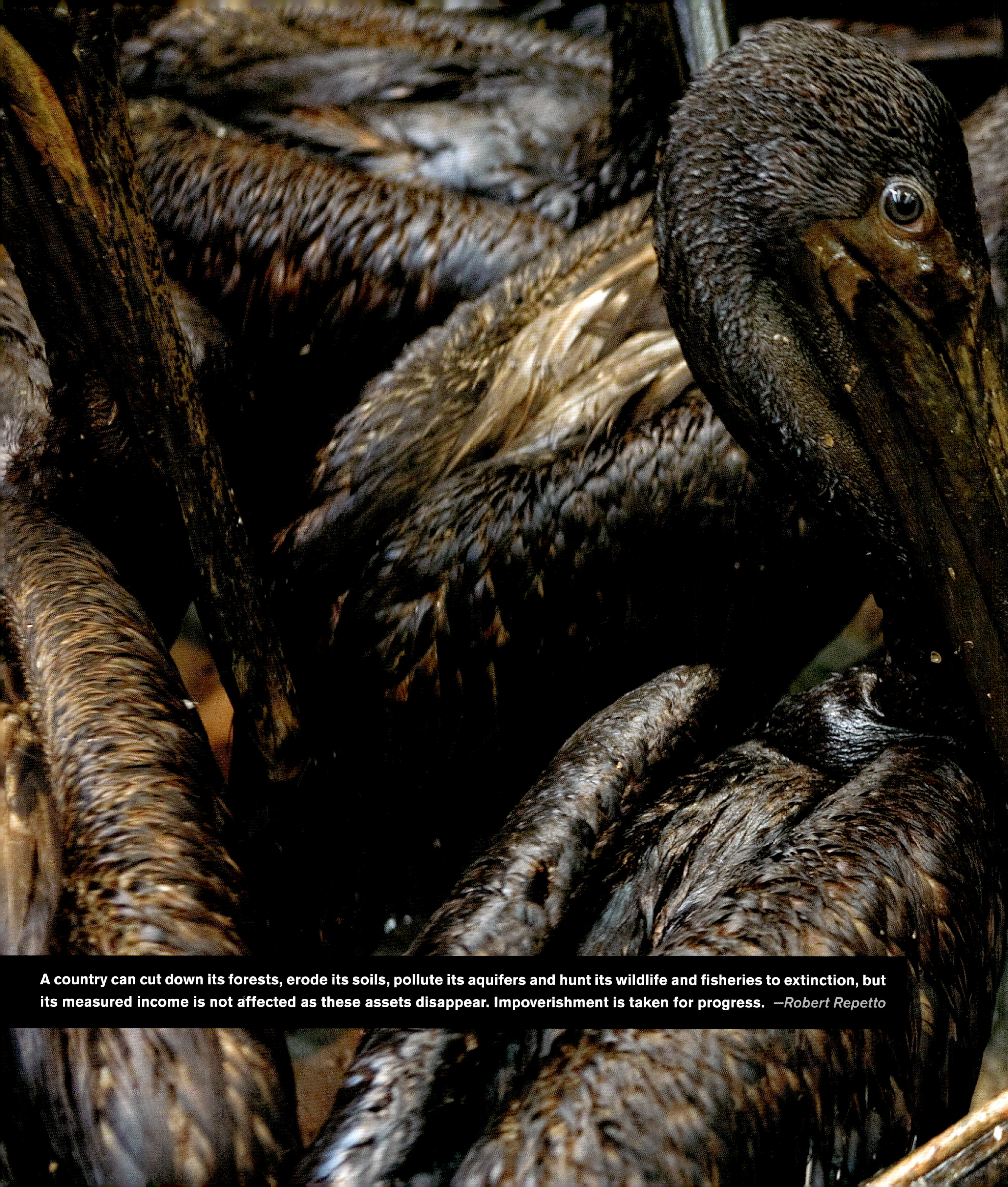

A country can cut down its forests, erode its soils, pollute its aquifers and hunt its wildlife and fisheries to extinction, but its measured income is not affected as these assets disappear. Impoverishment is taken for progress. *—Robert Repetto*

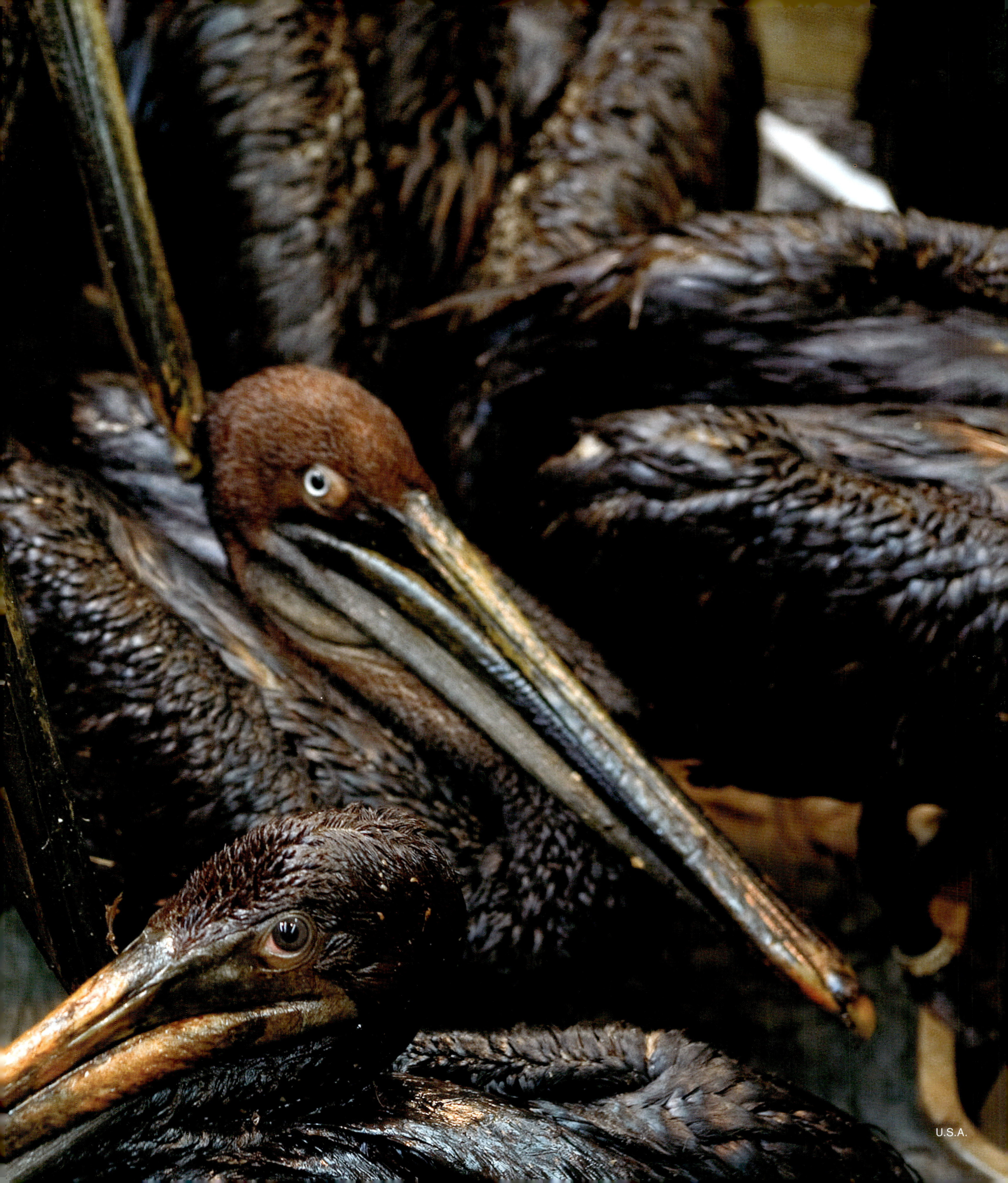
U.S.A.

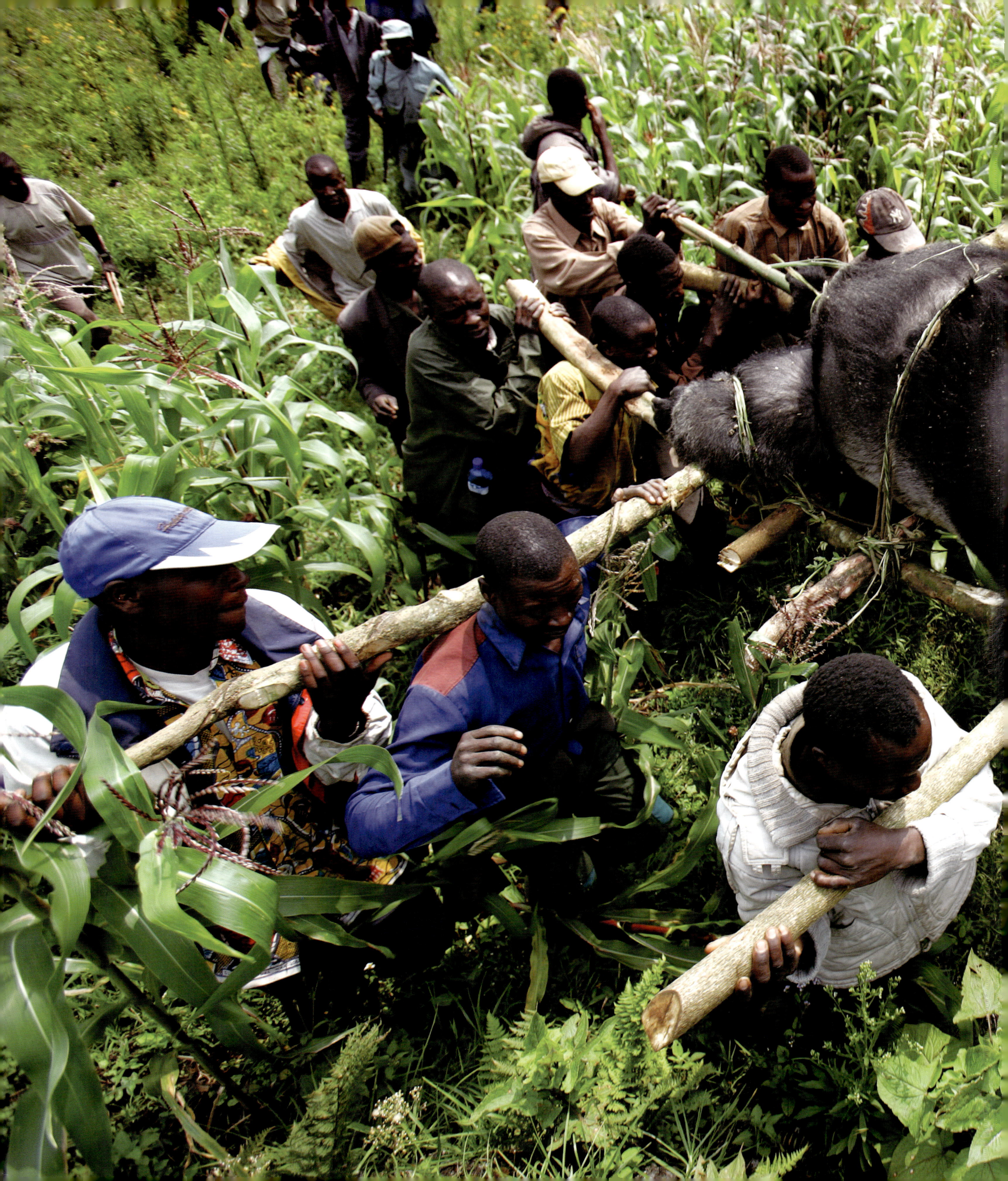

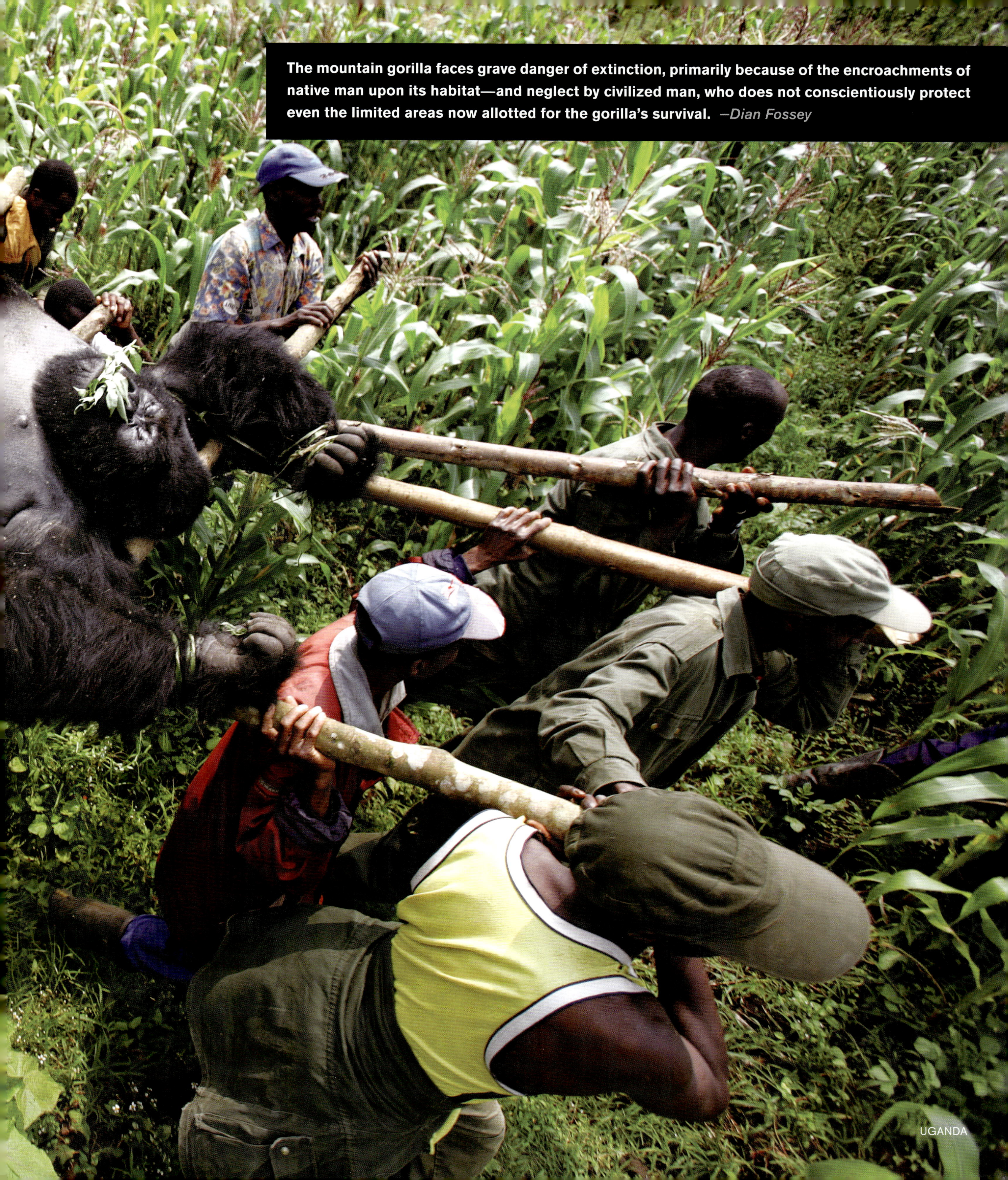

The mountain gorilla faces grave danger of extinction, primarily because of the encroachments of native man upon its habitat—and neglect by civilized man, who does not conscientiously protect even the limited areas now allotted for the gorilla's survival. —Dian Fossey
UGANDA

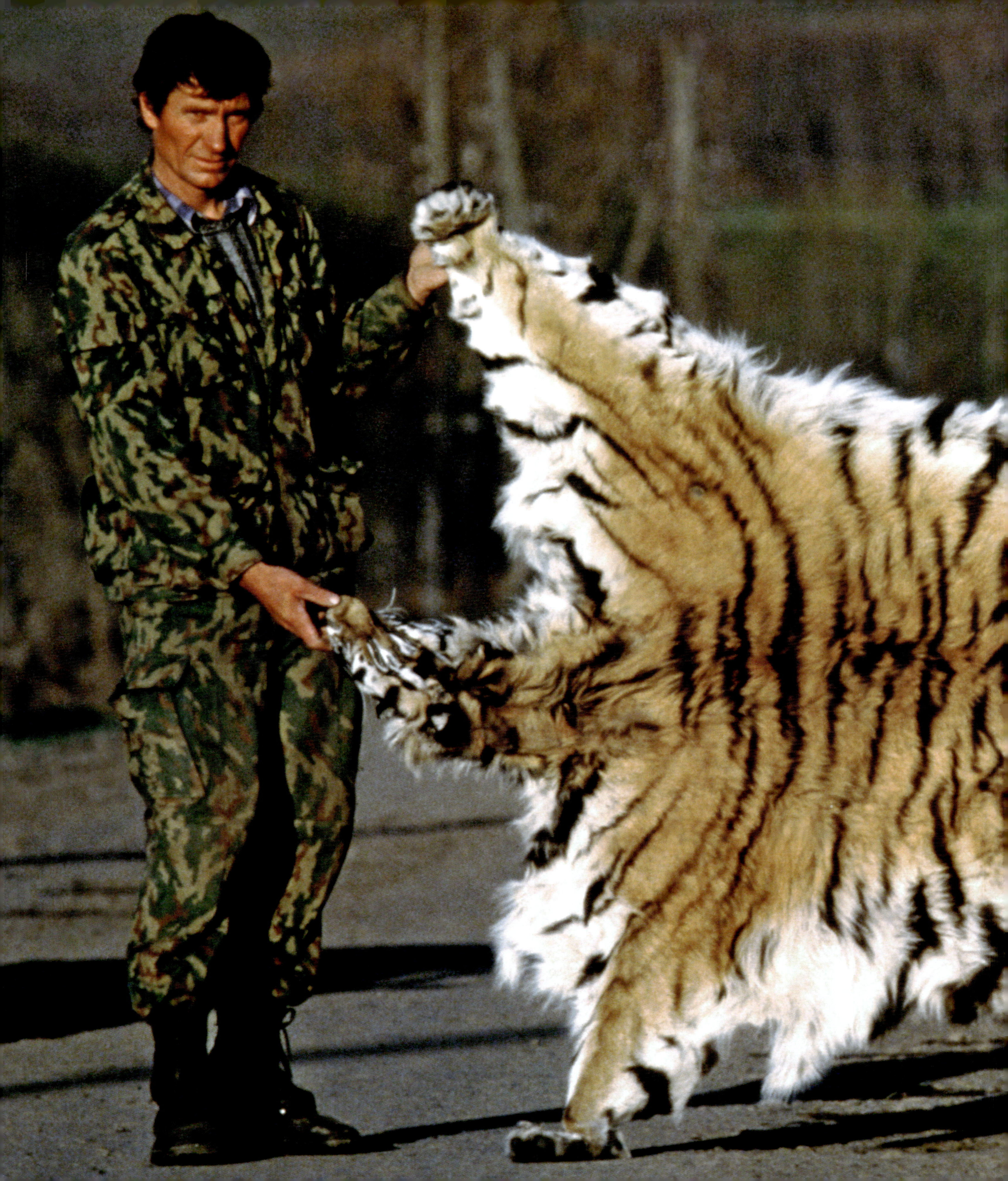

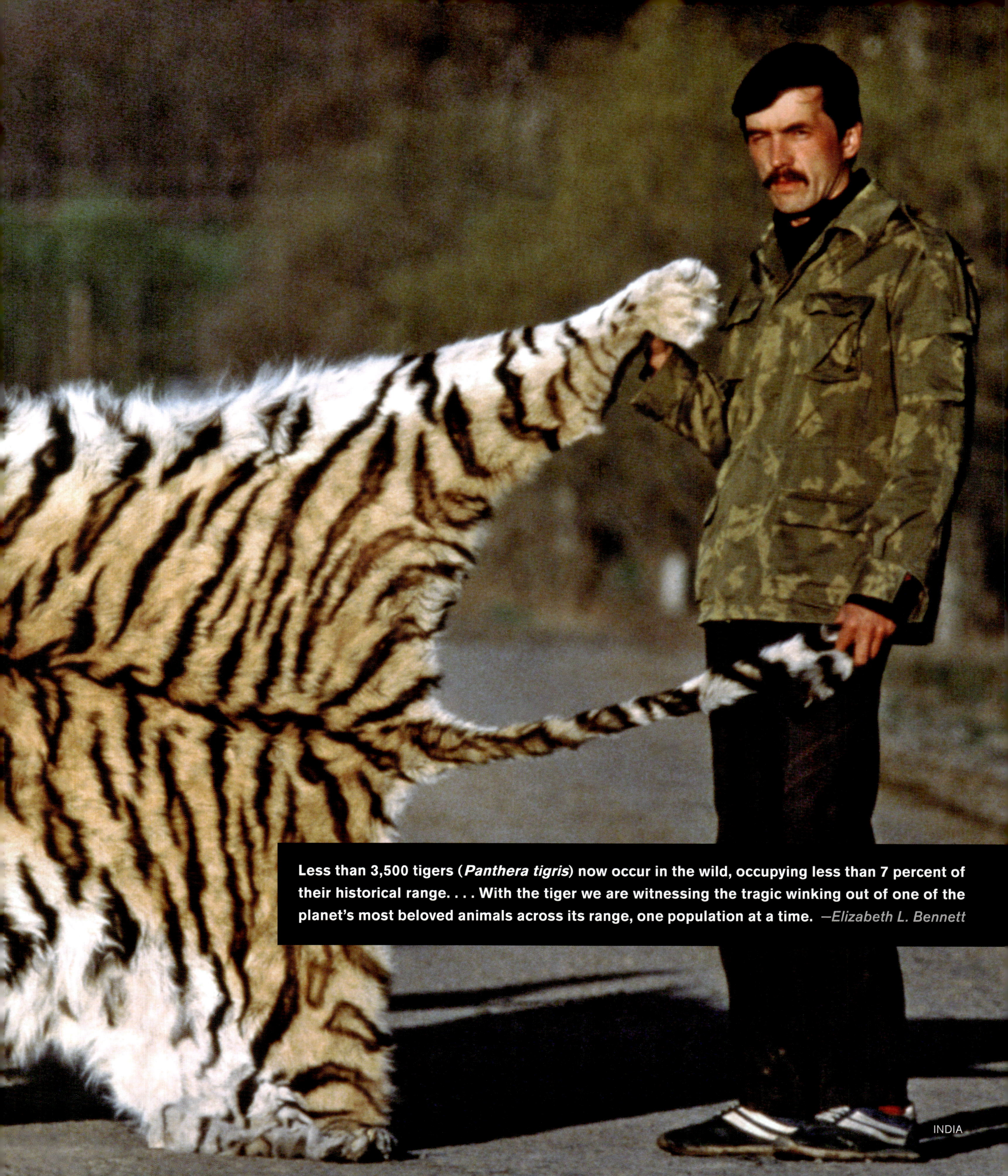

Less than 3,500 tigers (*Panthera tigris*) now occur in the wild, occupying less than 7 percent of their historical range. . . . With the tiger we are witnessing the tragic winking out of one of the planet's most beloved animals across its range, one population at a time. —*Elizabeth L. Bennett*

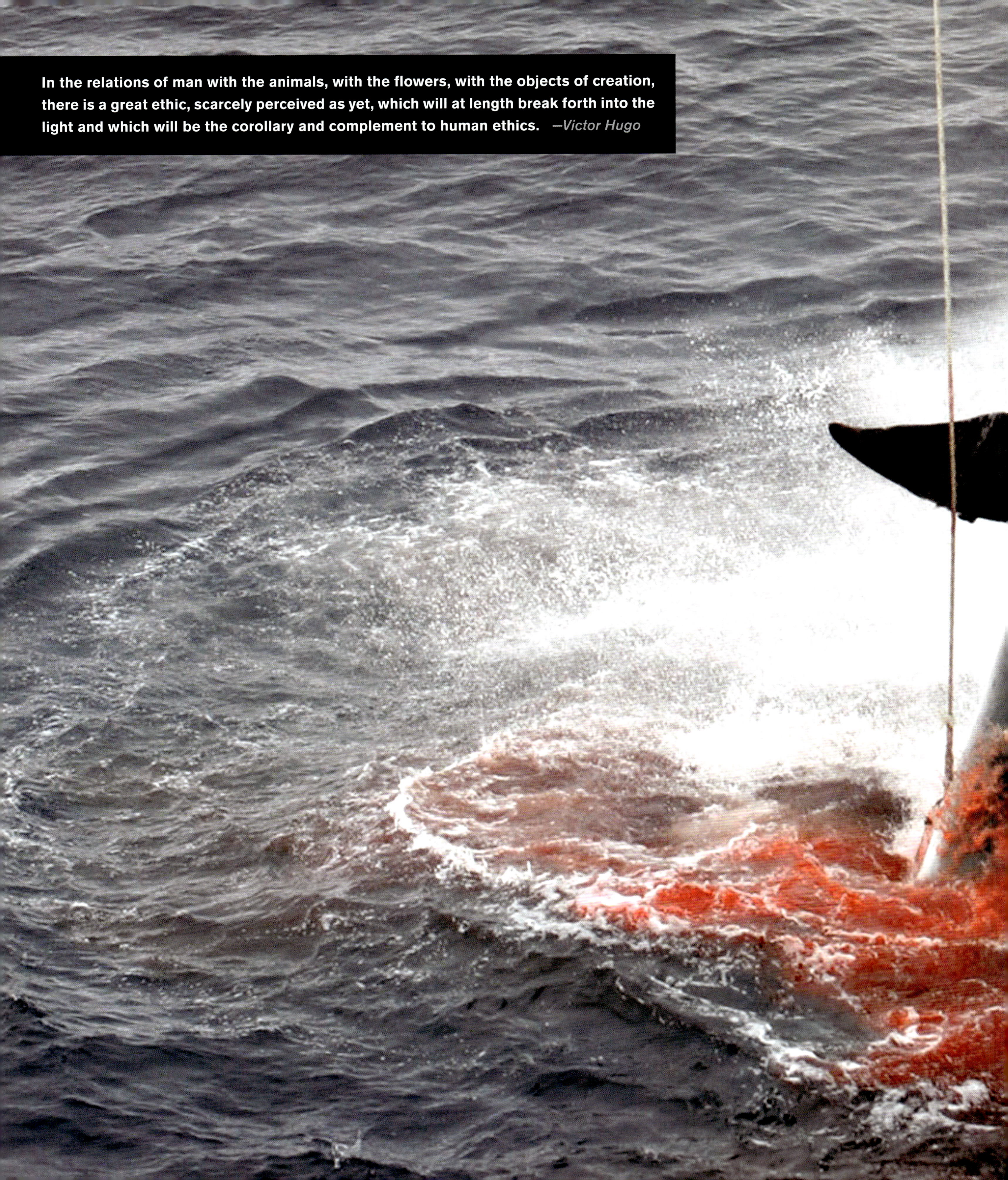

In the relations of man with the animals, with the flowers, with the objects of creation, there is a great ethic, scarcely perceived as yet, which will at length break forth into the light and which will be the corollary and complement to human ethics. —Victor Hugo

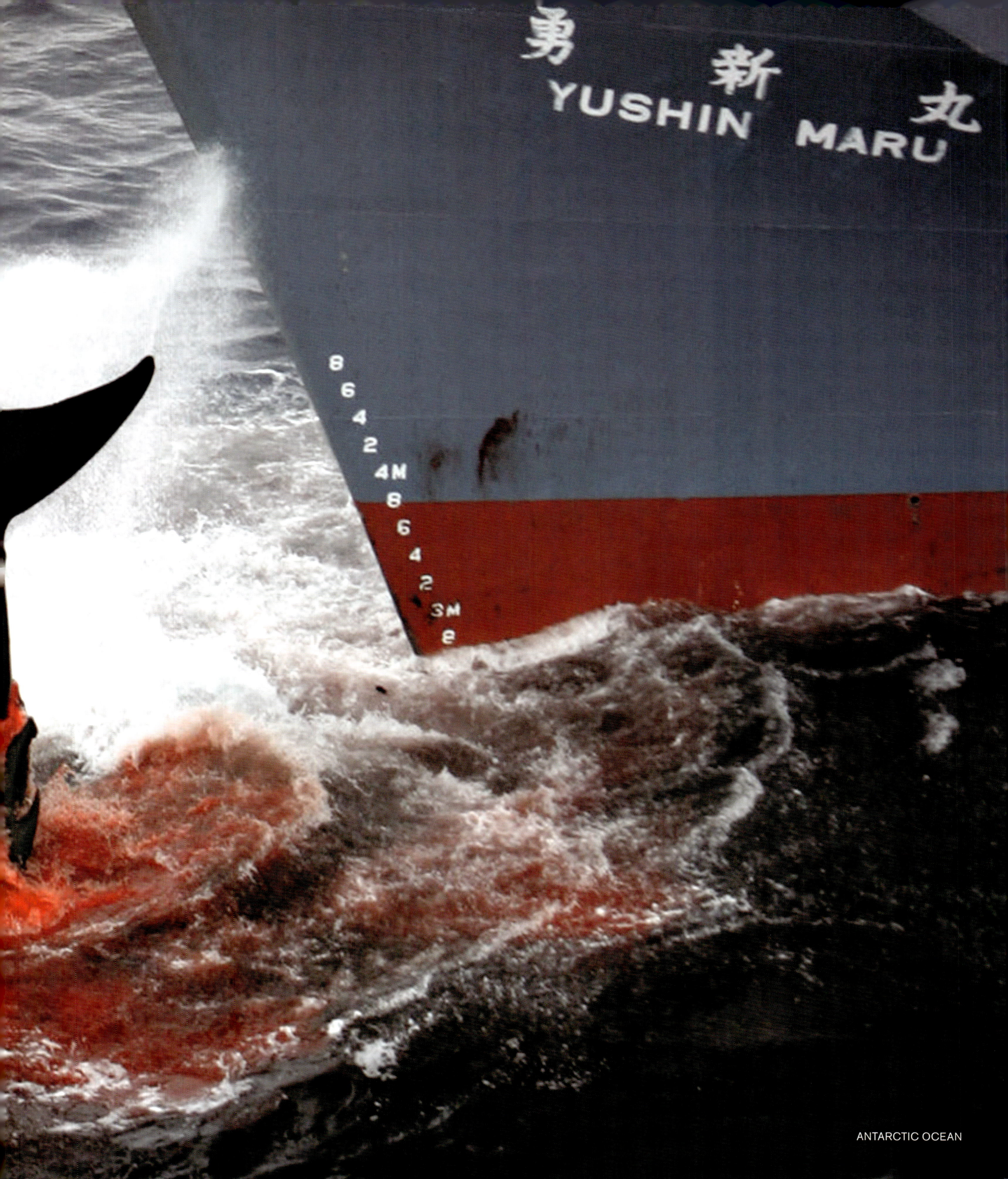

勇 新 丸
YUSHIN MARU
ANTARCTIC OCEAN

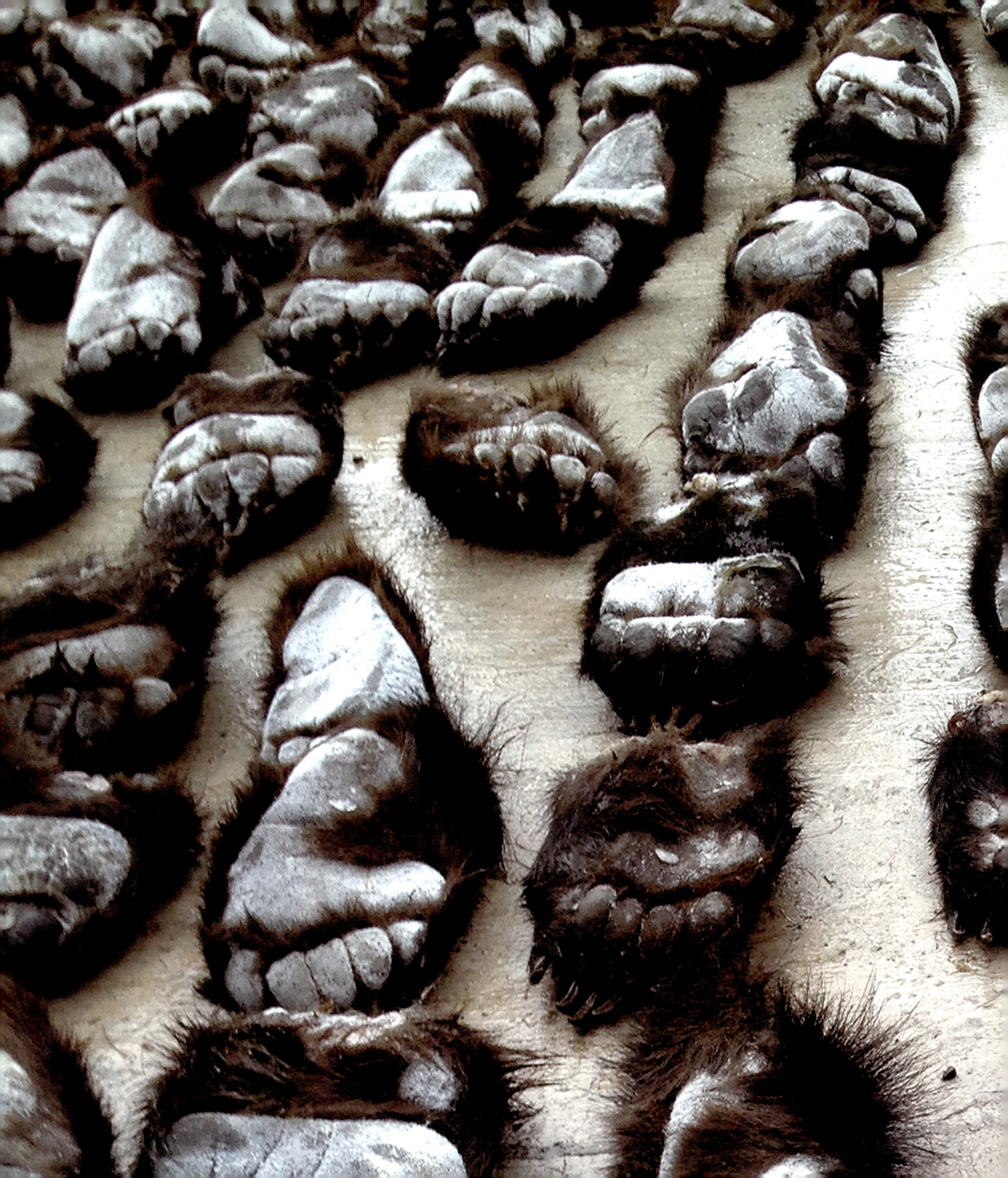

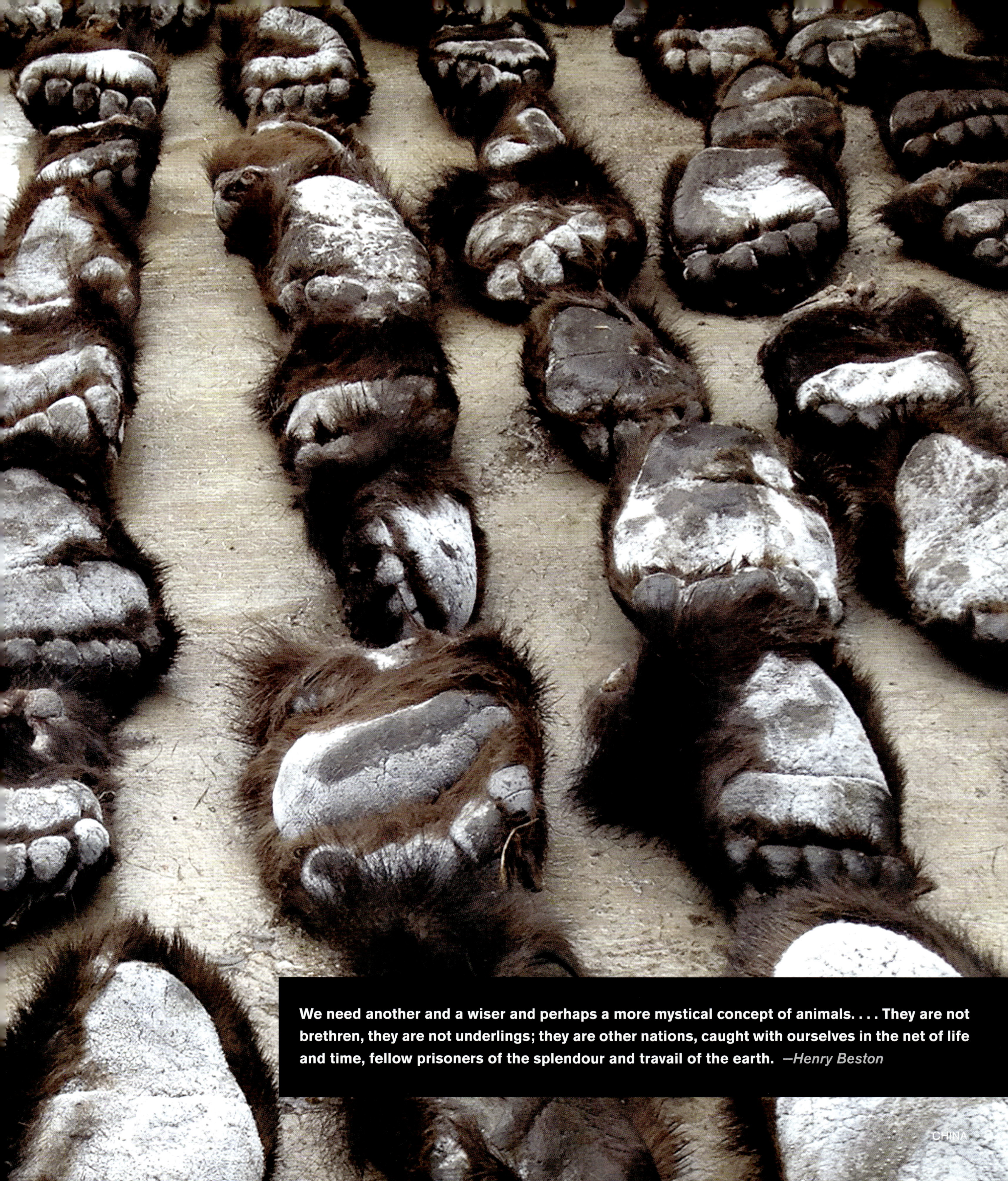

We need another and a wiser and perhaps a more mystical concept of animals. . . . They are not brethren, they are not underlings; they are other nations, caught with ourselves in the net of life and time, fellow prisoners of the splendour and travail of the earth. —Henry Beston
CHINA

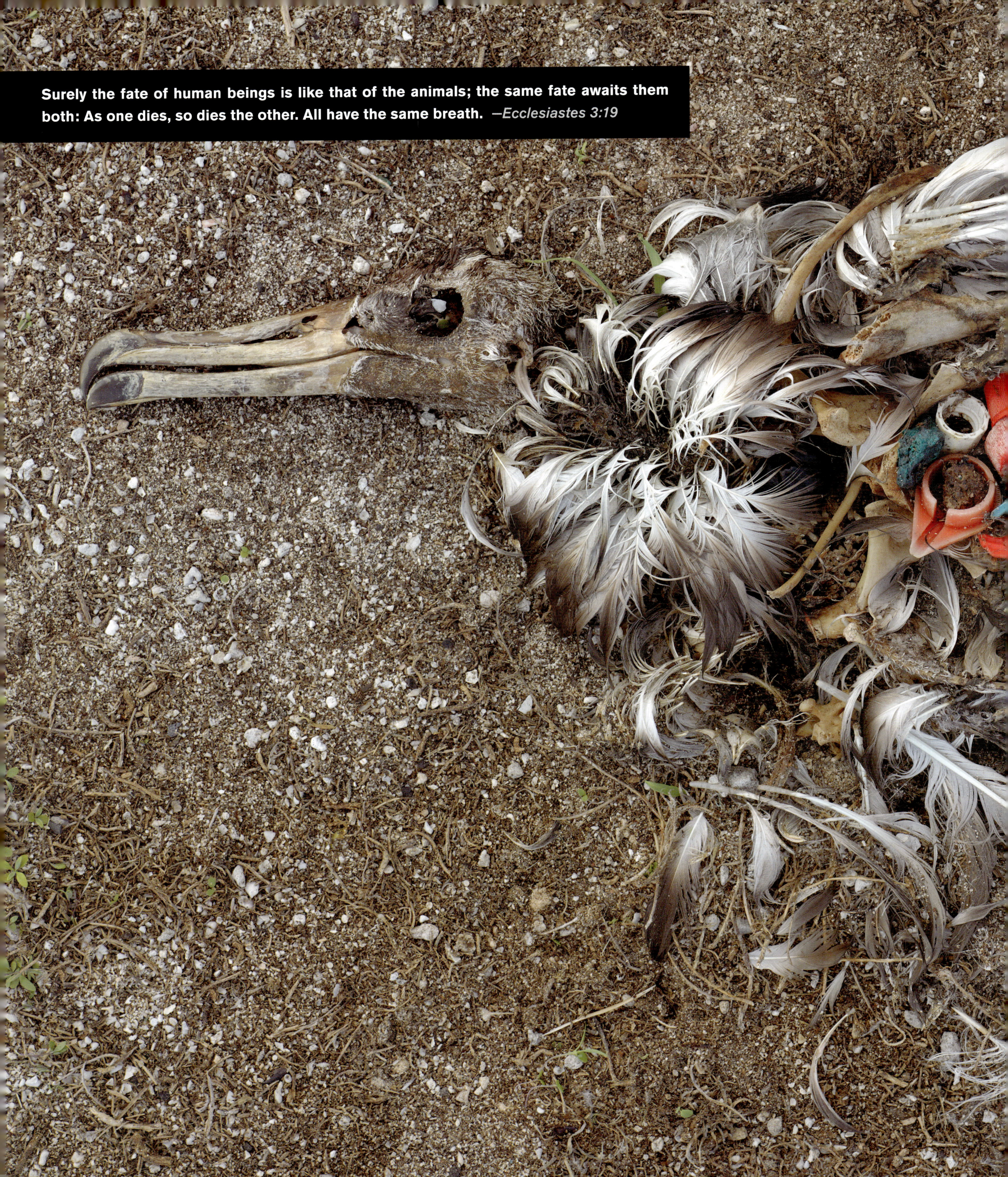

Surely the fate of human beings is like that of the animals; the same fate awaits them both: As one dies, so dies the other. All have the same breath. —Ecclesiastes 3:19

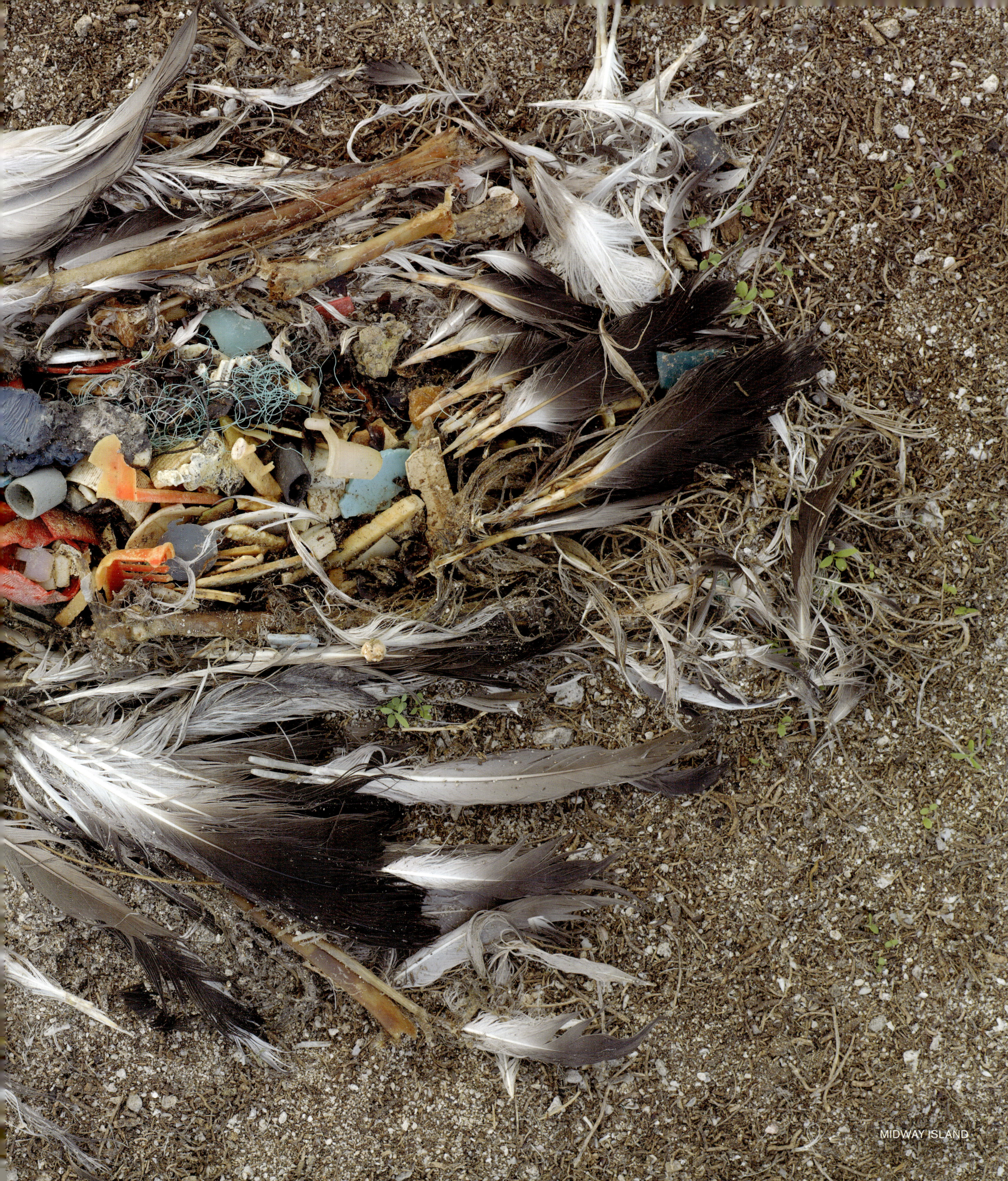
MIDWAY ISLAND

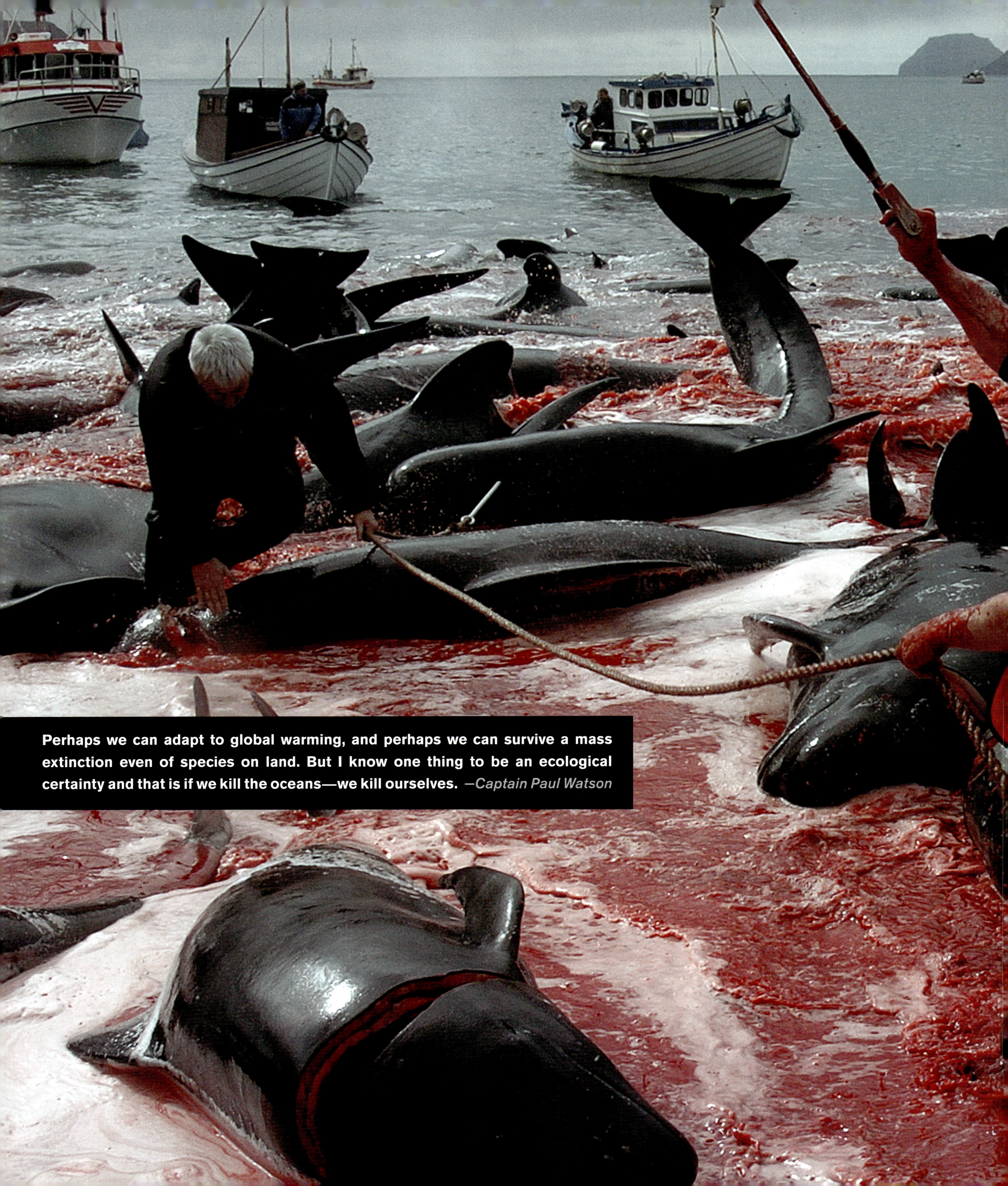
Perhaps we can adapt to global warming, and perhaps we can survive a mass extinction even of species on land. But I know one thing to be an ecological certainty and that is if we kill the oceans—we kill ourselves. —Captain Paul Watson

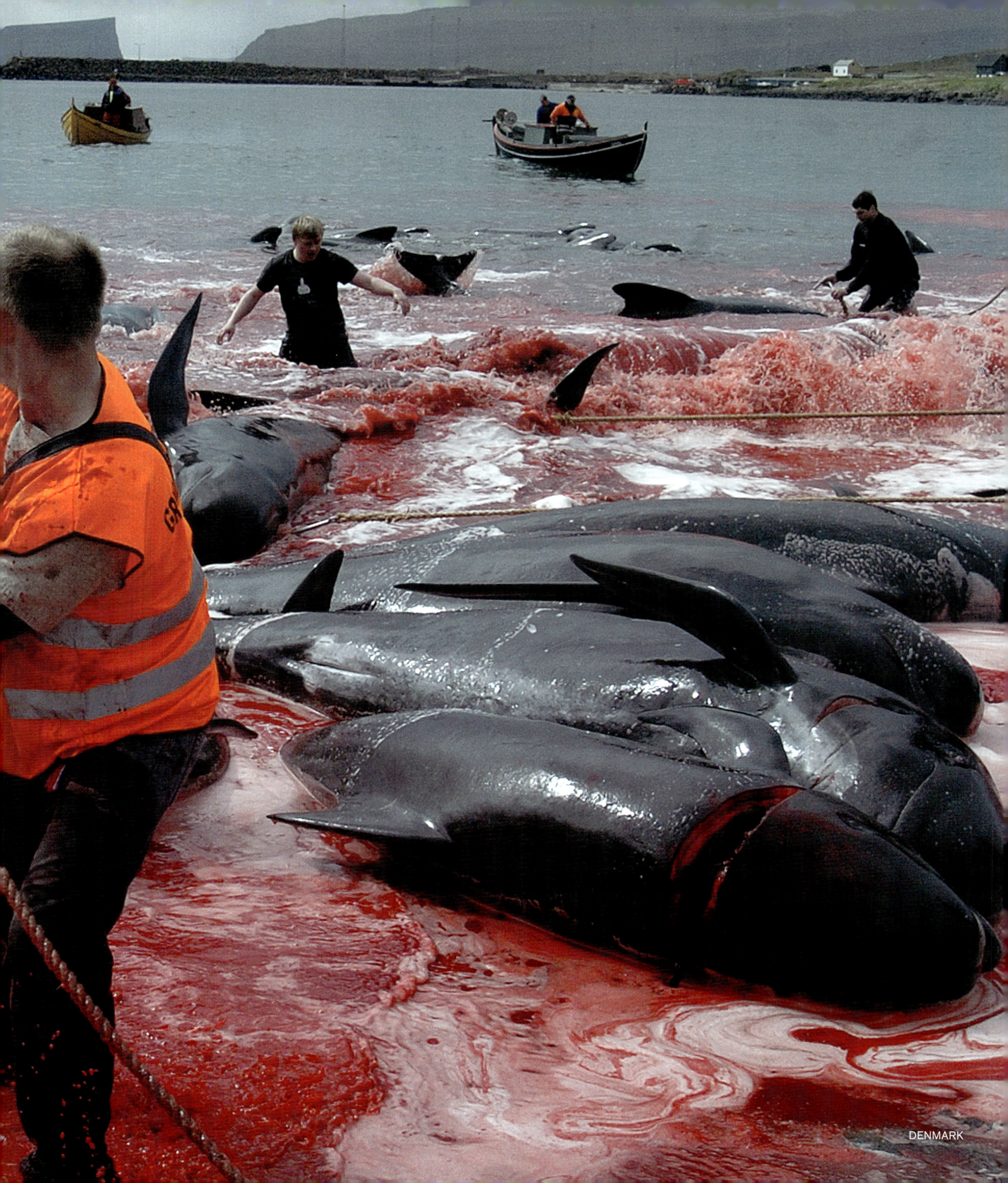
DENMARK

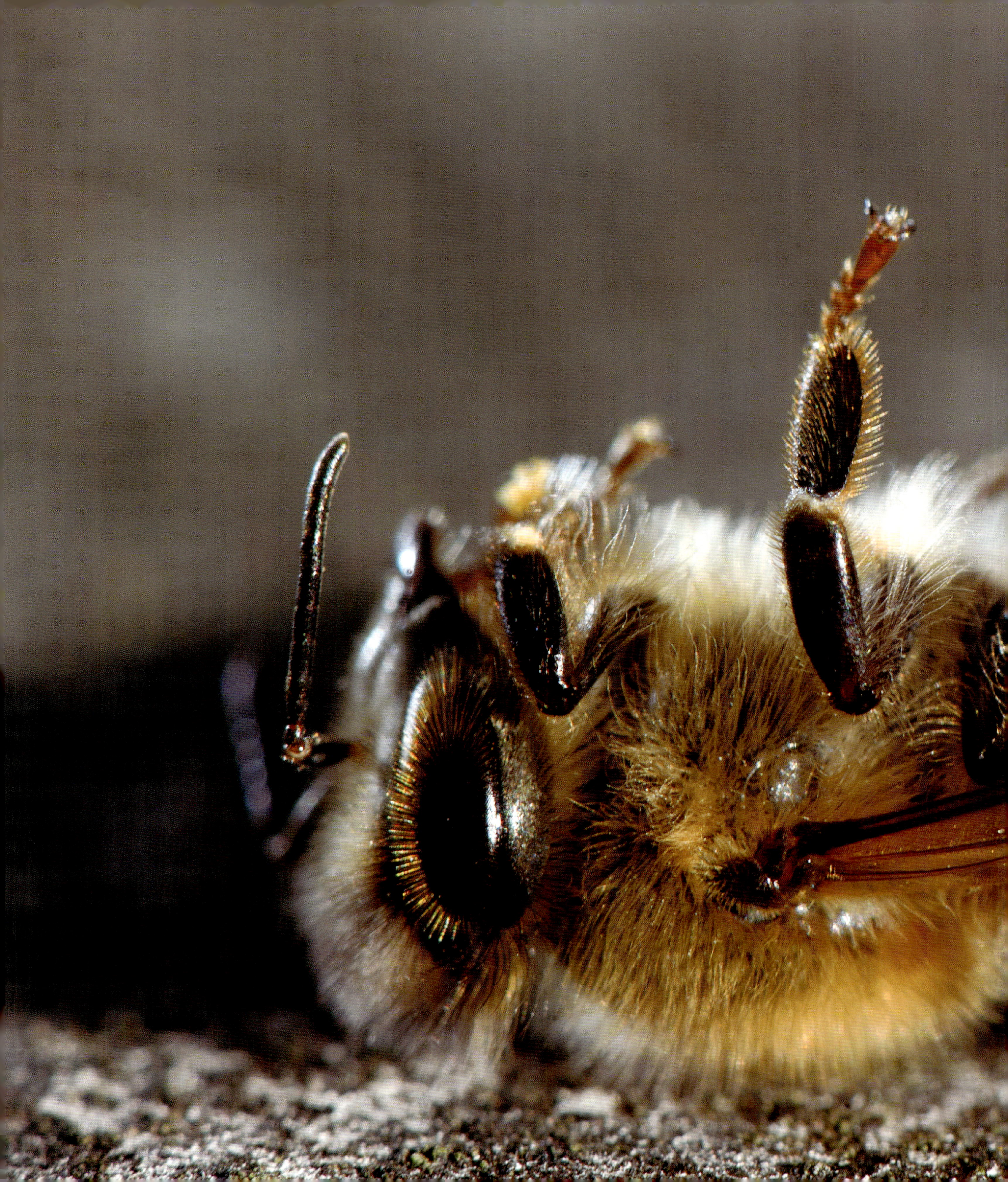

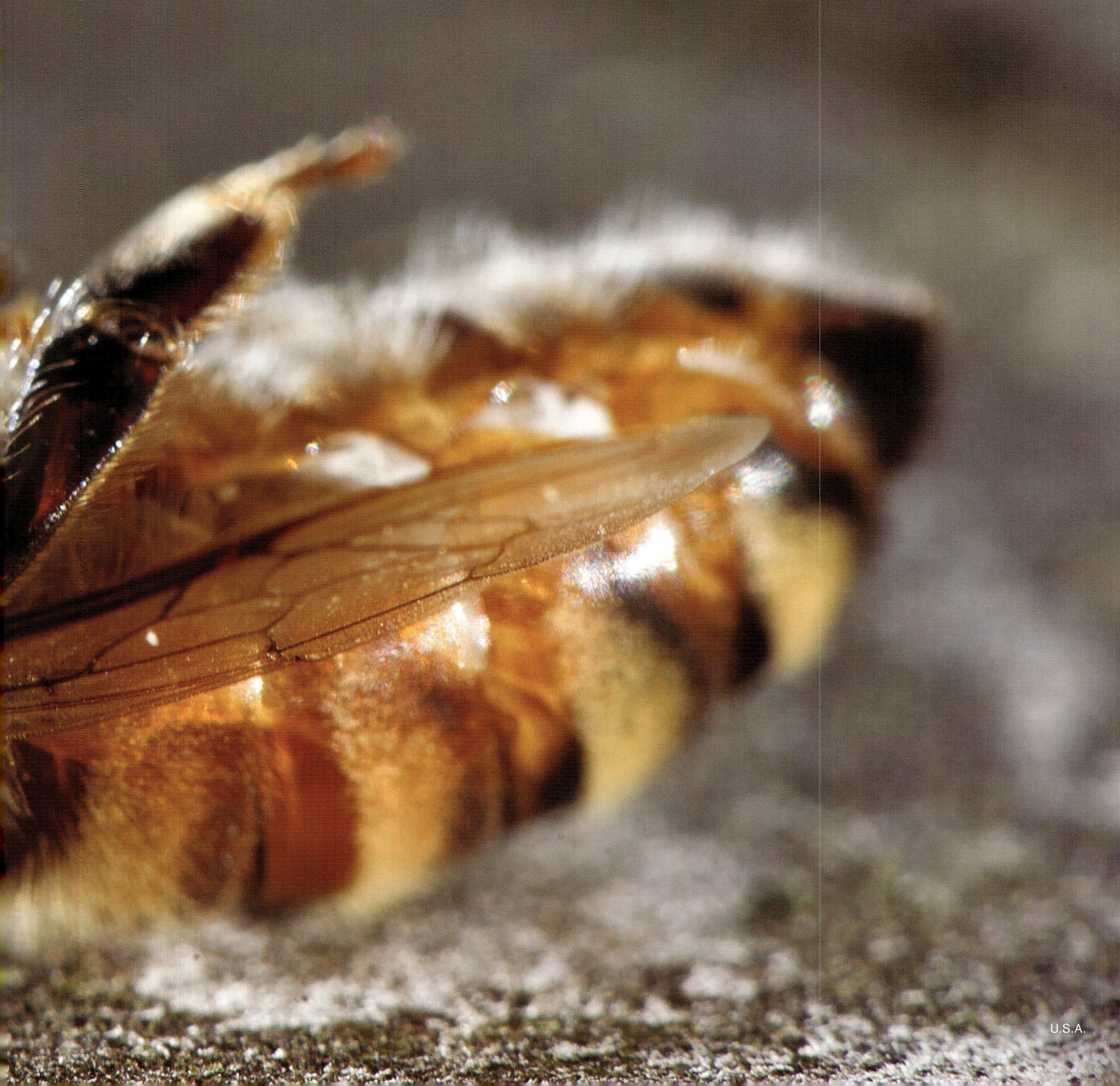

Of the 250,000 species of plants that share our world, three quarters rely on wild pollinators to reproduce. Wherever you live, look around and see a world engineered by these pollinators. Then look around and see a world in distress. Honeybees may have been filling in for wild pollinators to bolster our agriculture, but they can't do much for the other 249,900 species of flowering plants. That's up to the native bugs. And while evidence is hard to come by, many of these species are failing under the triple threats of habitat loss, pesticide poisoning, and exotics. —Rowan Jacobsen
U.S.A.

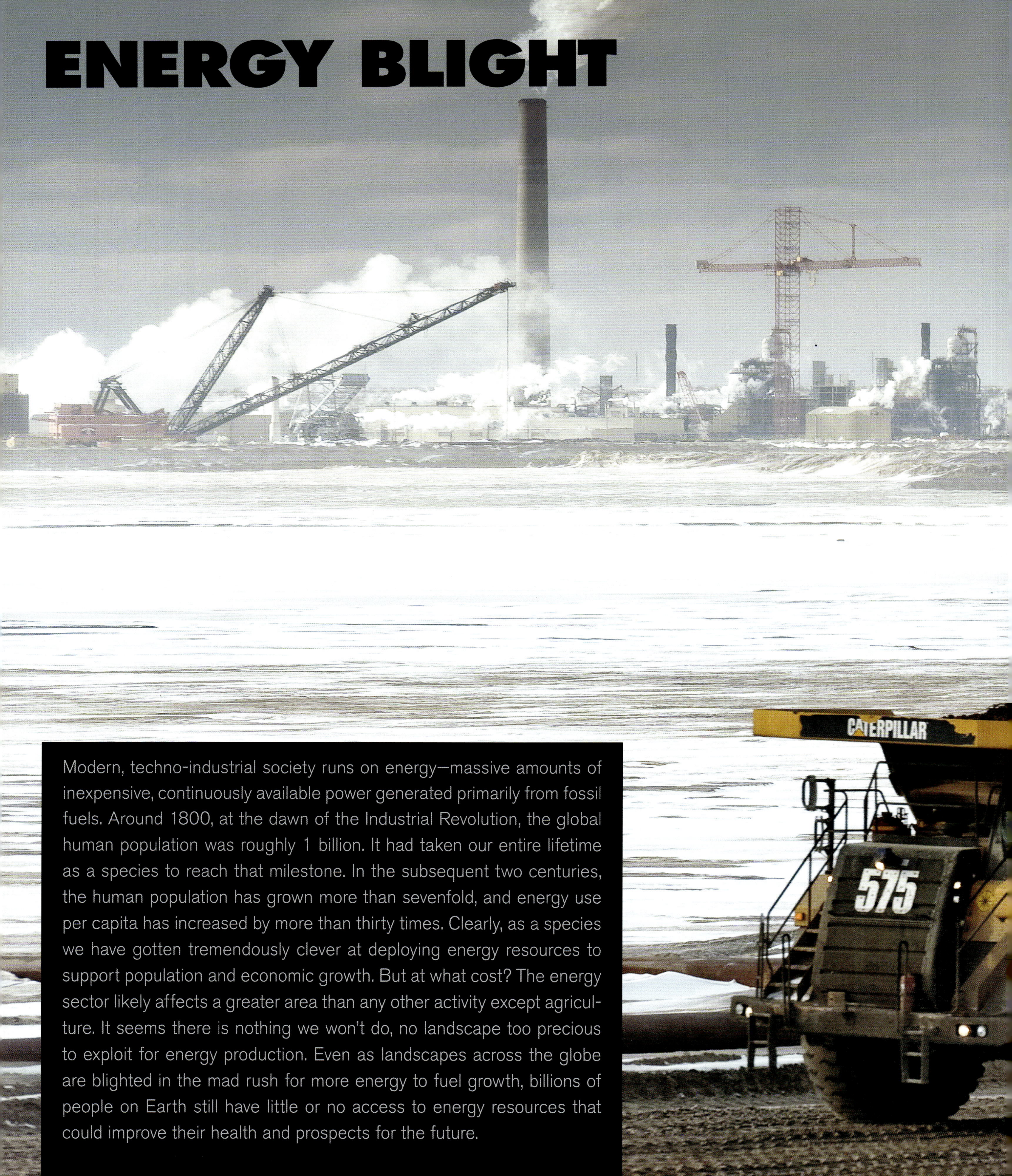

ENERGY BLIGHT

Modern, techno-industrial society runs on energy—massive amounts of inexpensive, continuously available power generated primarily from fossil fuels. Around 1800, at the dawn of the Industrial Revolution, the global human population was roughly 1 billion. It had taken our entire lifetime as a species to reach that milestone. In the subsequent two centuries, the human population has grown more than sevenfold, and energy use per capita has increased by more than thirty times. Clearly, as a species we have gotten tremendously clever at deploying energy resources to support population and economic growth. But at what cost? The energy sector likely affects a greater area than any other activity except agriculture. It seems there is nothing we won't do, no landscape too precious to exploit for energy production. Even as landscapes across the globe are blighted in the mad rush for more energy to fuel growth, billions of people on Earth still have little or no access to energy resources that could improve their health and prospects for the future.

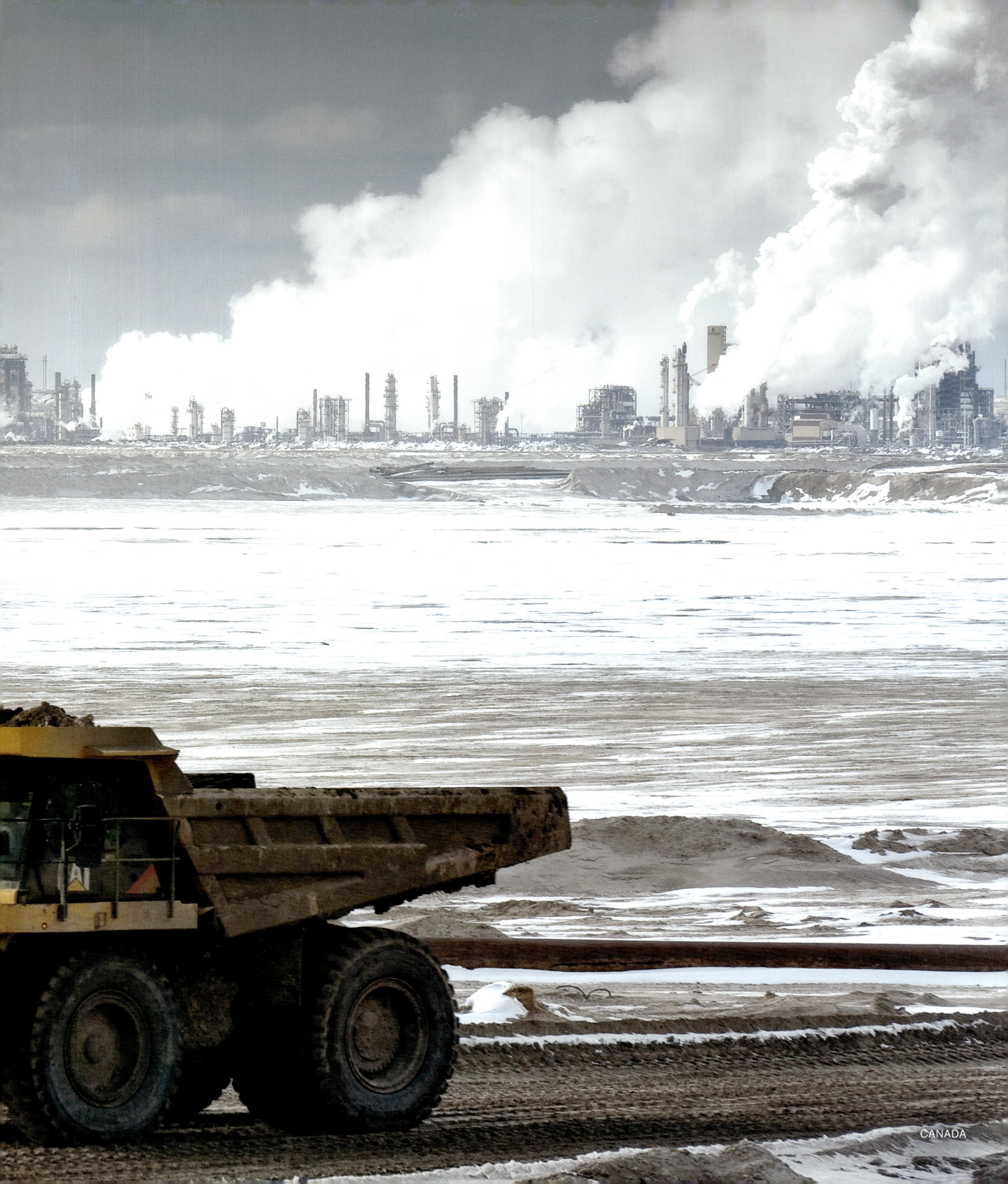
CANADA

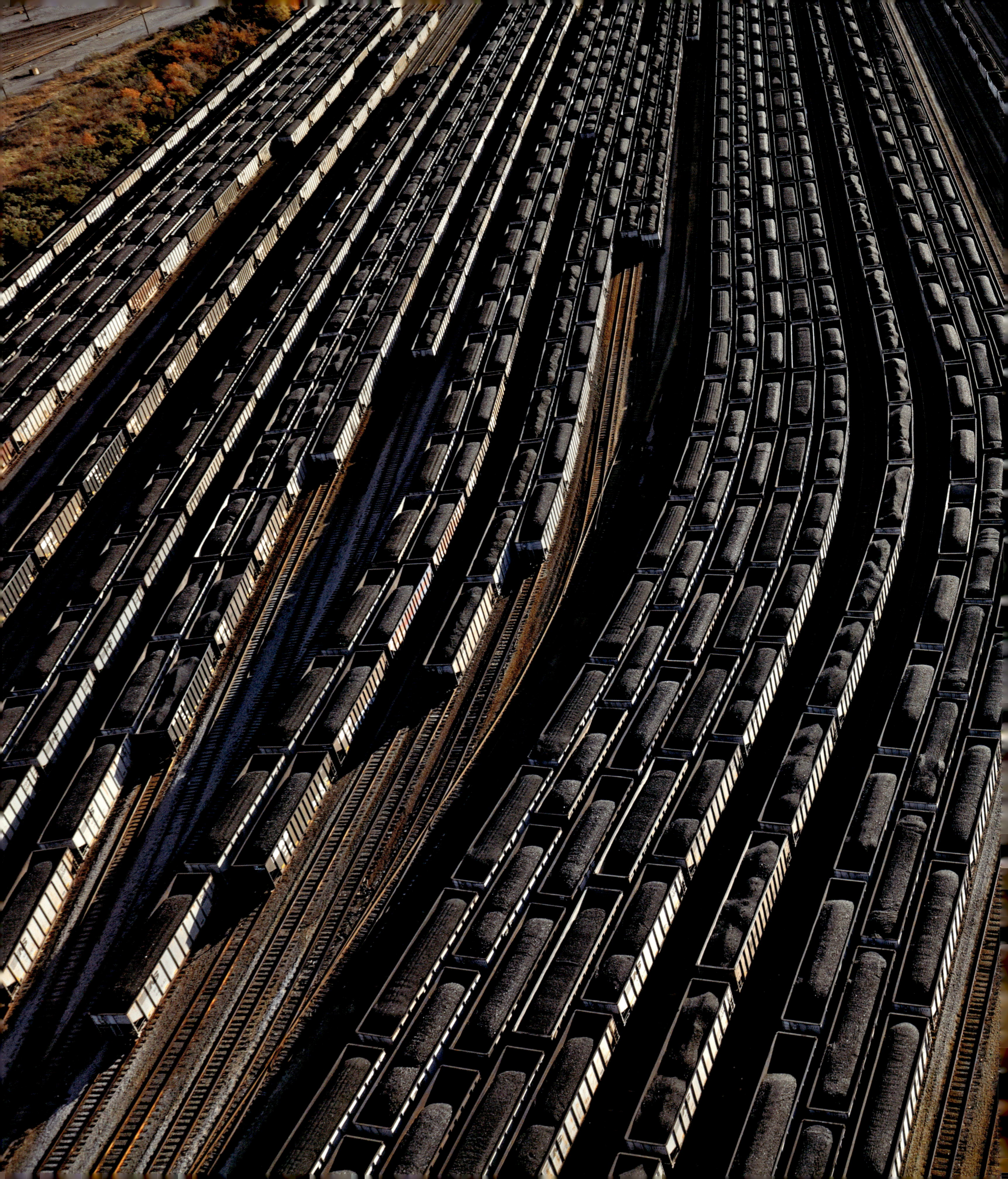

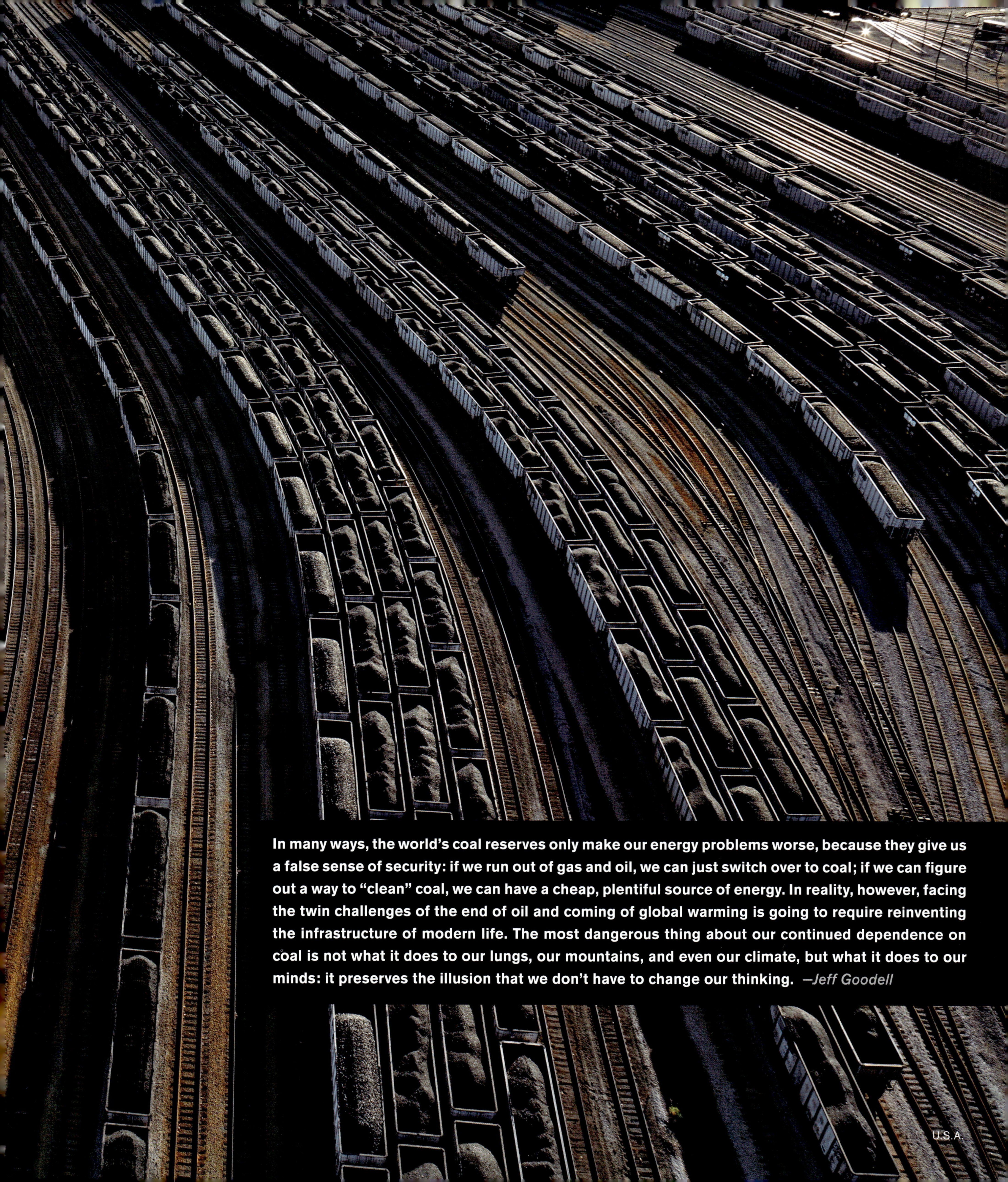

In many ways, the world's coal reserves only make our energy problems worse, because they give us a false sense of security: if we run out of gas and oil, we can just switch over to coal; if we can figure out a way to "clean" coal, we can have a cheap, plentiful source of energy. In reality, however, facing the twin challenges of the end of oil and coming of global warming is going to require reinventing the infrastructure of modern life. The most dangerous thing about our continued dependence on coal is not what it does to our lungs, our mountains, and even our climate, but what it does to our minds: it preserves the illusion that we don't have to change our thinking. —*Jeff Goodell*

U.S.A.

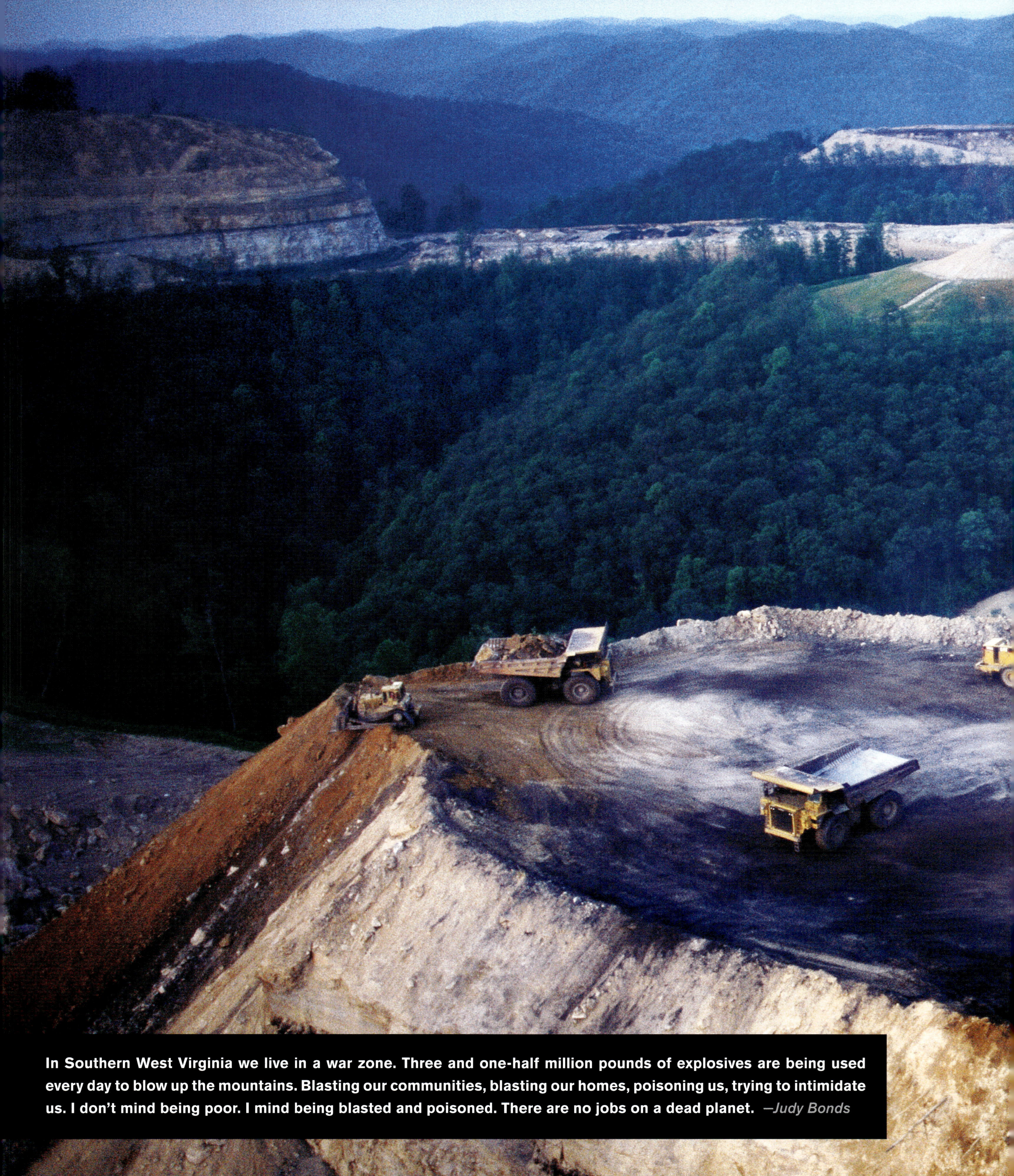

In Southern West Virginia we live in a war zone. Three and one-half million pounds of explosives are being used every day to blow up the mountains. Blasting our communities, blasting our homes, poisoning us, trying to intimidate us. I don't mind being poor. I mind being blasted and poisoned. There are no jobs on a dead planet. —*Judy Bonds*

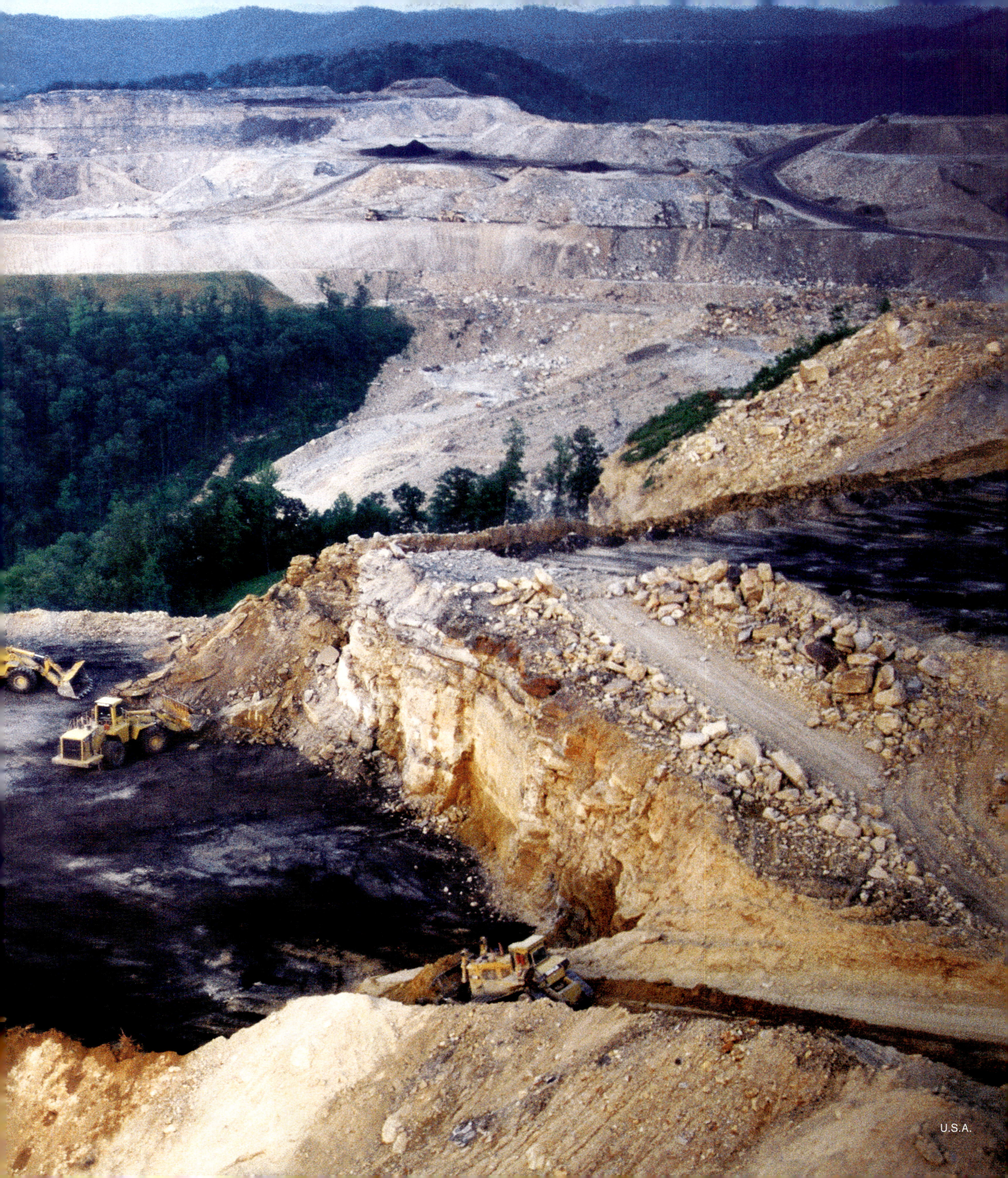
U.S.A.

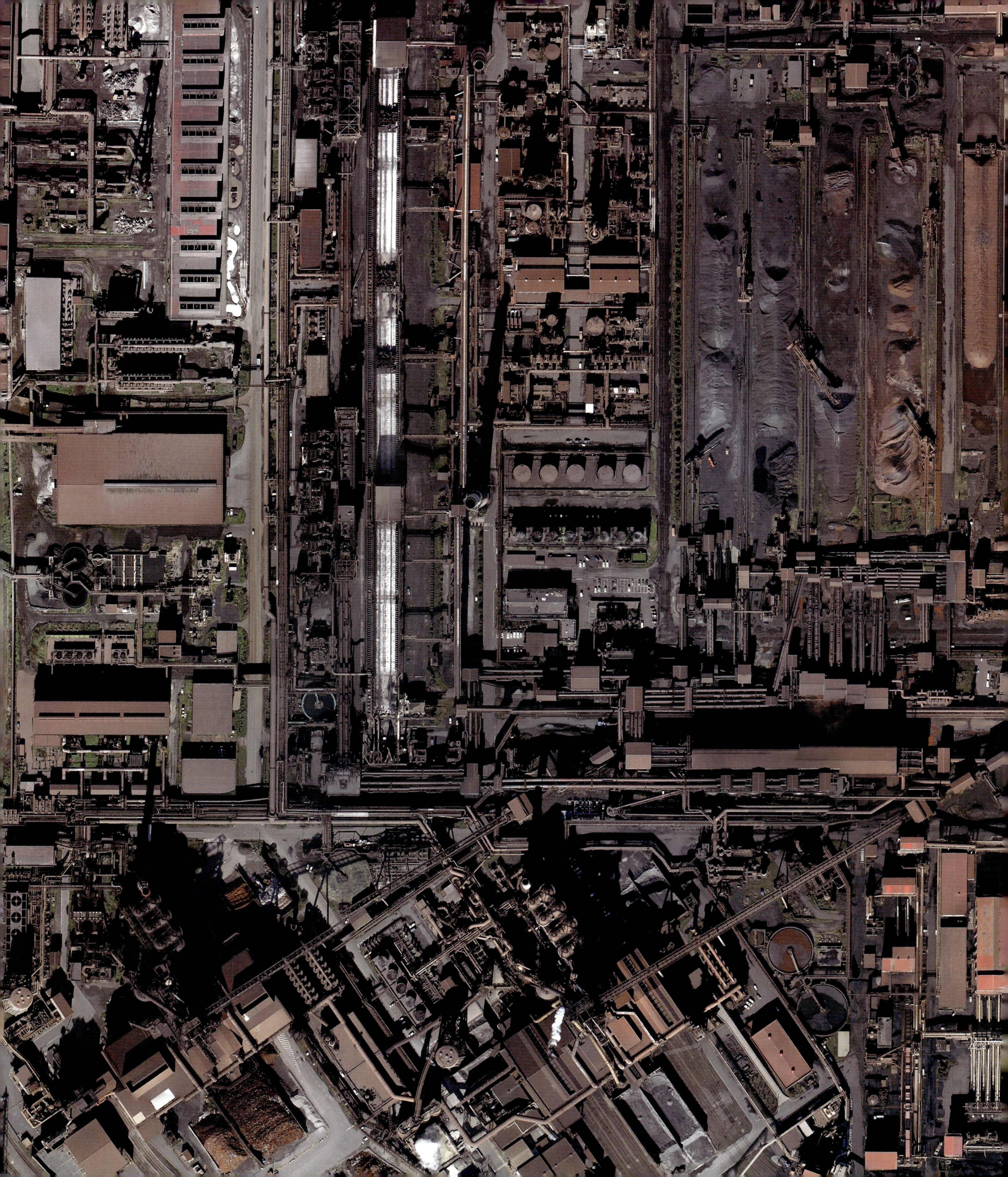

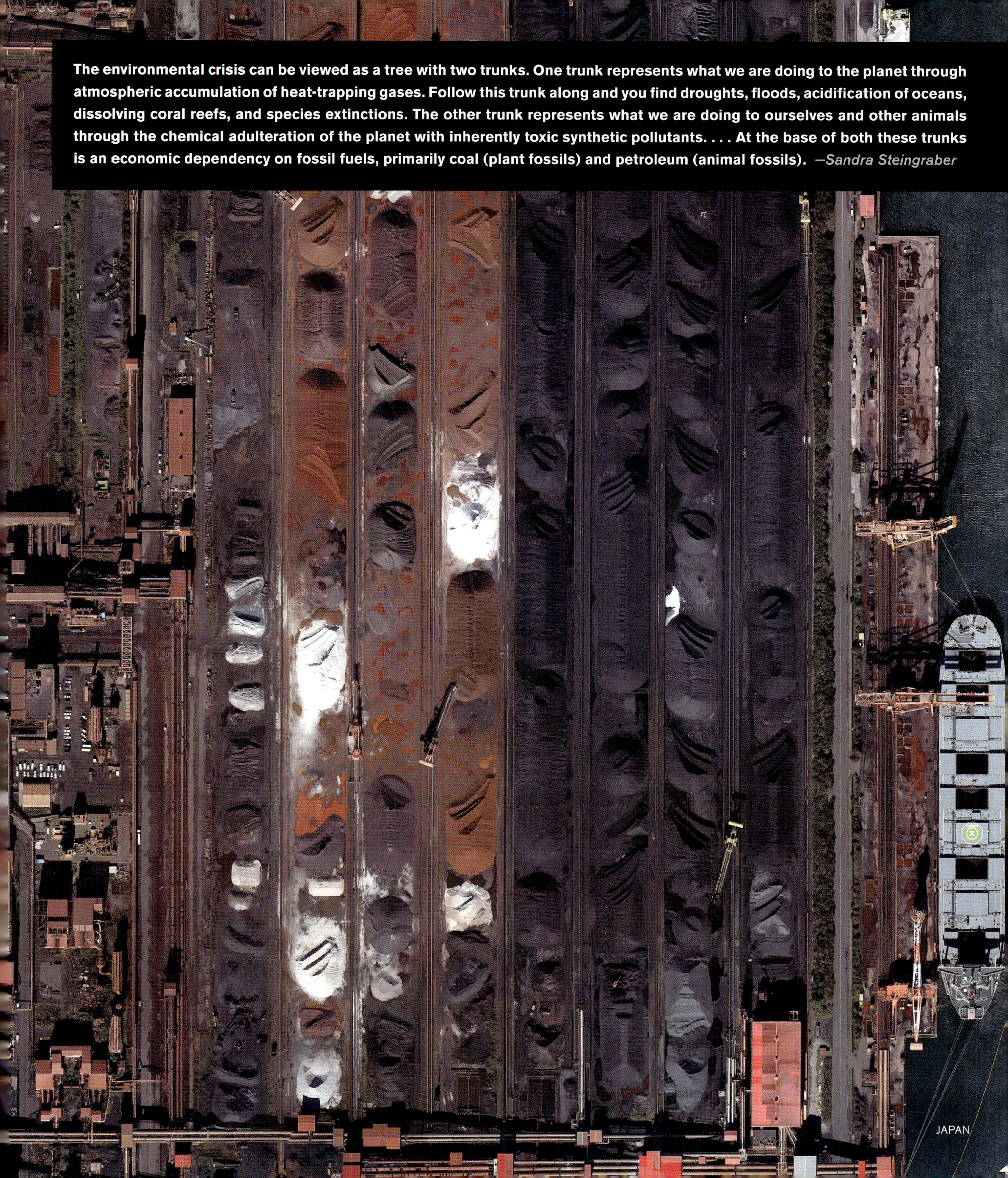

The environmental crisis can be viewed as a tree with two trunks. One trunk represents what we are doing to the planet through atmospheric accumulation of heat-trapping gases. Follow this trunk along and you find droughts, floods, acidification of oceans, dissolving coral reefs, and species extinctions. The other trunk represents what we are doing to ourselves and other animals through the chemical adulteration of the planet with inherently toxic synthetic pollutants. . . . At the base of both these trunks is an economic dependency on fossil fuels, primarily coal (plant fossils) and petroleum (animal fossils). —Sandra Steingraber

JAPAN

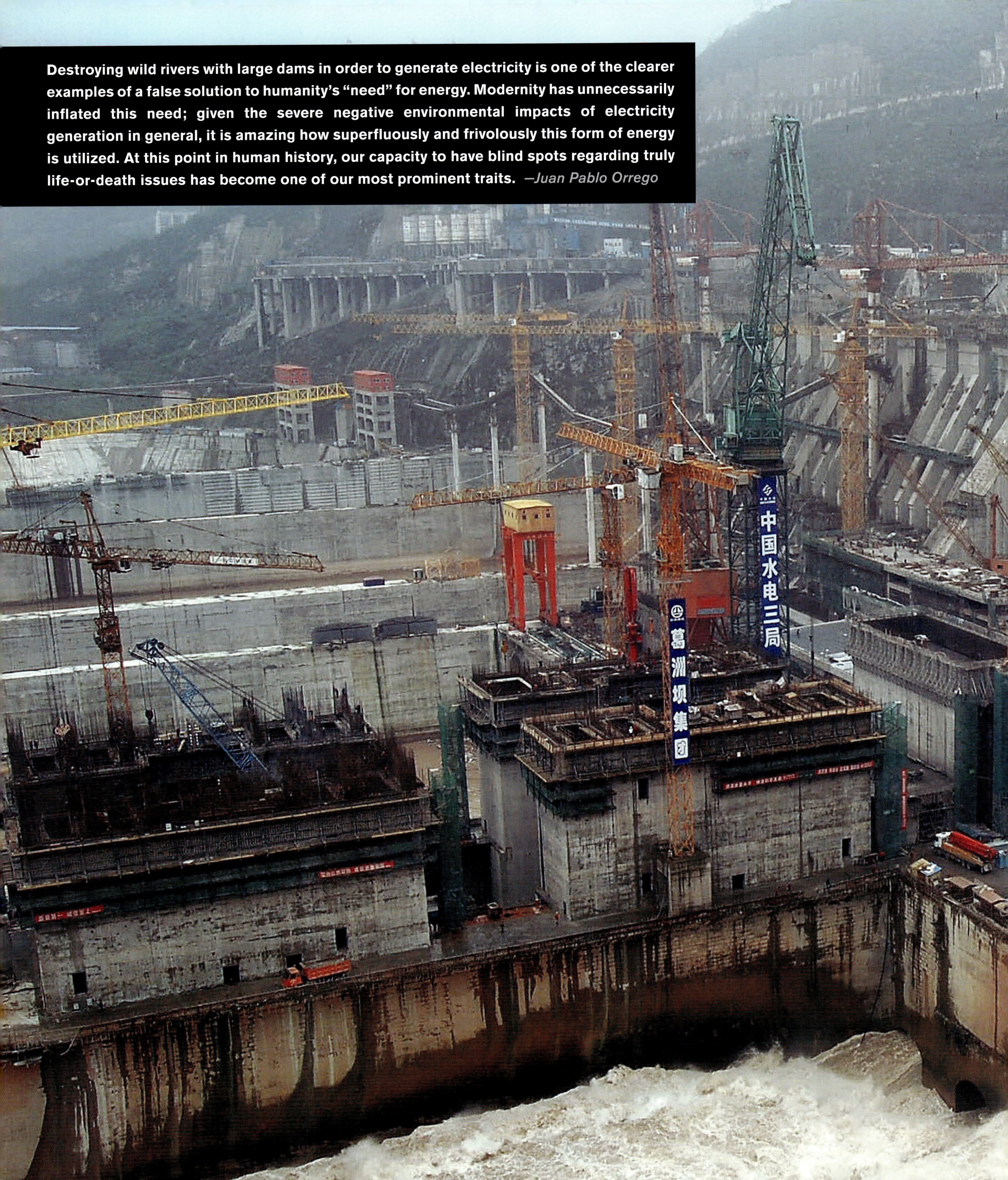

Destroying wild rivers with large dams in order to generate electricity is one of the clearer examples of a false solution to humanity's "need" for energy. Modernity has unnecessarily inflated this need; given the severe negative environmental impacts of electricity generation in general, it is amazing how superfluously and frivolously this form of energy is utilized. At this point in human history, our capacity to have blind spots regarding truly life-or-death issues has become one of our most prominent traits. —Juan Pablo Orrego

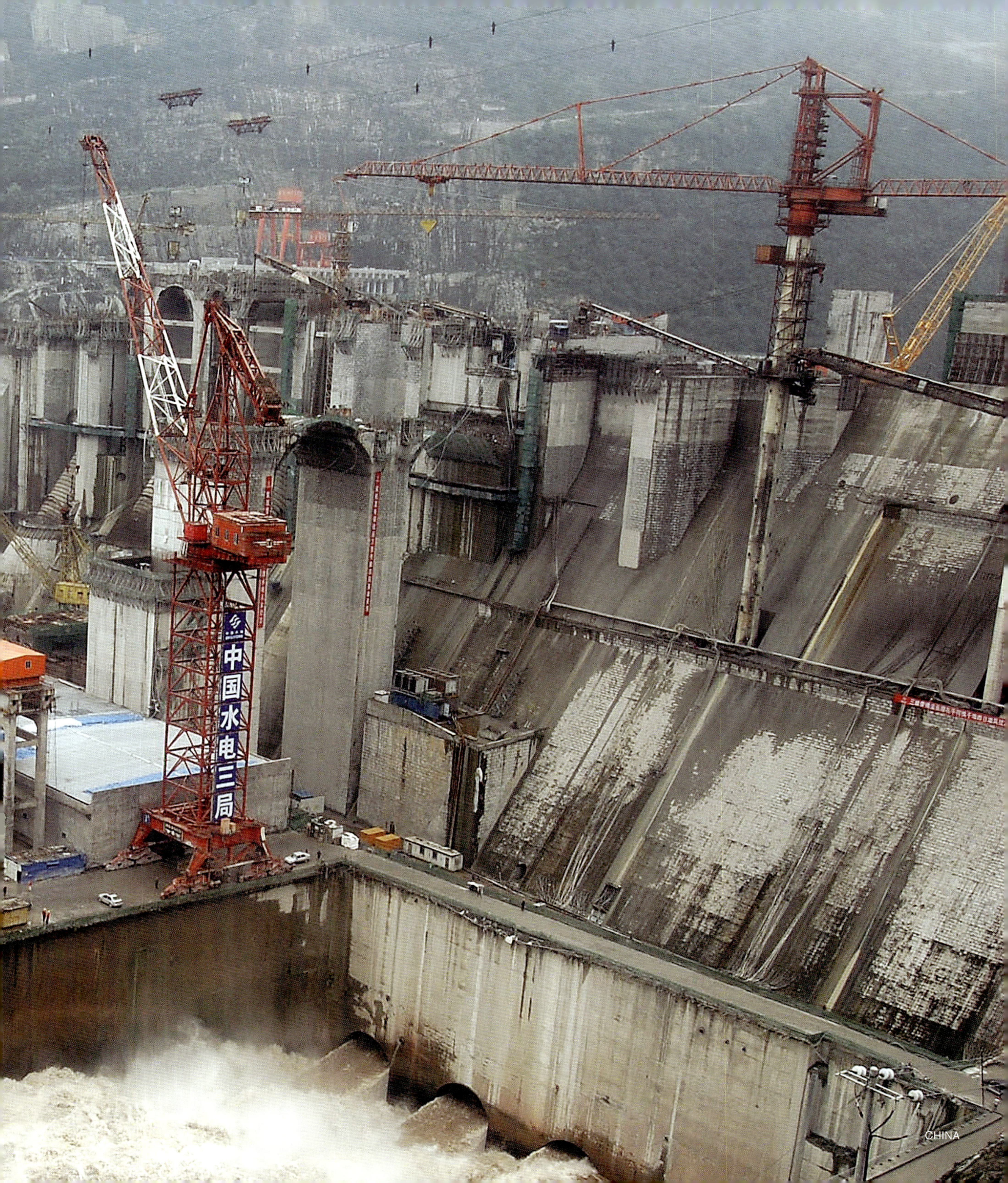

中国水电三局
CHINA

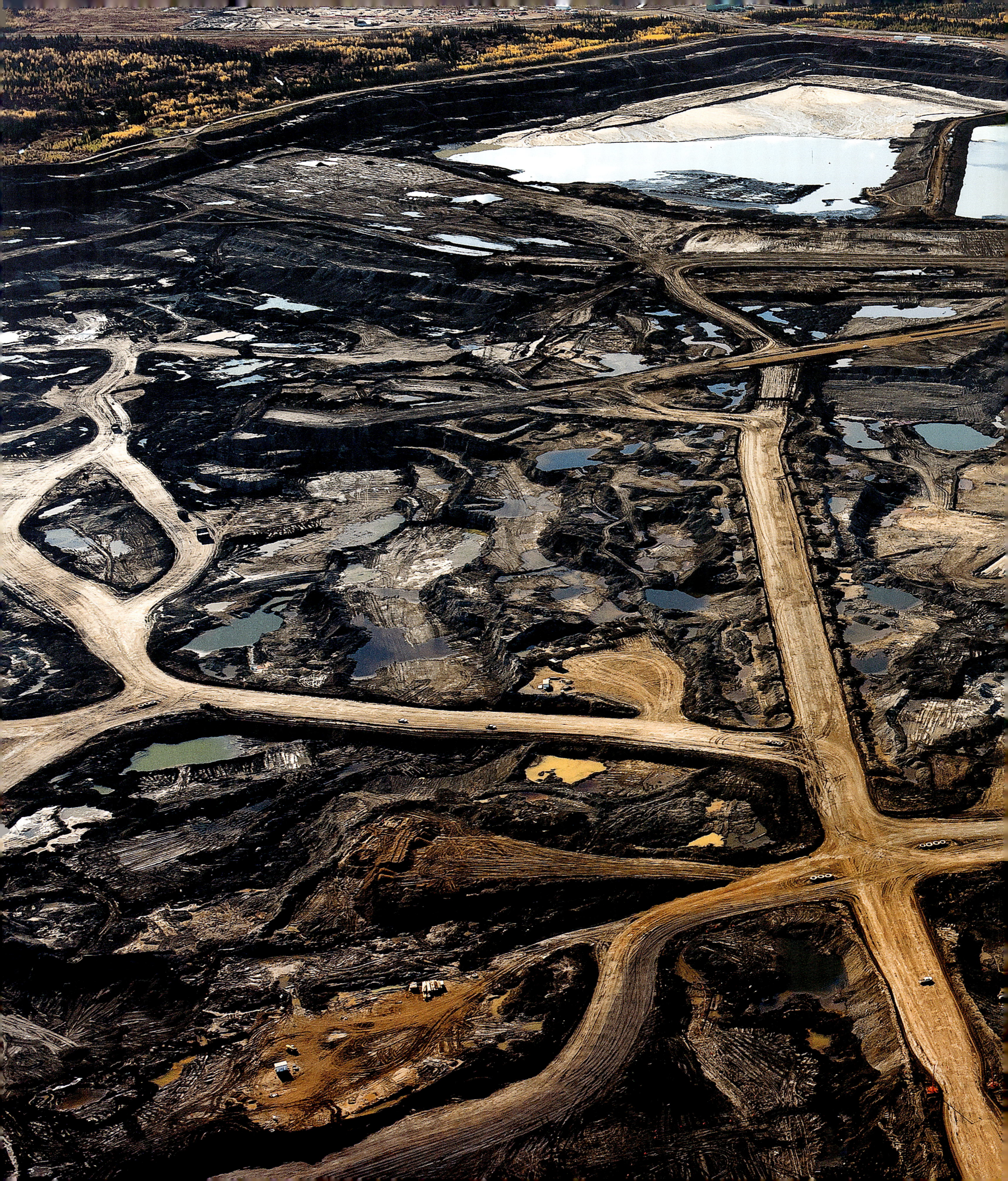

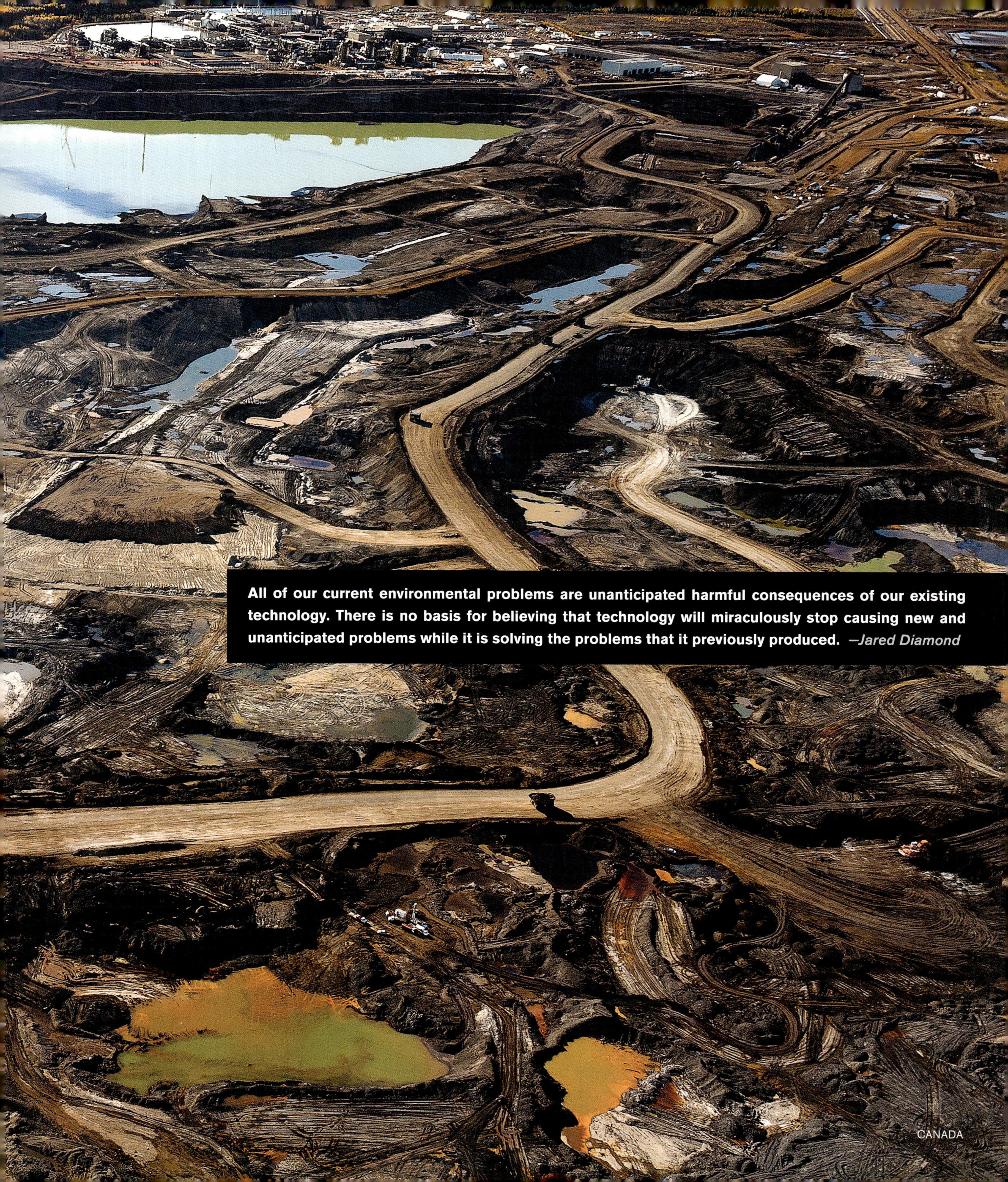

All of our current environmental problems are unanticipated harmful consequences of our existing technology. There is no basis for believing that technology will miraculously stop causing new and unanticipated problems while it is solving the problems that it previously produced. —Jared Diamond
CANADA

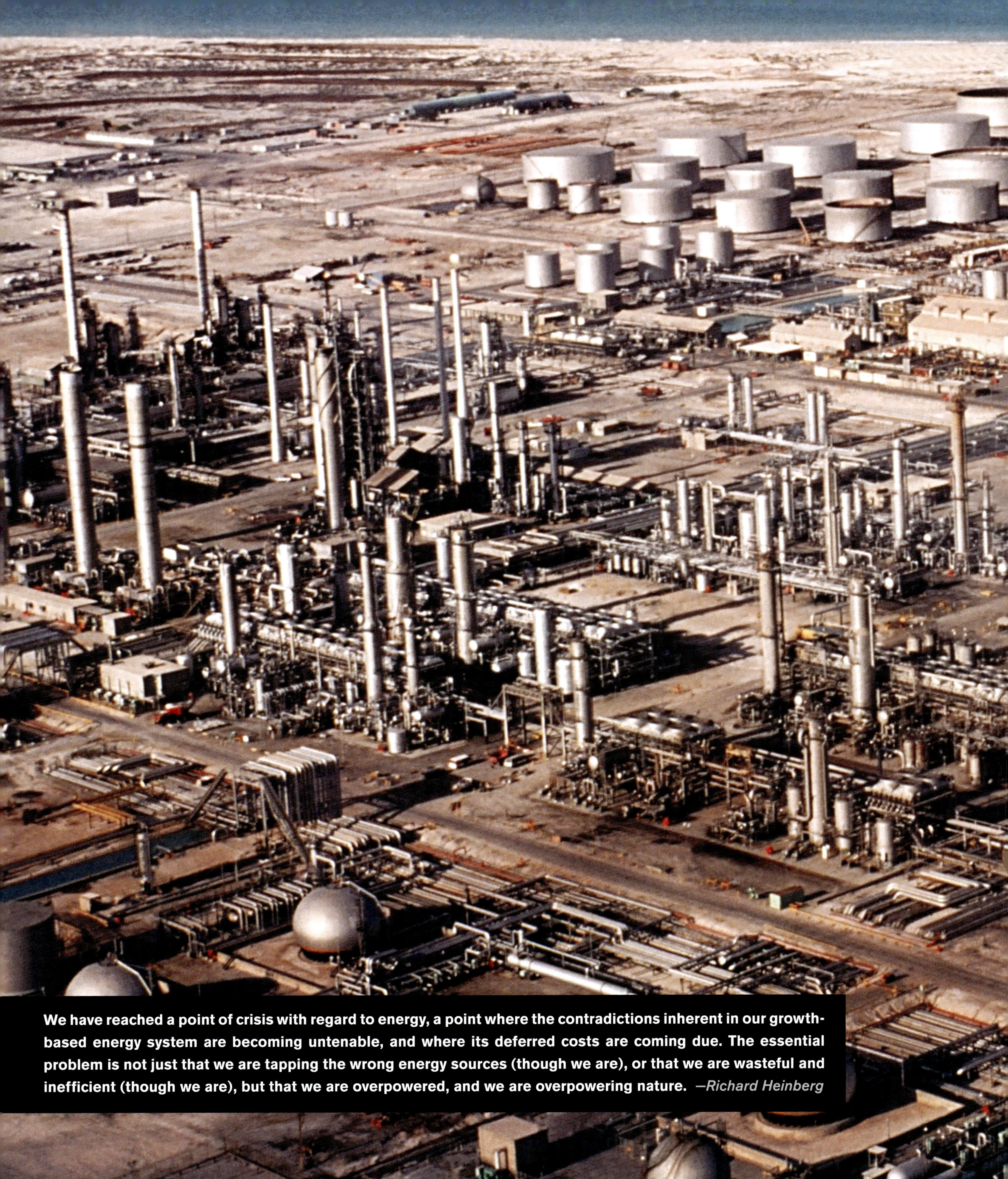

We have reached a point of crisis with regard to energy, a point where the contradictions inherent in our growth-based energy system are becoming untenable, and where its deferred costs are coming due. The essential problem is not just that we are tapping the wrong energy sources (though we are), or that we are wasteful and inefficient (though we are), but that we are overpowered, and we are overpowering nature. —Richard Heinberg

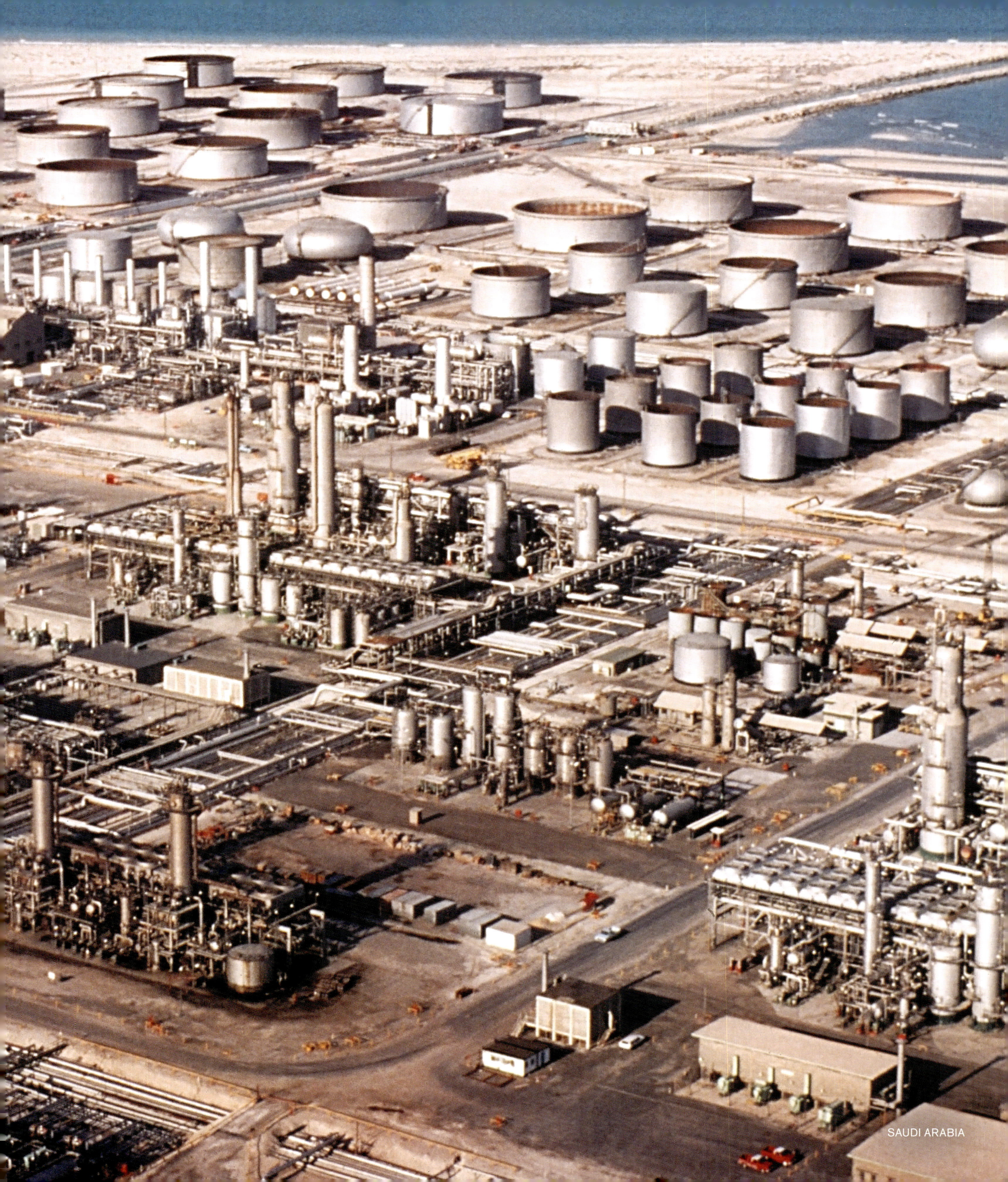
SAUDI ARABIA

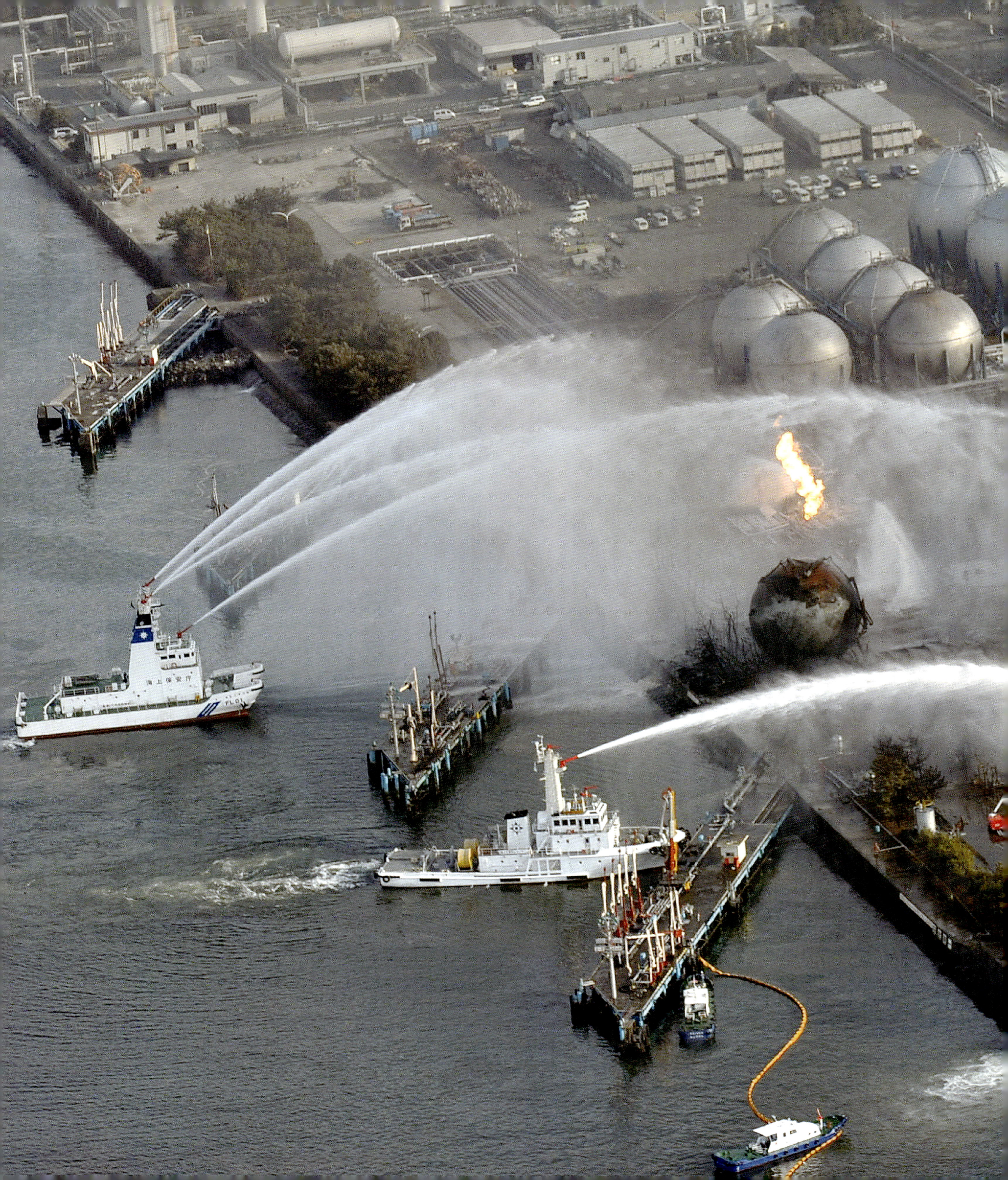

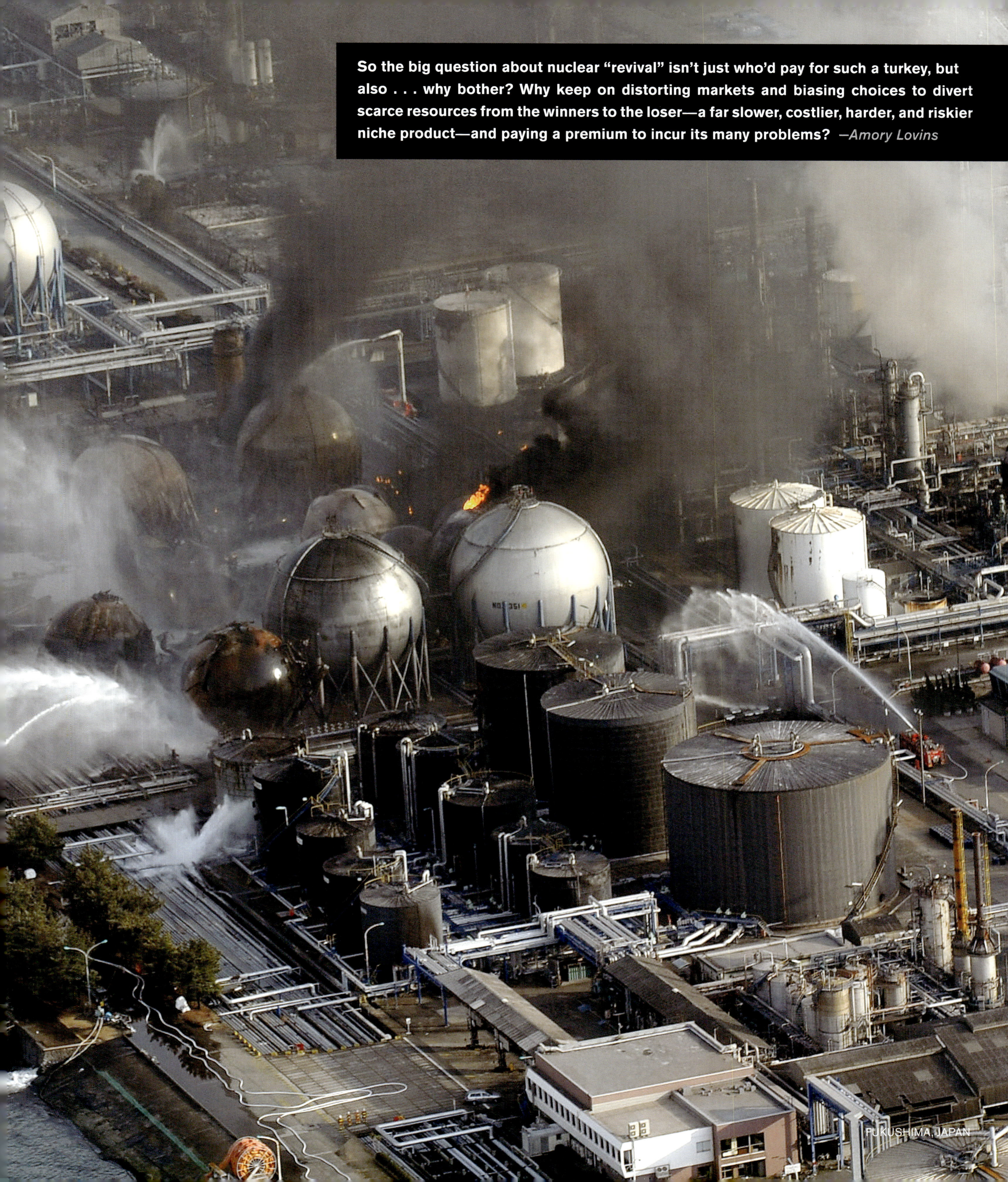

So the big question about nuclear "revival" isn't just who'd pay for such a turkey, but also . . . why bother? Why keep on distorting markets and biasing choices to divert scarce resources from the winners to the loser—a far slower, costlier, harder, and riskier niche product—and paying a premium to incur its many problems? —Amory Lovins
FUKUSHIMA, JAPAN

FOUL WATER

Water is life. Without it, we perish. Clean water and healthy seas are fundamental to the future welfare of humanity, and yet the industrial growth economy wastes copious amounts of water and treats the living oceans as a dumping ground for our effluent. This distain may stem, at least in part, from an old myth—the idea that the seas were limitless and their abundance immune to the efforts of people. That may have been true when we were few and our tools were simple. It is not true today when we are overabundant and no part of the ocean is out of reach of industrial fishers. The aggregate demands of a bloated humanity on marine and freshwater systems leave less and less room for nature and put billions of people at risk of having inadequate drinking water.

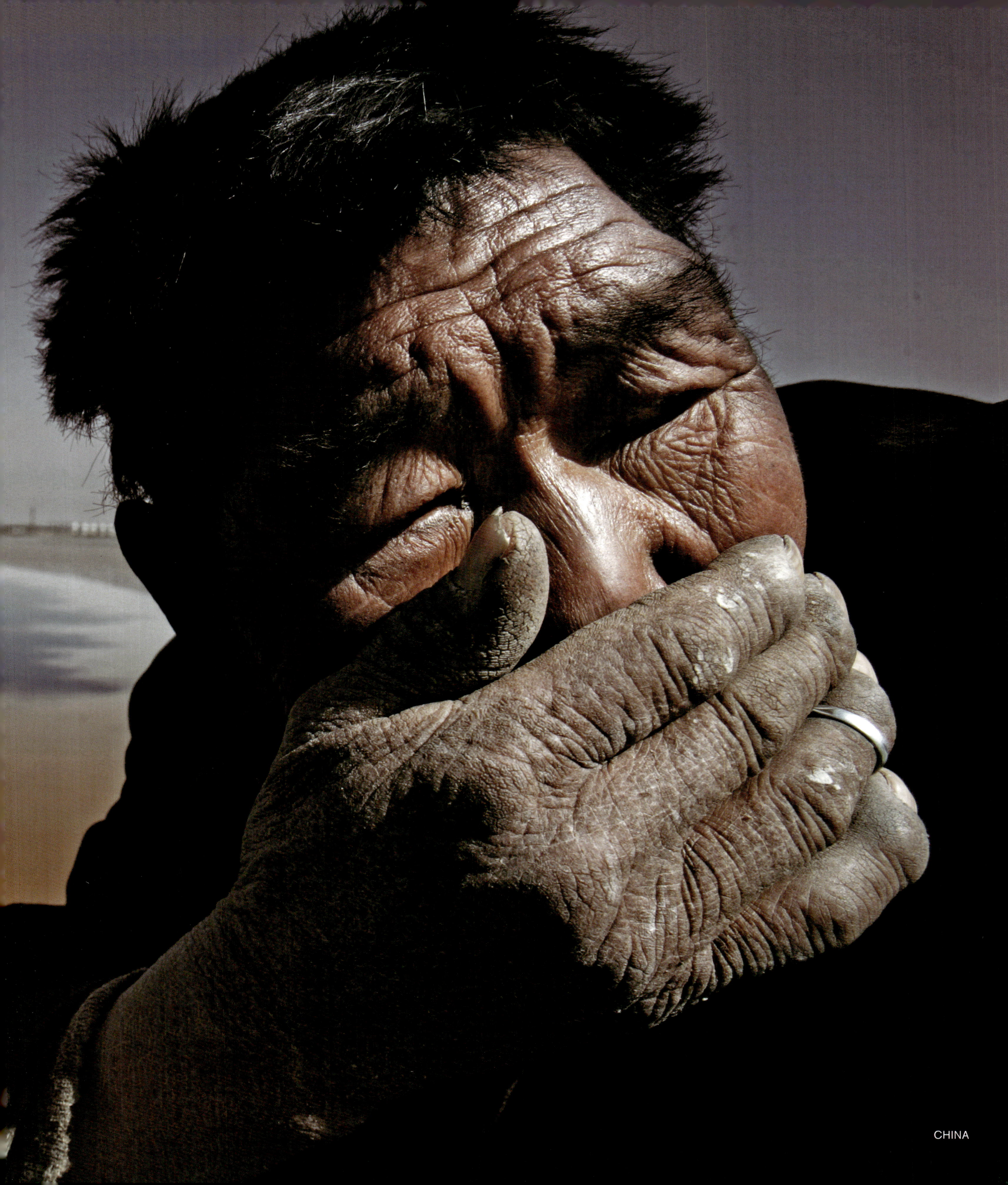

CHINA

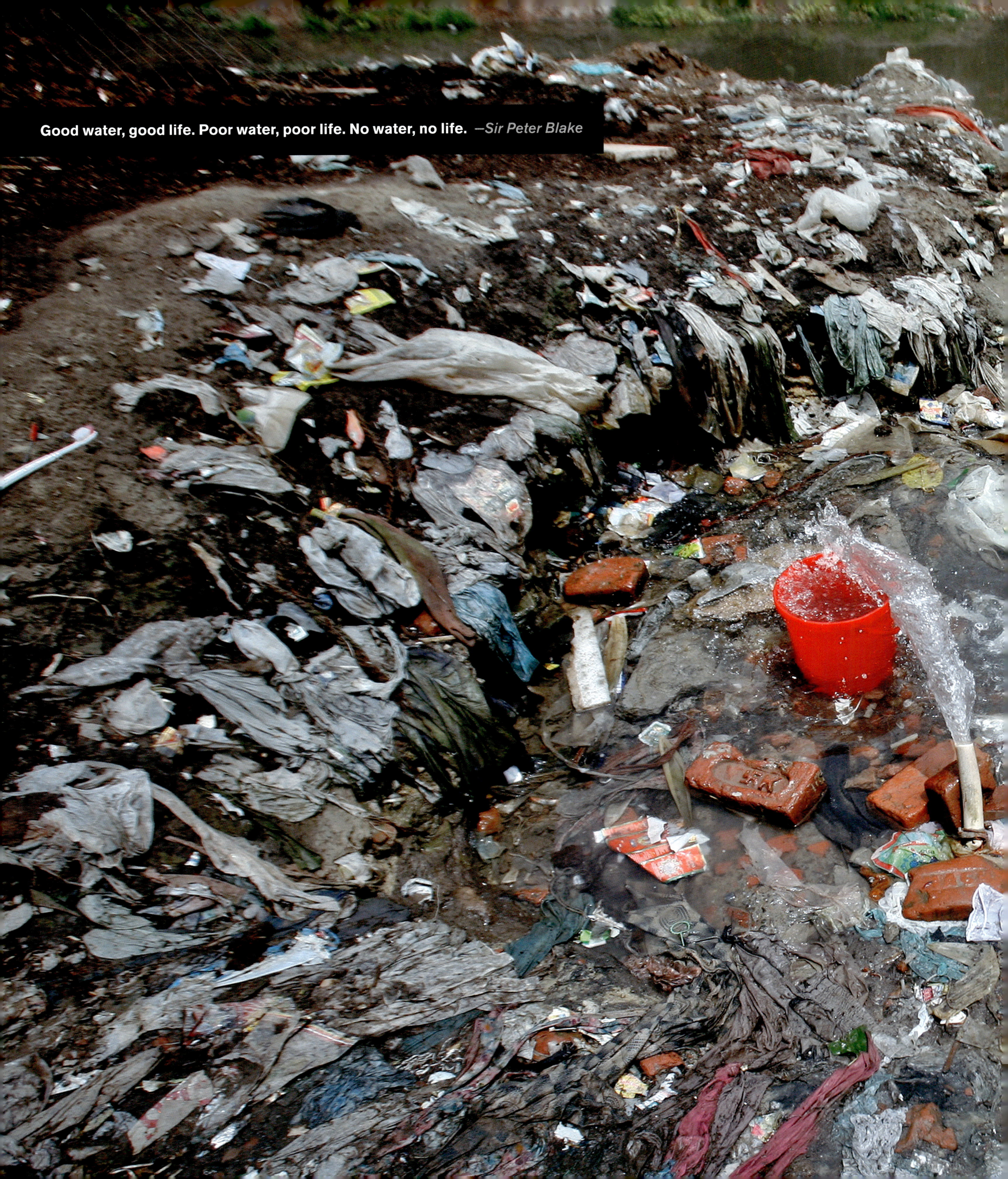
Good water, good life. Poor water, poor life. No water, no life. —Sir Peter Blake

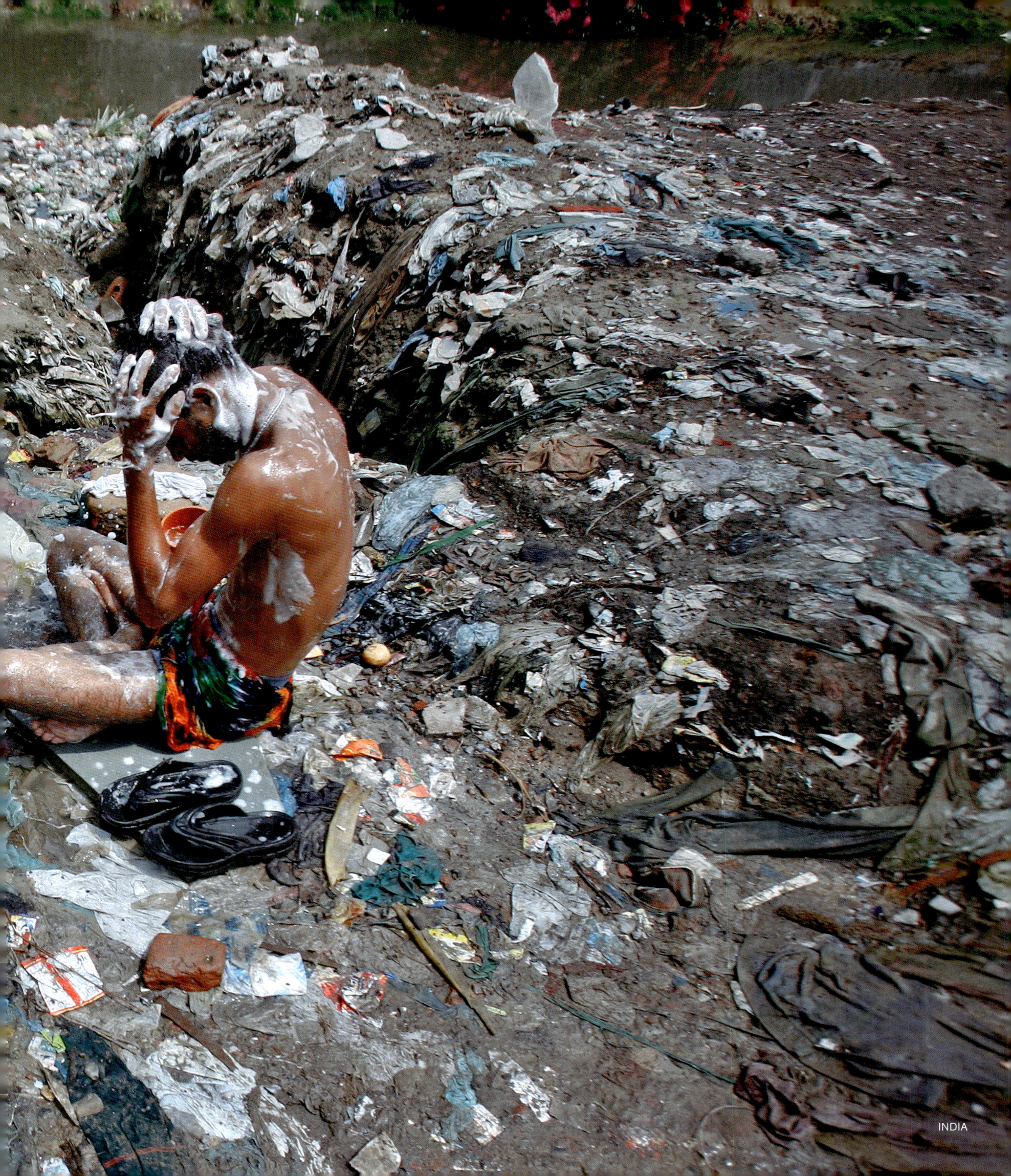
INDIA

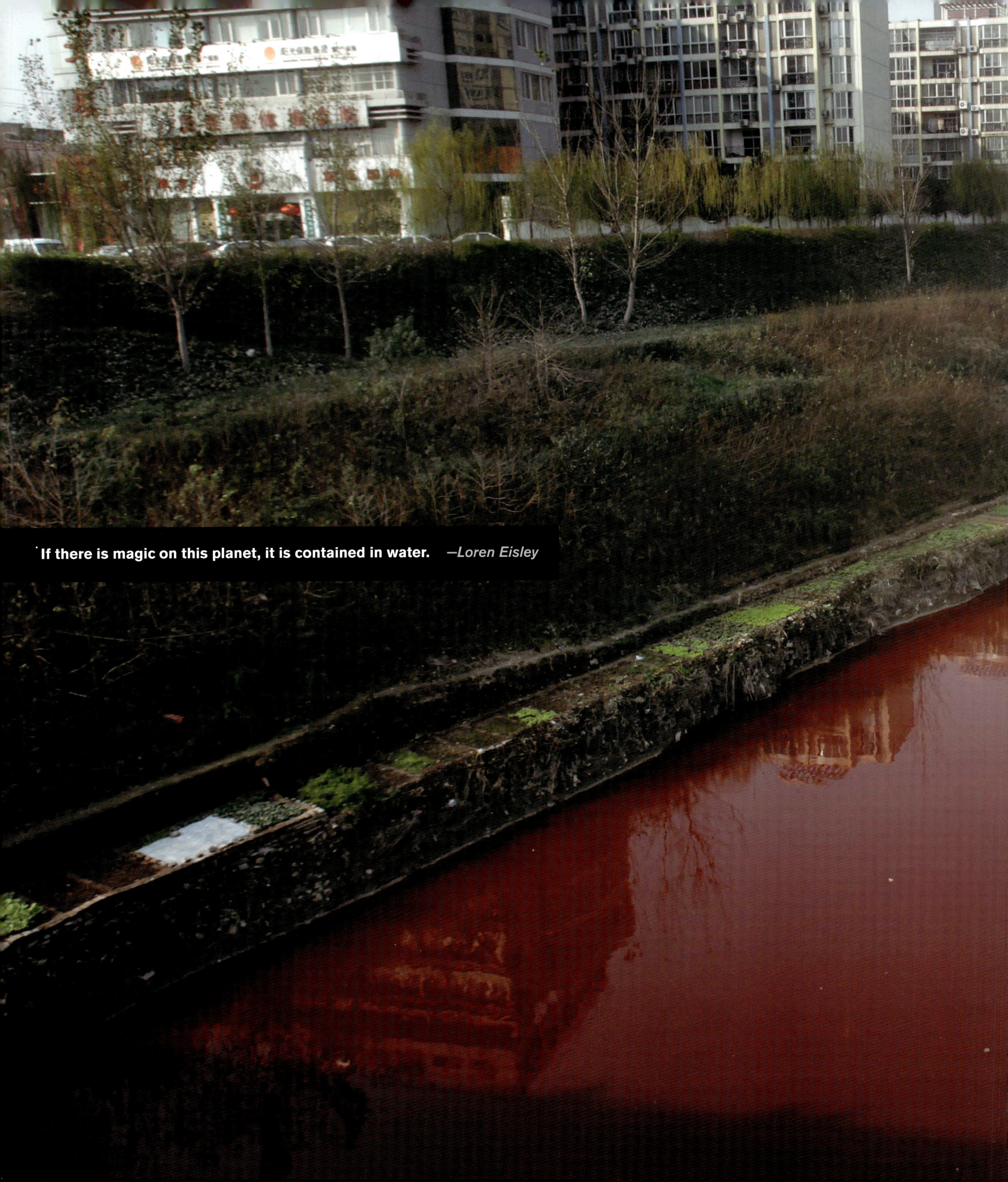

If there is magic on this planet, it is contained in water.   —Loren Eisley

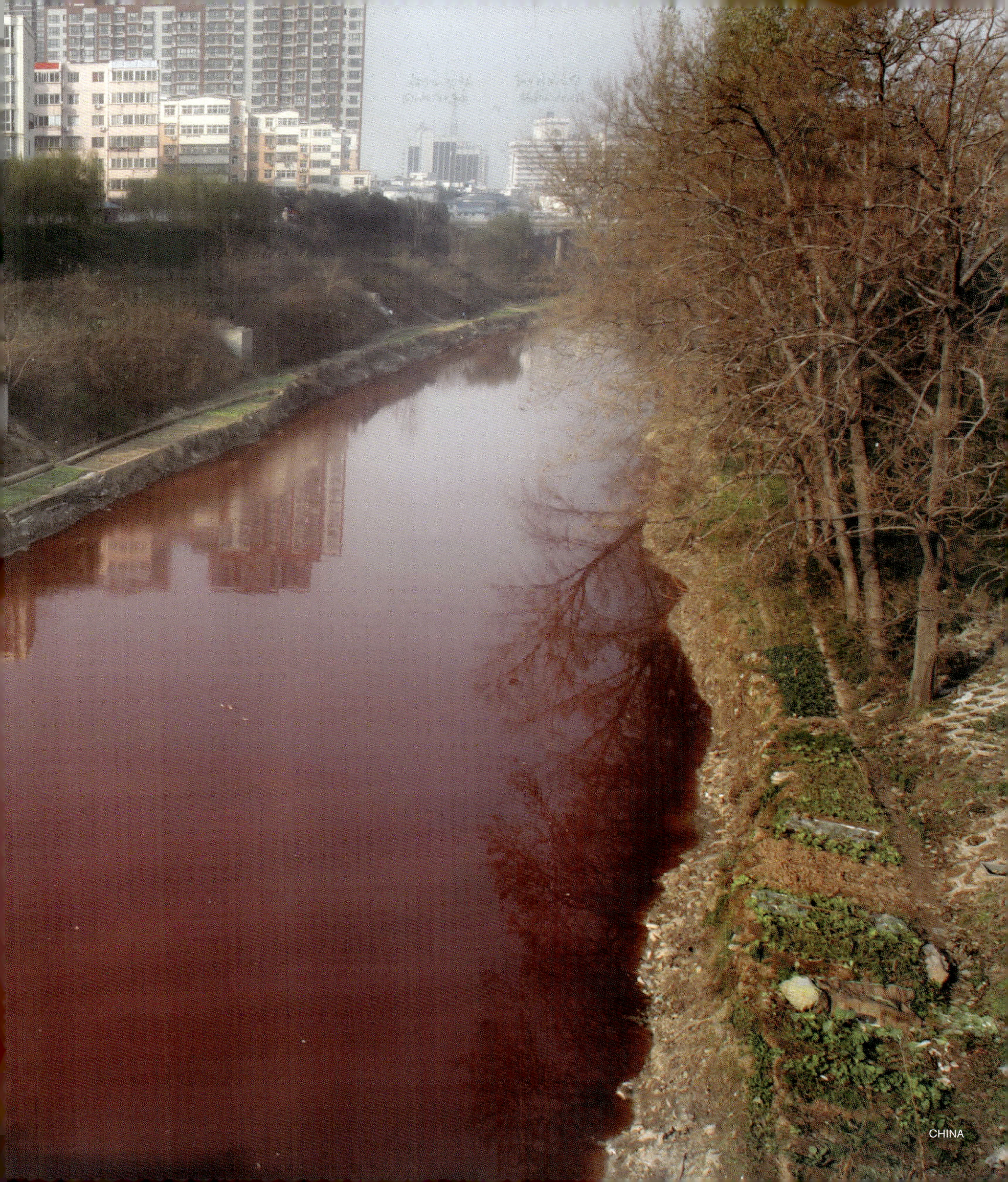
CHINA

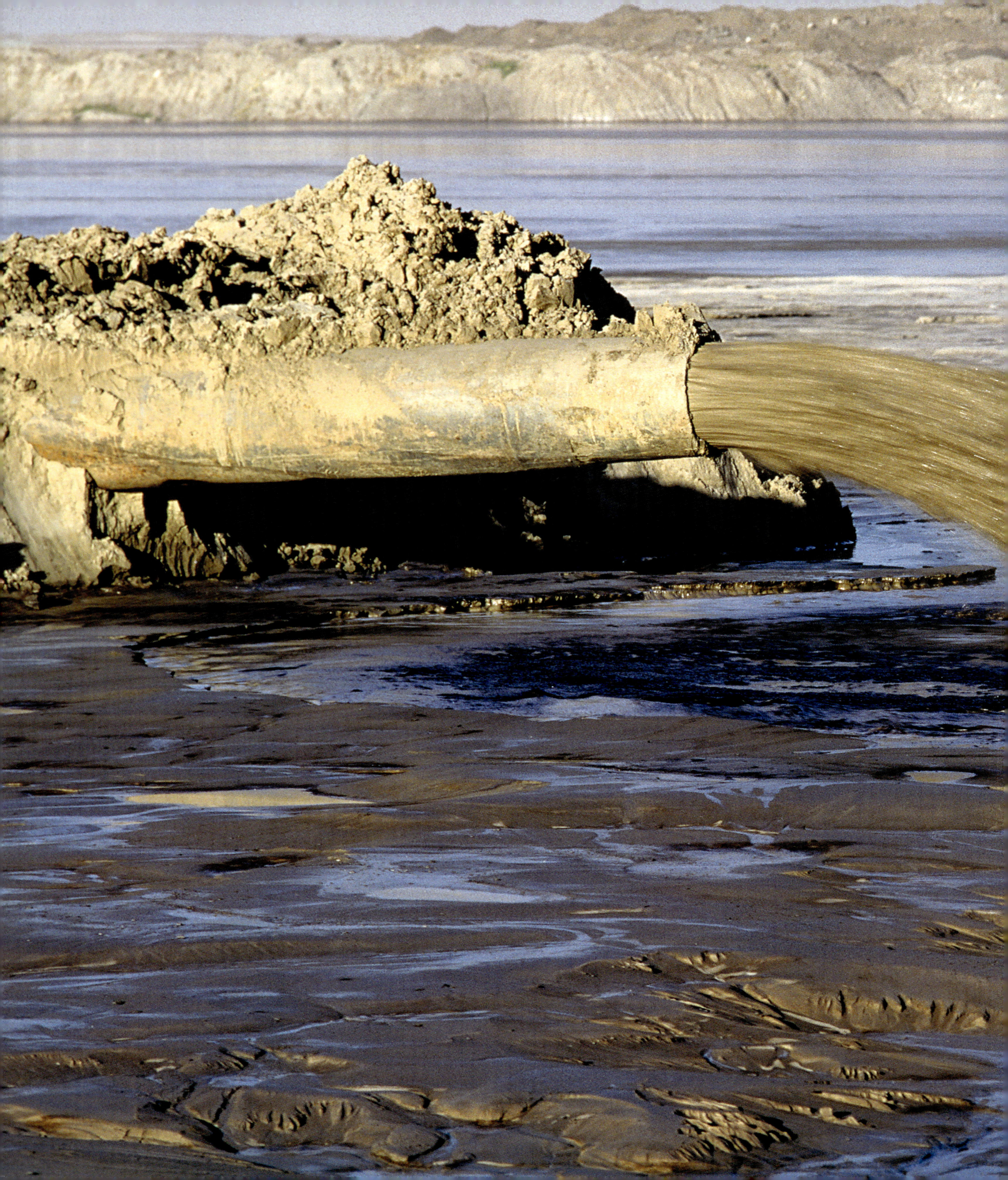

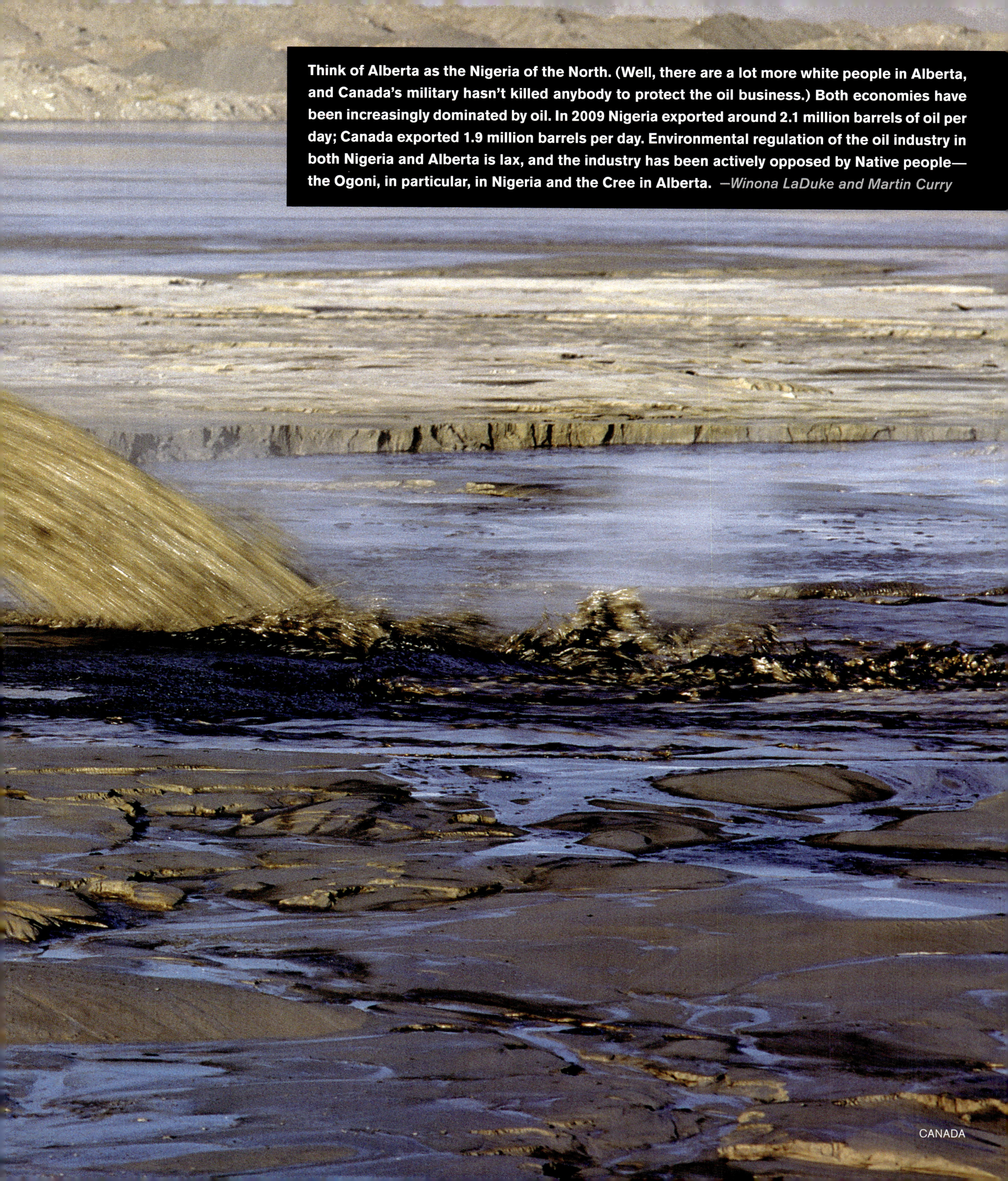

Think of Alberta as the Nigeria of the North. (Well, there are a lot more white people in Alberta, and Canada's military hasn't killed anybody to protect the oil business.) Both economies have been increasingly dominated by oil. In 2009 Nigeria exported around 2.1 million barrels of oil per day; Canada exported 1.9 million barrels per day. Environmental regulation of the oil industry in both Nigeria and Alberta is lax, and the industry has been actively opposed by Native people—the Ogoni, in particular, in Nigeria and the Cree in Alberta. —*Winona LaDuke and Martin Curry*
CANADA

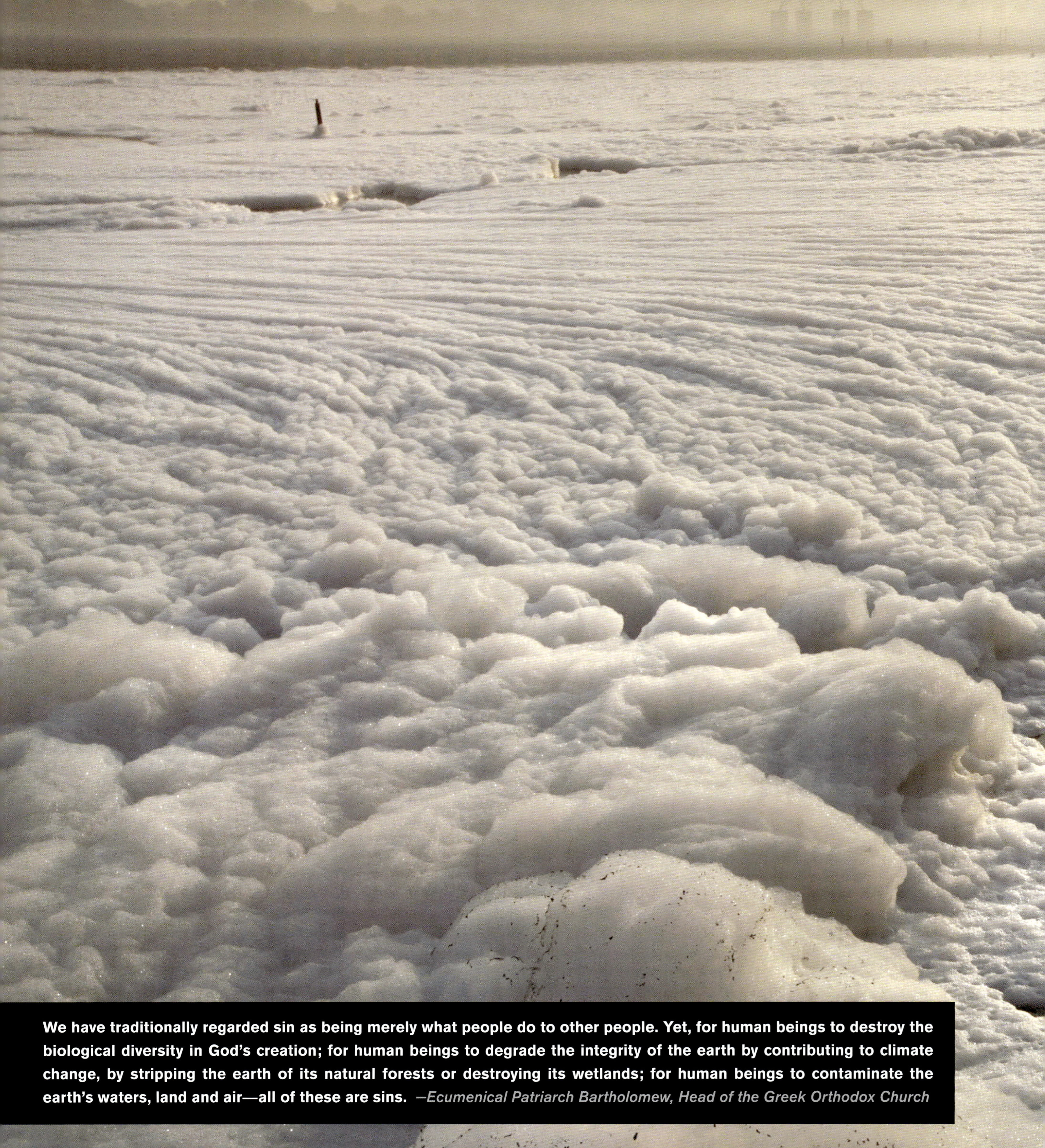

We have traditionally regarded sin as being merely what people do to other people. Yet, for human beings to destroy the biological diversity in God's creation; for human beings to degrade the integrity of the earth by contributing to climate change, by stripping the earth of its natural forests or destroying its wetlands; for human beings to contaminate the earth's waters, land and air—all of these are sins. —Ecumenical Patriarch Bartholomew, Head of the Greek Orthodox Church

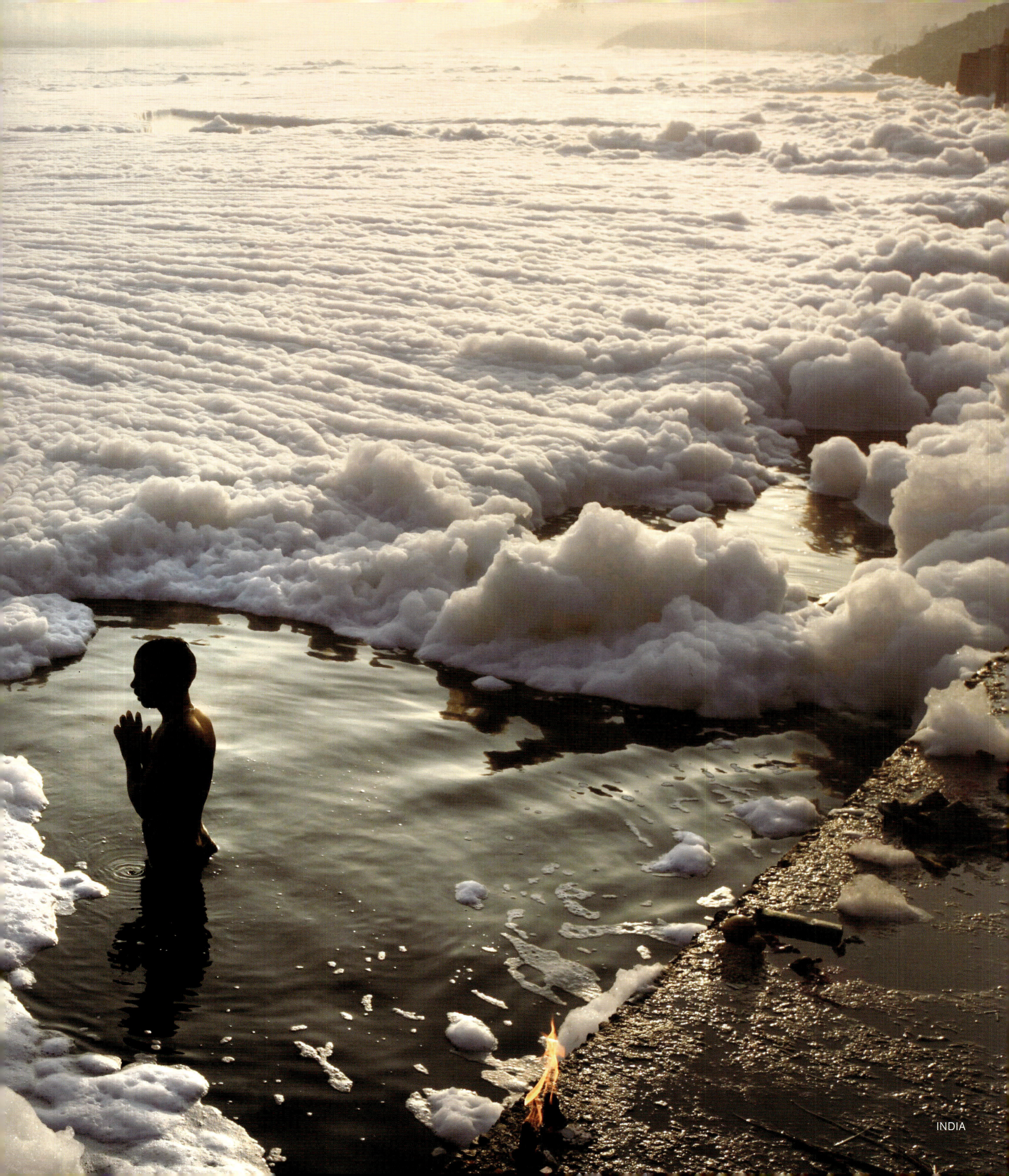
INDIA

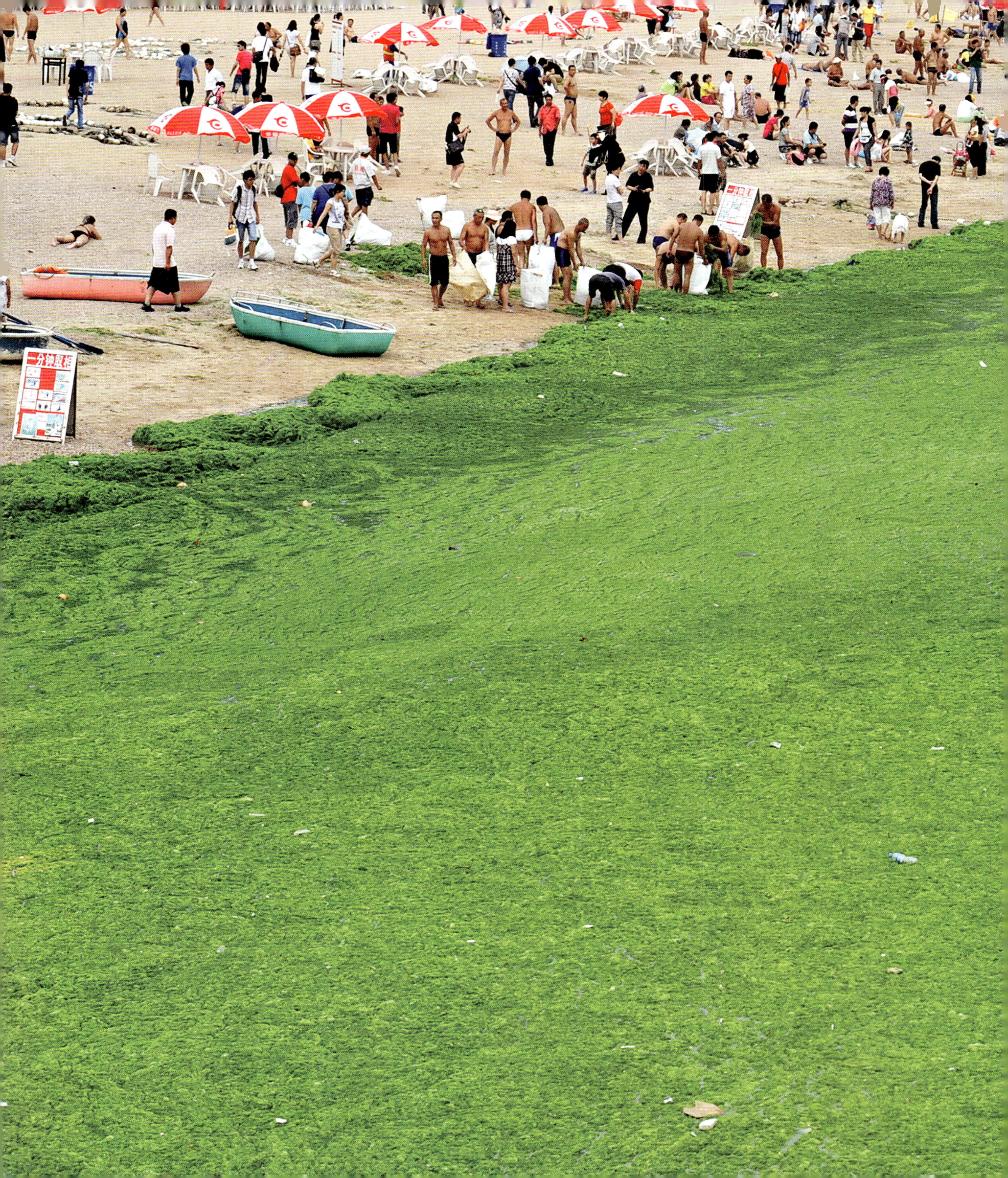

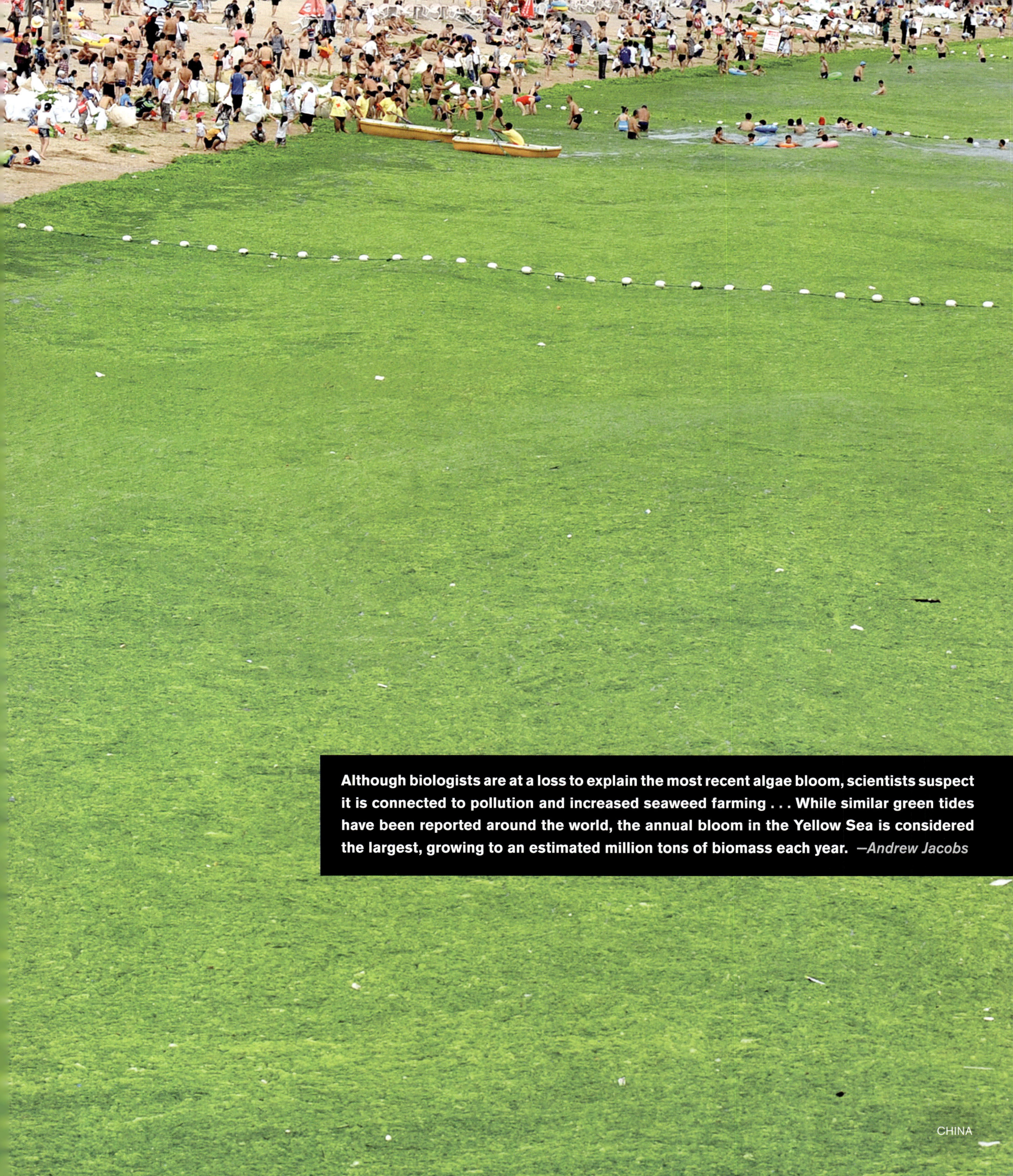

Although biologists are at a loss to explain the most recent algae bloom, scientists suspect it is connected to pollution and increased seaweed farming . . . While similar green tides have been reported around the world, the annual bloom in the Yellow Sea is considered the largest, growing to an estimated million tons of biomass each year. —Andrew Jacobs

CHINA

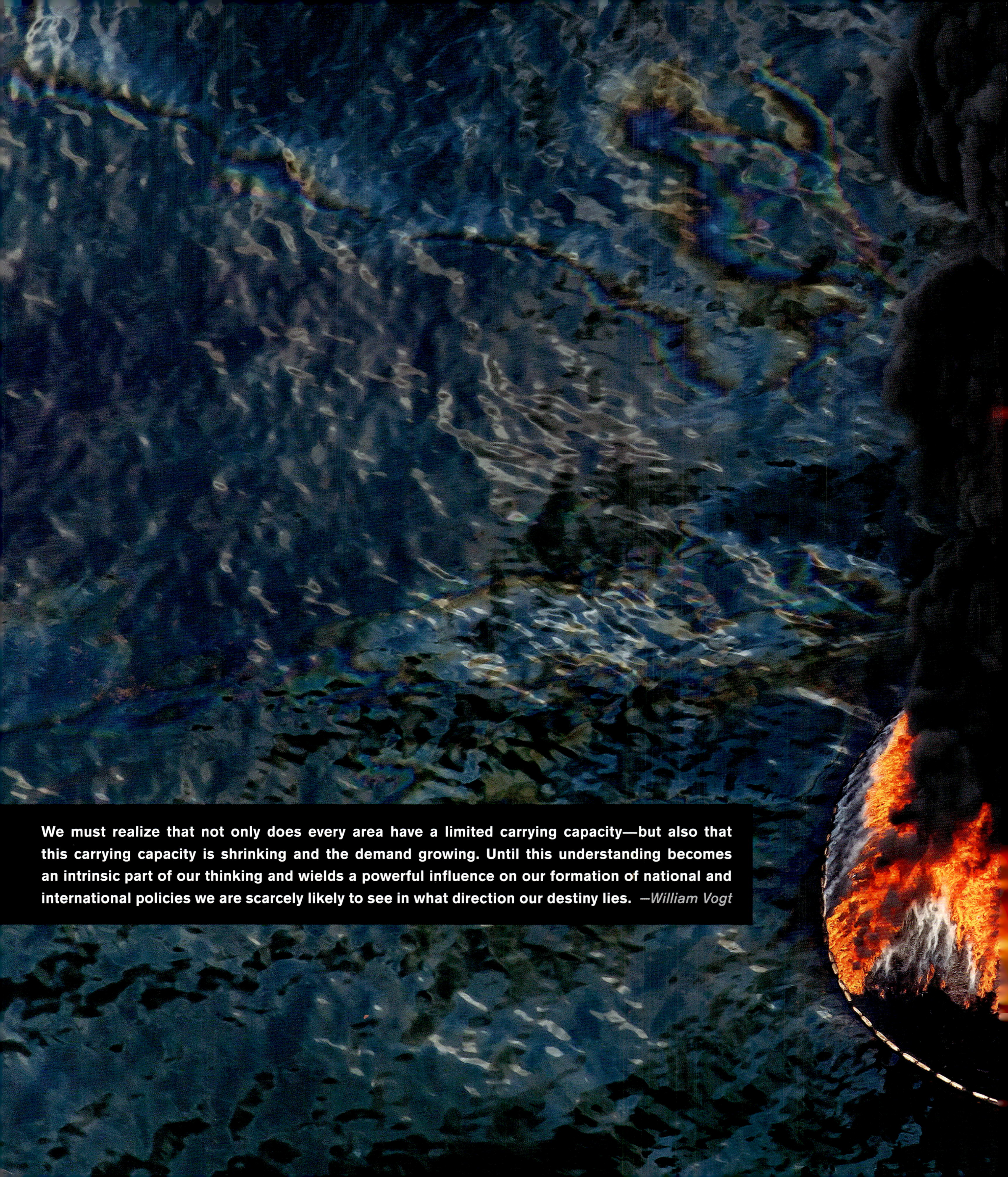

We must realize that not only does every area have a limited carrying capacity—but also that this carrying capacity is shrinking and the demand growing. Until this understanding becomes an intrinsic part of our thinking and wields a powerful influence on our formation of national and international policies we are scarcely likely to see in what direction our destiny lies. —William Vogt

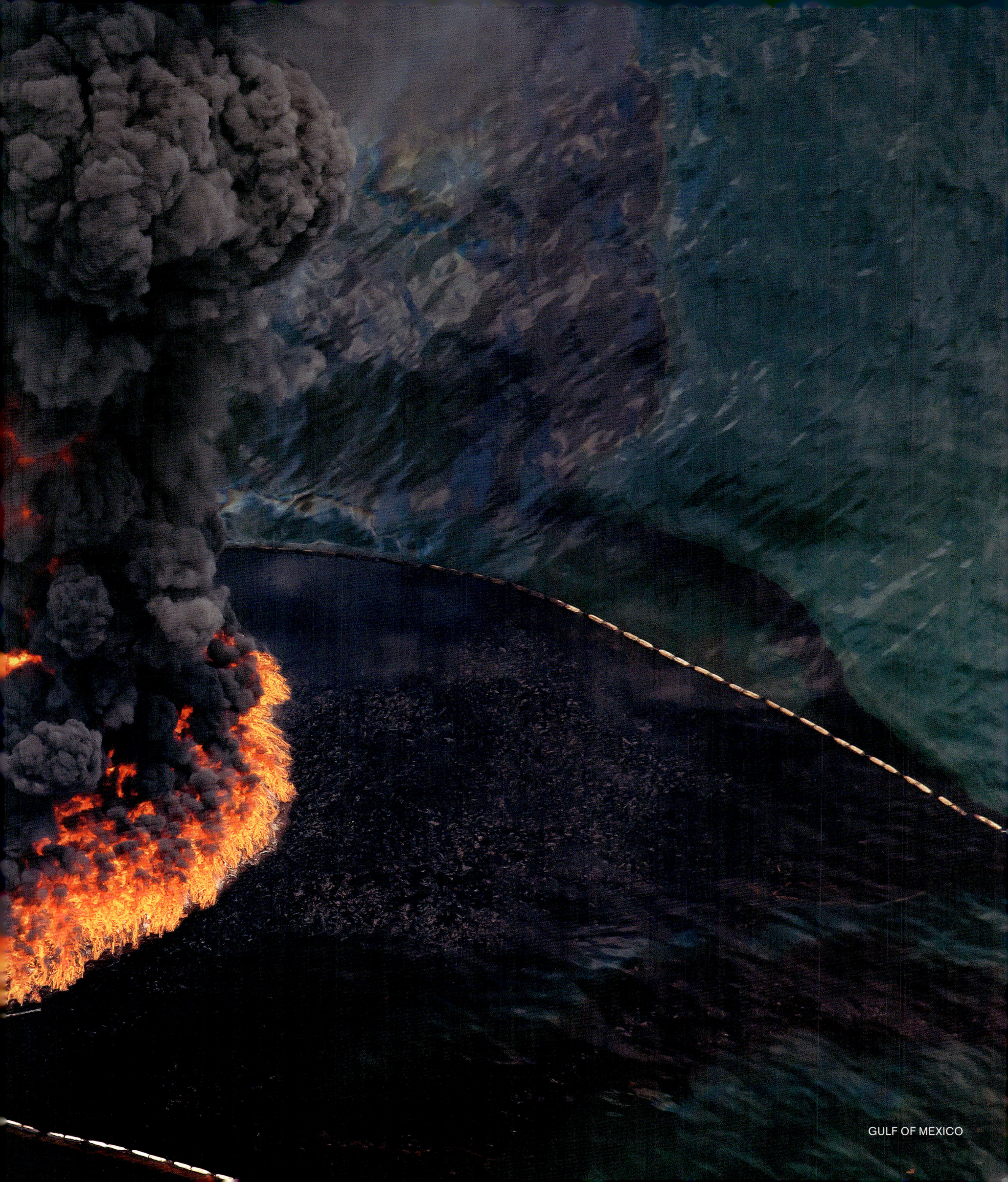
GULF OF MEXICO

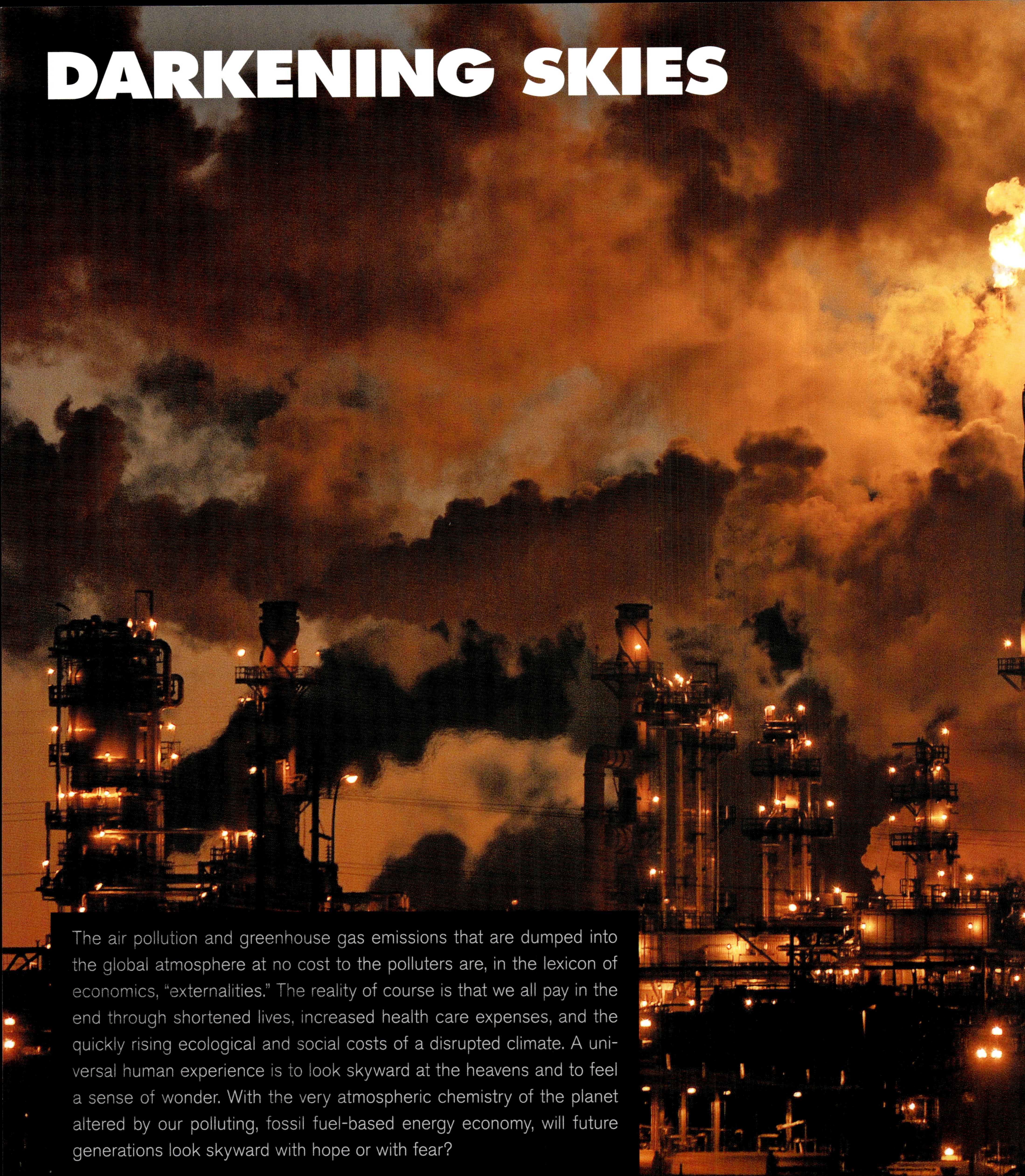

DARKENING SKIES

The air pollution and greenhouse gas emissions that are dumped into the global atmosphere at no cost to the polluters are, in the lexicon of economics, "externalities." The reality of course is that we all pay in the end through shortened lives, increased health care expenses, and the quickly rising ecological and social costs of a disrupted climate. A universal human experience is to look skyward at the heavens and to feel a sense of wonder. With the very atmospheric chemistry of the planet altered by our polluting, fossil fuel-based energy economy, will future generations look skyward with hope or with fear?

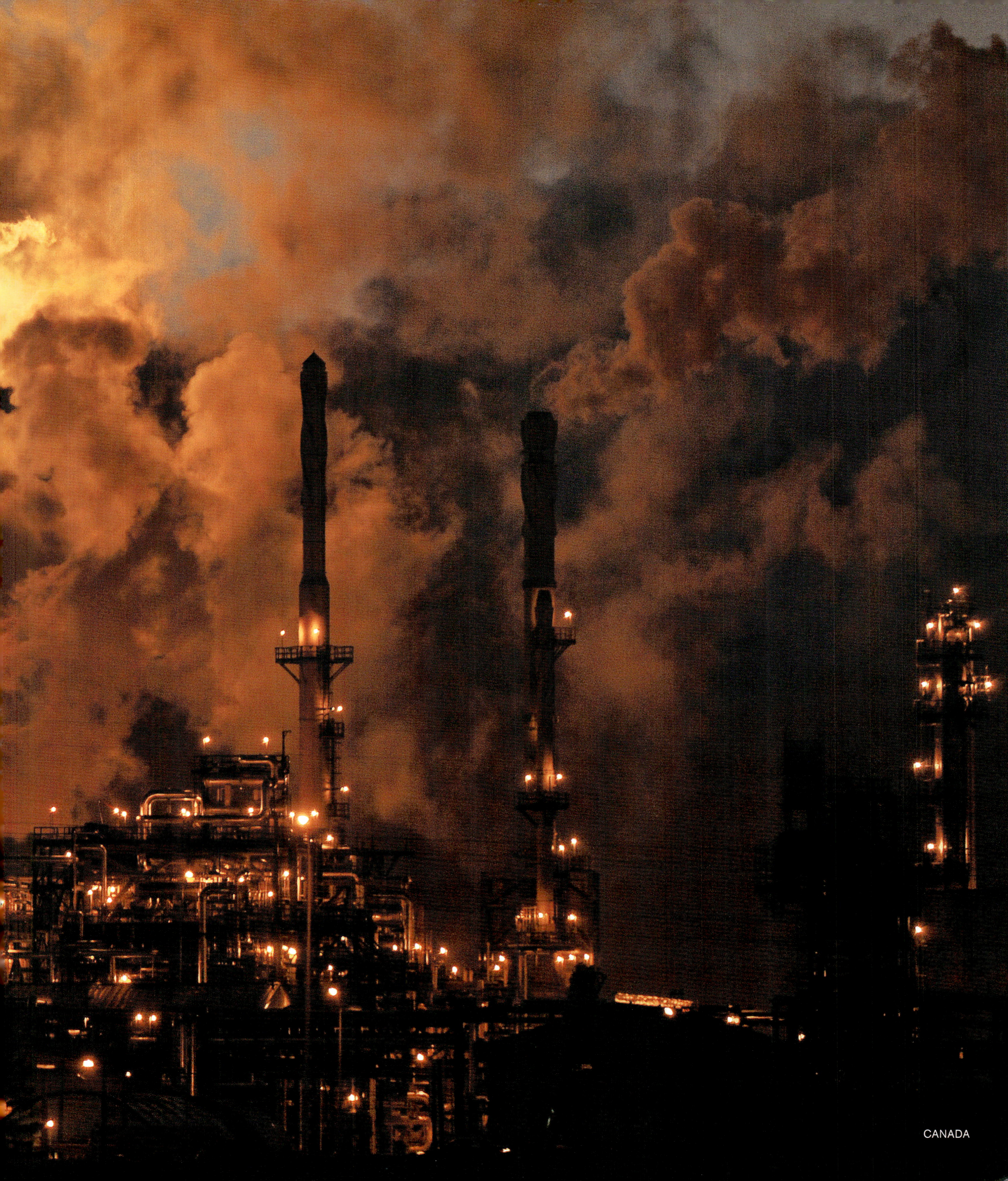
CANADA

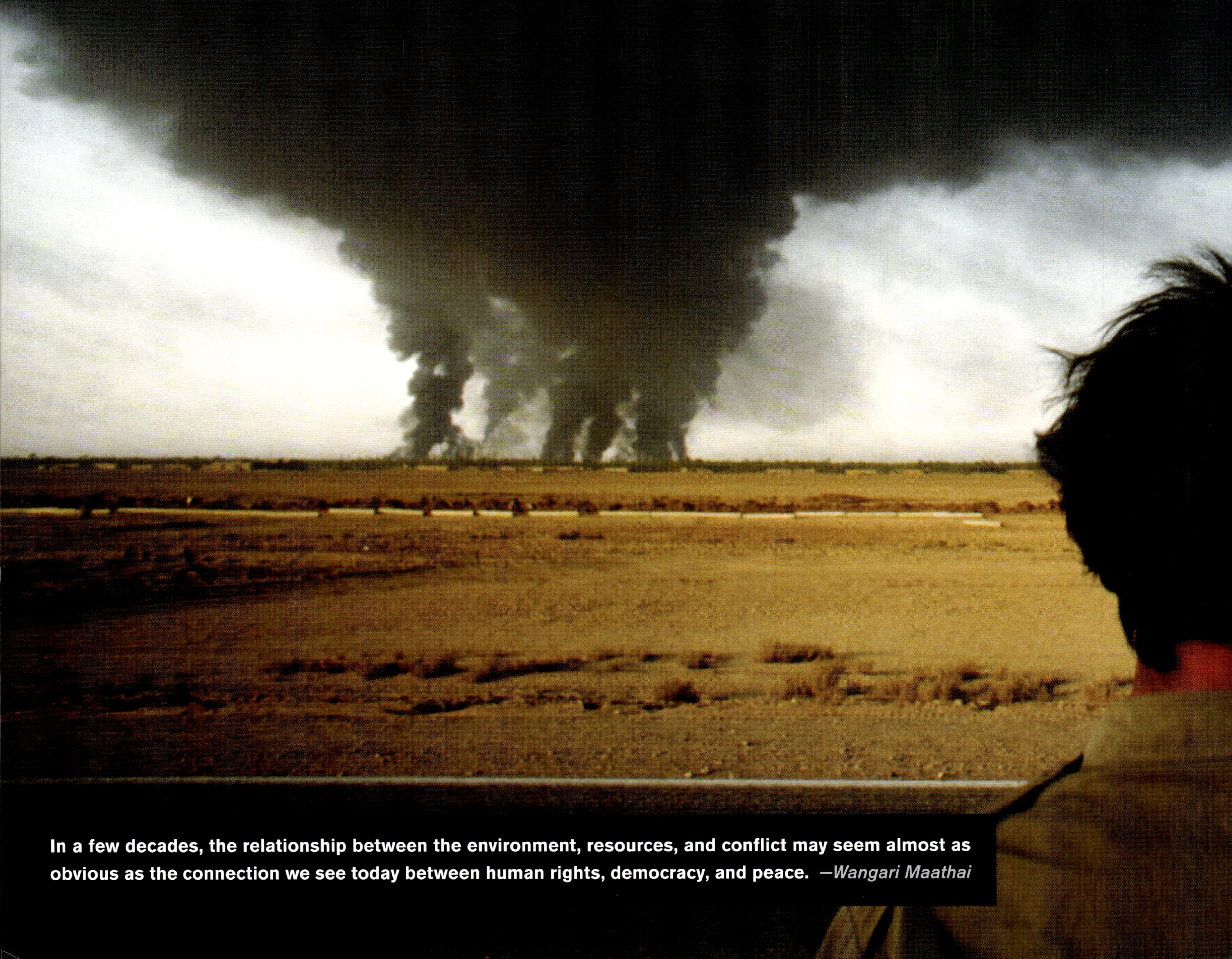
In a few decades, the relationship between the environment, resources, and conflict may seem almost as obvious as the connection we see today between human rights, democracy, and peace. —Wangari Maathai

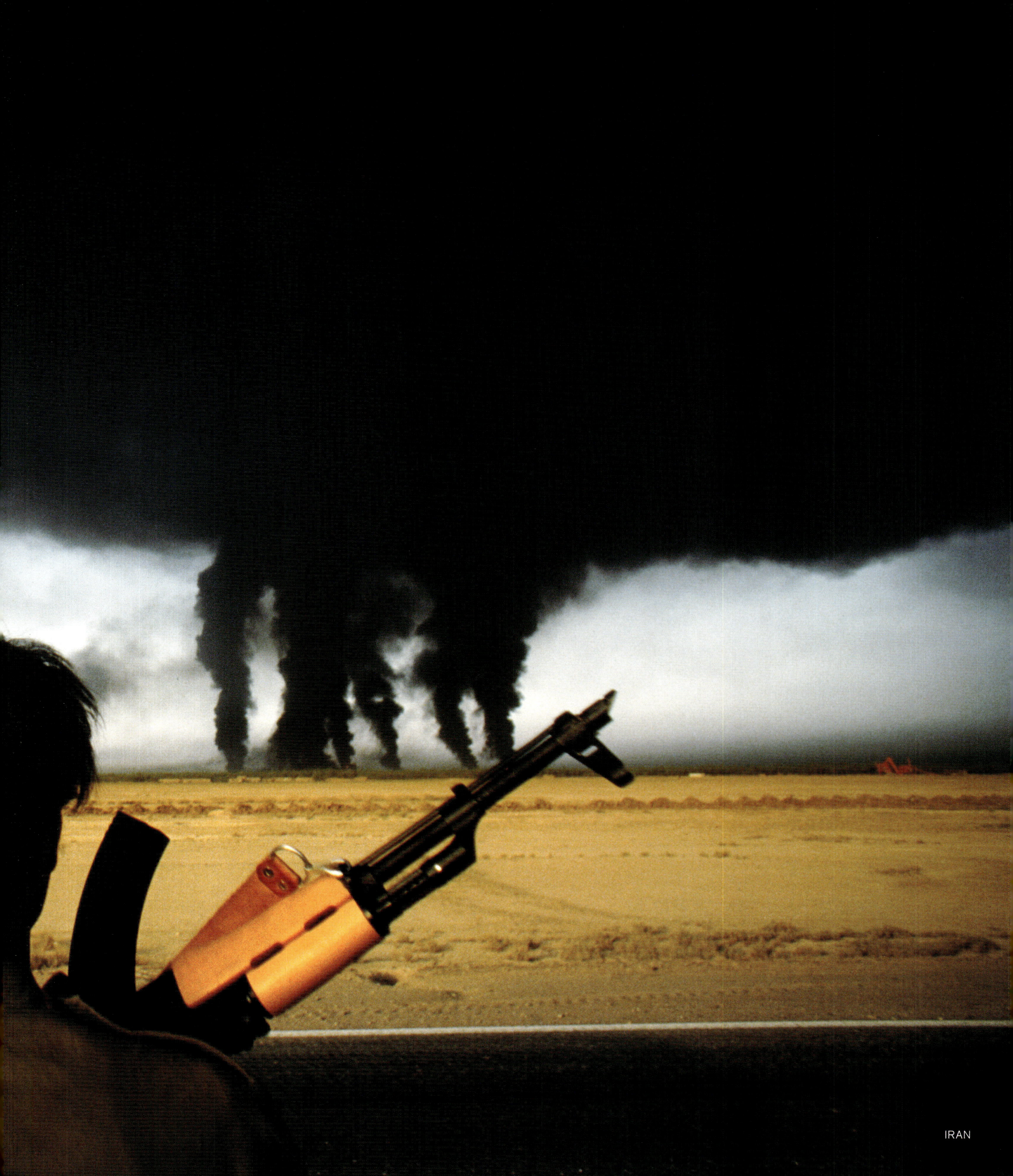
IRAN

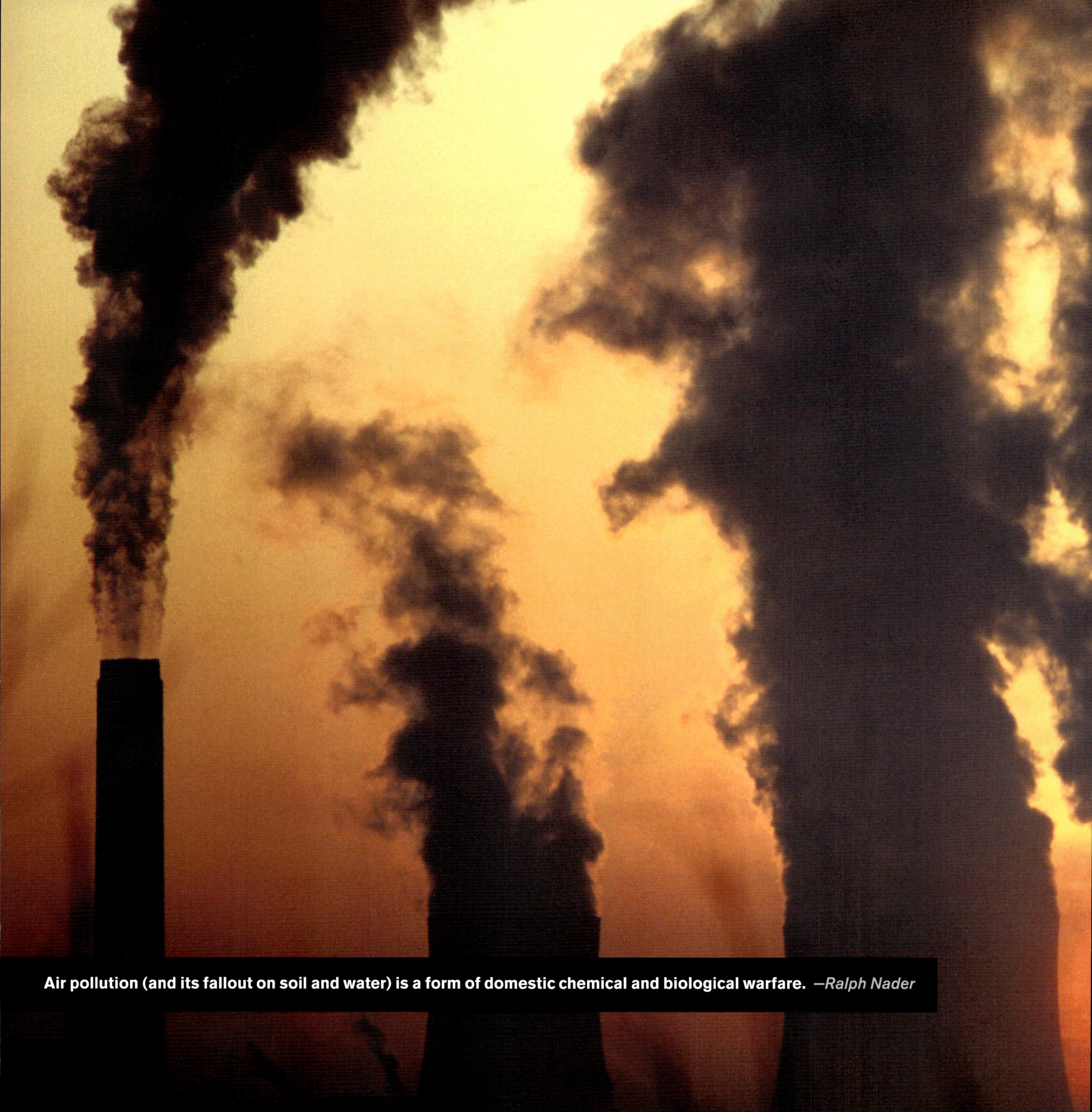
Air pollution (and its fallout on soil and water) is a form of domestic chemical and biological warfare. —Ralph Nader

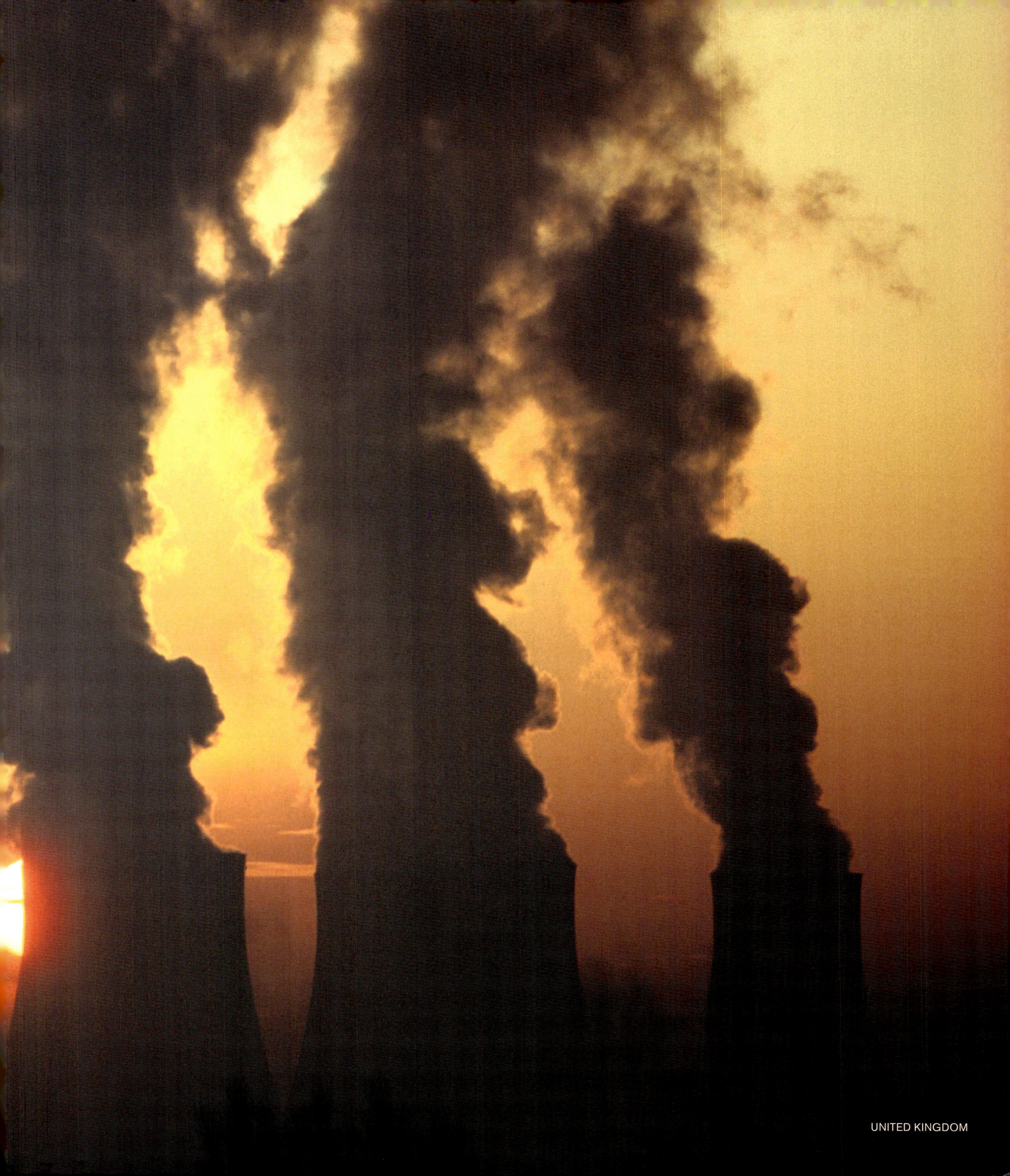
UNITED KINGDOM

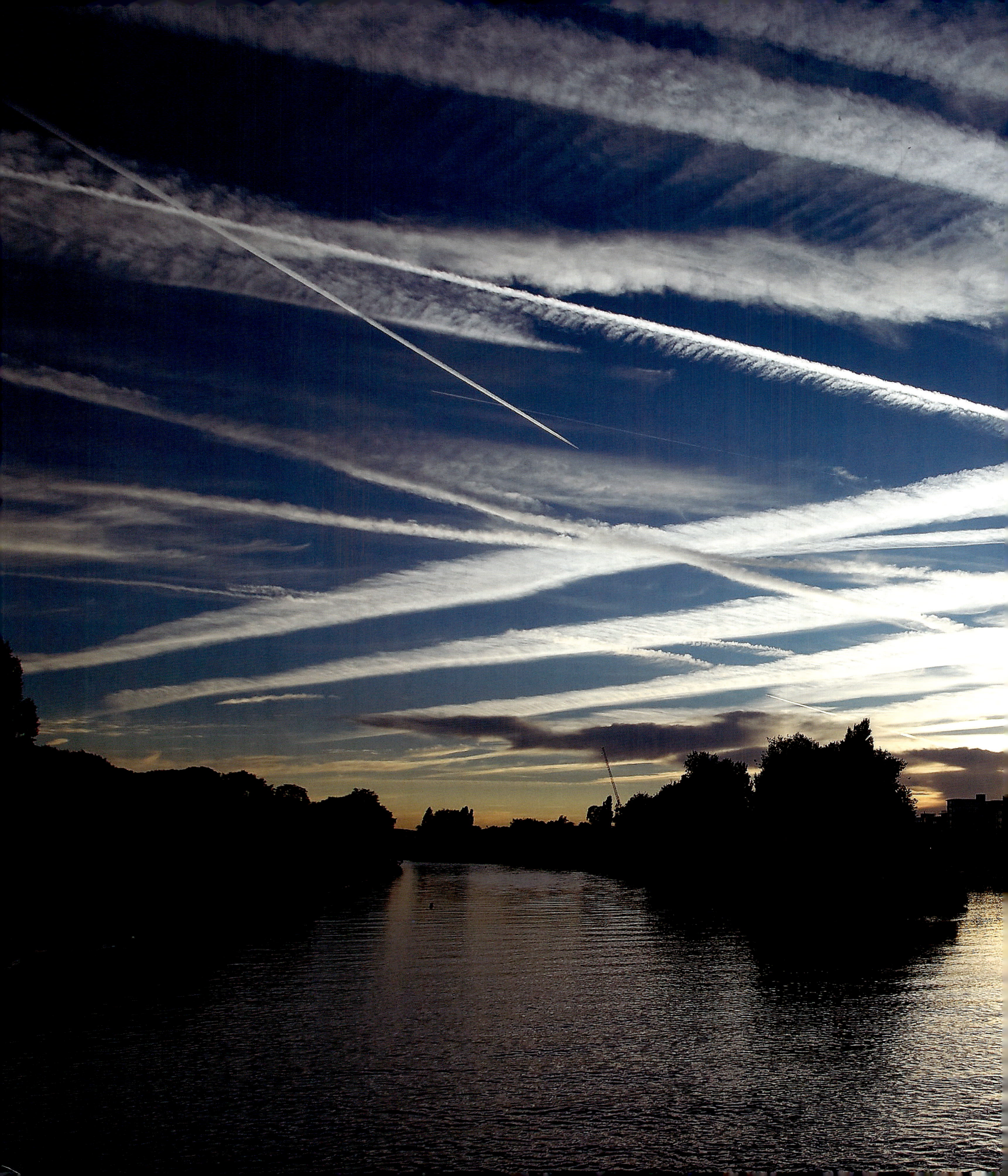

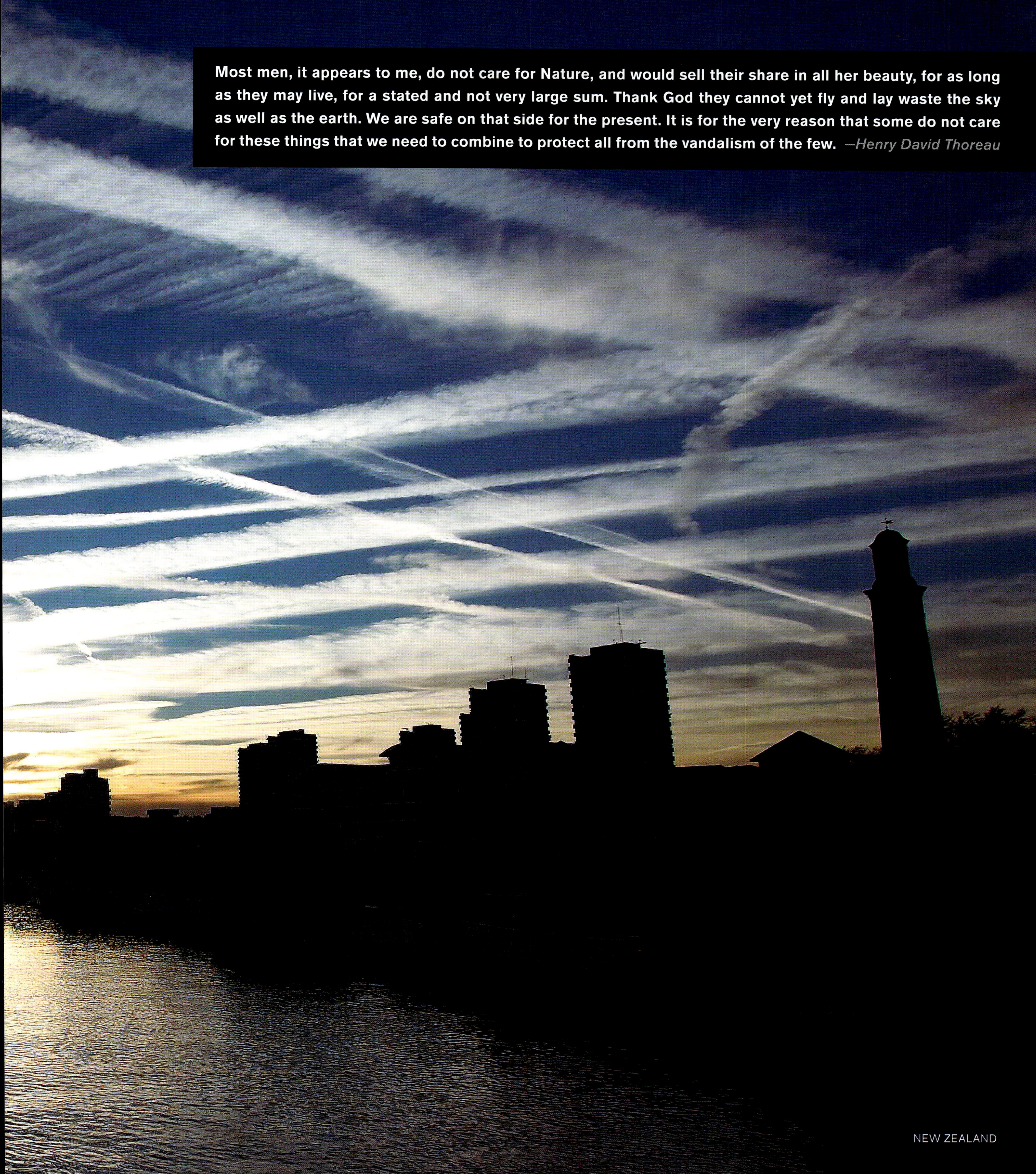

Most men, it appears to me, do not care for Nature, and would sell their share in all her beauty, for as long as they may live, for a stated and not very large sum. Thank God they cannot yet fly and lay waste the sky as well as the earth. We are safe on that side for the present. It is for the very reason that some do not care for these things that we need to combine to protect all from the vandalism of the few. —Henry David Thoreau
NEW ZEALAND

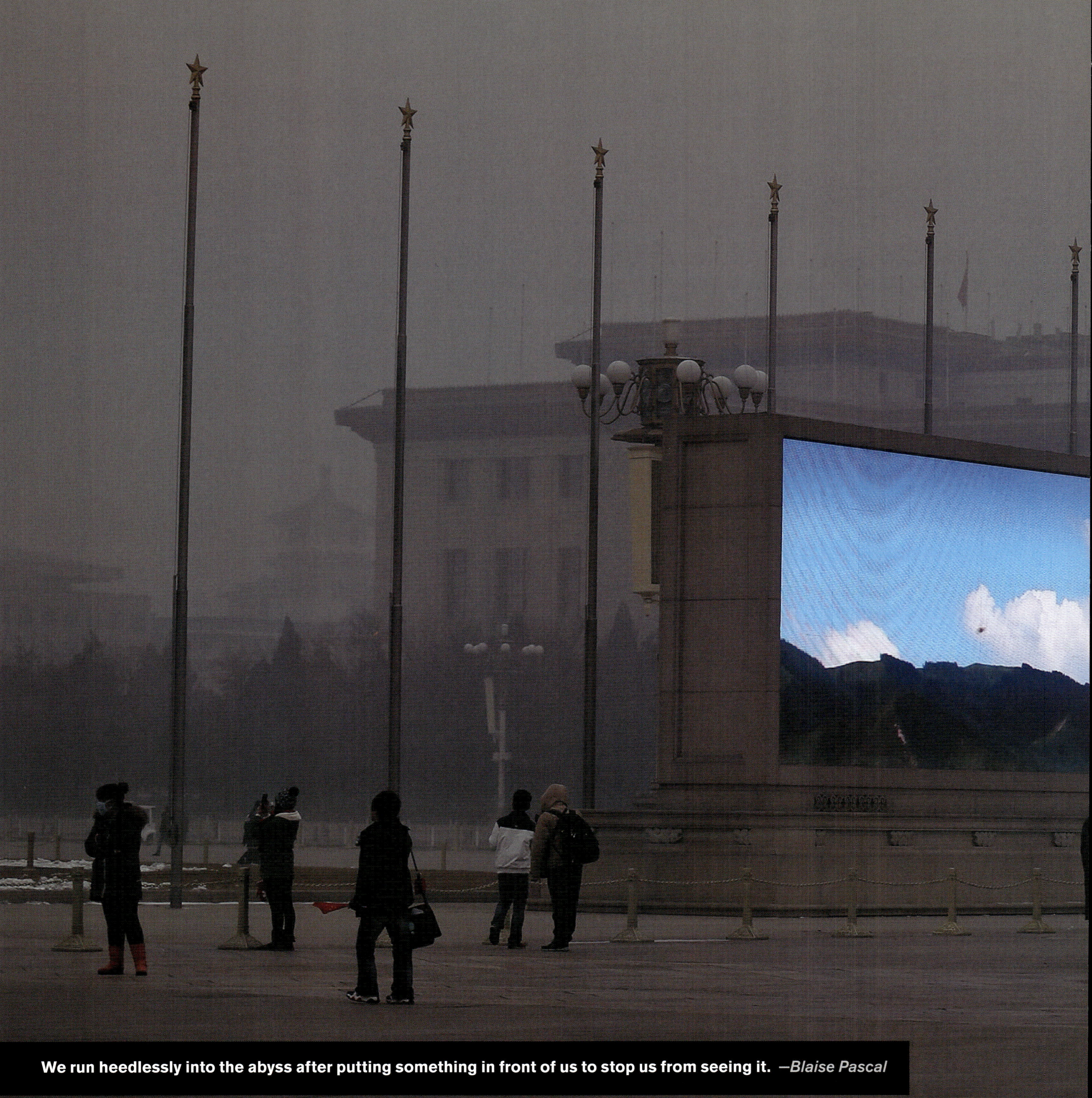

We run heedlessly into the abyss after putting something in front of us to stop us from seeing it. —*Blaise Pascal*

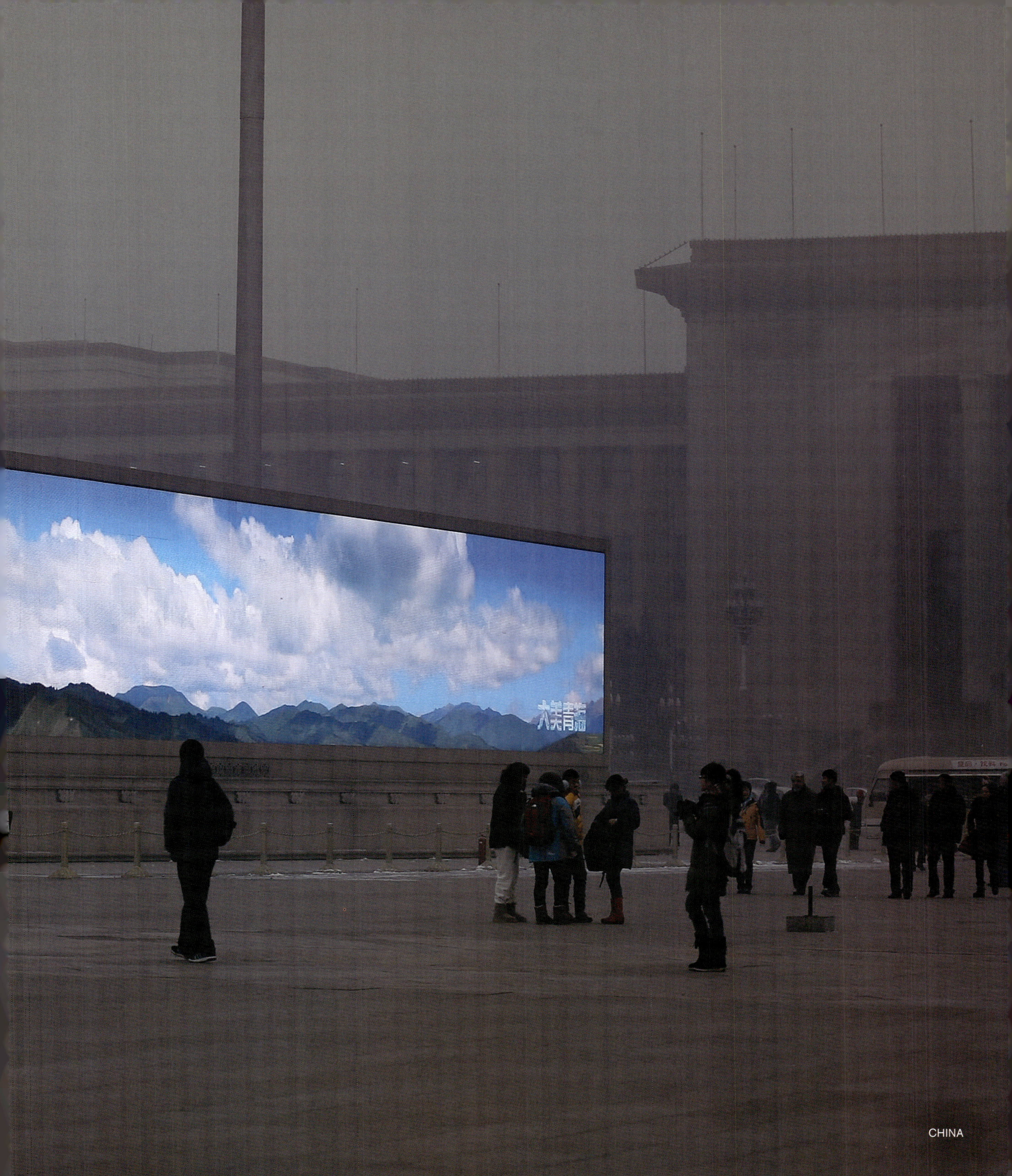

CHINA

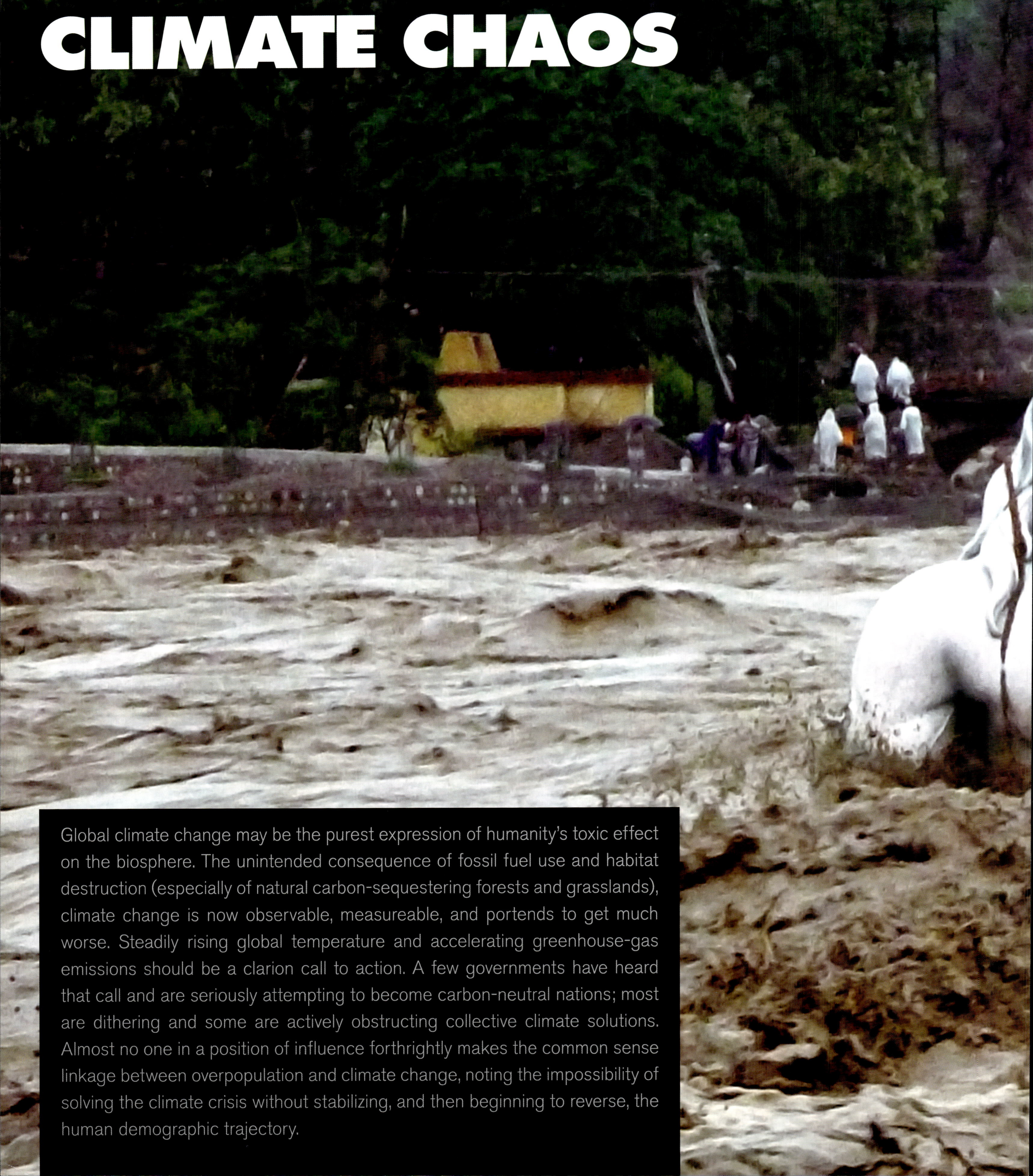

CLIMATE CHAOS

Global climate change may be the purest expression of humanity's toxic effect on the biosphere. The unintended consequence of fossil fuel use and habitat destruction (especially of natural carbon-sequestering forests and grasslands), climate change is now observable, measureable, and portends to get much worse. Steadily rising global temperature and accelerating greenhouse-gas emissions should be a clarion call to action. A few governments have heard that call and are seriously attempting to become carbon-neutral nations; most are dithering and some are actively obstructing collective climate solutions. Almost no one in a position of influence forthrightly makes the common sense linkage between overpopulation and climate change, noting the impossibility of solving the climate crisis without stabilizing, and then beginning to reverse, the human demographic trajectory.

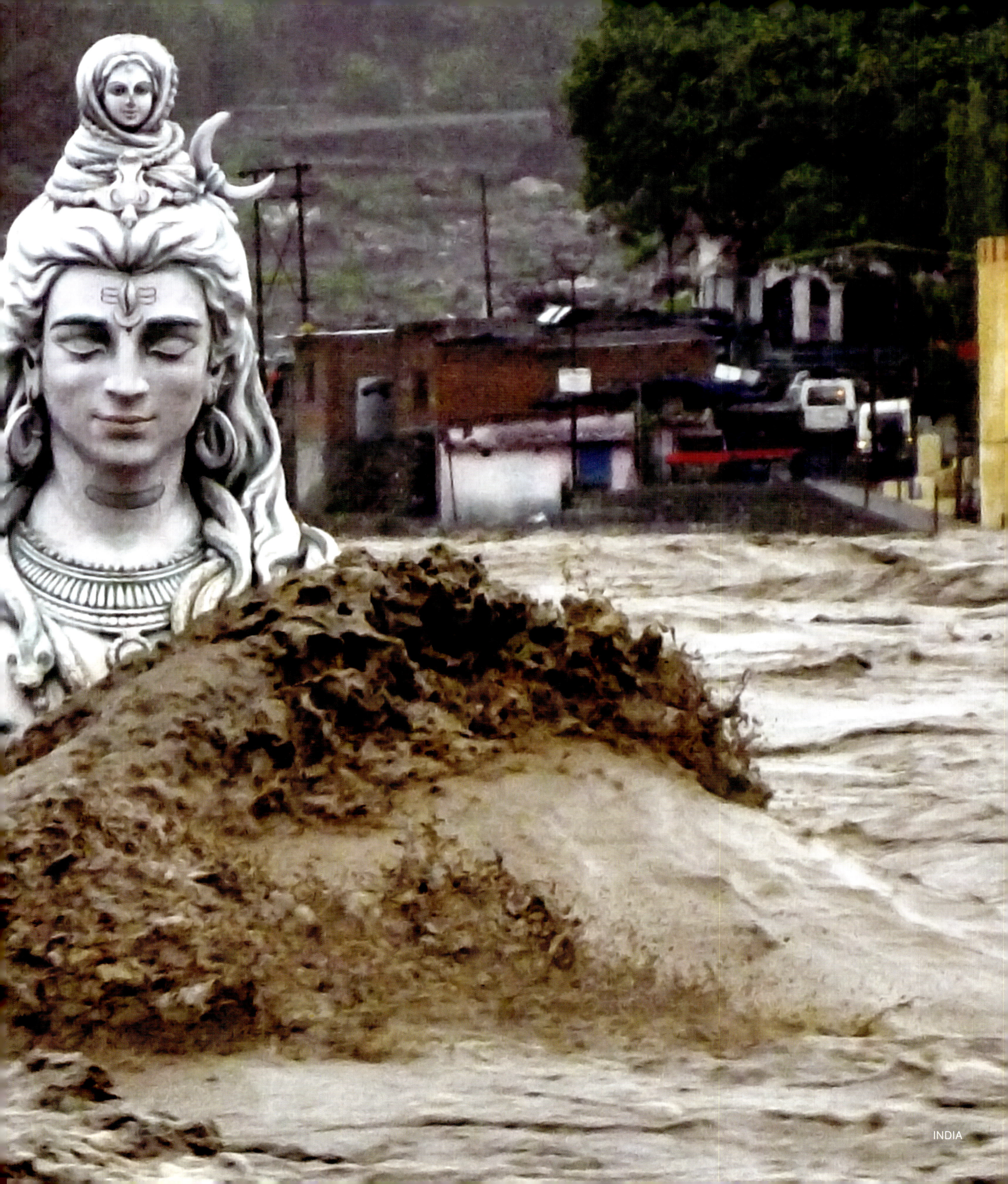
INDIA

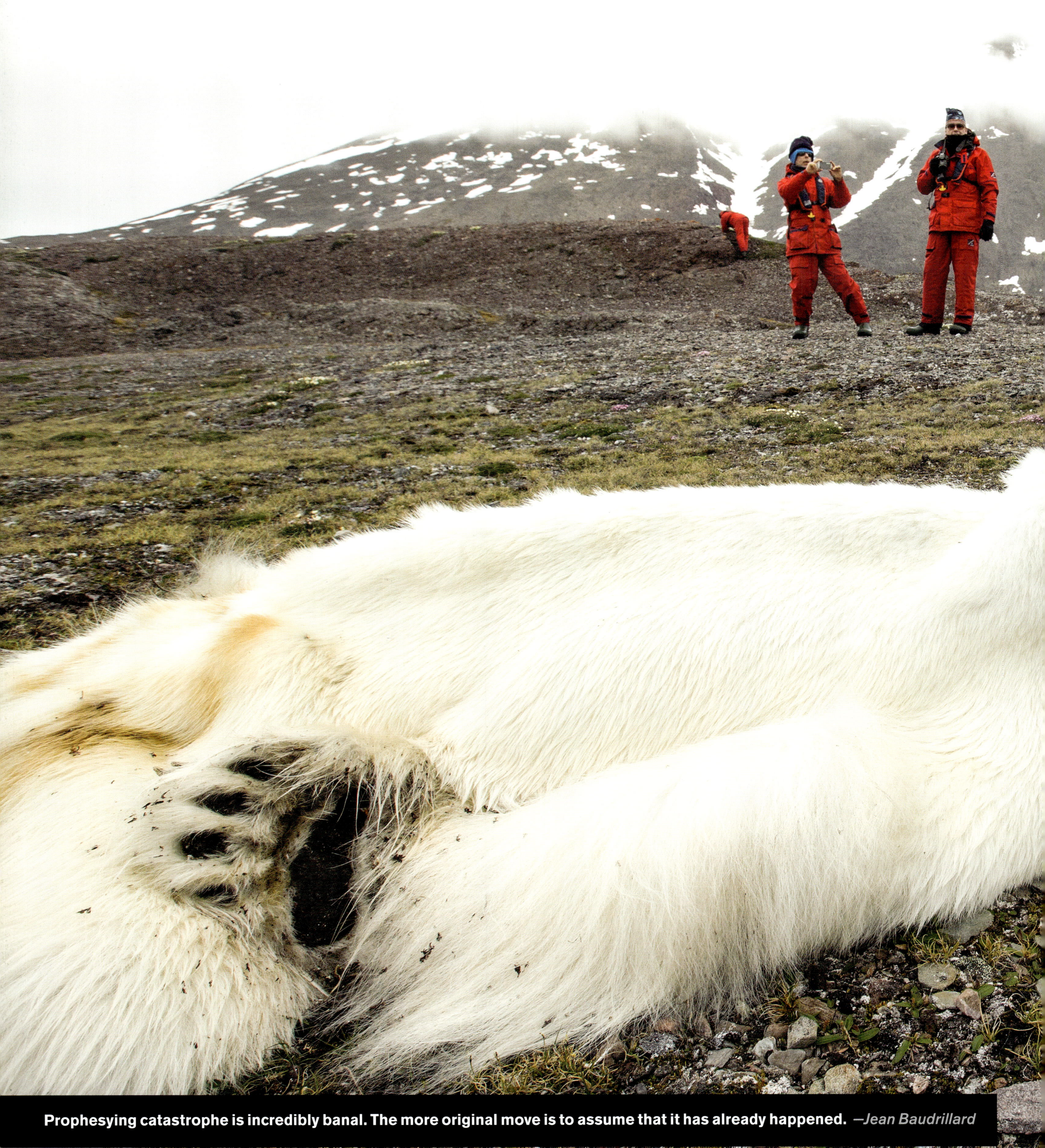

Prophesying catastrophe is incredibly banal. The more original move is to assume that it has already happened. *—Jean Baudrillard*

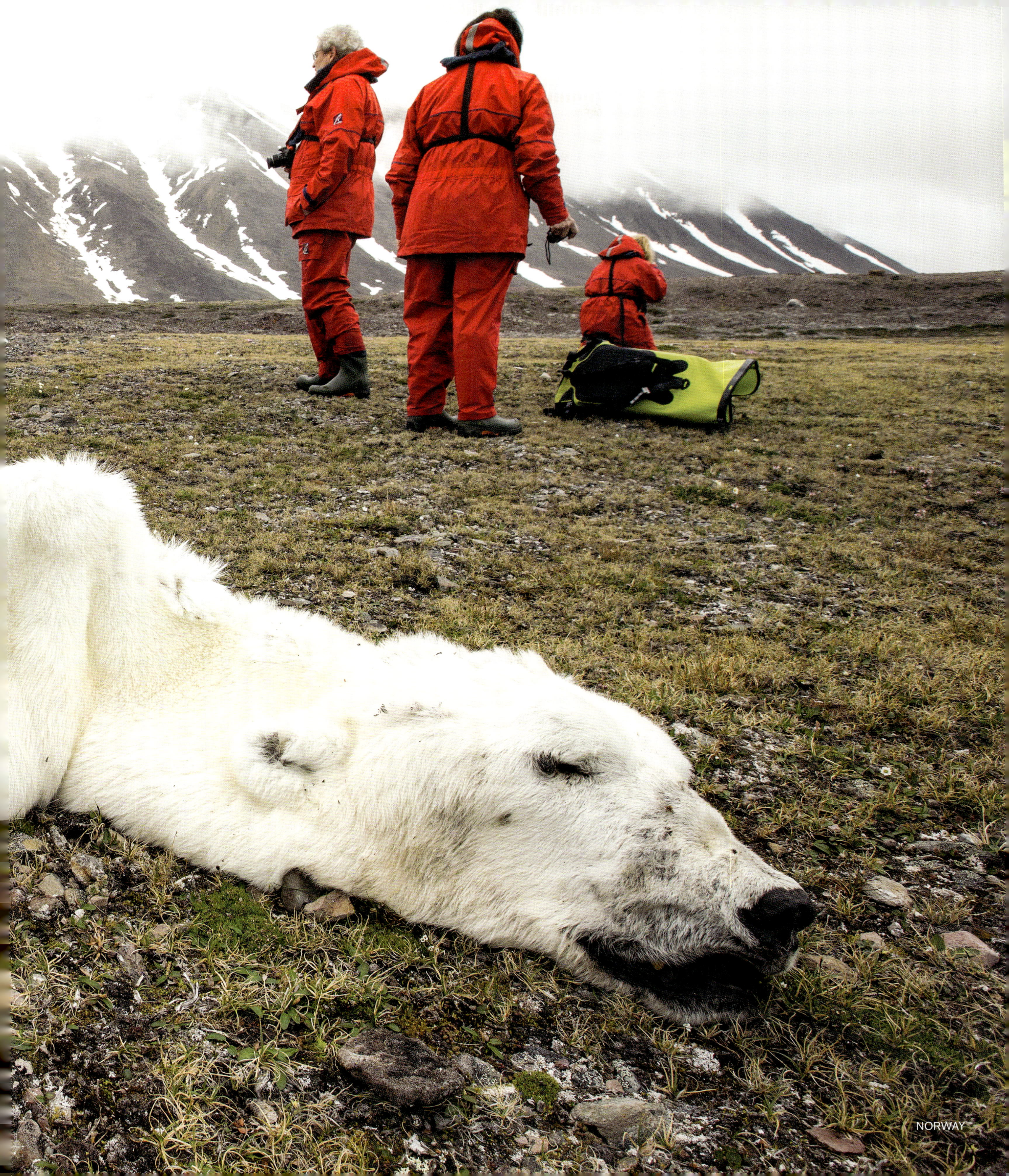

NORWAY

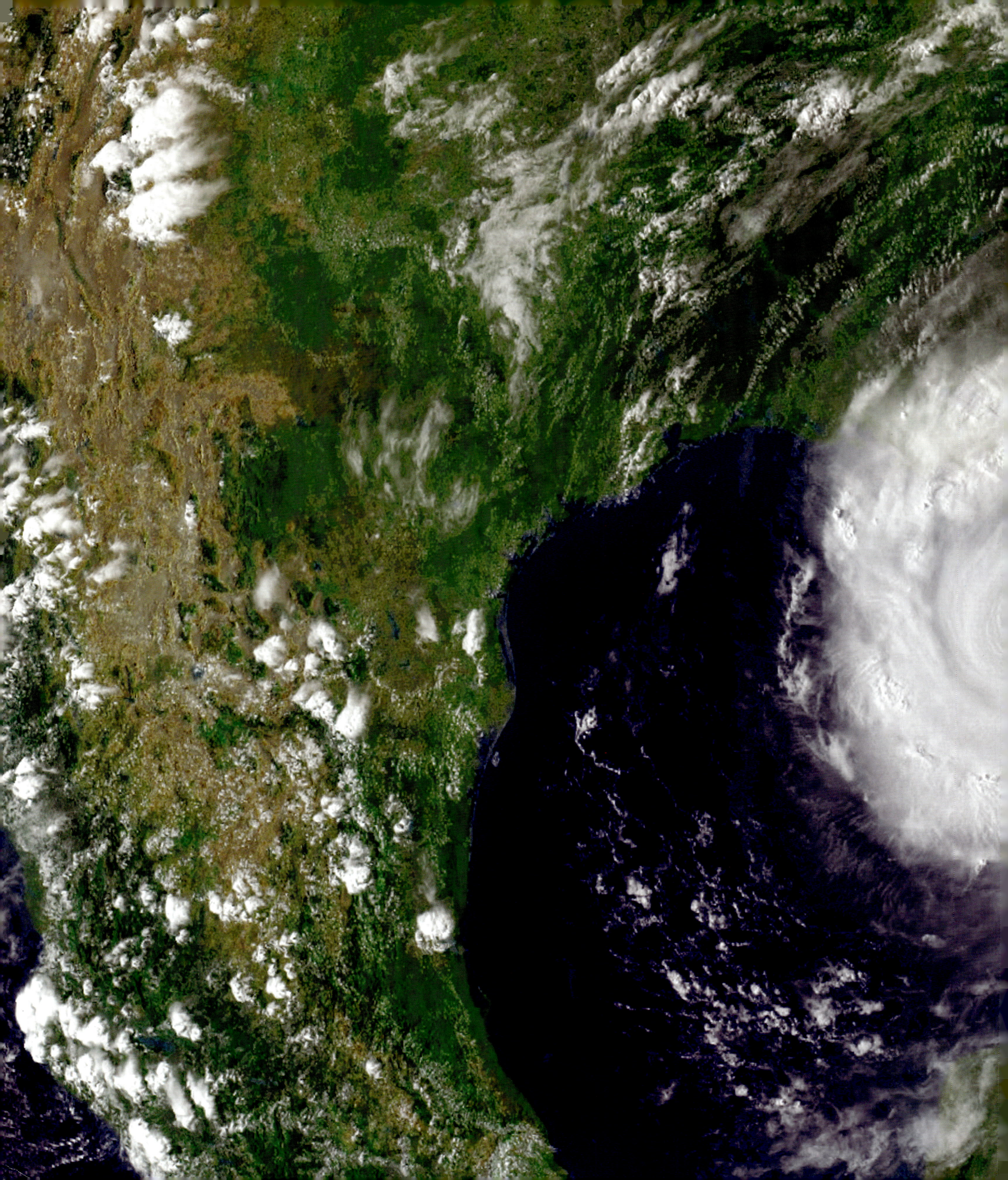

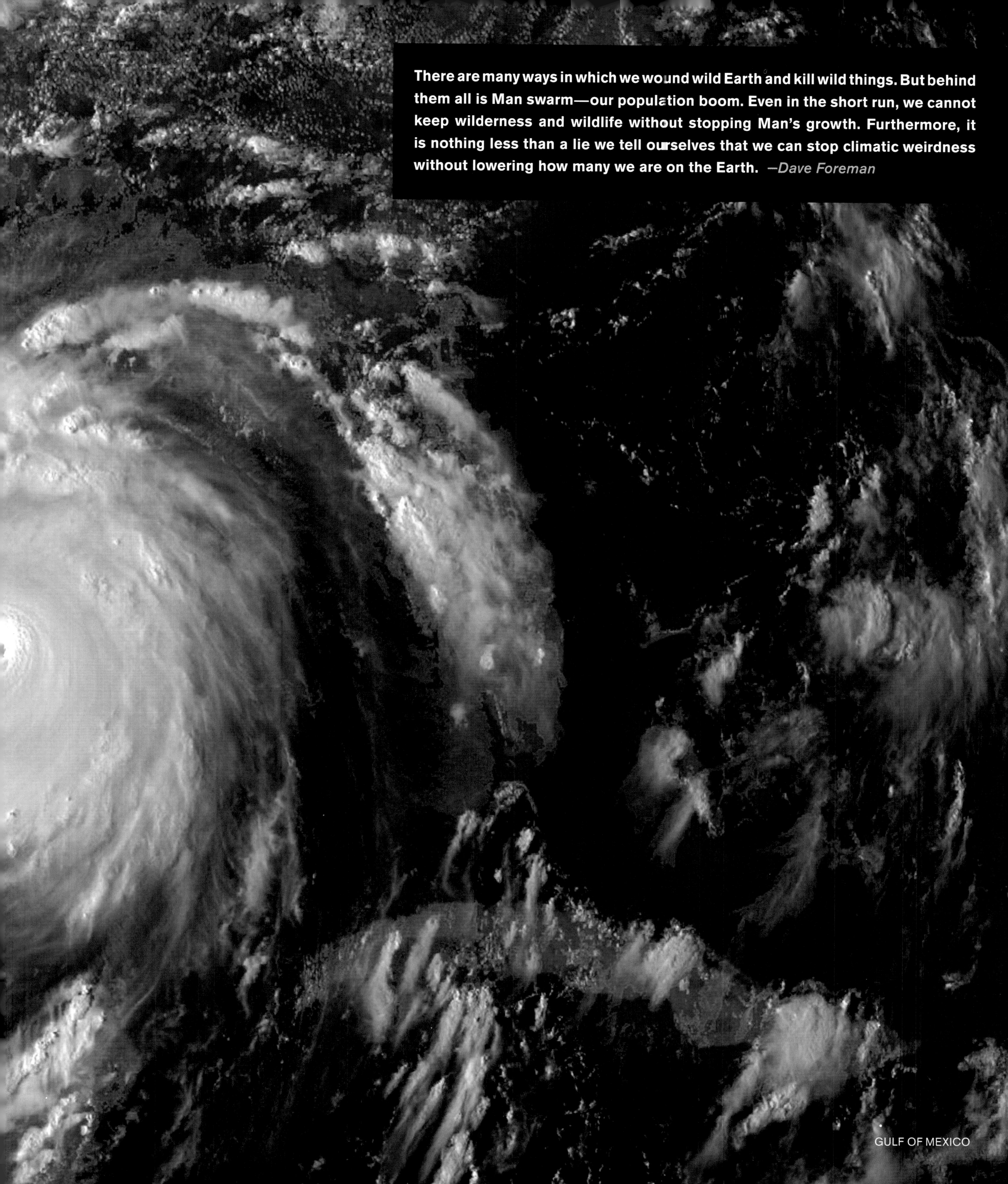

There are many ways in which we wound wild Earth and kill wild things. But behind them all is Man swarm—our population boom. Even in the short run, we cannot keep wilderness and wildlife without stopping Man's growth. Furthermore, it is nothing less than a lie we tell ourselves that we can stop climatic weirdness without lowering how many we are on the Earth. —Dave Foreman

GULF OF MEXICO

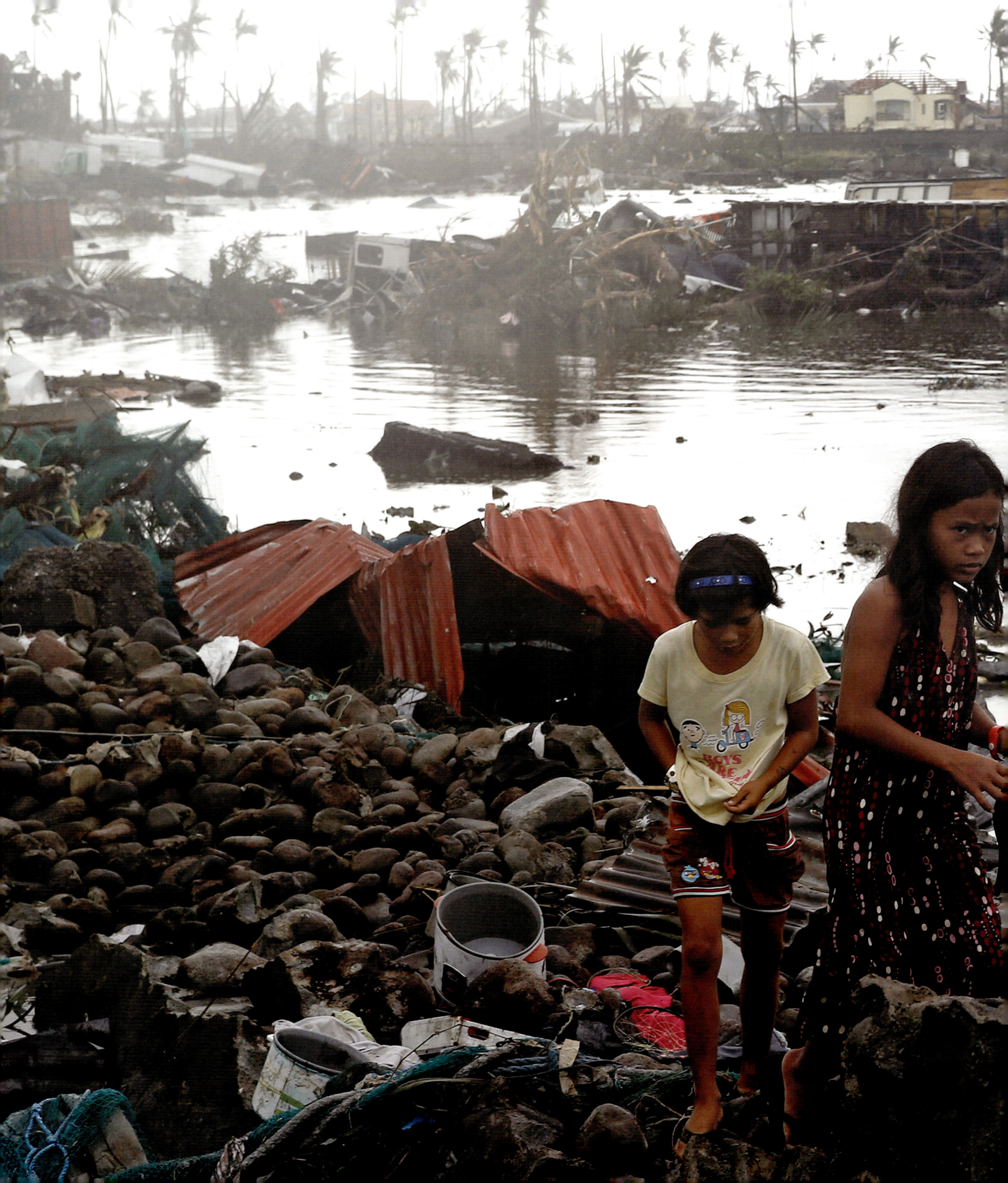

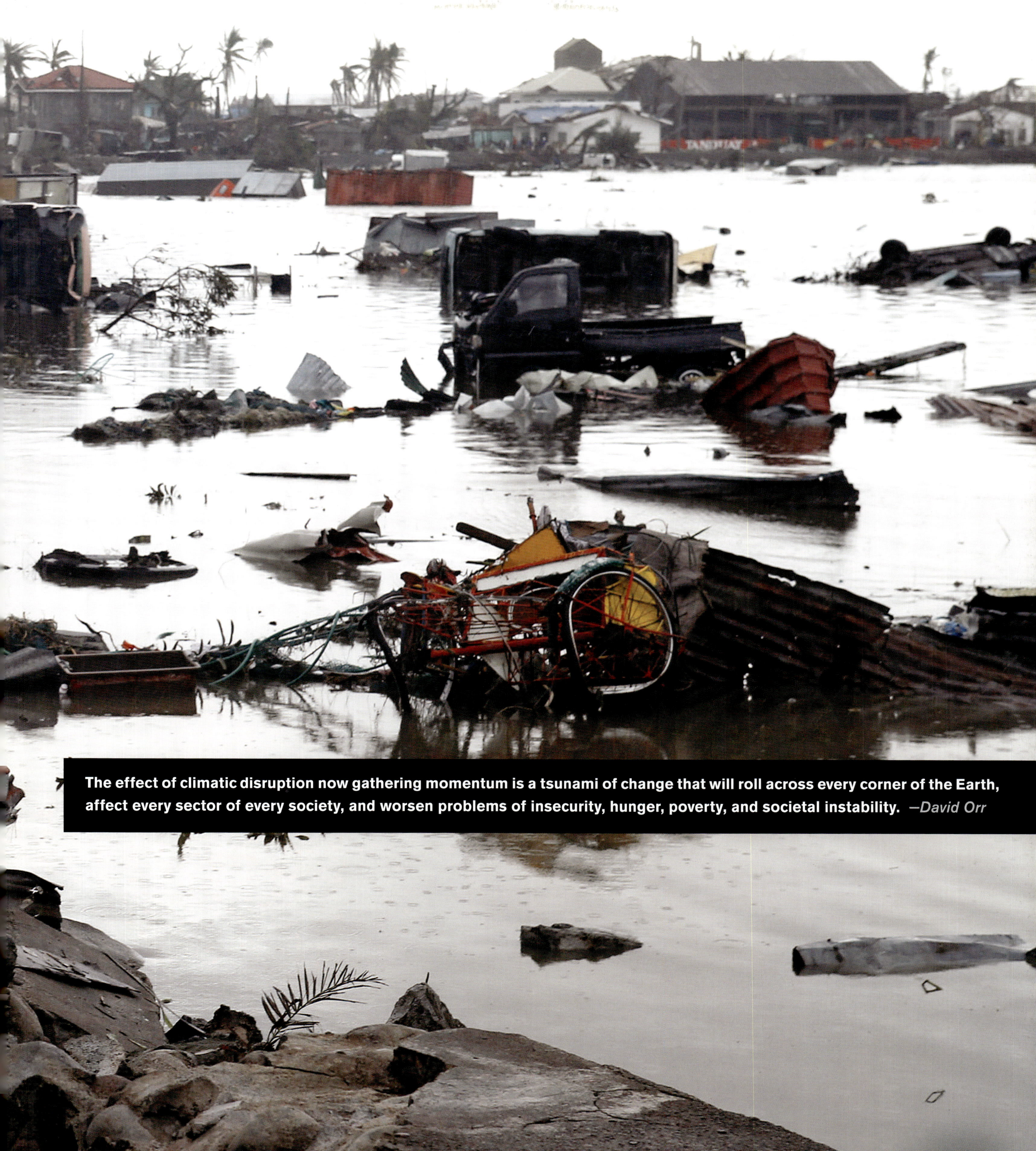
The effect of climatic disruption now gathering momentum is a tsunami of change that will roll across every corner of the Earth, affect every sector of every society, and worsen problems of insecurity, hunger, poverty, and societal instability. —David Orr

PHILIPPINES

The island is full of holes and seawater is coming through these, flooding areas that weren't normally flooded 10 or 15 years ago. There are projections of about 50 years before the islands disappear. After this, we will be drowned. —Paani Laupepa

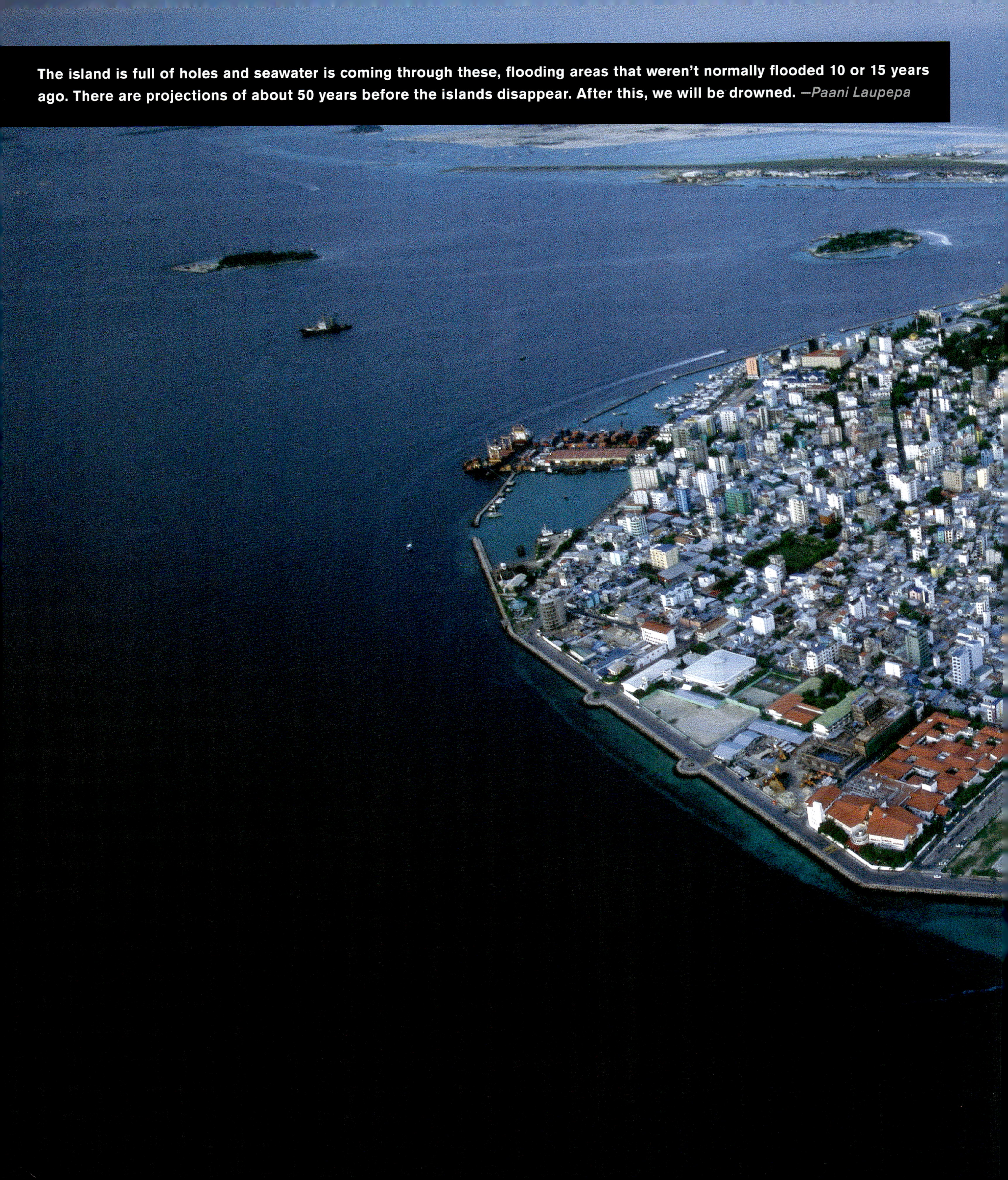

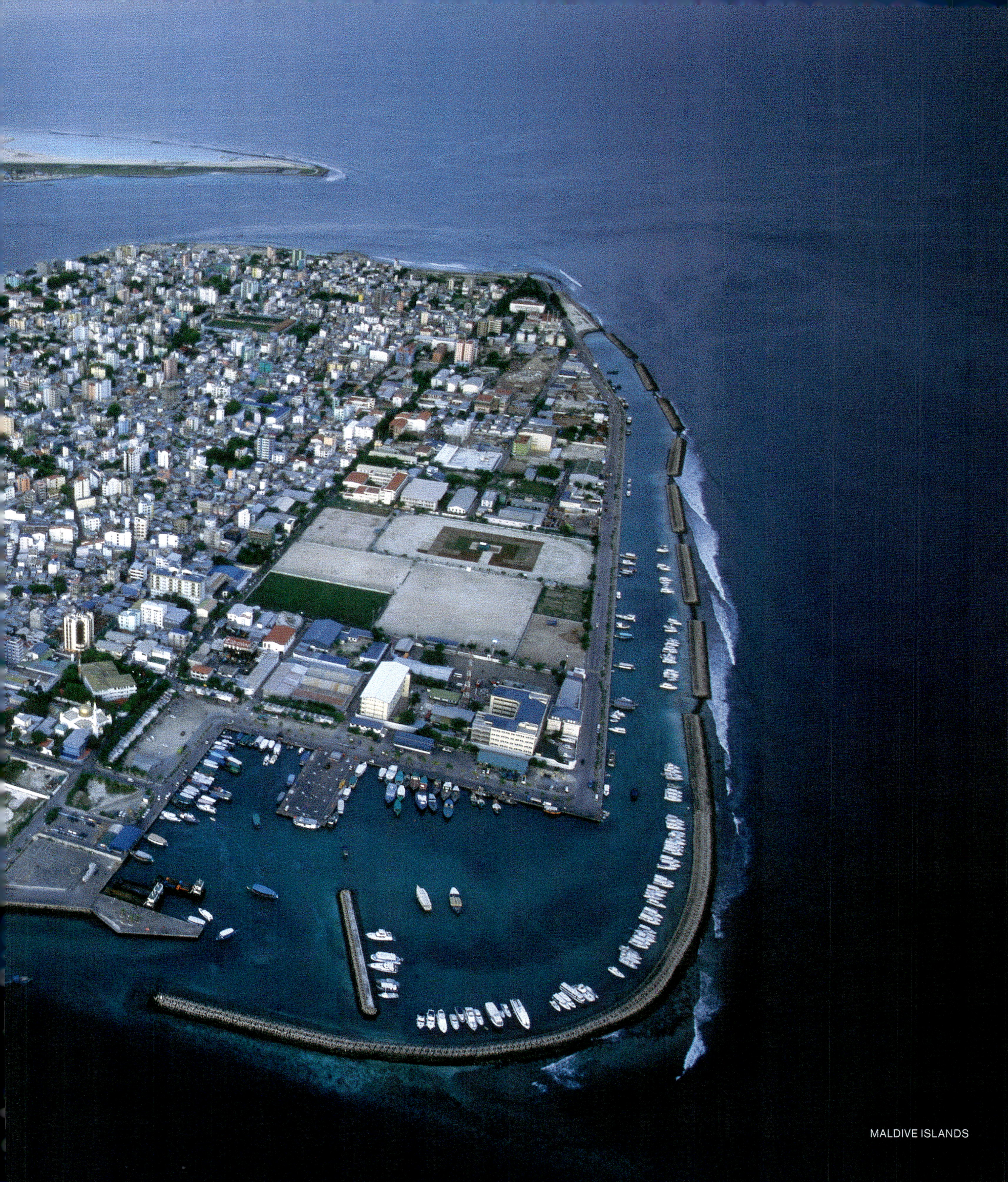
MALDIVE ISLANDS

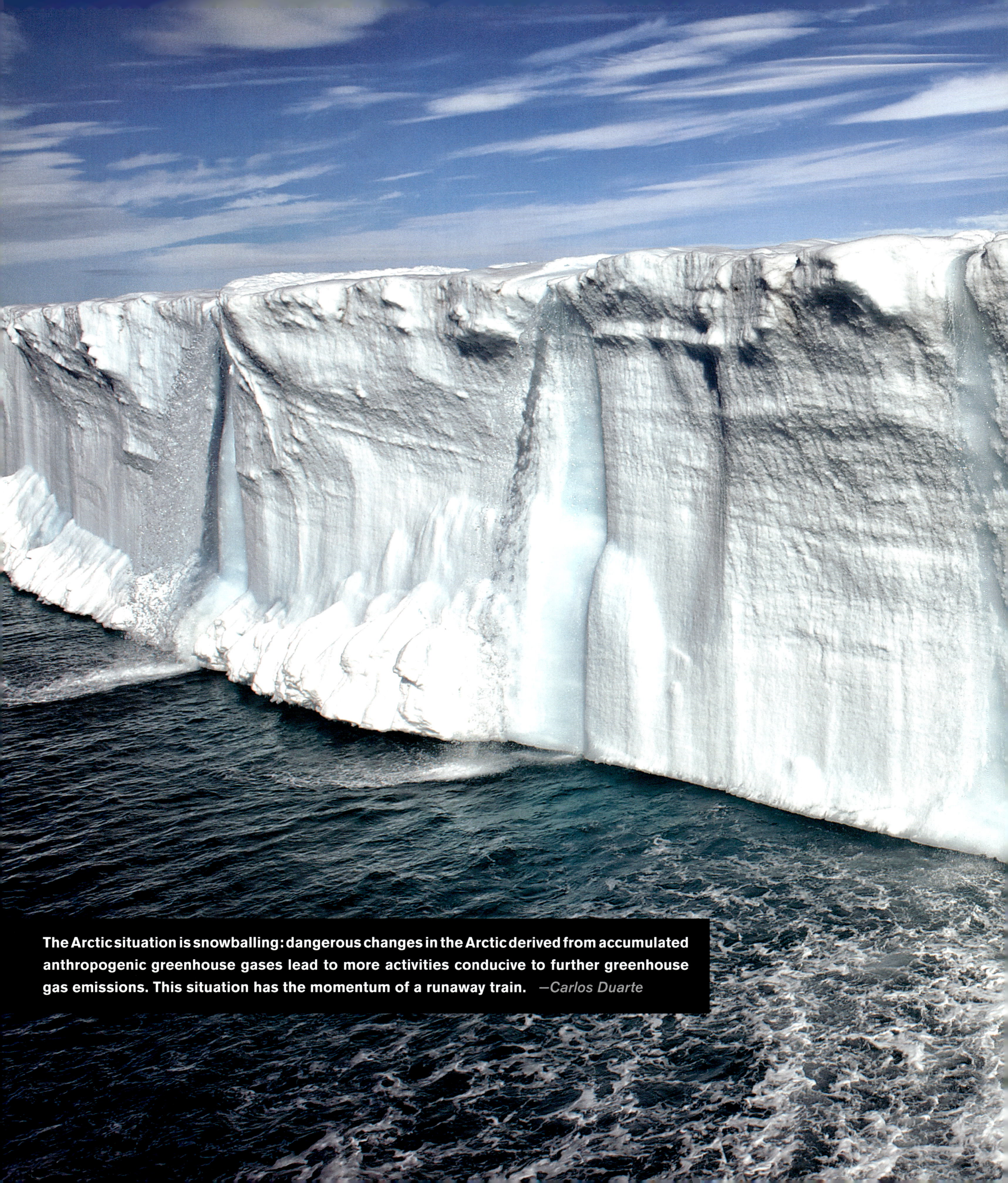

The Arctic situation is snowballing: dangerous changes in the Arctic derived from accumulated anthropogenic greenhouse gases lead to more activities conducive to further greenhouse gas emissions. This situation has the momentum of a runaway train. —Carlos Duarte

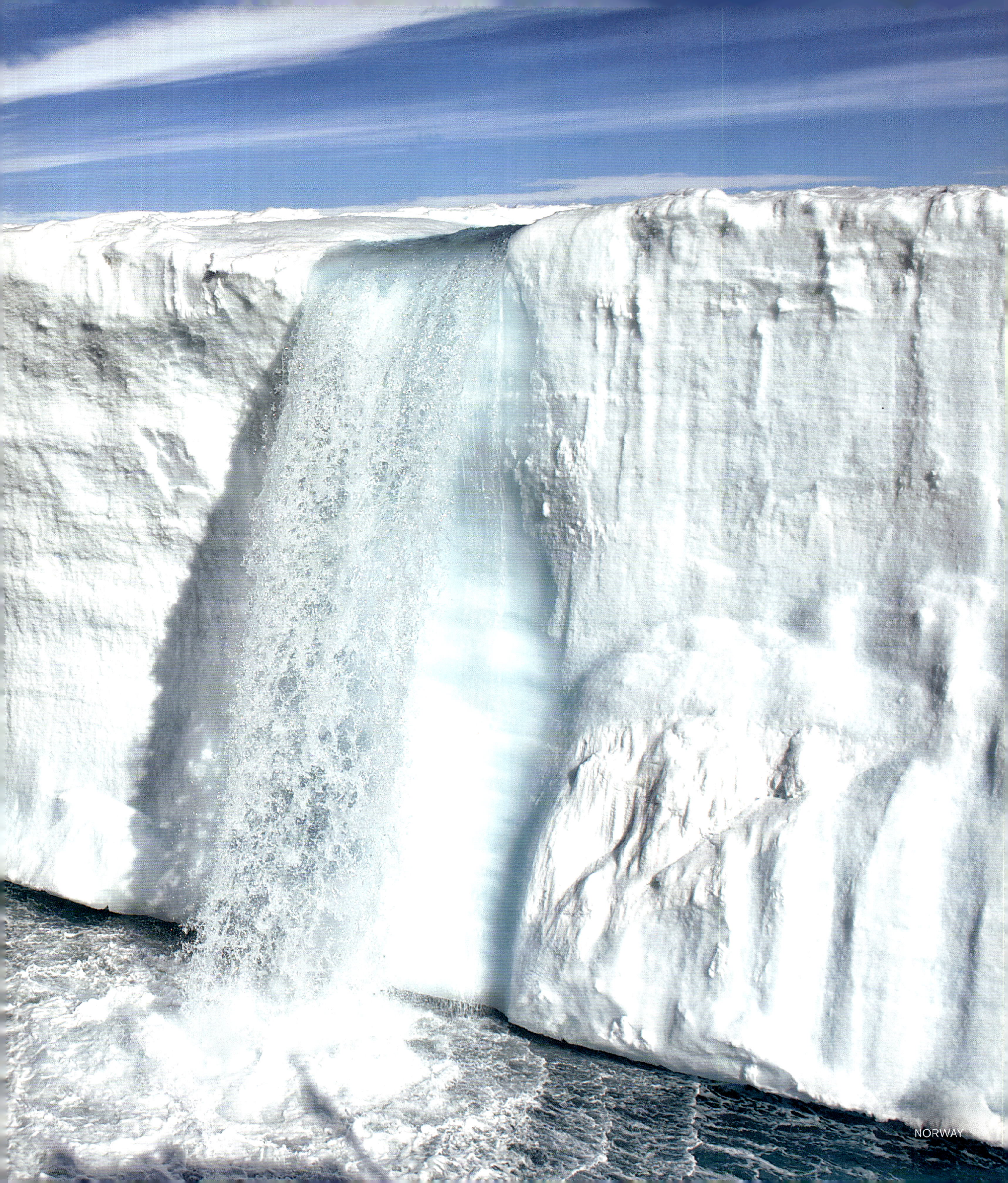
NORWAY

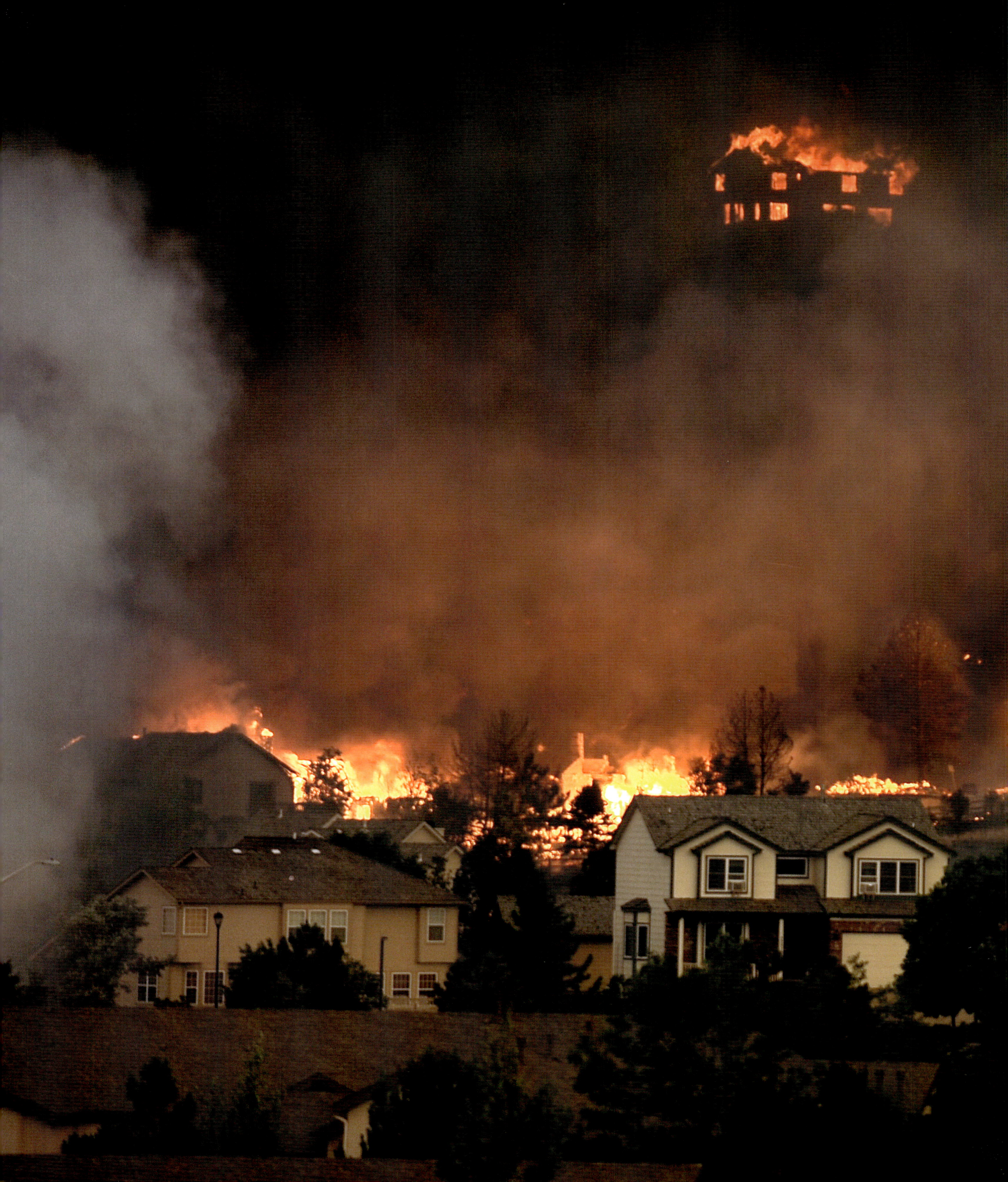

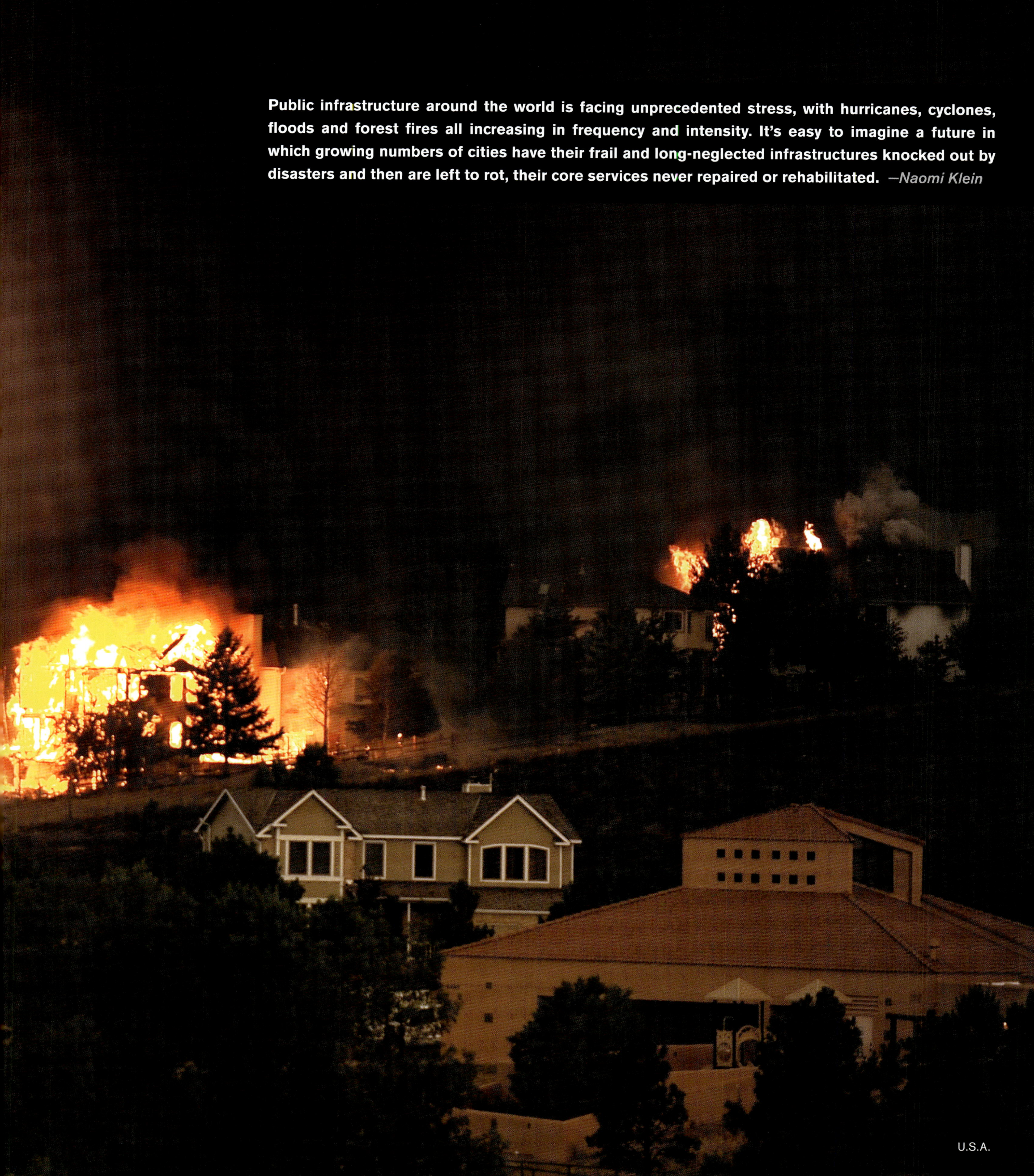

Public infrastructure around the world is facing unprecedented stress, with hurricanes, cyclones, floods and forest fires all increasing in frequency and intensity. It's easy to imagine a future in which growing numbers of cities have their frail and long-neglected infrastructures knocked out by disasters and then are left to rot, their core services never repaired or rehabilitated. —Naomi Klein

U.S.A.

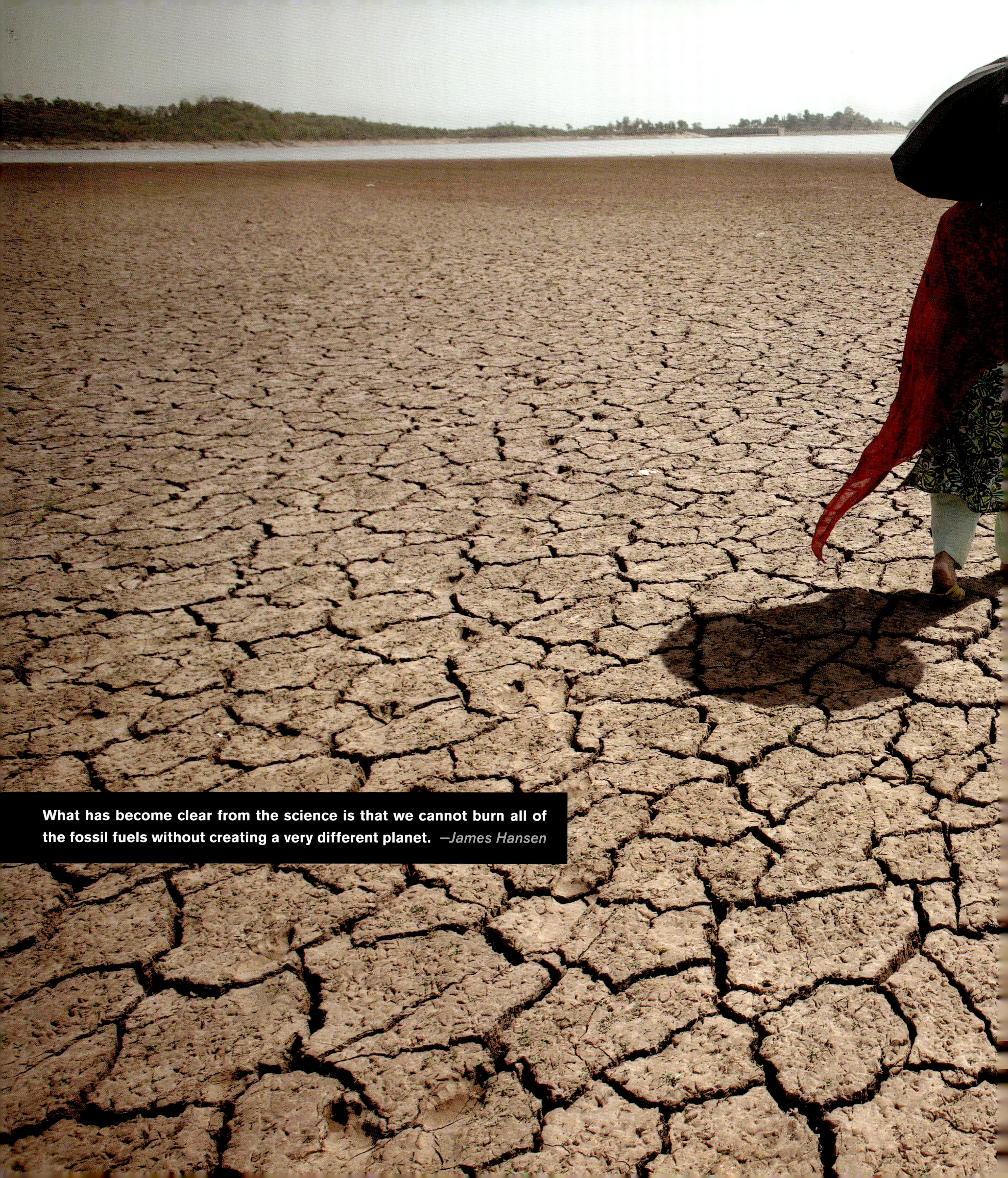

What has become clear from the science is that we cannot burn all of
the fossil fuels without creating a very different planet. —James Hansen

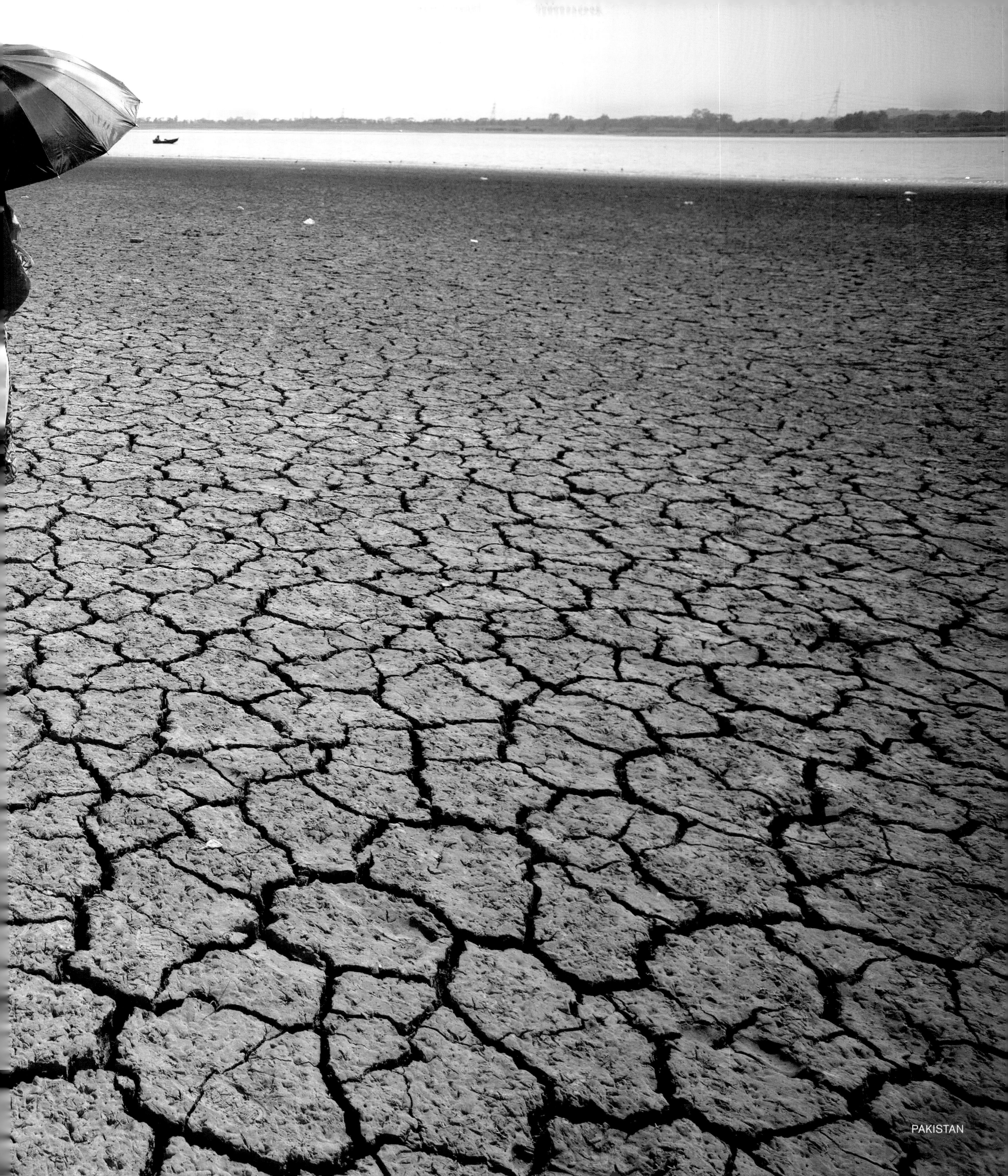
PAKISTAN

LORD MAN

PARABLE REDUX

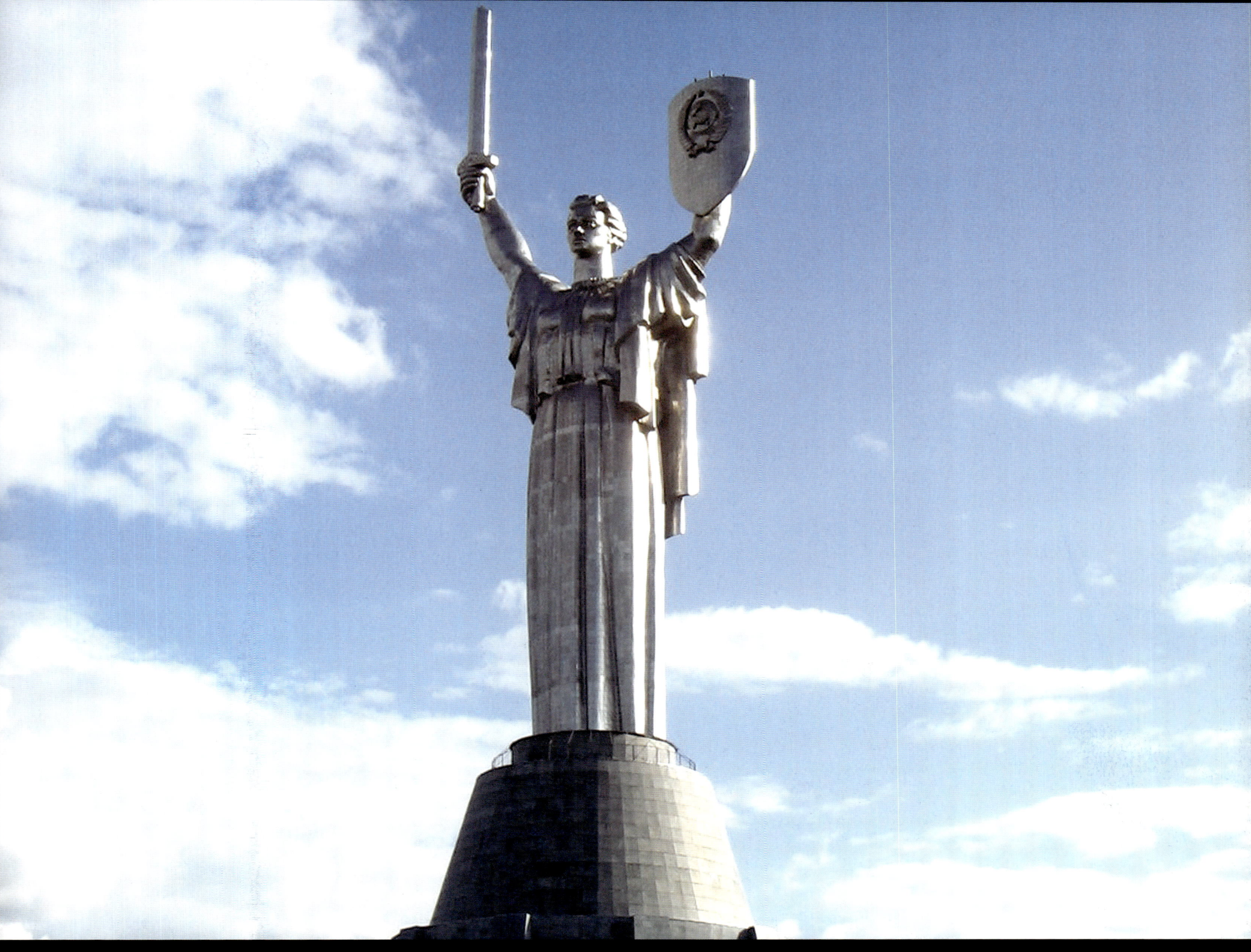

Facing the stark choice to continue his futile quest to bend the Earth to his will . . .

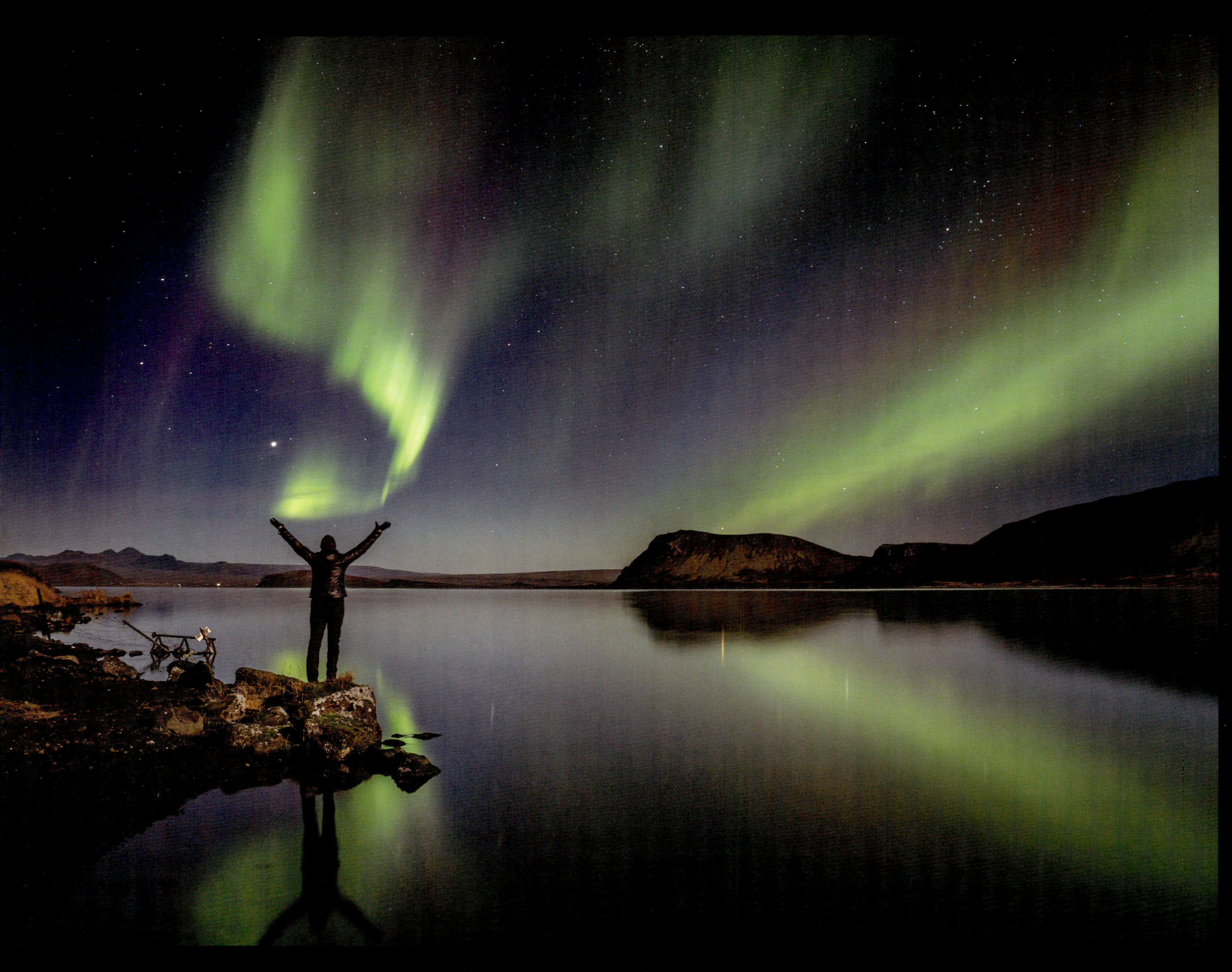

or to rejoin the community of life, Lord Man renounced the goal of empire, and was no more.

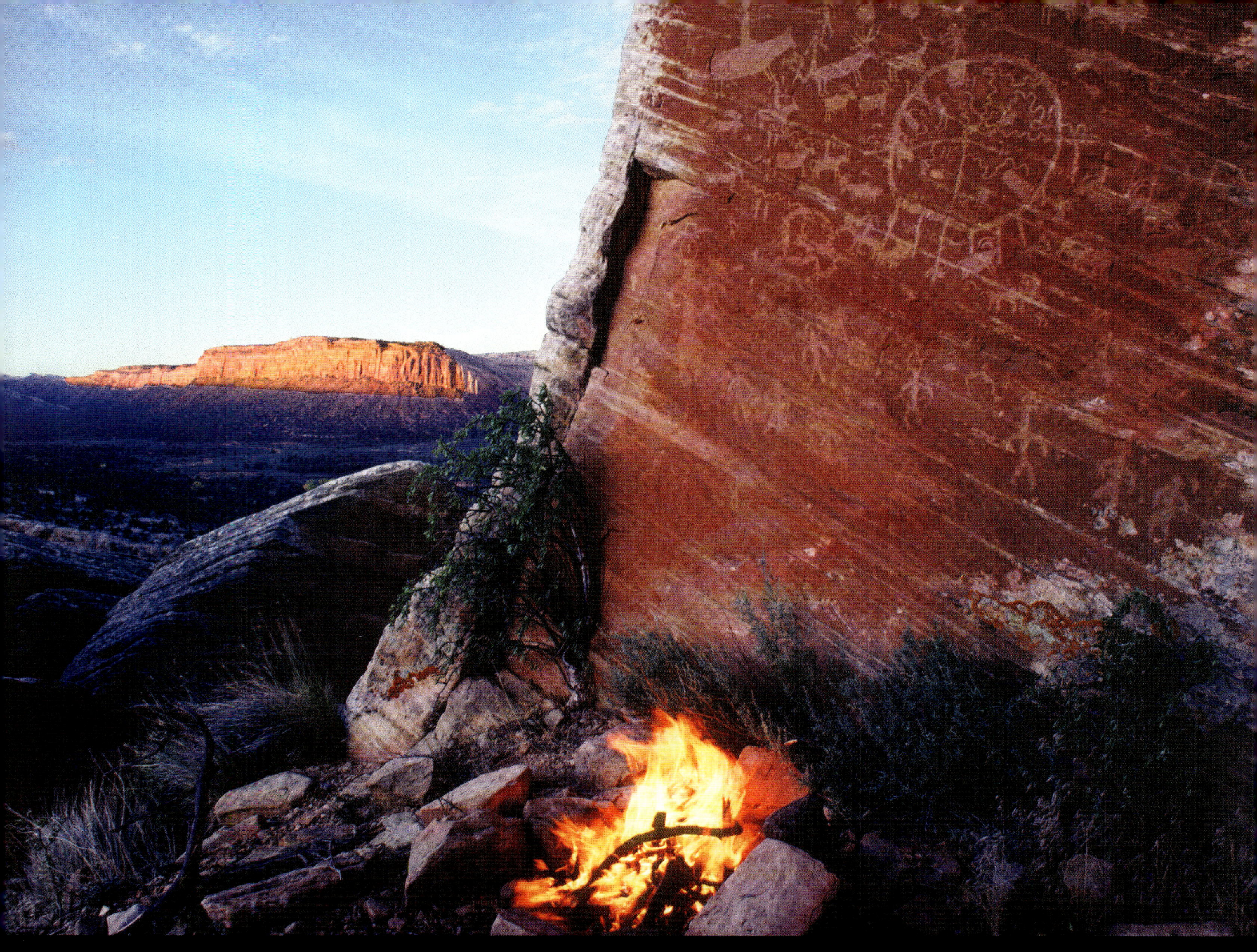

People around the globe began to remember the old ways, before humans behaved
as if the Earth was merely a storehouse of resources for them.

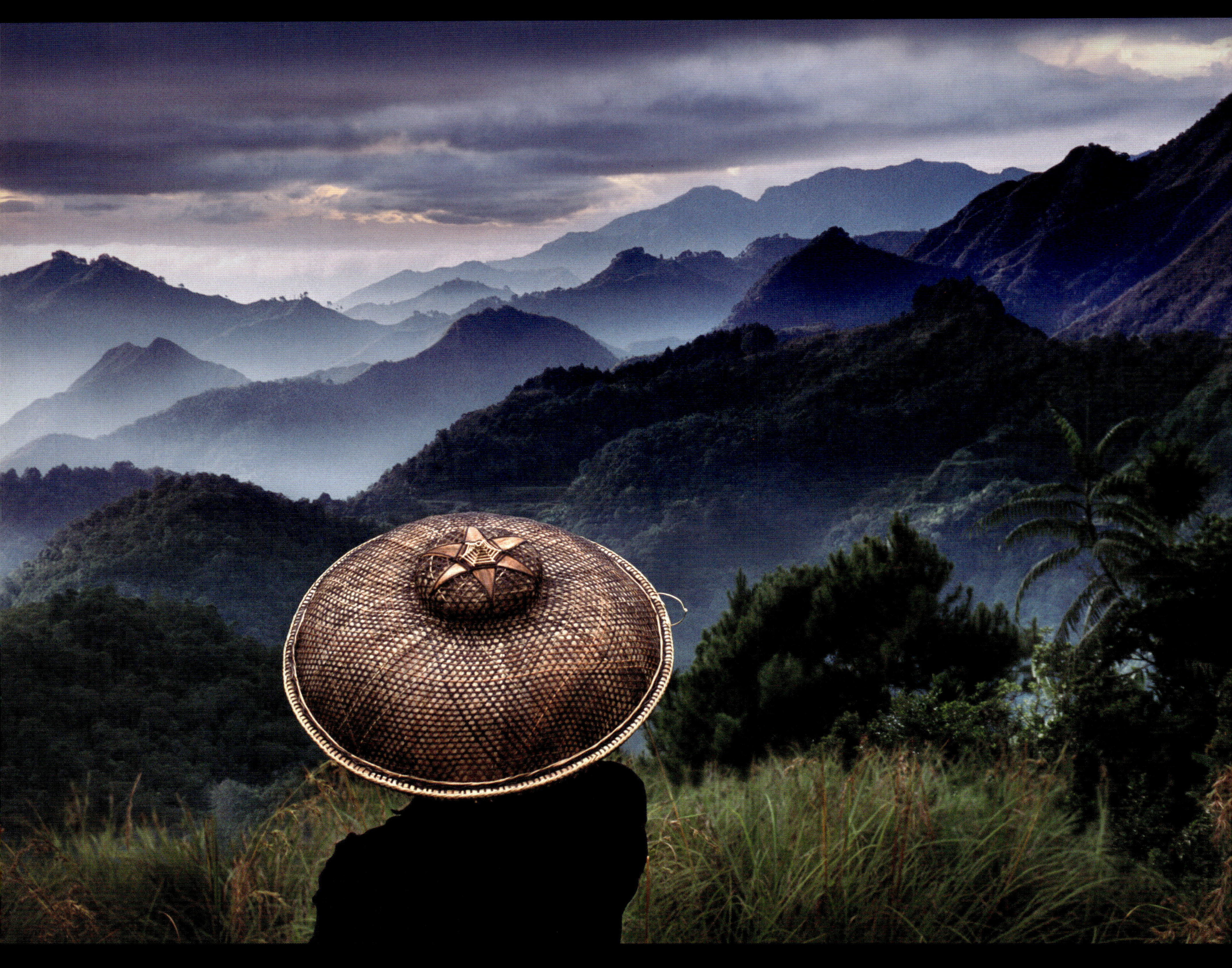

The people looked to the landscapes around them to inform their culture and shape their ways of living

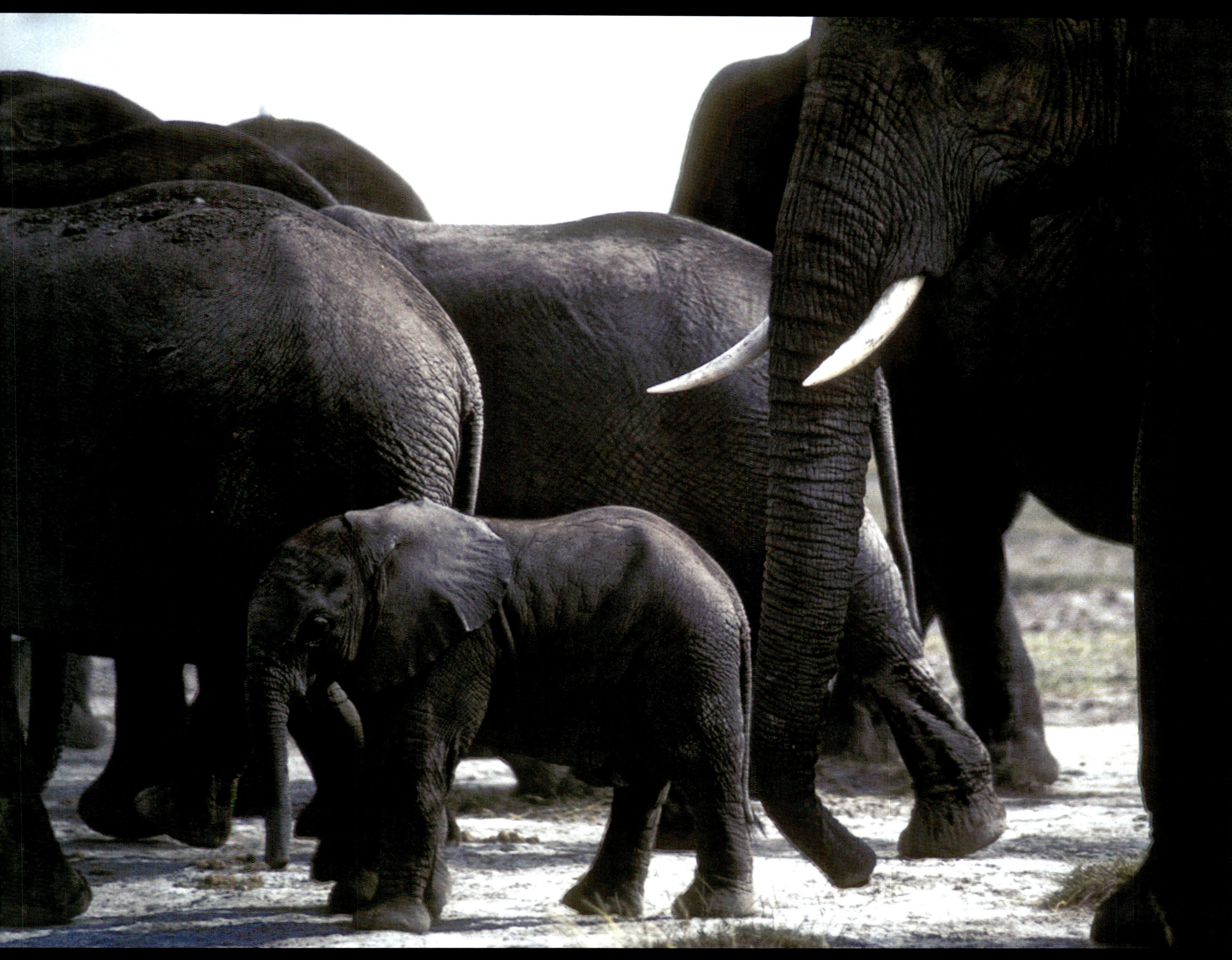

They valued the other members in the land community, giving them space enough to flourish in their own ways.

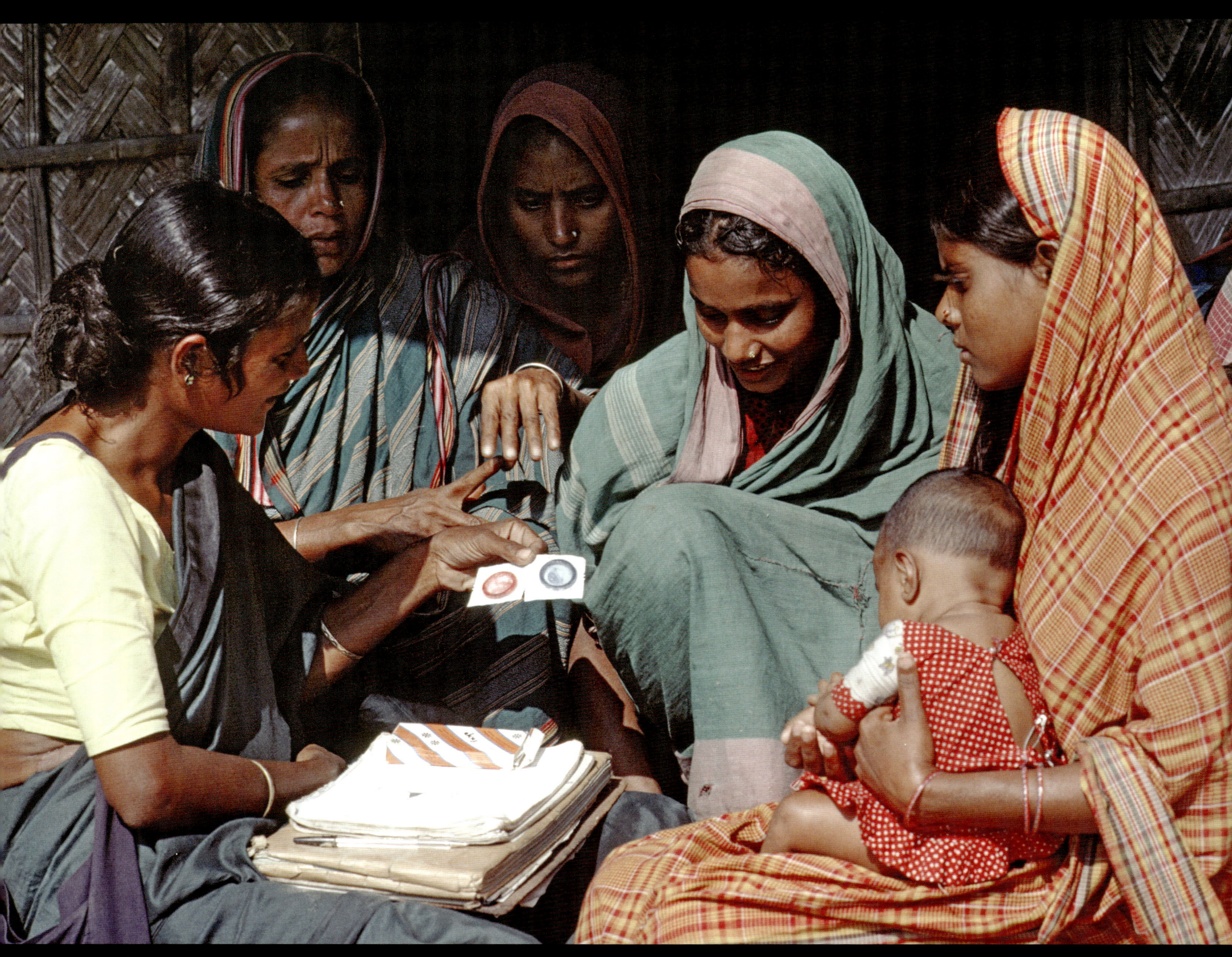

Knowledge and tools for family planning were universally shared

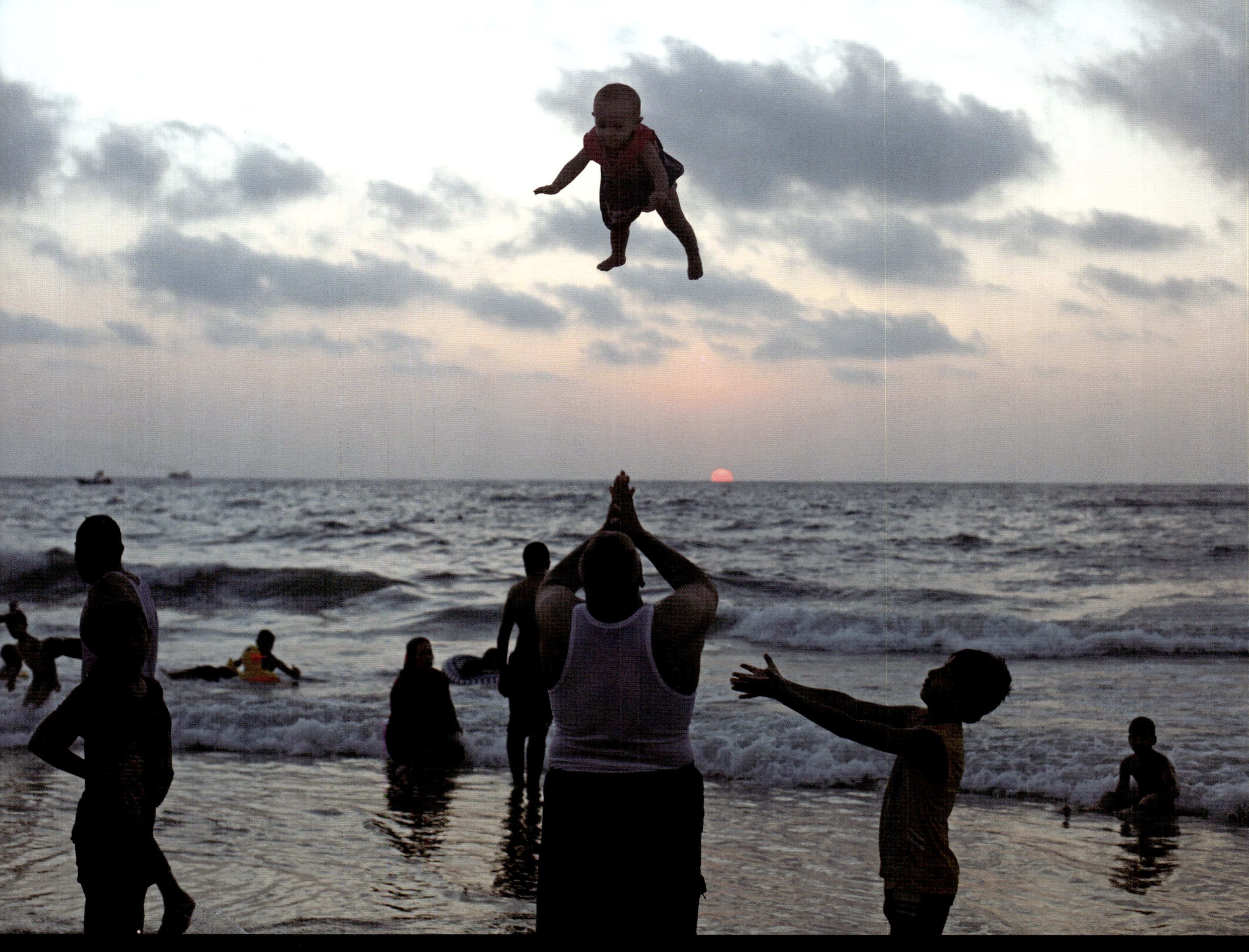

People began to restrain their numbers, with smaller families becoming a key to societal well-being.

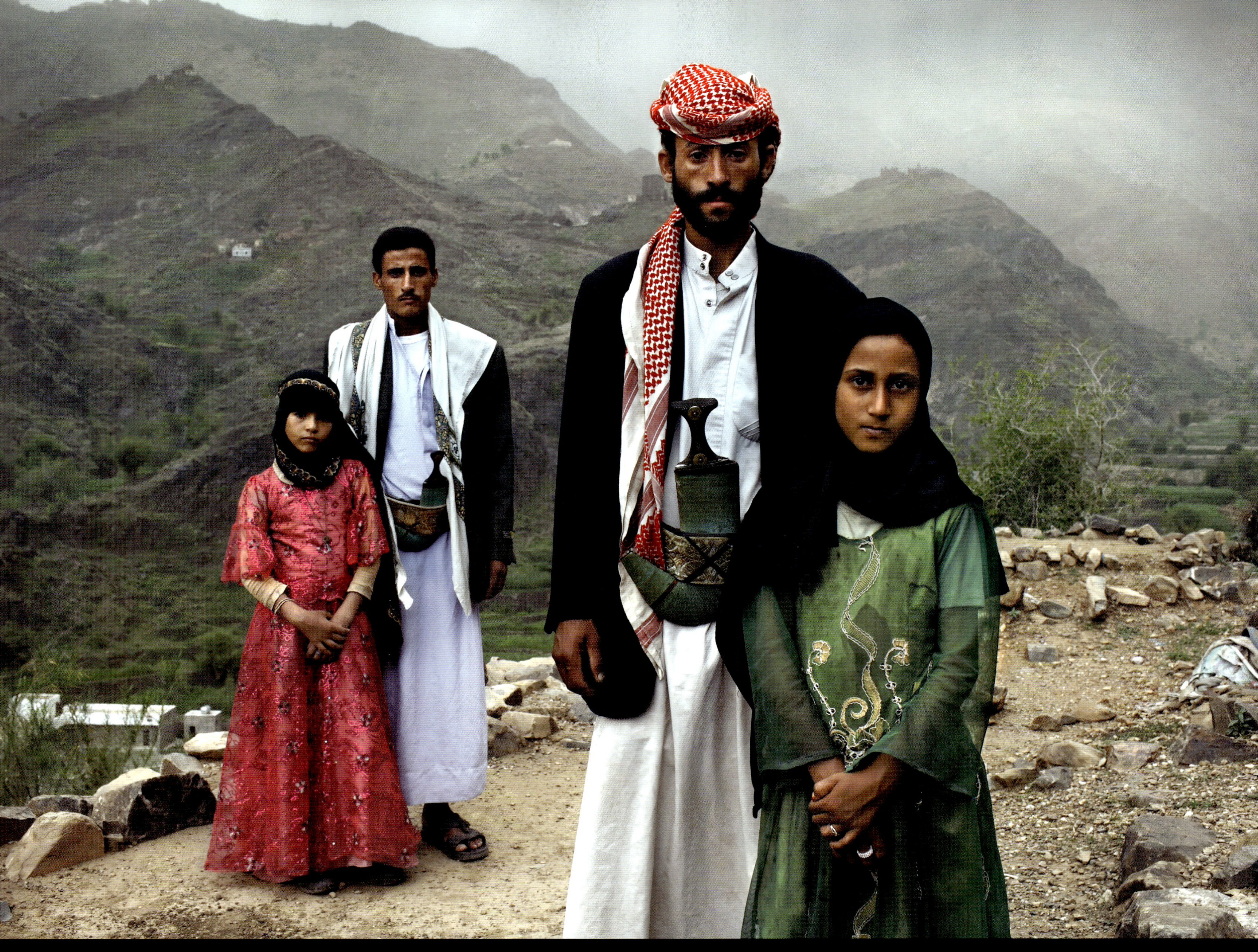

Children were allowed to be children, not forced into marriages with adults

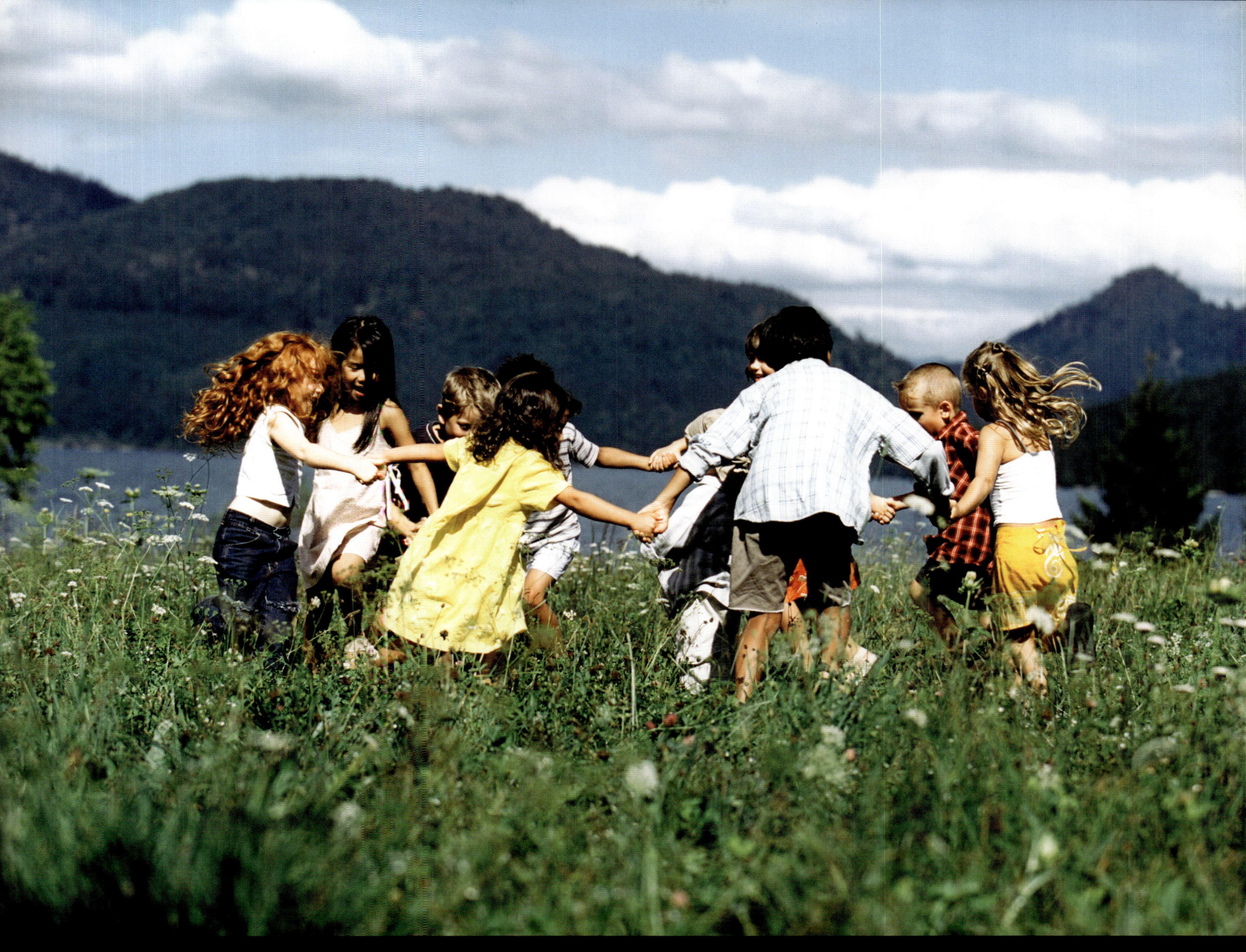

All the children were loved, and all were encouraged to follow their dreams.

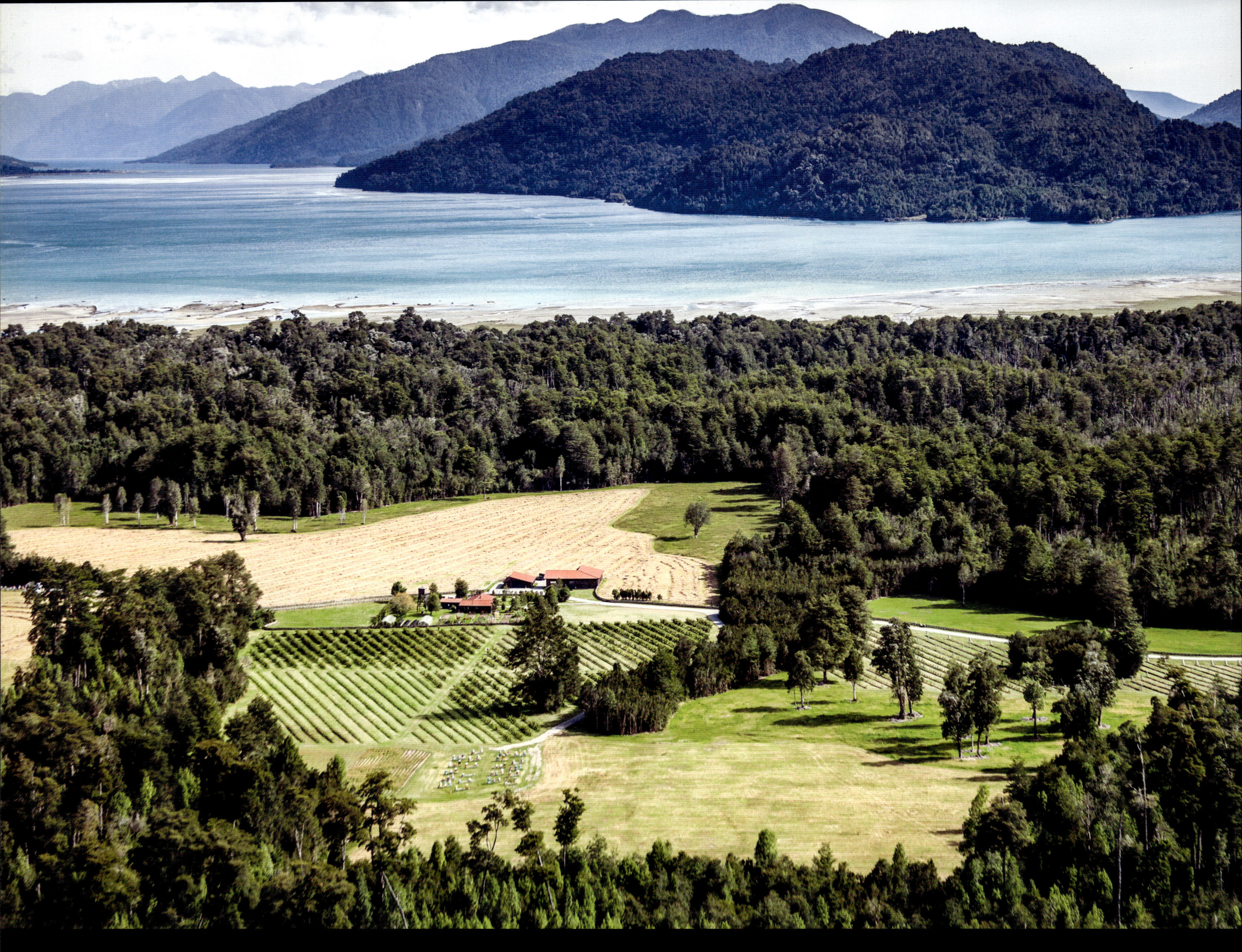
Development priorities began to shift from *more* to *better.

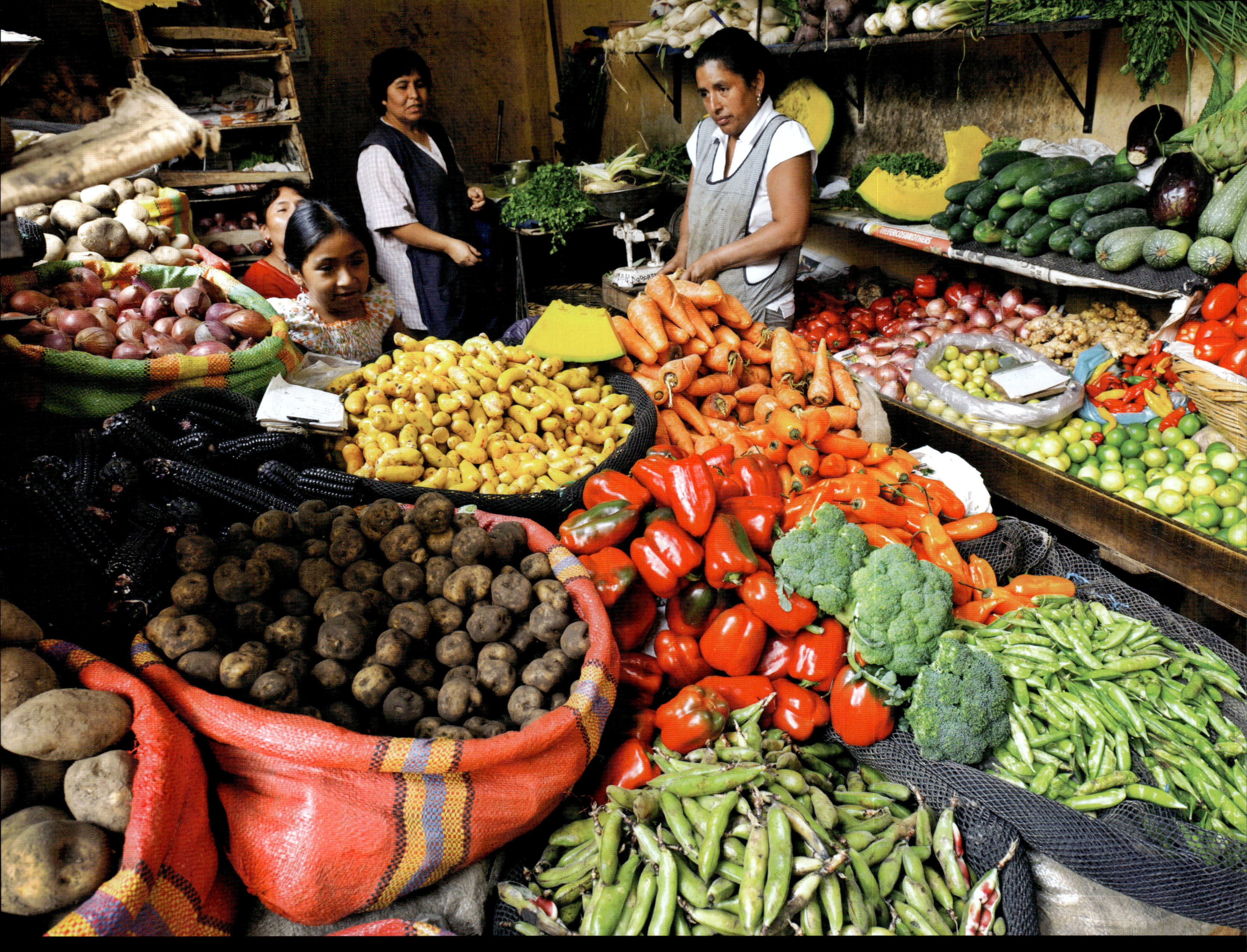

Economic objectives shifted toward sustainability, sufficiency, and resilience.

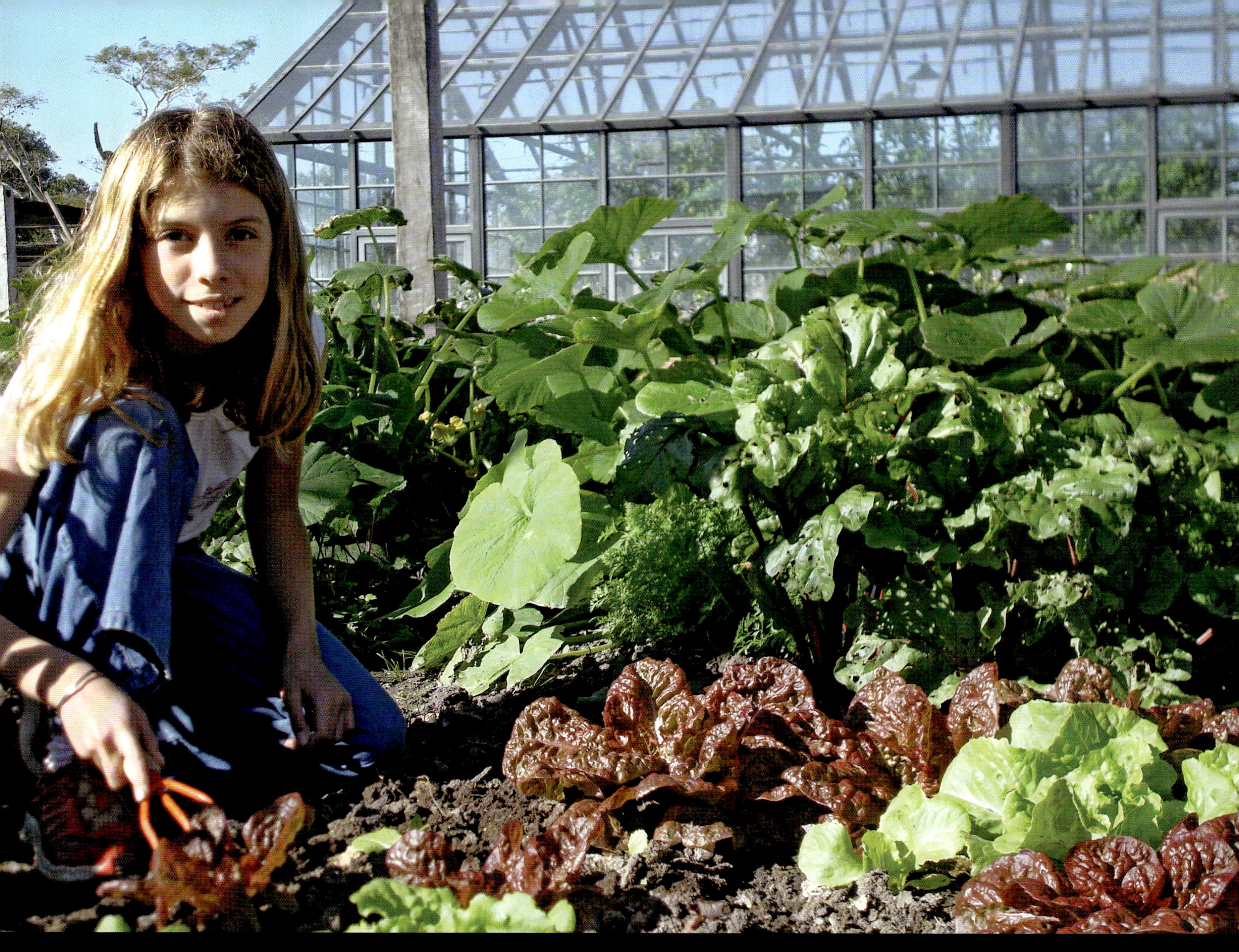

Actions were judged ethical or not by whether they helped sustain beauty, biodiversity, and health.

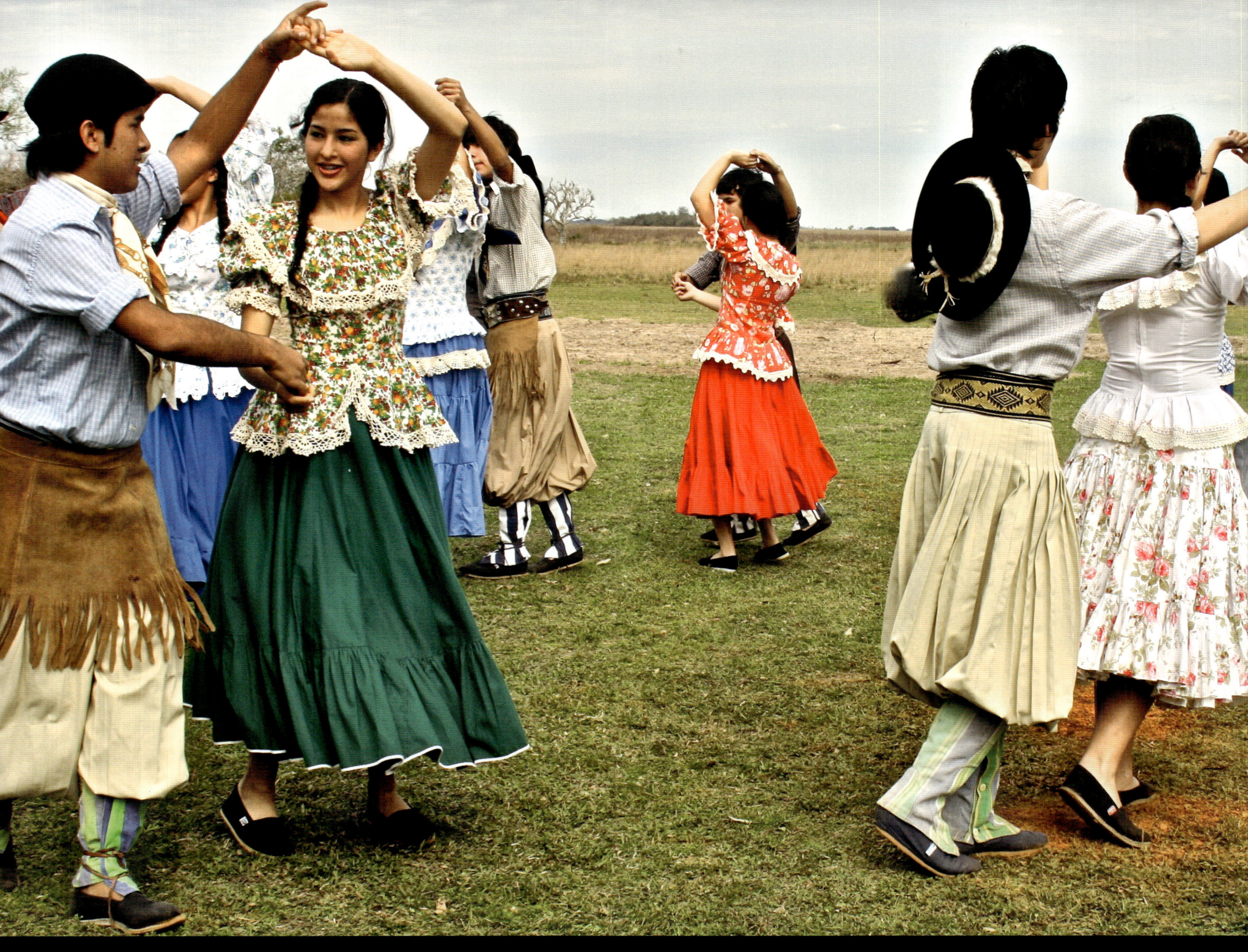

The people chose lives of quality, with sufficient time for the activities and relationships that gave them joy.

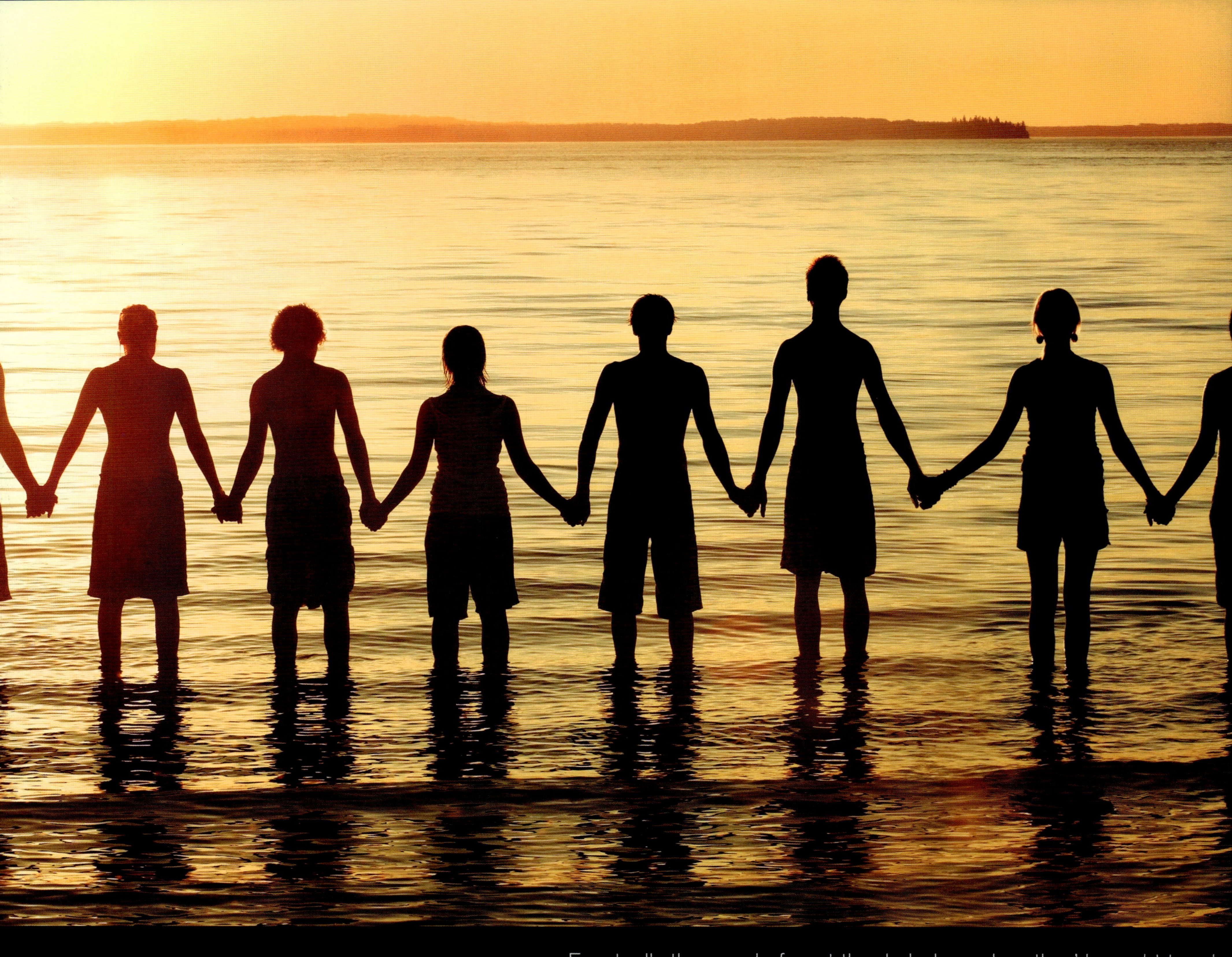

Eventually the people forgot the dark days when they'd sought to rule,

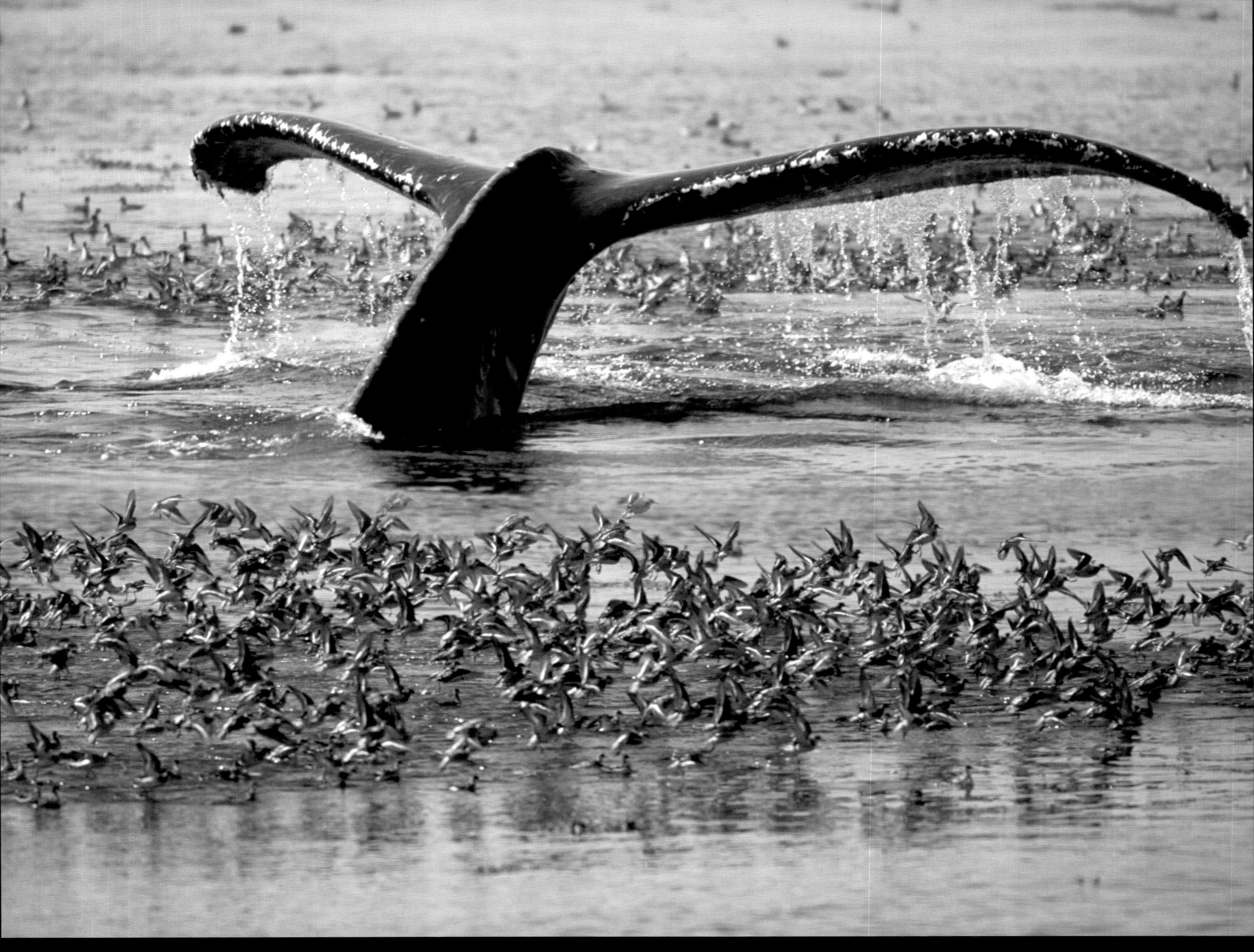

Whales, unmolested, sang in the deep.

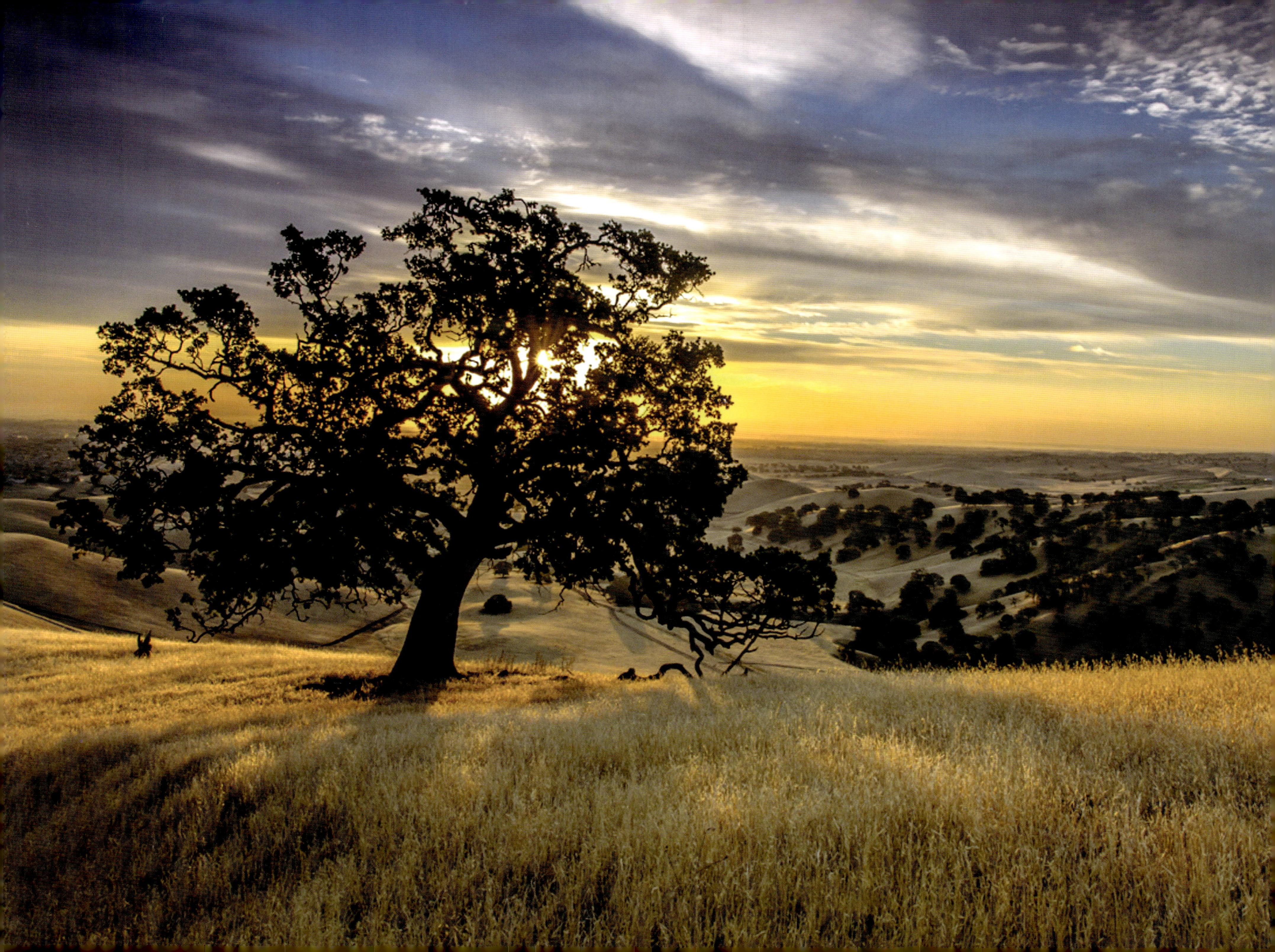

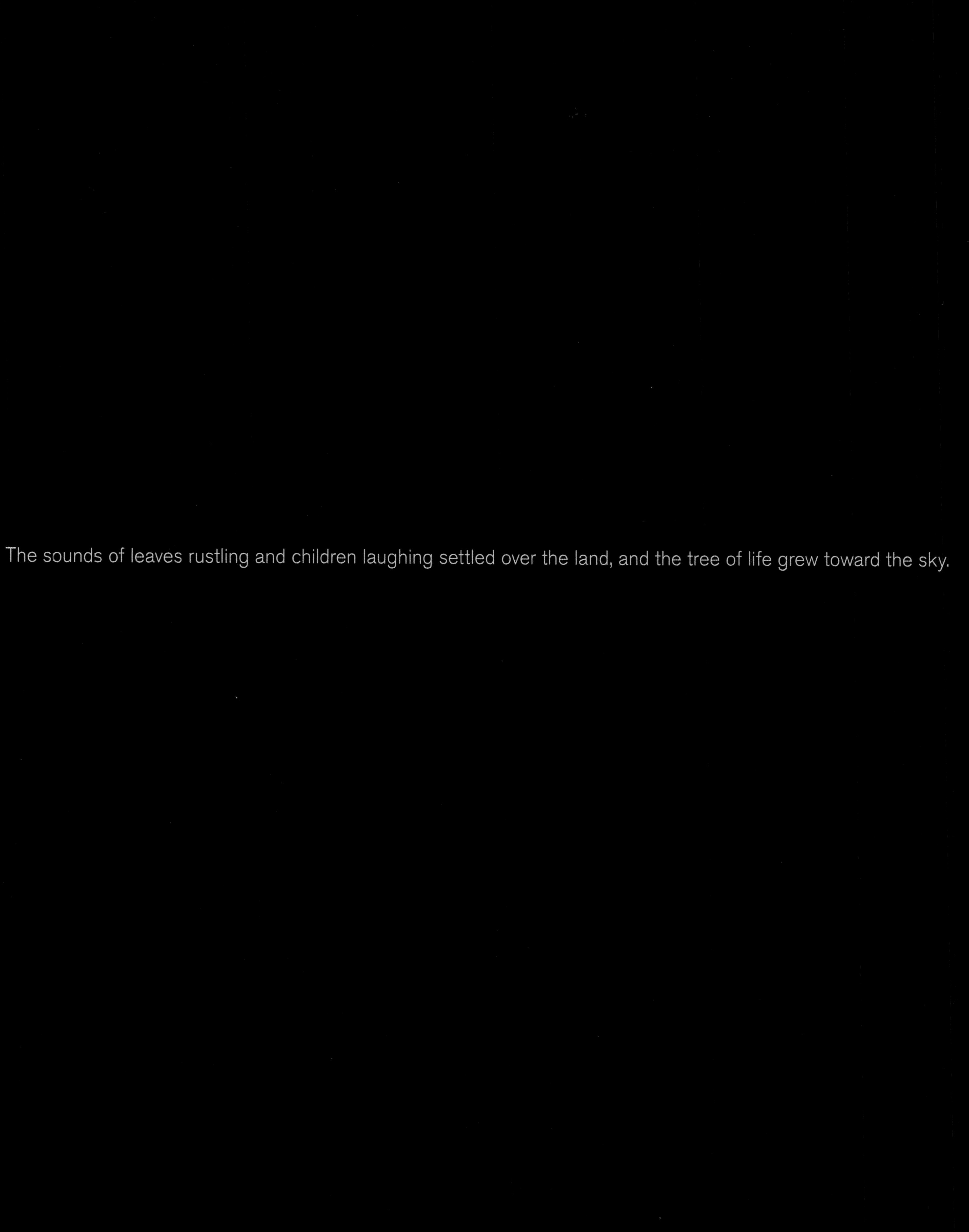

The sounds of leaves rustling and children laughing settled over the land, and the tree of life grew toward the sky.

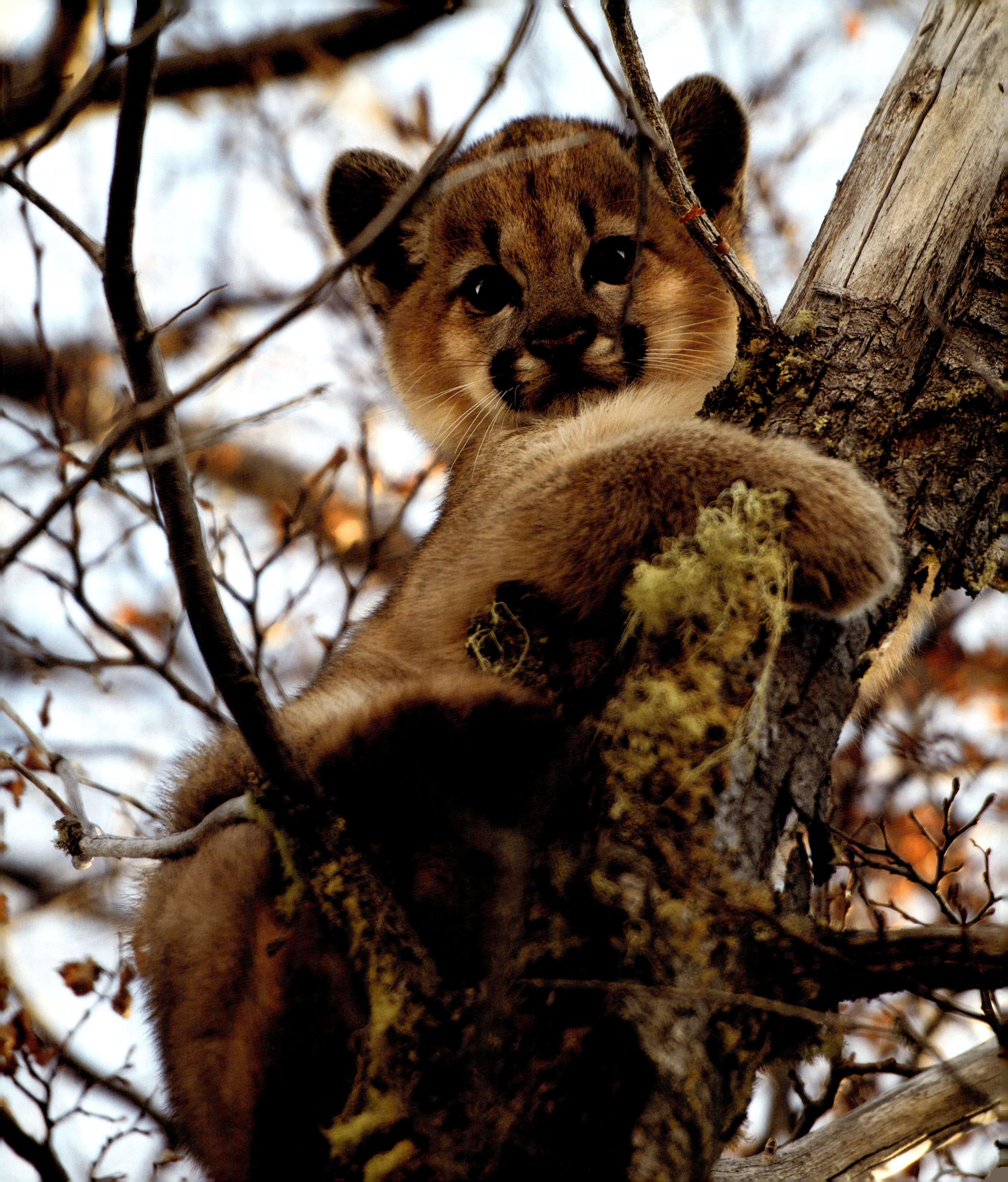

Choosing A Planet of Life

Eileen Crist

ONE OF THE COMMONPLACES of environmental writing these days is a population forecast of 10 billion (or more) people by century's end. Indeed, this projection is endlessly repeated, as if it were as inevitable as the calculable trajectory of an asteroid hurtling through space. Besides being a facile meme amenable to replication, this recurrent demographic report signals a widely shared fatalism: The coming growth has too much inertia behind it, and is far too politically sensitive, to question. At the same time, the projection reinforces a collective impression that nothing can be done to change it. Ironically, the incantation of "10 billion" seems at work as self-fulfilling prophecy, for without concerted, proactive intervention it is roughly the number to be expected; so do we hypnotize and propel ourselves in the predicted direction.

Environmental analysts have divergent responses to this particular figure (which is the latest United Nations estimate). Some are incredulous that such a number can be approached— let alone sustained—and contend that the consequences of moving in that direction will be disastrous; a catastrophe or combination of catastrophes is bound to derail professional demographers' expectations, and humanity (after enduring much suffering, or perhaps experiencing some kind of wake-up call) will stabilize at lower numbers. But other environmental observers, describing themselves as more optimistic, are endeavoring to figure out strategies that might sustain the expected billions. They hope that with the right developments and innovations in crop genetics, irrigation technologies, fertilizer application ("responsible nutrient management"), efficiency gains (including closing "yield gaps" and curbing food waste), requisite energy transitions, and other advances, the planet might feed, provide water for, house, educate, and medicate—at an acceptable standard of living for all—the coming 10 billion. There is reason to wager, they maintain, that humanity might succeed at the task, since people are resourceful, determined, and apt to get out of a tight spot even in the nick of time.

Thus where some see disaster on the immediate horizon, others submit that with another techno-managerial turn of the screw humanity might avert grim penalties to population growth. Yet despite considerable divergence in outlook, all environmental analysts agree that (even as our global numbers continue to climb) we face grueling challenges, each immense in its own right but dizzying in their unpredictable synergies: biodiversity destruction, climate change, freshwater depletion, ceilings on agricultural productivity, all manner of pollution, topsoil loss, and ocean acidification to mention some prominent examples.

Rather than taking sides between the forecast of impending tragedy versus optimism about "feeding the world," there is another way to tell the near future's story. On that telling, the issue is not whether it is possible for 10 billion people to eat industrial food, commune with iPhones, and make a decent living on planet Earth (an outlying scenario, in my view, but perhaps stranger things have happened in the universe). The point to focus on instead is that a world of so many billions does not, *in any case*, turn out well: Because such a world is only possible by taking a spellbindingly life-abundant planet and turning it into a human food plantation, gridded with industrial infrastructures, webbed densely by networks of high-traffic global trade and travel, in which remnants of natural areas—simulacra or residues of wilderness—are zoned for ecological services and ecotourism. In such a world, cruise ships with all-you-can-eat buffets will circumnavigate seas stripped of their plenitude of living beings, on waters awash with plastic refuse decomposing into bite-sized and eventually microscopic particles destined for incorporation into the worldwide food web.

What's more, a sustainable geopolitical status quo of 10 billion consumers will require comprehensive mega-technological support: offshore dike projects; more dams (already, according to a 2009 *Yale Environment 360* report, being constructed at "a furious pace"); desalinization plant construction with accompanying transport infrastructures; scaling-up of industrial aquaculture; genetic modification of crops and animals to adapt to climatic and consumer demands; cultivating so-called marginal lands to grow grasses and other plants for biofuels; the spread of the fracking scourge (globalizing "the oil and shale-gas boom"); climate engineering at global and regional scales; and the spread and normalization of factory farms. (*The Economist* praises the efficiency of the latter institution over traditional husbandry, calling it—in apparent oblivion of the term's Orwellian malodor— "the livestock revolution.")

In such a world corporations are likely to continue reigning supreme, for the coming technological gigantism (not to mention the escalation of mass consumption) will make them indispensable. Corporate expertise and products will be required to keep the biosphere on permanent "dialysis," to borrow a fitting metaphor from James Lovelock. Corporations will continue generating enormous revenues, via tax-based subsidies for their "public works" and by catering their products to huge numbers of people. (Any doubt regarding the relationship between private-sector opulence and consumer population size is dispelled by taking note of the correlation between today's wealthiest companies and their bulging middle-class client base. Indeed, capitalism is quite partial to the twin perks of population growth: cheap labor and mass clientele.) Whatever relatively natural places remain will be slated as the real estate and vacation destinations of the most affluent—as they are to a large degree today. But regardless of whether or not corporations and the gilded class entrench their reign, everyone (including the rich) will be wretchedly dispossessed, hustling for happiness on a planet degraded to serve a bloated user-species.

In such a world—whatever it augurs for humanity, which seems bleak to say the least—the exuberance of Life will suffer a tremendous blow. This Life is barely hanging on in the present world; it will not survive a world that is a magnified version of the one we live in. I use the word Life, with capital L, to mean something akin to what life scientists call "biodiversity"; unfortunately, though, the latter term is often mistakenly conflated with numbers of species on Earth. While numbers of species are a significant dimension of Life's fecundity, Life is far greater than a total species inventory—as extravagant as that inventory

may be. Life is be*wild*ering in its creative expressions, its beauty, strangeness, and unexpectedness, its variety of physical types and kinds of awareness, and its dynamic, burgeoning, and interweaving world-making.

Earth's story is about Life, whose phenomena emerge in each place uniquely and over the whole planet diversely, always contiguous and interconnected at local, regional, and global levels. Life fills niches and also creates them; life-forms accommodate other life-forms via niche construction and by their edible, breathable, or otherwise consumable waste by-products (including, ultimately, their own corpses). With the exception of mass extinction events, Life is always enabling more of itself to surge. There's ceaseless feeding on one another and on each other's by-products, as well as a co-molding of a physical and chemical environment in which more life is supported to flourish. Importantly, a vast array of life-forms—from all five kingdoms of life—are involved in building soil, which is not only Life's foundation but *itself* a living phenomenon. Through organism-mediated processes, the land brings nutrients to the seas, and the seas (through organism-mediated processes) return nutrients to the land. Forest canopies feed the life in the understory, and life in the forest understory feeds the trees and all who live in their canopies. Beings in the seas' upper layers sustain the strange menagerie of abyssal creatures, and organism-created nutrients in the depths well up and nourish fellow beings in the upper zones.

In the "interdisciplinary" dance of Life—where phenomena of physics, organismal biology, biochemistry, behavior, awareness, and chaos jostle in established and spontaneous patterns—Life creates abundance. For example, hundreds of millions of eggs wash to the sea's edge, feeding multitudes before a fraction develop into the organisms that spawned them. Prey species proliferate wildly in response to the pressure of their predators— incalculable numbers of marine creatures once sustained the tens (and perhaps hundreds) of millions of sharks and whales who existed before their concerted extermination began. Enormous, ever-on-the-move ungulate herds do not decimate the lush grasslands that feed them, but on the contrary the grasses grow because of them, and the animals and grasses (with other life-forms) together create more soil. Freely moving, pristine rivers teemed with fish even in recent history. Great flocks of birds graced skies, wetlands, and seashores. And land, sea, and air animal migrations have not only told the seasons' stories but contributed to bringing the seasons into being. The intermingled manifestations of Life on Earth—when Earth is allowed to manifest them—have no finitude.

As for a popularized claim that, alas, life is all about struggle, competition, and selfishness, it is best to turn away from such claptrap: for it is only within a planet of Life, a Life-world, that phenomena of struggle, competition, and selfishness arise and pass away in their relevant contexts. The Life-world itself is far more encompassing in the kinds of phenomena it manifests and cannot be reduced to a one-dimensional schema — except for *the one thing* we know in the marrow of our bones and in our hearts: that the Life-world is *All*-good.

And here's the crux of the matter: Humanity can choose to live on a planet of Life instead of haplessly plunging toward a human-colonized planet on dialysis ("wisely managed"). To live on a planet of Life it is necessary to *limit* ourselves so as to allow the biosphere freedom to express its ecological and evolutionary arts. For that, we in turn need to cultivate the breadth of imagination to give the concept of freedom wider scope—pushing its territory beyond the sheath of human exclusivity. In the name of a higher freedom that encompasses Earth and its entire community of beings, we can choose to let the world be the magnificence and wealth it was and still can be. Borrowing words from nature writer Julia Whitty's *Deep Blue Home*, this path is about cultivating intimacy with the natural world, taking as our lover the way things really are and finding our way home.

But the wisdom of limitations—of our numbers, economies, and places of habitation—is rarely entertained in mainstream thought for what it is: the elegant way home and the surest means for addressing the deepening (and likely self-endangering) problems of extinctions, ecosystem destruction and simplification, rapid climate change, freshwater and topsoil depletions, as well as (relatedly) mounting concerns about "feeding the world." The path of limitations is rarely entertained, for it is assumed to be unrealistic and thus politically inexpedient. But knowledge of the multiple stresses on the biosphere, along with an understanding of the adverse, volatile ways these may compound one another,

yield the recognition that drastically scaling down the human project is the most realistic approach to imminent catastrophes. If political expediency cannot see that, then political expediency and those who speak for it need to be deposed so we can get on with *the real work*.

IN THE MEANTIME, even as the available option of limitations is bypassed as ostensibly unrealistic, the prevailing question voiced with increasingly shrill urgency is: *Can the Earth feed 10 billion people?* By most expert accounts, because of population growth along with the rise of meat and animal product consumption, food production will have to double by 2050 to meet demand—and the big question is: *Can it be done?* There is an effort under way to figure this out, by experimenting in research and development labs, working in research stations, and analyzing agricultural databases. And because it is well known that most (and certainly the most fertile) arable lands are already in cultivation, and that the areas where wild creatures live are already pushed to their limits, the effort to increase food production (to double it in about forty years and triple it by century's end) is invariably escorted by the caveat that it must be done without "further damage to biodiversity" or "taking over more uncultivated lands."

Since at least the early 2000s, this "ecologically correct" sound bite has been activated in environmental writings, journalistic reports, and corporate web pages: We must produce more crops (for food, feed, and fuel), as well as more meat and animal products, by means of careful planning and management, with minimal additional ecological impacts. Oddly, the latter disclaimer is stated as if tropical forests are not today giving way to soybean monocultures, cattle ranches, and oil palm, sugar, tea, and other plantations; as if large-scale acquisitions recruiting land in Africa and elsewhere are not already under way in the name of "food security"; as if marine life is not being chewed up by the industrial machine; and as if rivers are not today so taxed by damming, extraction, diversion, and pollution that the crisis of freshwater Life may well be the gravest extinction site on Earth (a big nonevent as far as the public and its elected officials are concerned). Despite all these things happening already today (in a global economy of 7.3 billion), those at work to figure out if food production can be doubled and eventually tripled (to serve a world of 9, 10, or more billion in an intensified global economy) always add that it must be done without additional ecological damage. When we encounter such pious declarations of intent we'd do well to recall Hamlet's sardonic response to the question, "What do you read?" *Words, words, words.*

Those endeavoring to figure out how to increase food production without more harms to nature may well be sincere; but they appear to be in the throes of wishful thinking. For even if for a moment we ignore the fact that present-day industrial agriculture, industrial aquaculture, and industrial fishing constitute a mounting planet-wide disaster—which goes largely unremarked only because it is nigh equaled by planet-wide unawareness—simply *saying* that we need to grow more food without further ecological destruction is not going to stop hungry and acquisitive people from taking what they need and think they need: clearing more forests and grasslands, moving up slopes, overgrazing pastures and rangelands, decimating sea creatures, replacing mangrove forests with shrimp operations, or killing wild animals for cash or food.

Even so, the most pernicious thing about this formulaic mandate-plus-caveat—grow more food, don't damage more nature—has yet to be stated: namely, that it insinuates that the current damage our food system inflicts is acceptable and irreversible. Hands down, however, food production is the most ecologically devastating enterprise on Earth. (More on this shortly.) Yet mainstream discourses do not tend to flag the food system's earth-shattering demands on the biosphere. Instead, the *current* ability to produce ample amounts of food—enough for all, including those not yet at the table—appears to merit a different cluster of conclusions: that humanity's food-producing capacity is not constrained by natural limits; that we may be able to stretch that productivity even further via managerial and technological innovations; and that *Homo sapiens* is unlike all other species, who are checked by nature whenever their numbers exceed the capacity of the environment to sustain them. Indeed, the belief that humans are *exempt* from any natural "carrying capacity" is a cornerstone of the mission to continue expanding food production to support the coming billions.

The demographic idea of *carrying capacity* refers to the maximal population of a species that its environment can support, without that environment becoming too degraded to support the species in the future. If a species, for some reason or other, does exceed its carrying capacity—with numbers mounting beyond what the natural setting can sustain—the consequences are implacable: Starvation, disease, and death follow, until the population is brought back within a supportable range. While this natural law of the relationship between population size and sustenance appears broadly applicable in the animal kingdom, here's the key point regarding human exemption: It is widely believed that history has shown that it does *not* apply to us.

In the early nineteenth century the Reverend Thomas Robert Malthus, in his *Essay on the Principle of Population*, endeavored to apply the logic of natural limits, and the severe costs of transgressing them, to humanity. He predicted that because population grows faster than food production, human numbers would outstrip the available food supply and people would reap the woes of famine, disease, and war. But the two centuries following his analysis did not see a human population crash, as food production kept up with mounting numbers of people; in fact, during the last half of the twentieth century the rate of food production even *outpaced* the rate of population growth. So Malthus's thesis came to be viewed as repudiated, and the doctrine of human exemptionalism from natural limits received a victorious boost.

Indeed, the foreboding forecast that the human population would inevitably exceed the amount of available food to (at least in principle) feed everyone did not come to pass. It was refuted by converting Earth's most fertile lands for agriculture (after being denuded of their Life-rich forests, grasslands, and wetlands); by taking over extensive swaths of natural areas for domestic animal grazing; by appropriating half the world's freshwater—with the biggest share diverted for agriculture; by applying enormous quantities of synthetic chemical and fertilizer pollutants; and by plundering untold numbers of wild fish. In other words, the prediction of human tribulation in the wake of unsustainable numbers was refuted by means of the near conversion of the biosphere into a human-food pantry.

The seemingly "winning argument" that humanity is uniquely capable of keeping food production apace with (or ahead of) demographic growth reveals a profound lack of insight into the bigger picture of what stretching our food-producing capacity has really portended. It reveals an inability to appreciate—or even to entertain as a passing thought—that human carrying capacity (how many people the Earth can support) has been extended not simply because we are so clever at manipulating natural processes and inventing stuff, but through forcefully taking over the carrying capacity of other life-forms and, in the process, wiping them out regionally or globally. Moreover, the exemptionalism thereby displayed—that we are not bound by natural conditions like other species—beyond the superficial "fact" that it seems to be, serves *conveniently* as an ideological handmaiden of human expansionism. For what the doctrine of exemptionalism tacitly conveys and inculcates is that because humanity is so special by comparison to all other creatures, it is proportionately that much more entitled; and thus the acts of war on the natural world that undergird our expansionism (for food production in particular) become unrecognizable *as* acts of war.

The question of whether ultimately there are (or not) natural limits to our food-producing ability, which will (or not) check human demographic growth, is not so interesting; *the experiment required for the final verdict is an ugly one either way.* Instead, I along with other deep ecologists invite consideration of something far more enticing: that by choosing the wisdom of limitations and humility, humanity can reject life on a planet converted into a human food factory and allow for the rewilding of vast expanses of the biosphere's landscapes and seascapes. To drive home why the latter option is much more beautiful (as well as more prudent), I turn to the highlights of how food production is contributing the lion's share of anthropogenic ecological havoc.

Cropland uses a portion of the planet the size of South America, while land for grazing farm animals eats up an even larger share—an area the size of Africa. Effectively, humanity has seized the temperate zone for agriculture, wiping out all or most former nonhumans and ecologies in order to mine the soil. ("How did *they* get on top of *our* soil?") The raising of tens of billions of domestics has exacted the eradication or displacement of wild animals from their former habitats, the persecution and slaughter of carnivores viewed as threats to farm animals (themselves reduced to being "live-*stock*"), and the erosion and degradation

of lands from overgrazing. And the alternative to grazing—*The Economist*'s so-called livestock revolution—constitutes a pollution nightmare and an egregious violation of basic decency in the treatment of animals. (Yet factory farming is a production method that today both supplements grazing and is swiftly spreading.) Regarding the seas, the human food factory has demanded that 98 percent of them be fishable. This reign of terror for marine species is partly underwritten by an institution called, without the slightest irony, "the freedom of the seas." As a consequence, only about 10 percent of the big fish are left and there is no end in sight to the demand on everything from krill to sharks. In the literal and figurative industrial mowing of the world's oceans, the countless beings who suffer and die in the name of mass consumption and profit are referred to as "catch" and "bycatch."

Furthermore, food production contributes at least 30 percent of anthropogenic greenhouse gases; the latter are driving a climate change episode that—barring the energy transition everyone is still waiting for—*could* egg the planet to an average temperature increase in the ballpark of the Paleocene-Eocene Thermal Maximum. (If you have never heard of the Paleocene-Eocene Thermal Maximum, please wiki it.) The food factory—the one often touted as a miracle of ingenuity bestowing the badge of exemptionalism on *Homo sapiens*—consumes at least 70 percent of the freshwater taken from ecological watersheds, thus depriving the nonhumans who called that water home, and killing or driving them to extinction (in many cases even before we could meet them). Food production drives soil erosion and desertification, giving rise to ocean-spanning dust storms. It also depends on constant applications of pesticides, herbicides, and other biocides: Indeed, many consumers and growers, alike, have been duped by corporate salesmen (and their government allies) into believing that it is normal and necessary to poison the biosphere for the purpose of producing human nourishment. Streams, rivers, lakes, wetlands, and estuaries around the world are fouled or deadened by agricultural runoff and farm animal excrement—all just "how things have to be" if we are to eat.

This unprecedented impact on the living world allows for the production of so much food as to seemingly demonstrate our ability to feed billions and, with some additional resourcefulness, perhaps feed even more. From a deep ecological perspective, however, the unprecedented ecological impact demanded for the production of so much food has demonstrated our capacity to take a magnificent planet—second to none in the known universe—and turn it into, or use it as, a human feedlot, and then muster the arrogance to call this act of pilfering and degradation an "achievement."

In his latest work, *Countdown*, author Alan Weisman sums our current Green Revolution food system as involving "fossil fuel gluttony," "river fouling fertilizers," "dependence on poisons," and a "monocultural menace to biodiversity." So how is the amount of food we produce to be doubled or more without additional damage? Remarkably, one of the strategies being considered is to extend the productivity of Green Revolution methodologies to places they have not yet reached. Indeed, as the global population continues to grow, spreading the Green Revolution in order to "feed the world" will be the likely tack of the present-day policy framework, which is beholden to (in no particular order) corporate interests, institutional inertia, and acute anthropocentrism. Predictably, the call to extend the Green Revolution is cushioned by all the ecologically correct pleas for wiser uses of water, more efficient application of fertilizers, prudent deployment of pesticides and herbicides, inclusion of no-till agriculture, and so forth: an appeal to "greening" the Green Revolution that not only is politic but also constitutes necessary retooling in a time of potential phosphate shortages, water wars, and fossil fuel price hikes. But making a destructive food model more efficient does not the model make good. At best it yields a world—as Rachel Carson so cuttingly put it—that is *not quite lethal*.

I HAVE DIGRESSED into the ecological discontents of humanity's current food production in order to submit the following: that the social mission to double or triple it is madness. But the proposal to move deliberately in the direction of more than halving our global population, and simultaneously radically changing our food system, is not.

If women (and their partners) today were voluntarily to choose having an average of one child (meaning many would choose none, many one, and others no more than two), then the world's population—instead of climbing toward 10 billion—would stabilize and then begin descending toward 2. Were the current generation of childbearing women to embrace this voluntary mandate for the sake of a living planet and the quality of life (perhaps even survival) of future people, how could this possibly be construed as a sacrifice? It is an intelligent and compassionate action that many people would be willing to take if they became properly informed and knowledgeable about the planetary emergency we are in. As for those who hear "coercion" in such a proposal—and respond by defending "human reproductive rights"—they should at least take a moment to acknowledge a fact that population experts are well aware of: that some of the grossest violations of human rights are perpetrated in societies that force women to start (involuntarily) having children when they are barely beyond childhood themselves, and to continue reproducing until their bodies give way or they are no longer fertile. The population question is indeed pressing in countries where patriarchic, polygamous, fundamentalist, and military cultures are keeping women handcuffed and thus adding roadblocks to a restored future.

Yet population size is not strictly a "developing world" problem but a global issue and task. One of the most effective and tangible ways to address climate disruption, as well as to curb the excessive consumption of everything (including food), is to move toward the substantial reduction of the number of consumers worldwide, meaning both the populations of the developed world and of "emerging economies" in Asia, Southeast Asia, and Latin America. Concerning the developed world's responsibility in addressing overpopulation, it is also reasonable to insist that monetarily affluent nations and institutions should provision the financial backing and expertise for bringing state-of-the-art reproductive health services around the world—including their home territories. For example, half the pregnancies that occur in the United States are unintended—a statistic that speaks to a social, cultural, and educational failure, not just to a weakness of human nature. The important work of demographic expert Robert Engelman has shown that if unintended pregnancies (everywhere) were reduced to a humanly possible minimal, this would lead to a reduction in *both* population size and numbers of abortions.

Wherever concerted policies to lower birthrates have been implemented, birthrates have declined with alacrity. By concerted policies I include the following: prominent, unembarrassed public discourse and campaigning on the issue; prioritizing the education of girls and women; establishing reproductive clinics that are accessible and affordable to all; training large numbers of health workers for grassroots education and support; making marriage counseling widely available; bringing sex education to school curricula; providing the full array of modern contraceptive methods for free or at minimal cost; and instituting legal, safe abortion services. On the latter controversial point, it needs to be added that implementing all the above measures would significantly lower the number of abortions worldwide as well as the number of deaths from slipshod, illicit abortions.

The combination of heightened public awareness, the empowerment of women, and the availability and affordability of up-to-date reproductive information and services yields swift declines in birthrates. Such declines have nothing to do with the imposition of some top-down coercion; rather, they follow from a straightforward bio-cultural cause: that the vast majority of women, when they attain free choice, rarely want more than one or two children, because numerous offspring are hard on the female organism and also take time away from other personal pursuits. As the peerless work of population analyst Martha Campbell has shown, this natural female propensity for few offspring surfaces straight away, once barriers to reproductive services are removed and freedom of choice becomes reality. If, additionally, today's fertile women were presented with the beautiful and compassionate mandate to help alleviate the world's most pressing ecological and social problems, then the average fertility rate might well shrink even further. Does this sound unreasonable? Certainly not more so than the unthinkable mission to double or triple food production, which augurs a colonized and ecologically impoverished biosphere, haunted by scarcity, and possibly marauded by nasty social mayhem to boot.

Bringing our global population down to, say, 2 billion will not be the magic bullet that solves every ecological and social problem. But we can rest assured that it will be *a* magic bullet for doing so. Significantly lowering our numbers facilitates a more harmonious way of life on Earth in at least two ways. First, many problems— from traffic jams, to health care budgets, to climate change— become more tractable as the dimension that magnifies them is curtailed. Lowering our numbers, in other words, helps *downscale*

harms: For example, there is a yawning difference between a world of 1 billion vehicles (causing damage enough) versus a world of 2, 3, or 4 billion vehicles (the direction we are headed). There is also a vast difference between urban settlements beautified and balanced by an abundance of open, green spaces versus the nightmare of unending road, housing, and strip-mall construction to serve the glutton of sprawl.

The second way in which significantly lowering our global population supports the turn to what we might call "beautiful human habitation" involves food production: A lower population will make possible the radical transformation of an industrial food regime that is currently bludgeoning ecologies, wild and domestic animals, and human wellness. (Four leading causes of disease and death are linked to industrial food, and especially to the consumption of mass-produced, low-quality animal products: heart disease, diabetes, cancer, and stroke.) The whole world can indeed be fed: with organically grown, nutritious food; by prioritizing local and regional food economies; without mining, polluting, and dispersing the soil but by caring for it and building it; through diversified, smaller-scale farm operations modeled on natural ecosystems; in lovely and fecund interfaces with wild nature ("farming with the wild"); and by forsaking high quantities of animal foods, for the occasional consumption of such foods produced with due consideration to ethical and nutritional values. This wholesome turn only becomes possible if our global numbers are far lower than today's.

We need an authentic green revolution. Instead of holding demographic growth as given, and a biosphere-wrecking food system as normal, let's imagine what the world might look like if we renounced both. Such a world would be dramatically more beautiful and sane following expansive rewilding—with abundant food, ecologically and ethically produced; with streams, rivers, lakes, and estuaries returned to being living waters; with deforestation halted and grassland ecologies reinstated; with the extinction crisis arrested and seas thriving again with Life; and with climate change made more manageable via carbon-sequestering forests and grasslands and decelerated emissions. If all these things can be achieved, what is keeping us from pursuing such a world? Indeed, what is detaining us from creating a civilization in harmony with wild Earth?

DEDICATION

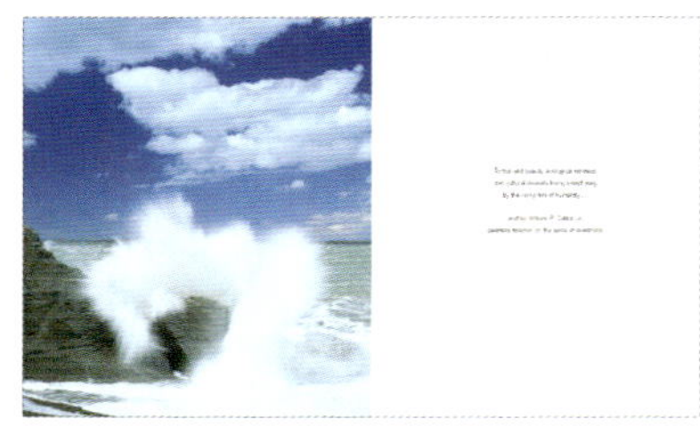

Patagonia coast: Monte León National Park, Argentina; photo © Antonio Vizcaíno/America Natural.

CONTENTS

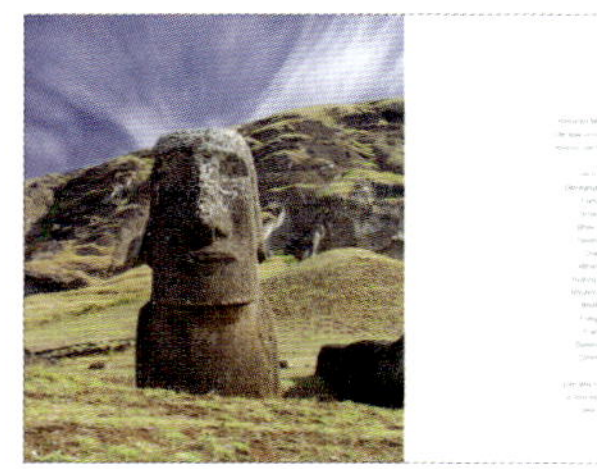

Easter Island: Giant statues, "moai," on Easter Island have become icons for a civilization that overshot the land's carrying capacity; © Christian Wilkinson.

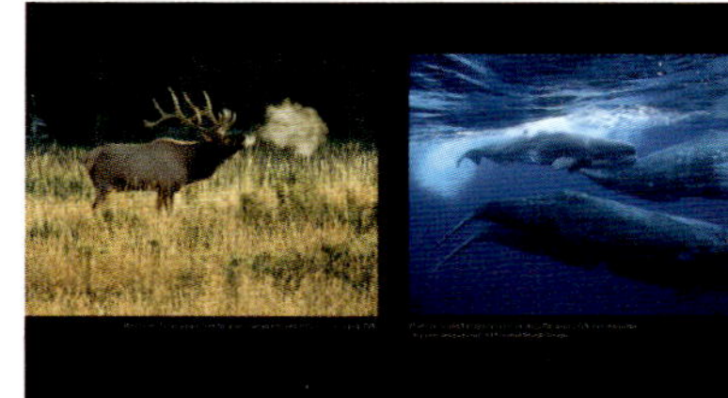

Elk: Bull elk bugling in Yellowstone National Park, Wyoming; © Royce Bair.

Whales: Sperm whale pod with calf, Azores Islands, Portugal; © Hiroya Minakuchi/National Geographic.

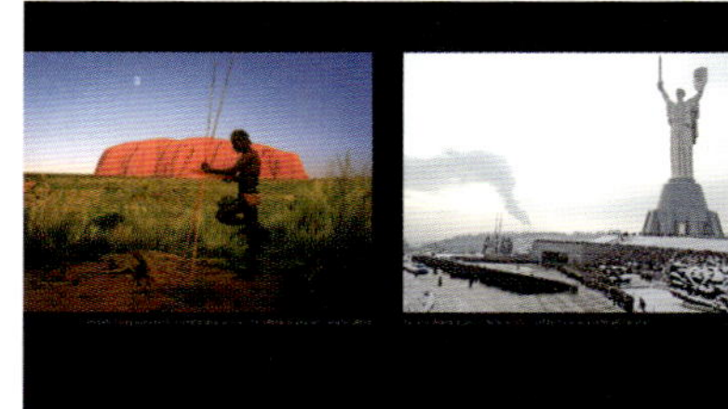

Man: Person connected to place—Aborigine near Ayers Rock, Australia; Alamy.

Statue: Motherland Monument, Kiev, Ukraine; © Anatolii Stephanov/Reuters.

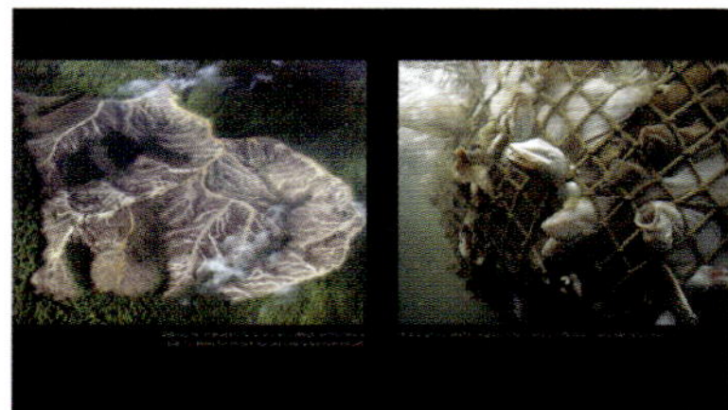

Deforestation: Logged-over area in the mountains of Jambi province, Indonesia; AFP/Getty.

Fish: Cod in net off the Gulf of Maine; © Bill Curtsinger/National Geographic.

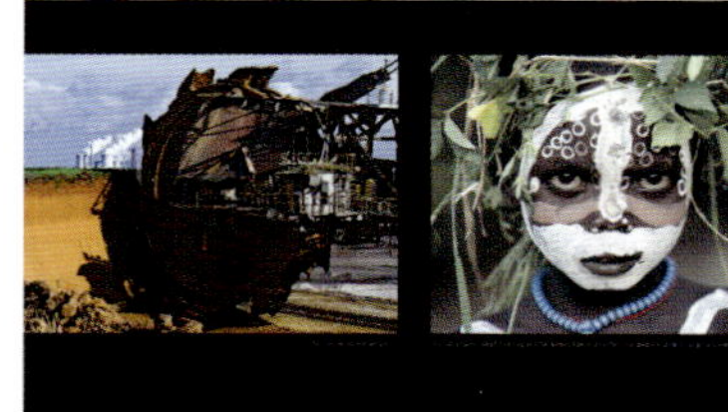

Giant Machine: The Bagger 288, largest nonstationary machine in the world, removing "overburden" prior to coal mining in Tagebau Hambach, Germany; © Achim Blum.

Child: Omo girl in Ethiopia; © Hans Sylvester.

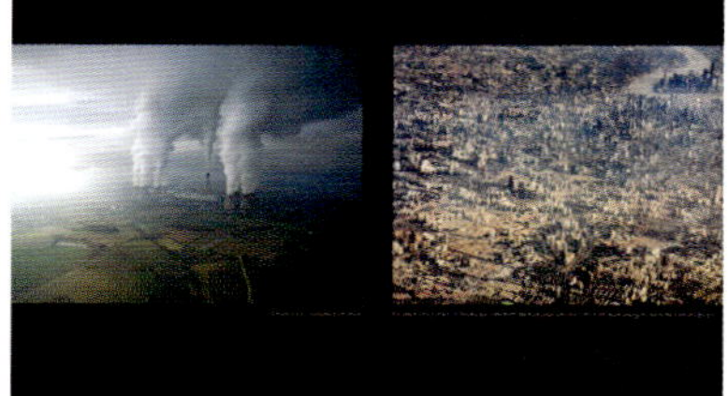

Darkening Skies: Coal-burning power plant, United Kingdom; © Jason Hawkes.

Megalopolis: Shanghai, China, a sprawling megacity of 24 million; © Mike Hedge.

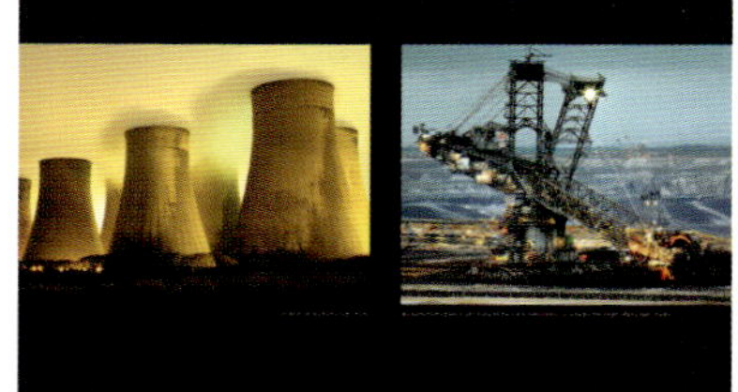

Cooling Towers: Nuclear power station in Scotland; Alamy.

Bucketwheel Excavator: Strip-mining coal in Germany; © Jorg Dickman.

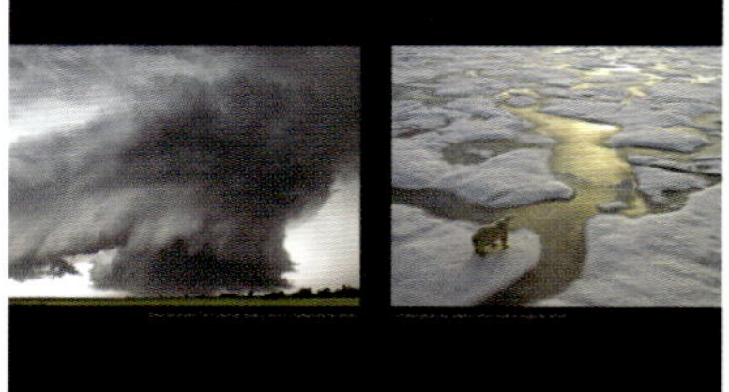

Tornado Storm Cloud: Supercell over Nebraska; © Mike Hollingshead/Extremeinstability.com.

Polar Bear: Poster child for disappearing Arctic ice and accelerating climate change; Getty Images.

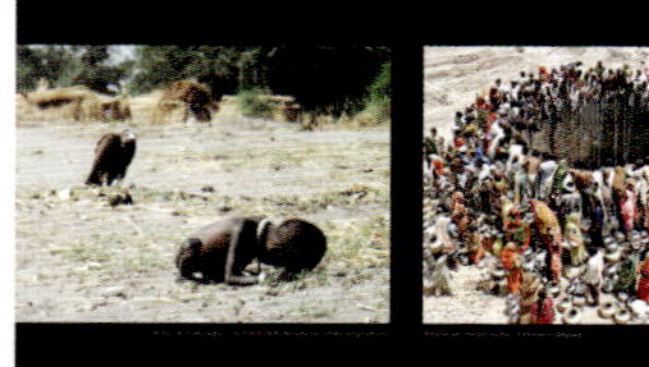

Child and Vulture: Pulitzer Prize-winning image of vulture and starving child, Sudan; © Kevin Carter.

Water Well: Crowding around a communal well in India; Reuters.

INTRODUCTION

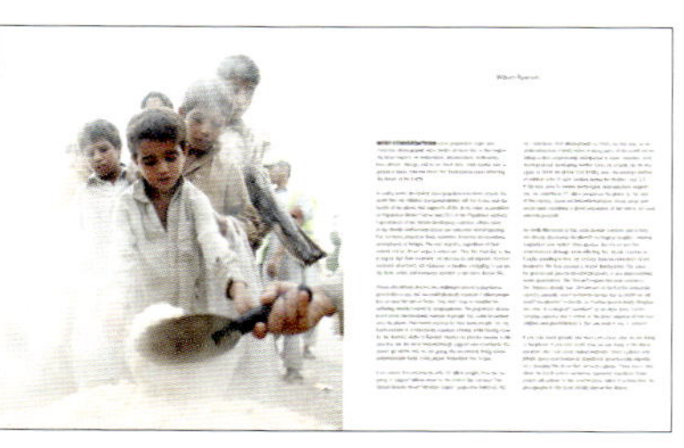

Boys: Flood-affected children in Pakistan stand in a queue to get food relief; Reuters.

DEMOGRAPHIC EXPLOSION

FOREWORD

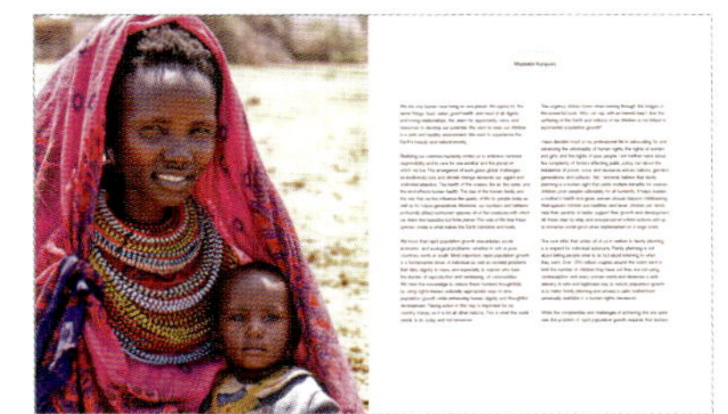

Mother and child, Kenya;
© Michele Burgess/Alamy.

PARABLE

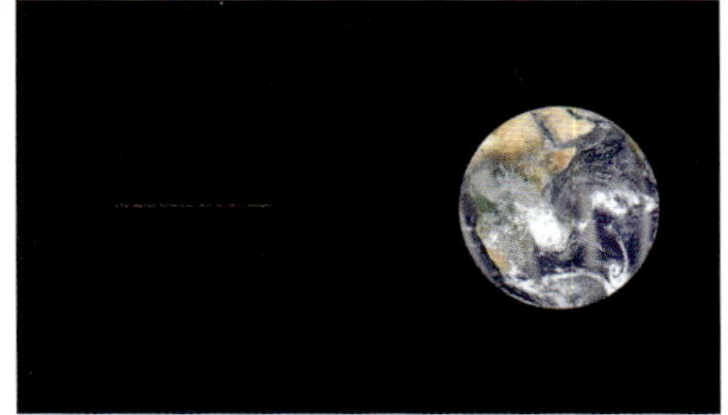

Earth: A fine little planet, roughly 4.5
billion years old, filled with life; NASA.

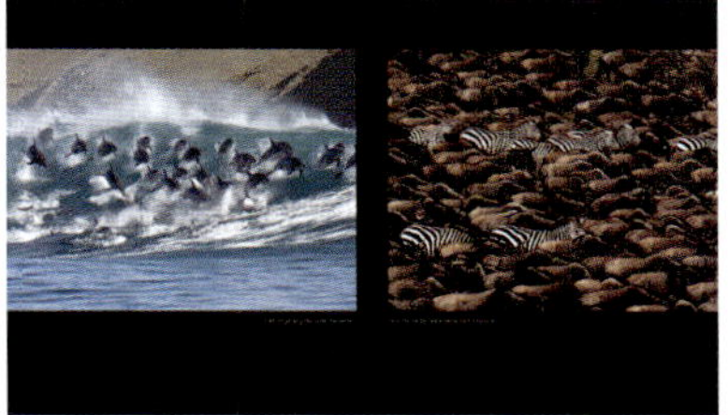

Dolphins: Off the South African coast;
© Mark Van Coller.

Zebras and Wildebeest: Animal migration
in Serengeti National Park, Tanzania;
© Mitsuaki Iwago /National Geographic.

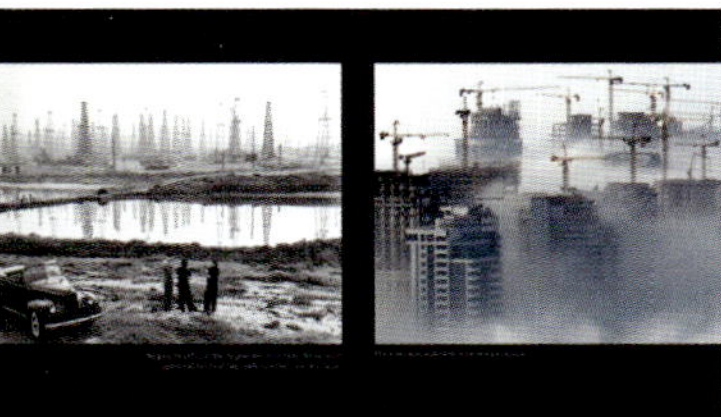

il Wells: Oil field at Signal Hill,
alifornia, 1941; © B. Anthony Stewart,
ational Geographic.

onstruction Cranes: Expanding urban
kyline in Dubai; Reuters.

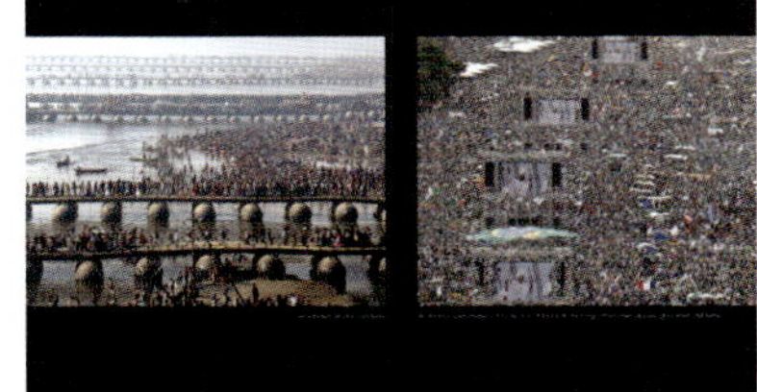

Multitudes on Bridges: Kumbh Mela
Hindu holy festival; AP.

Papal Visit: Pope Francis conducts
outdoor mass in Rio de Janeiro, Brazil;
Reuters.

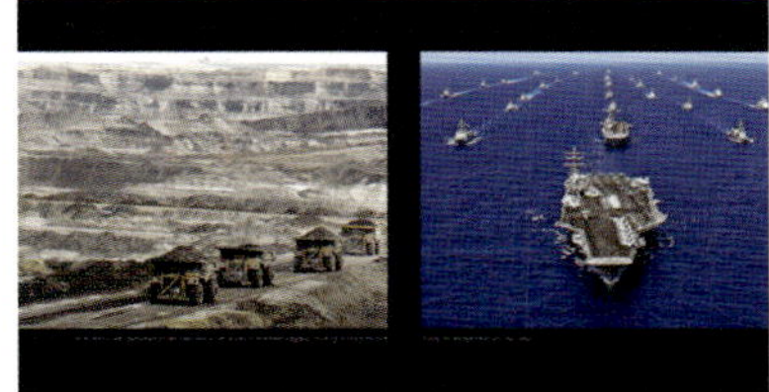

Trucks: Surface mining as part of tar
sands development, Alberta, Canada;
AP.

War Ships: USS Ronald Reagan aircraft
carrier in the Pacific Ocean; Alamy.

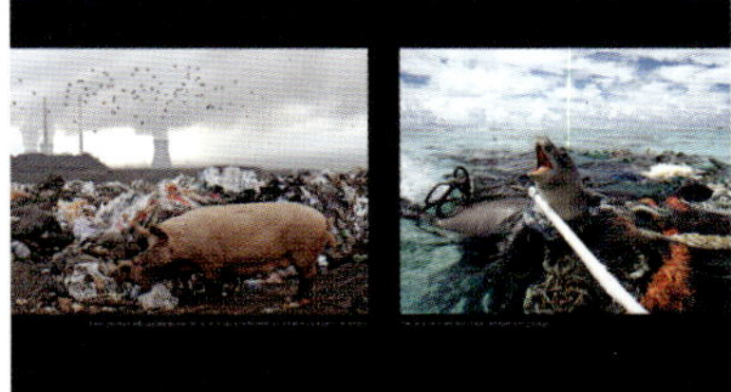

Pigs and Pollution: Pigs feed on waste
dumped near the cooling towers
of a coal power plant near Skopje,
Macedonia; © Robert Atanasovski, AP.

Seal in Net: Hawaiian monk seal caught
in fishing tackle off Kure Atoll, Pacific
Ocean; © Michael Pitts/NaturePL.

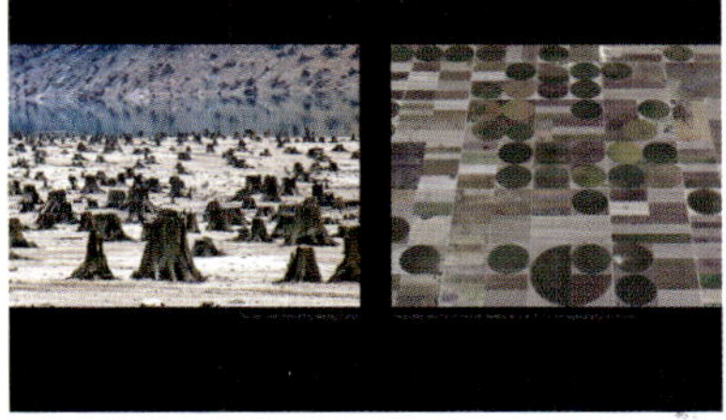

Clear-cut: Industrial forestry degrading
public lands, Willamette National Forest,
Oregon; © Daniel Dancer.

Irrigation Circles: Making arid lands
bloom, at least for a while. Irrigation
circles in Texas; Google Earth/Image
Landsat, 2013.

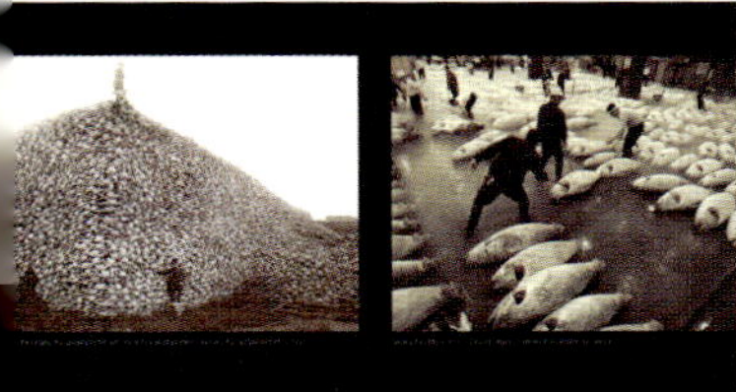

ison Bones: Mountain of collected
kulls to be ground into fertilizer; Burton
istorical Collection, Detroit Public
ibrary. Glenbow Archives NA-2242-2.

una Market: Atlantic bluefin tuna for sale
t the Tsukiji Market, Tokyo, Japan; Corbis.

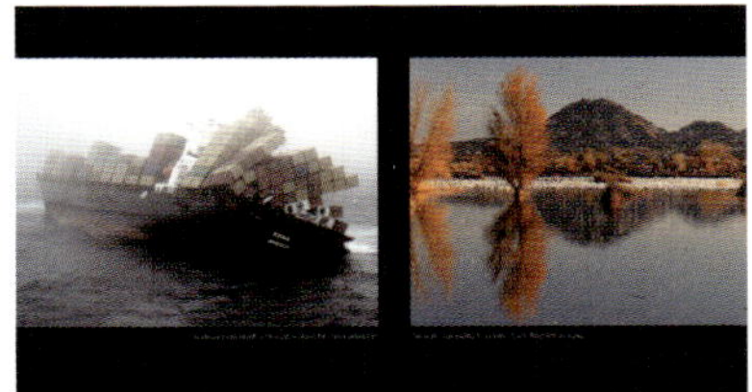

Listing Ship: Shipping containers wobble
as the MV Rena is battered by strong
seas on Astrolabe Reef, Tauranga, New
Zealand; Getty Images.

Lake and Birds: Gray Lodge Wildlife Area,
Butte County, California; © Daniel Dancer.

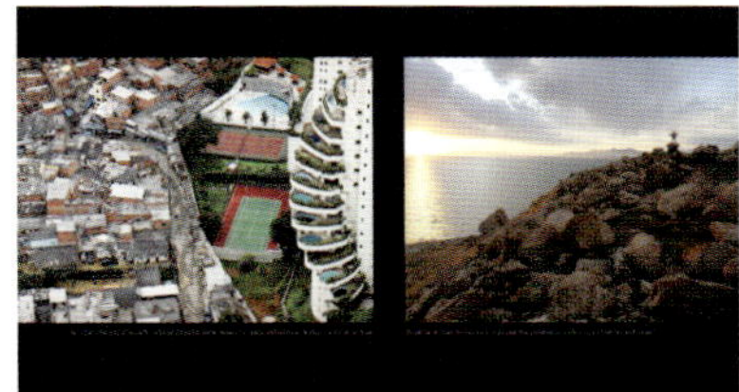

Sao Paulo, Brazil: The Paraisópolis
favela borders the affluent district of
Morumbi; Tuca Viera.

Human at Sunset: Antelope Island, Utah;
© Daniel Dancer.

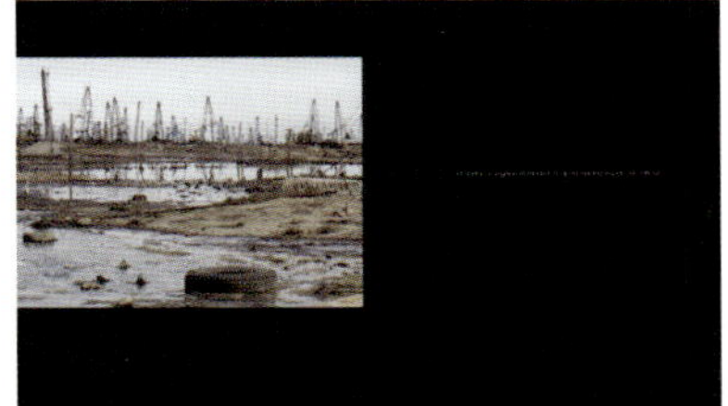

Oil Desolation: Detritus of the oil age,
Azerbajan; Corbis.

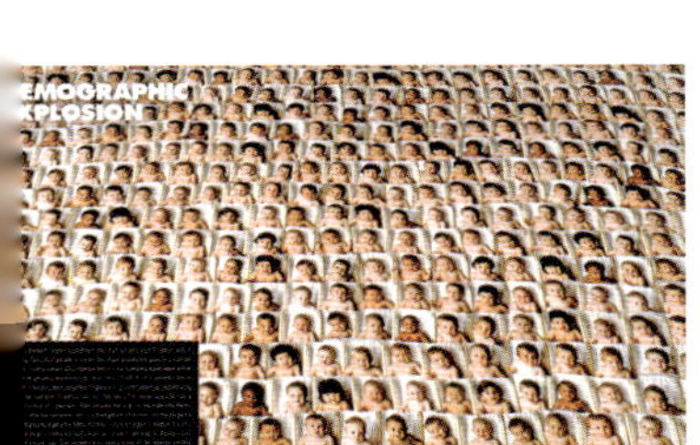

abies: Composite image;
Steve Cavalier/Alamy.

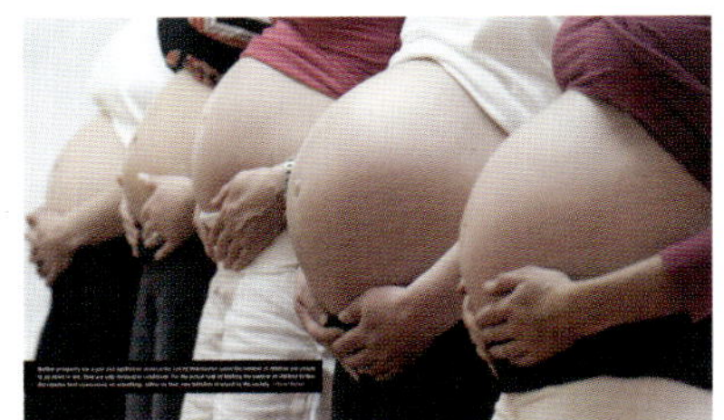

Pregnant Women: Every woman should
be able to choose when, and if, she
becomes pregnant; © 67photo/Alamy.

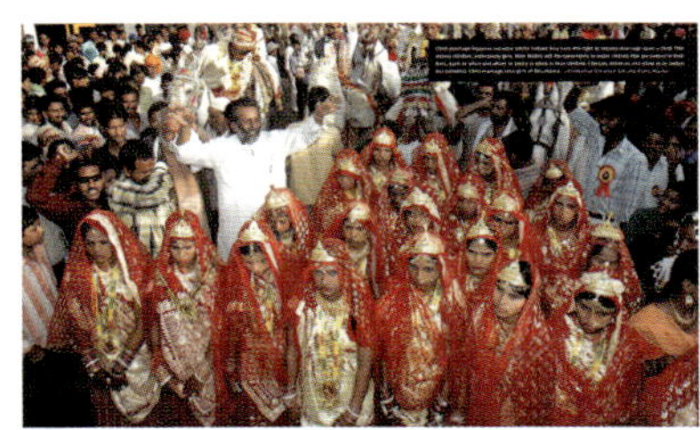

Child Brides: Girls gather for inspection
before mass wedding, India; Getty Images.

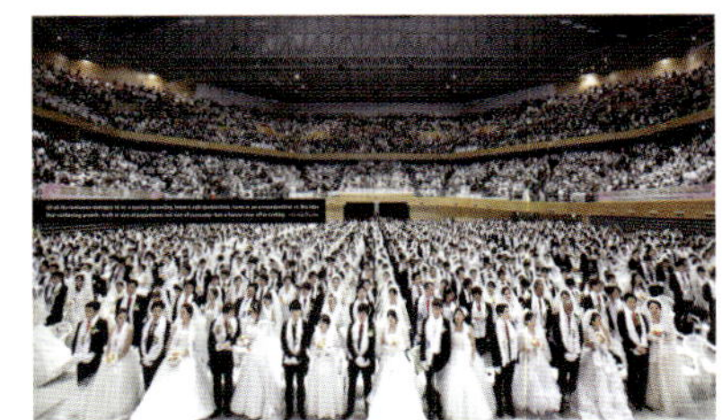

Wedding Party: Mass wedding in South
Korea; Kim Hong-Ji/Reuters.

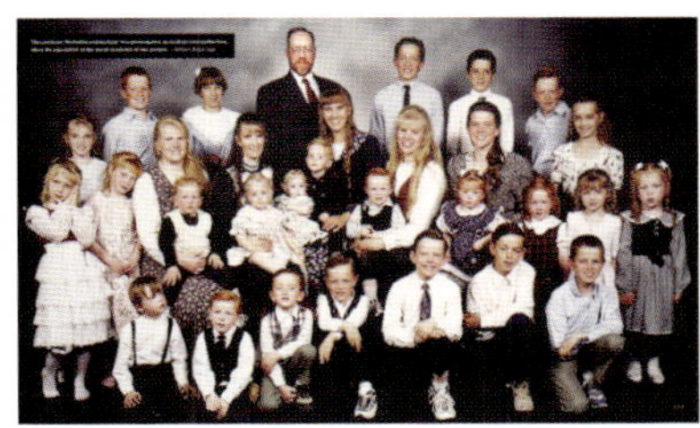

Growing Family: Polygamist Tom Green
with his wives and children, U.S.A;
Getty Images.

HUMAN TIDE

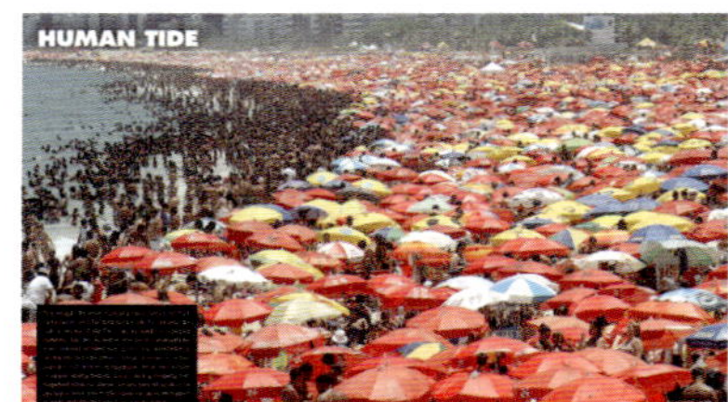

Beach: A rising tide of humanity covers the Ipanema Beach in Rio de Janeiro (population 12 million), Brazil; Picture Alliance.

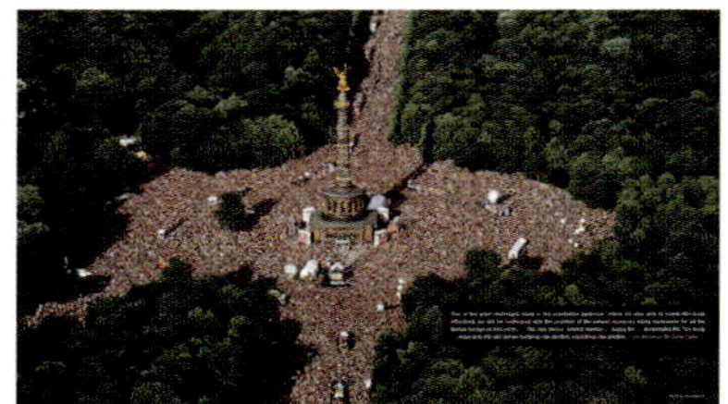

People and Trees: Mass rallies and other cultural events are only possible in a mass society. The "Love Parade" in Tiergarten Park, Berlin, Germany; © Yann Arthus-Bertrand.

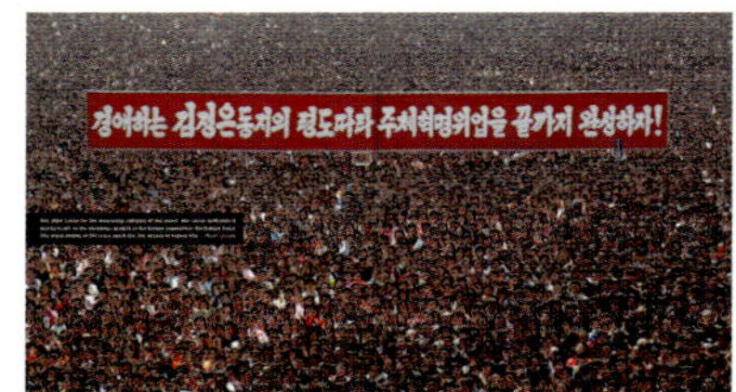

North Korean Gathering: Thousands gather at the inauguration of a mosaic portrait of Kim Jong-il in Pyongyang, North Korea; Bobby Yip/Reuters.

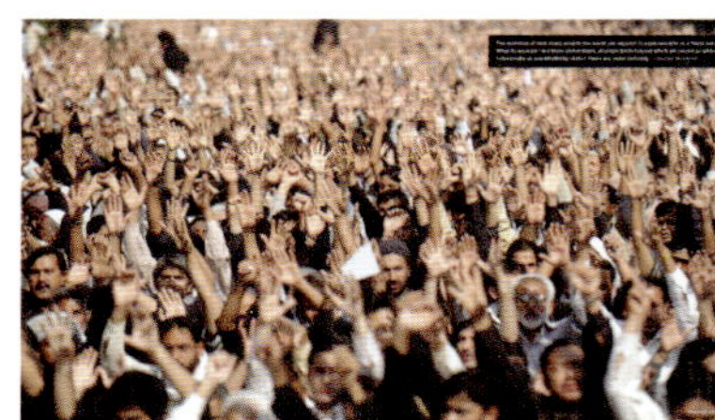

Raised Hands: Kashmiri Muslims gather at the Hazratbal Shrine in Srinagar, India; Tauseef Mustafa/Getty Images.

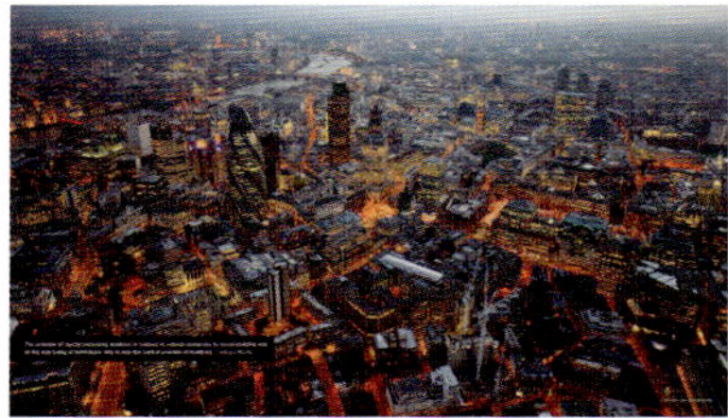

City Night: Large urban areas like London, U.K. (population 13 million), represent a huge amount of embodied energy in their infrastructure as well as require massive ongoing inputs of energy; © Jason Hawkes.

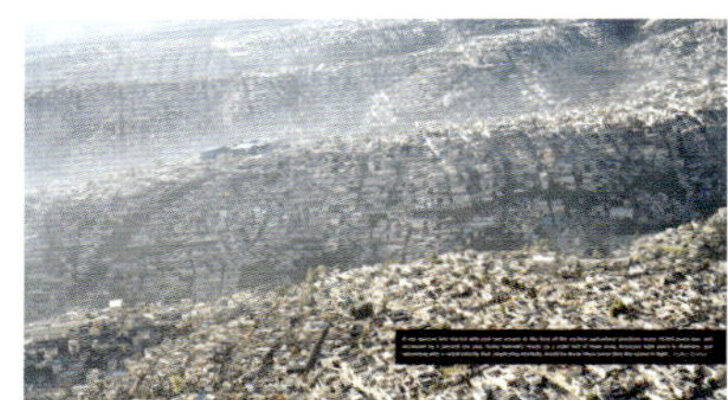

Waves of Humanity: Sprawling Mexico City, Mexico, population 20 million, density 24,600/mile (63,700/square kilometer), rolls across the landscape, displacing every scrap of natural habitat; © Pablo Lopez Luz.

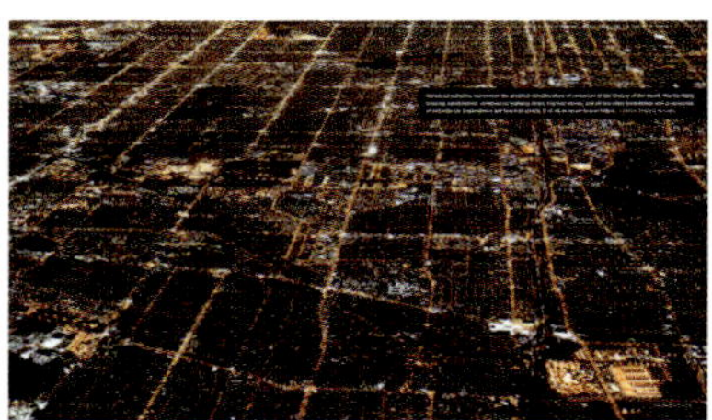

Nighttime Grid: Los Angeles, California, population 15 million, typifies America's consumption-oriented and car-dependent culture; © Mike Hedge.

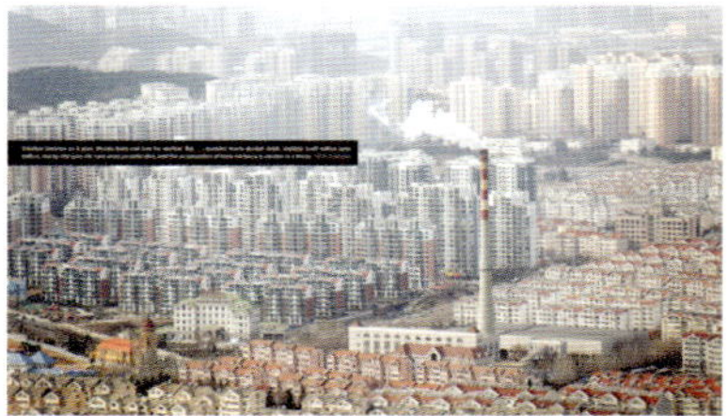

Boomtown: Second-tier Asian cities such as Qingdao, Shandong Province, China (population 8.7 million), are some of the fastest-growing urban zones on Earth; © Wu Hong/Corbis.

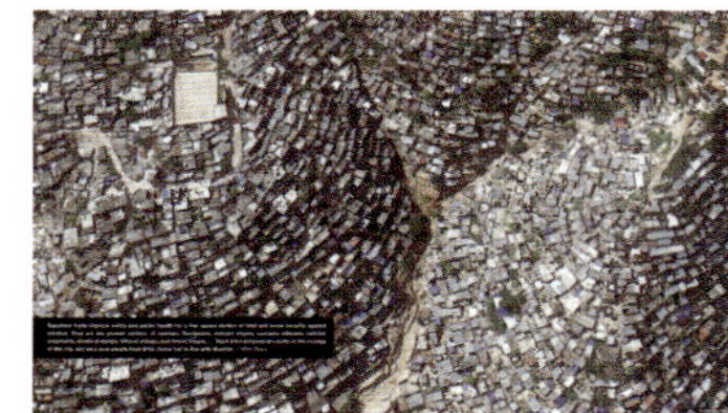

Hillside Slum: Slum-dwelling residents of Port-au-Prince, Haiti (population 4 million, density 50,000/mile [129,500 km²]) face bleak living conditions in the Western Hemisphere's poorest country; Google Earth/2014 Digital Globe.

ELBOW TO ELBOW

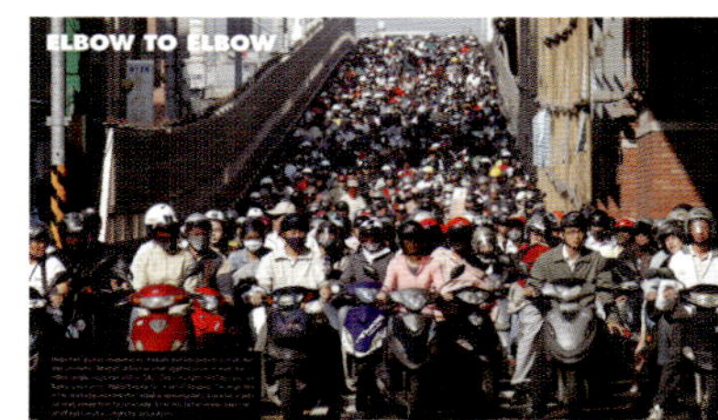

Motorcycles: Daily life in Taipei, Taiwan (metro area population 6.9 million), includes packed streets; © Nicky Loh/Reuters.

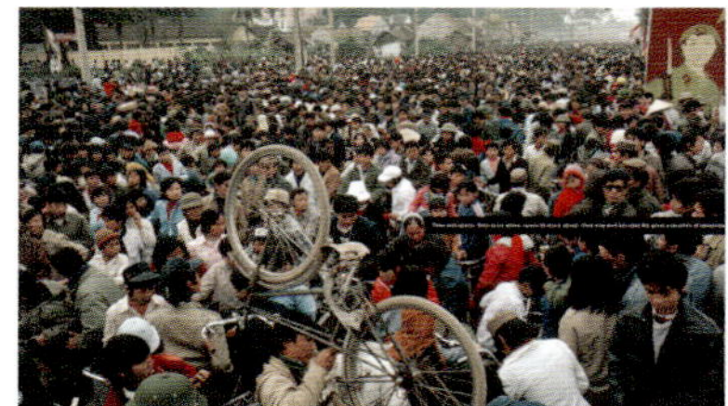

Bike and Crowd: A lone biker attempts to get through a crowd in Hanoi, Vietnam's second-most populous city; © David Alan Harvey/Magnum.

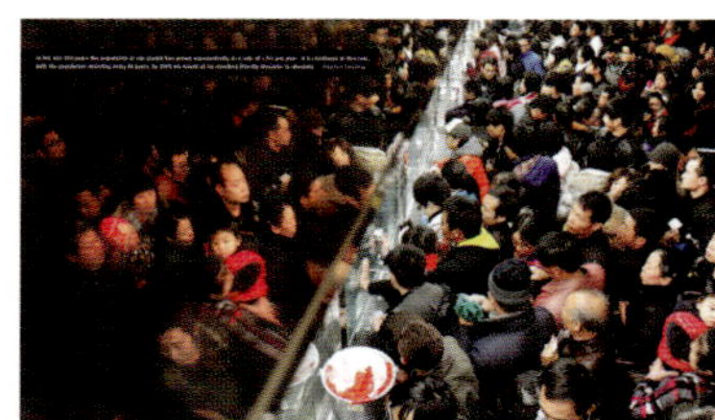

Crowded Line: Passengers queue to buy train tickets at Changsha Railway Station in Changsha, Hunan province, China; FotoPress/Corbis.

FEEDING FRENZY

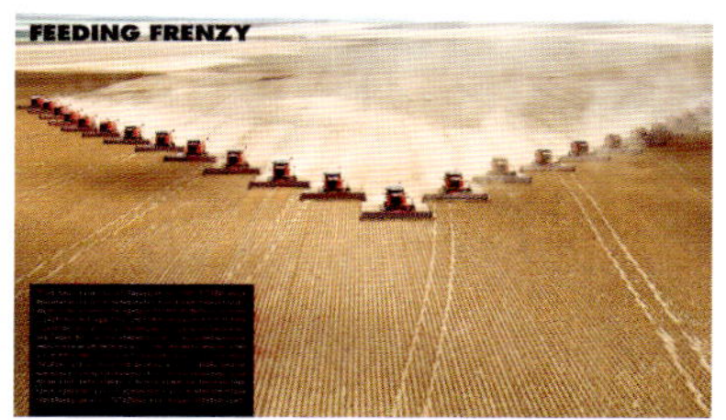

Harvesters: An army of machines harvest soybeans in Mato Grosso state, Brazil. Brazil is the second-largest soy producer worldwide; Photo © Henri Bureau/Sygma/Corbis.

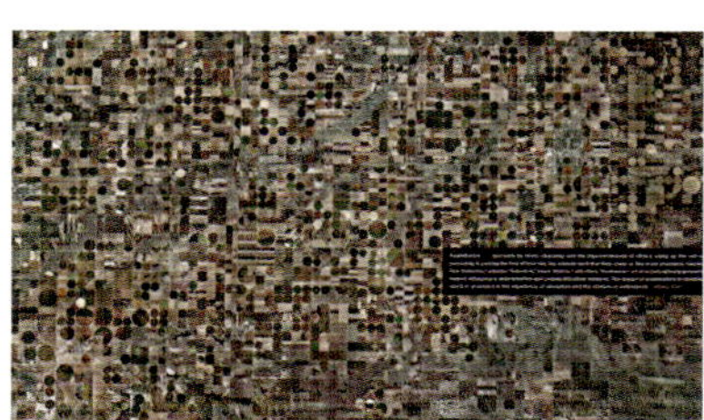

Circles and Squares: An industrialized landscape—center pivot irrigation grid amongst square fields in West Kansas, USA; Google Earth/Image Landsat.

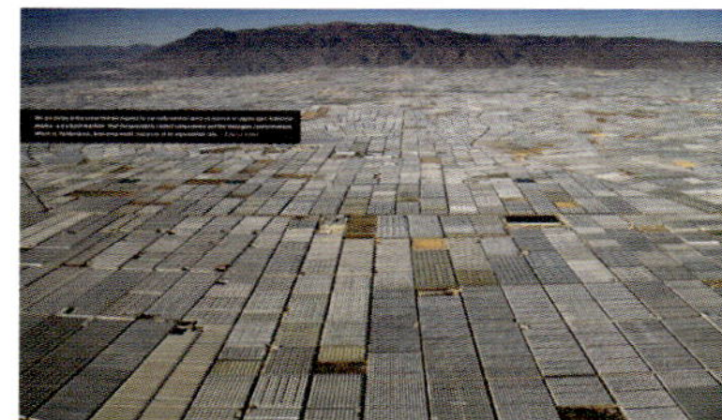

Greenhouses: As far as the eye can see, greenhouses cover the landscape in Almeria, Spain; © Yann Arthus-Bertrand.

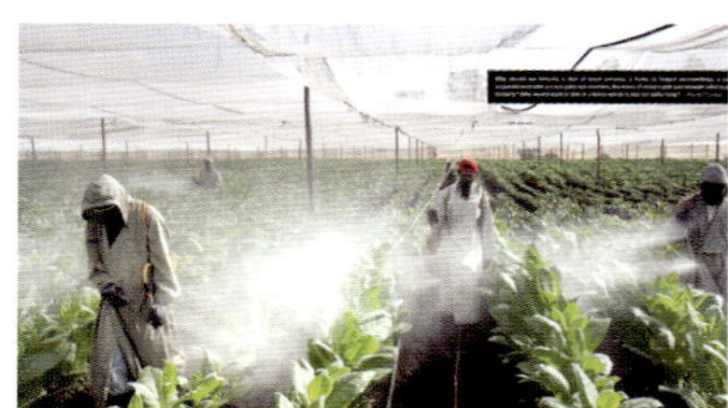

Sprayers: Industrial agriculture depends upon a regime of chemicals; pesticide spraying inside a greenhouse, Nicaragua; © Peter Essick.

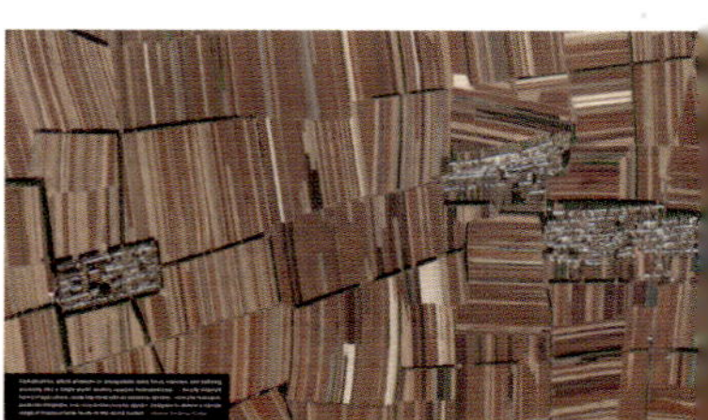

Rectangular Fields: No room for nature, the entire landscape is devoted to crop production, China; Google Earth/2014 Digital Globe.

OVERSHOOT

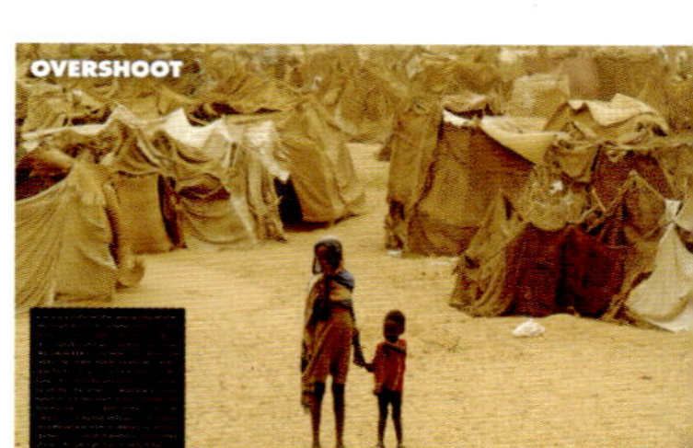

Kids in Refugee Camp: Wracked by famine and conflict, El Fasher Darfur Region of Sudan, 1985; © Brian Harris/Alamy.

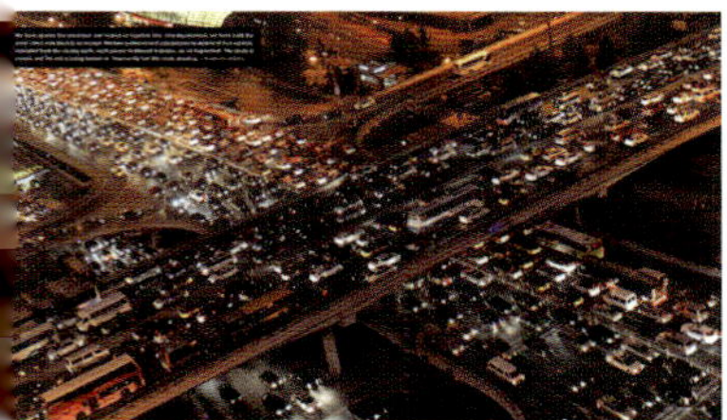

Crosstown Traffic: Rush hour commuters in Beijing, China (population roughly 20 million), face a maze of congested highways; © Xinhua/Gong Lei/Corbis.

URBAN ANIMAL

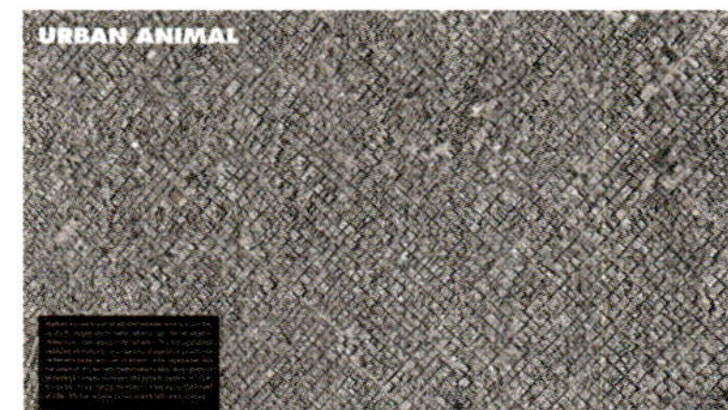

Delhi Grid: Aerial view of New Delhi, India, population 22 million, density 30,000 per square mile (77.700/km²); Google Earth/2014 Digital Globe.

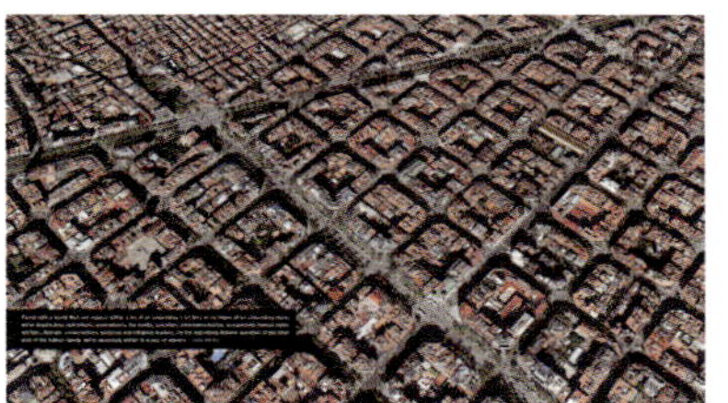

Urban Scene: Central area of Barcelona, Spain, population 5 million, density 16,000 per square mile (41,400/km²); Google Earth/NOAA, U.S. Navy, NGA, GEBCO.

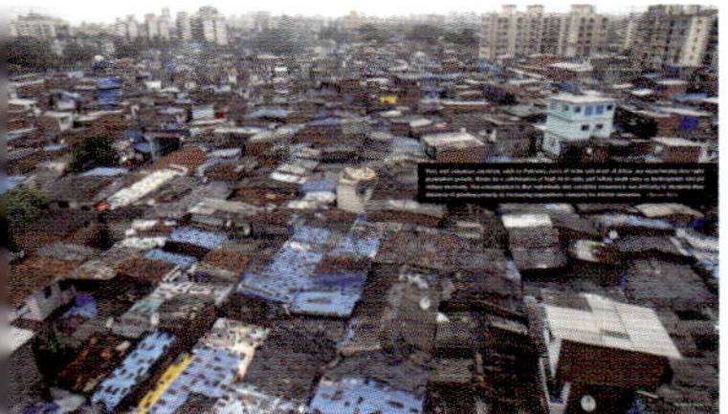

Slums: Dharavi, a major slum of Mumbai, India (population: 17 million, density: 30,000/mile), has almost no public sanitation facilities; Alamy.

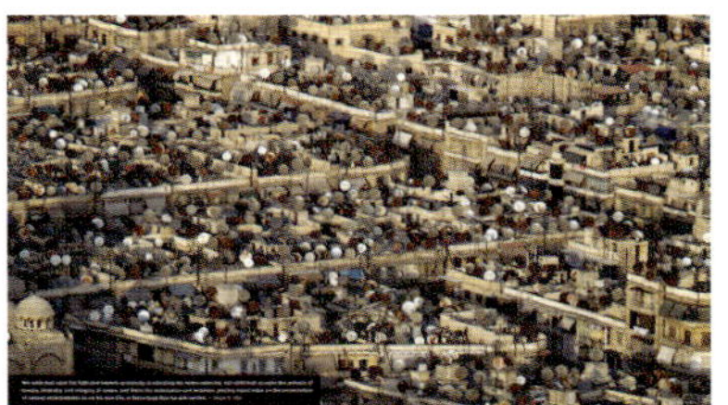

Satellite Dishes: The rooftops of Aleppo, Syria, one of the world's oldest cities, are covered with satellite dishes, linking residents to a globalized consumer culture; © Yann Arthus-Bertrand.

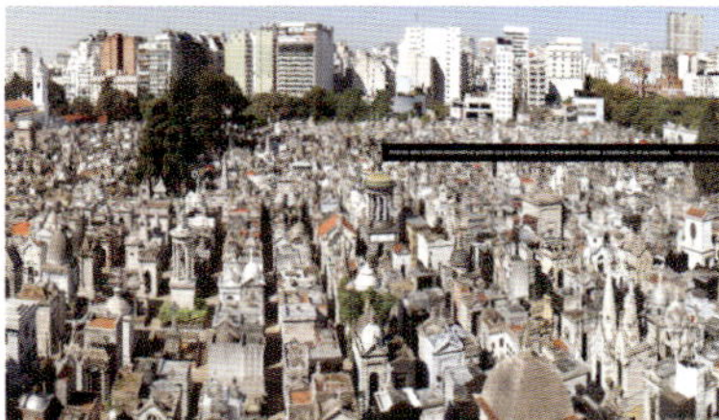

Cemetery: Crowded even in death, La Recoleta Cemetery in Buenos Aires, Argentina; © Gianni Muratore/Alamy.

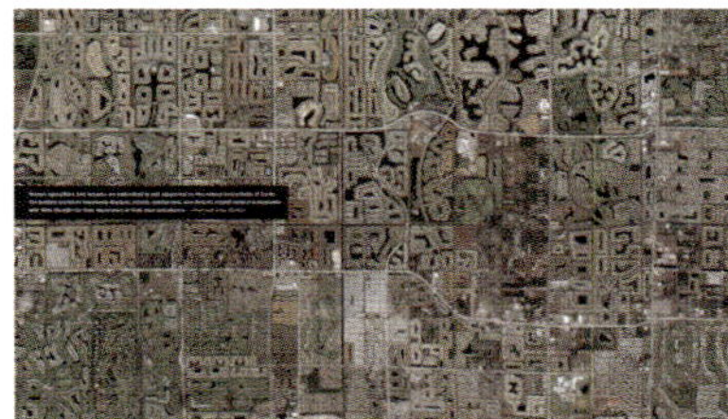

Suburban Sprawl: Aerial view of landscape outside Miami, Florida, shows 13 golf courses amongst track homes on the edge of the Everglades; Google Earth/NOAA, U.S. Navy, NGA, GEBCO.

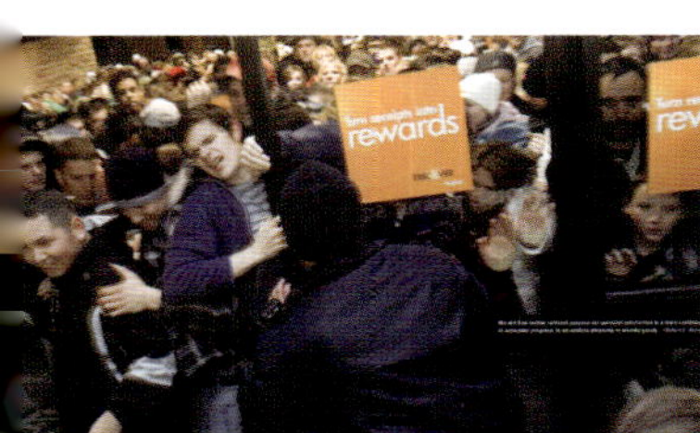

"Black Friday" Shoppers: Aggressive bargain hunters push through the front doors of the Boise Towne Square mall as they are opened at 1 a.m. Friday, Nov. 4, 2007, Boise, Idaho, USA; © Darin Oswald/Idaho Statesman.

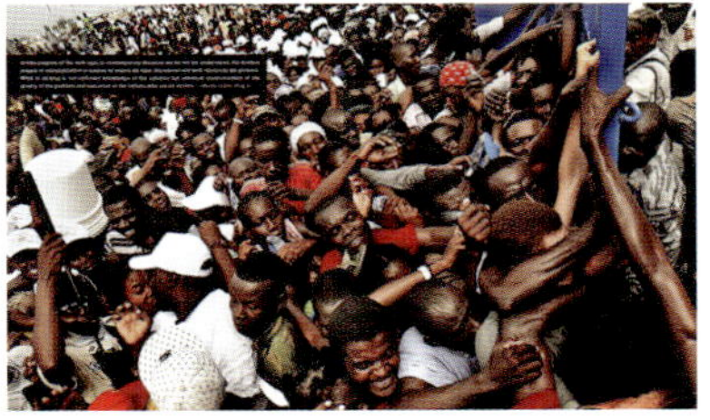

Angry Crowd: People jostle for food relief distribution following the 2010 earthquake in Haiti; © Carolyn Cole/LATimes.

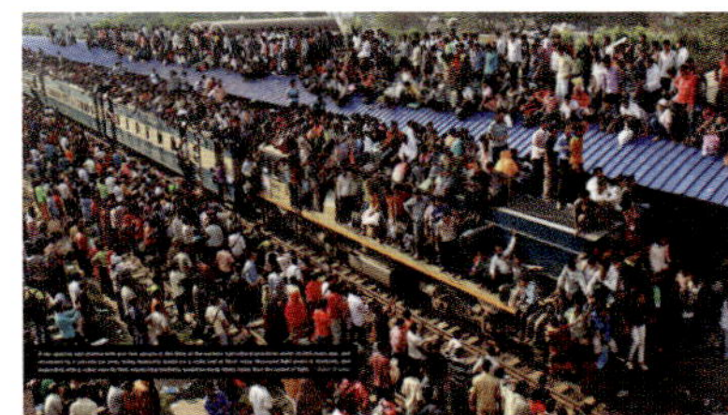

Train Station: Mass crowds are commonplace at the train station in Dhaka, Bangladesh; © Pavel Rahman/AP.

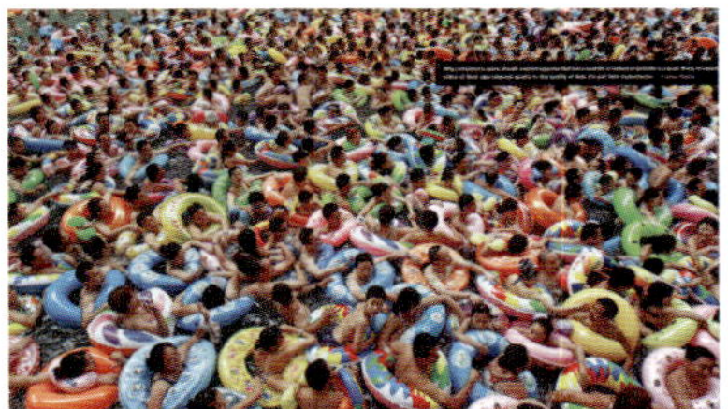

Swimming Pool: A relaxing day in the water, as swimmers wait for the gush of a man-made tide in a salty swimming pool dubbed China's Dead Sea, Sichuan province; Alamy.

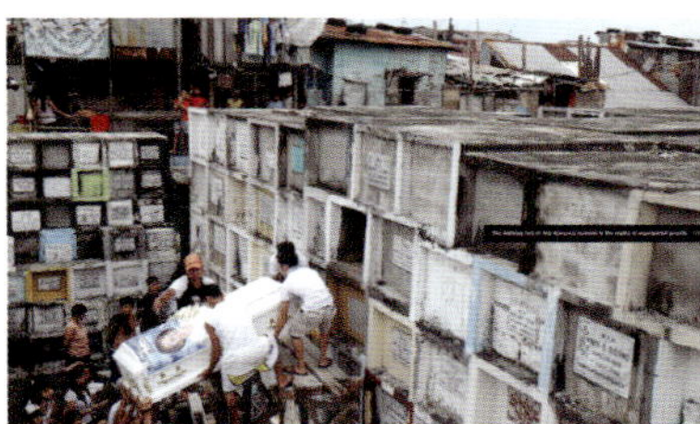

Coffin: Stacking the dead in a municipal cemetery, Manila, Phillipines, population 22 million, density 40,000/sq. mile (103,600/sq. km); © Noel Celis/Getty Images.

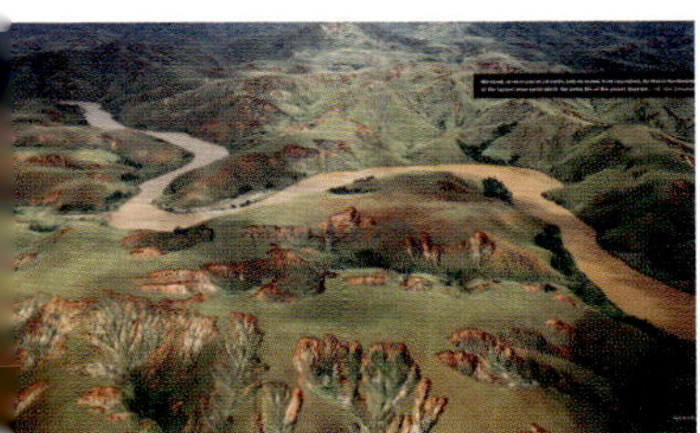

Eroded Hills and Silt-laden River: Deforestation along the Betsiboka River in Madagascar has caused massive erosion; Corbis.

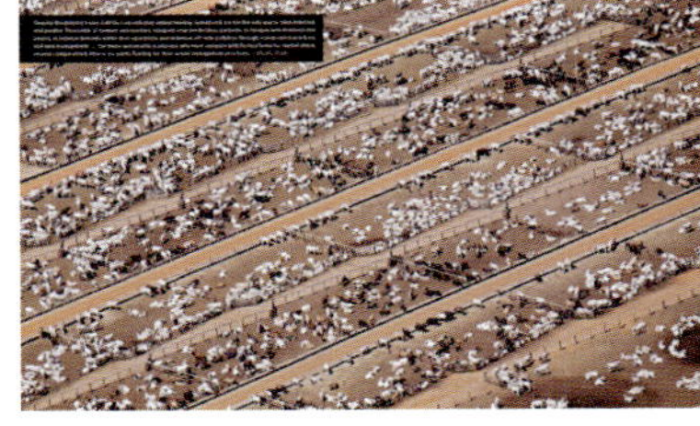

Feedlot: Industrial livestock production in Brazil; © Peter Beltra.

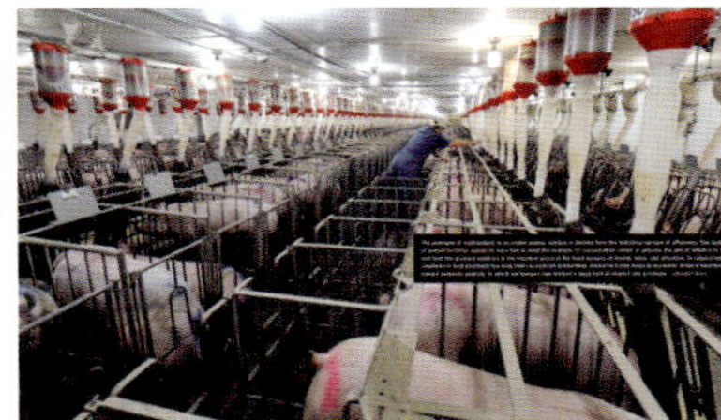

Pigs: Automatic feeding system in a hog barn in Southern Ontario, Canada; © Greg Taylor/Alamy.

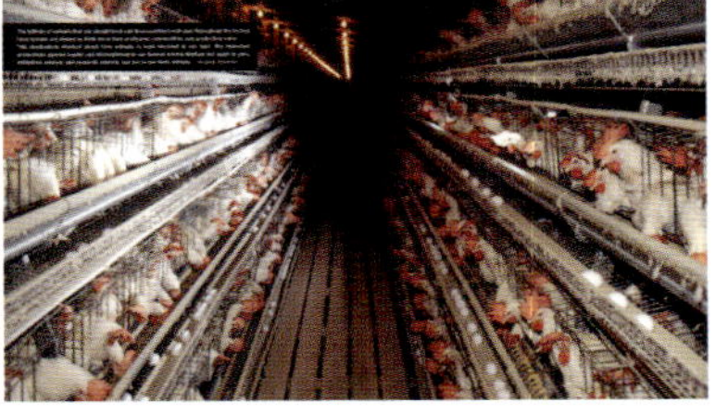

Chickens: "Efficiency" at work—a modern egg farm in Idaho; © David R. Frazier, Photolibrary, Inc./Alamy.

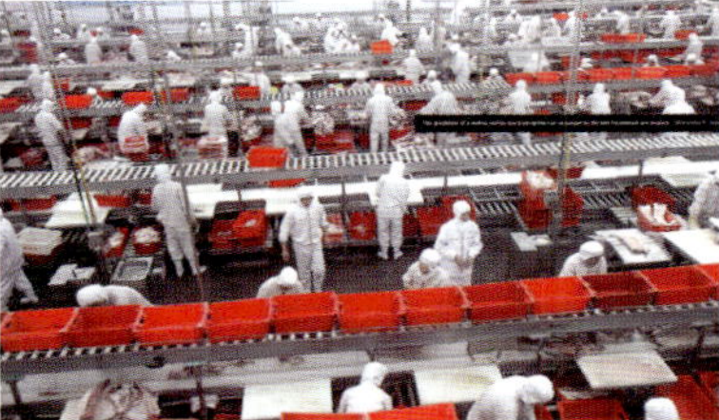

Animal Factory: Carcasses are disassembled for packaging at a pork-processing facility in Luohe City, China; Alamy.

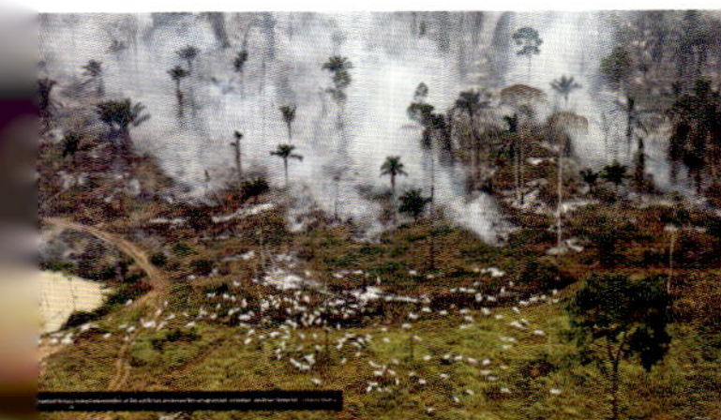

Cows and Smoke: Ground zero in the war on nature—cattle graze amongst burning Amazon jungle, Brazil; © Daniel Beltra.

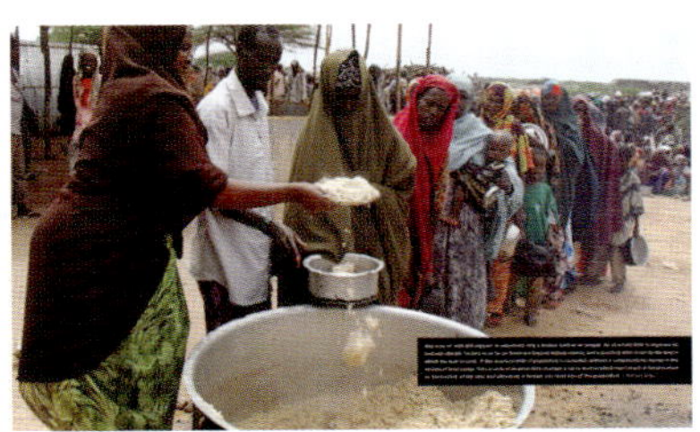

Food Line: Refugee camp in Kenya; © Farah Abdi Warsameh/AP.

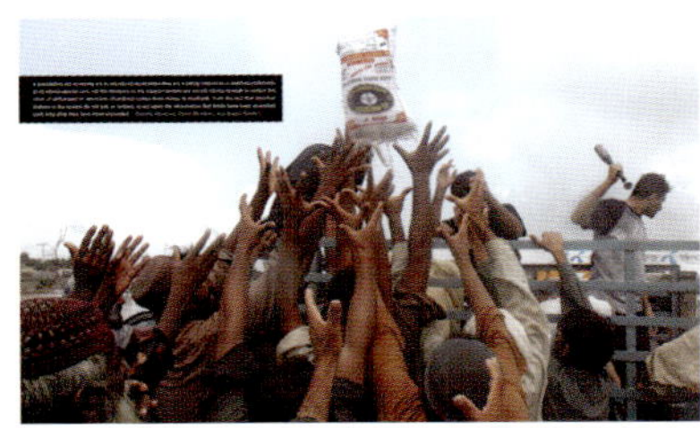

Food Fight: Pakistani flood victims jostle for food distributed by flood-relief workers, 2010; © Shakil Adil/AP.

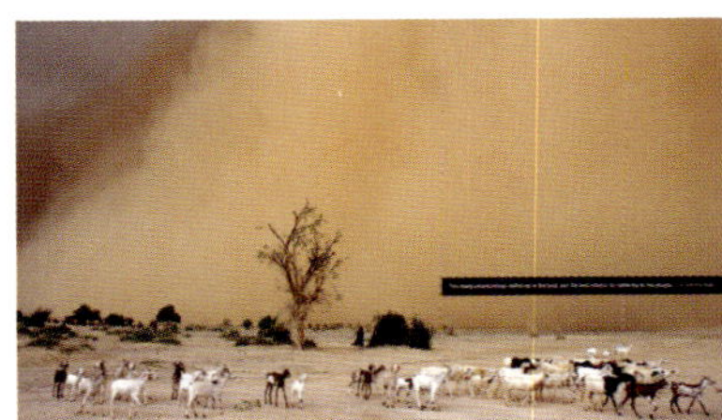

Goats: Approaching sand storm is as foreboding as the severely overgrazed landscape, Mali, Africa; © Remi Benali/Corbis.

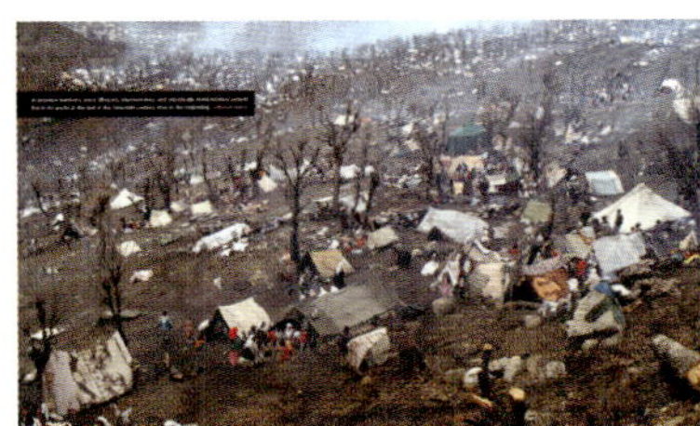

Refugee Camp: Following the Iraq war, hundreds of thousands of Kurds fled into Turkey where they were held by the Turkish army in a massive makeshift refugee camp near Isikveren; © Roger Hutchings/Corbis.

MATERIAL WORLD

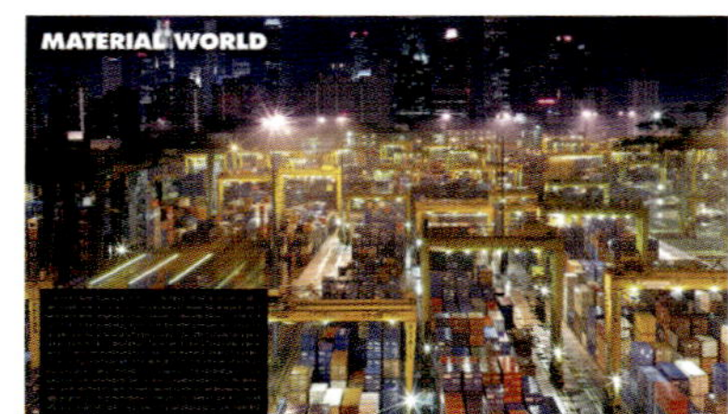

Container City: Shipping containers, indispensable tool of the globalized consumer economy, reflect the skyline in Singapore, one of the world's busiest ports; © John Stanmeyer.

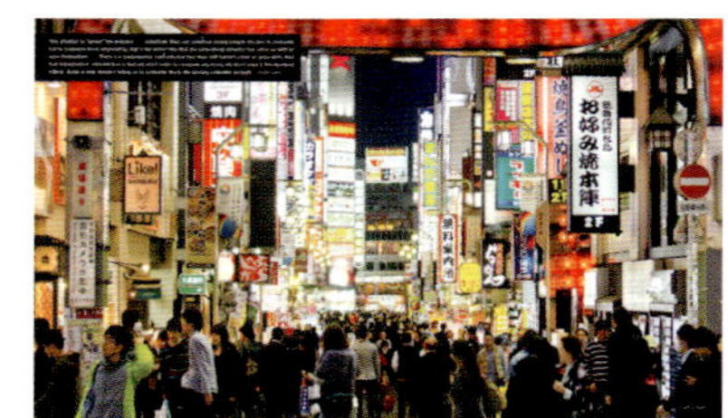

Signs and People: Advertising is ubiquitous in the overdeveloped world, helping to fuel a culture of hyperconsumption. Photo of Tokyo, Japan; © Mike Theiss/Corbis.

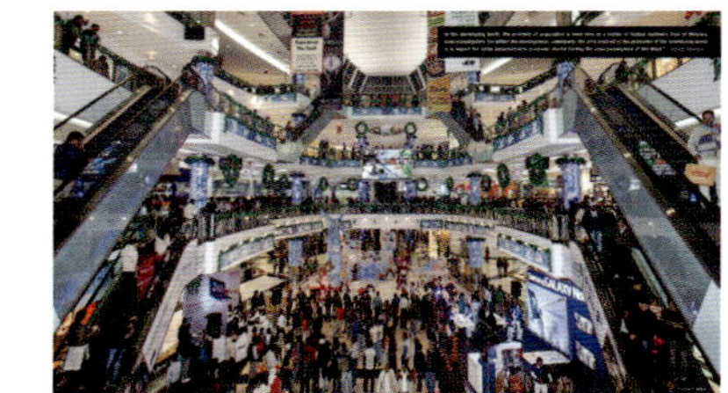

Shopping Mall: Consumer culture spreads to the "developing" world—South City Mall, Kolkata, India; © Brett Cole.

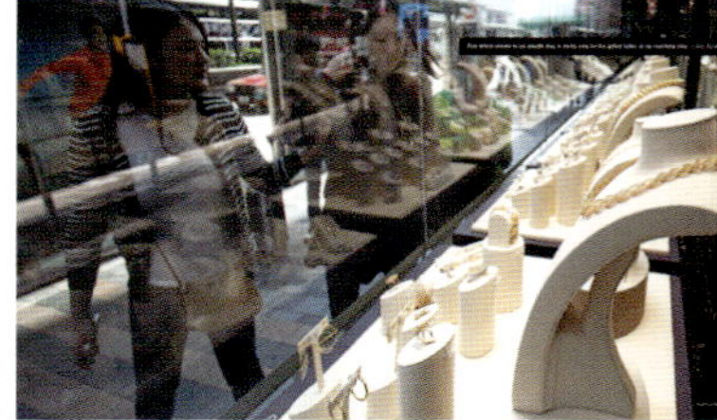

Jewelry Shoppers: Economic elites around the globe now have access to luxury goods, setting a cultural standard for others in the society to emulate; Getty Images.

TRASHING THE PLANET

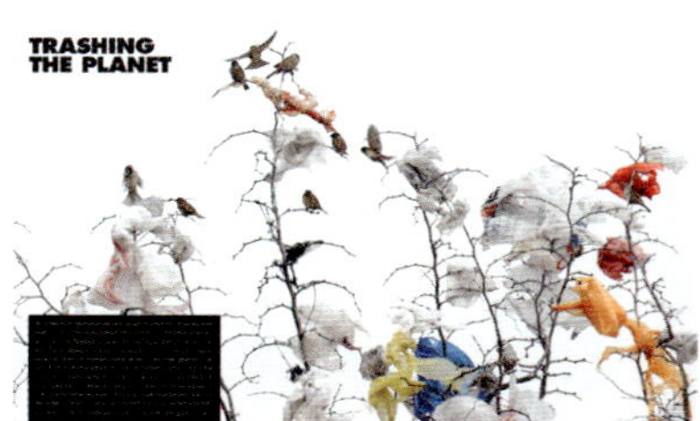

Birds and Plastic: Plastic bags adorn the trees near a garbage dump in Changzhi, Shanxi province, China; Reuters.

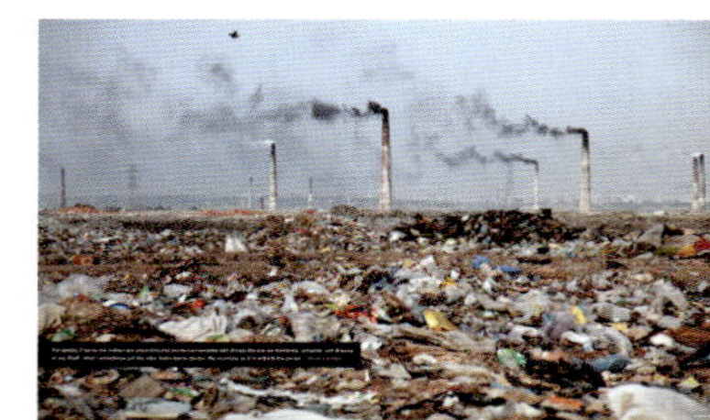

Smokestacks and Garbage: Brick kilns dot a dystopian landscape of trash in Bangladesh; © M.R. Hasasn.

NATURE'S UNRAVELING

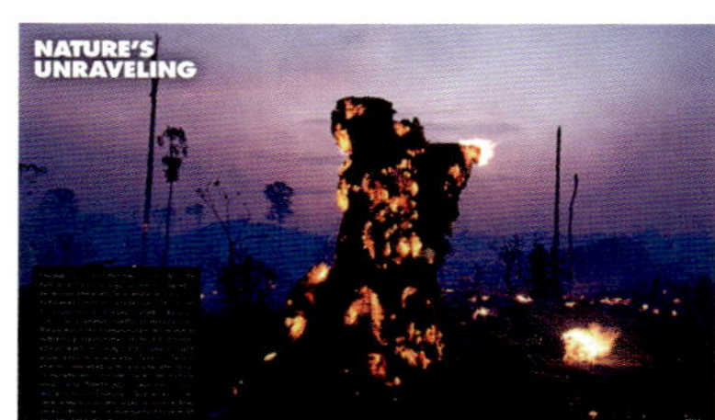

Burning Stump: Amazon rainforest destruction—lungs of the planet developing lung cancer; © John Maier, JR/Still Images.

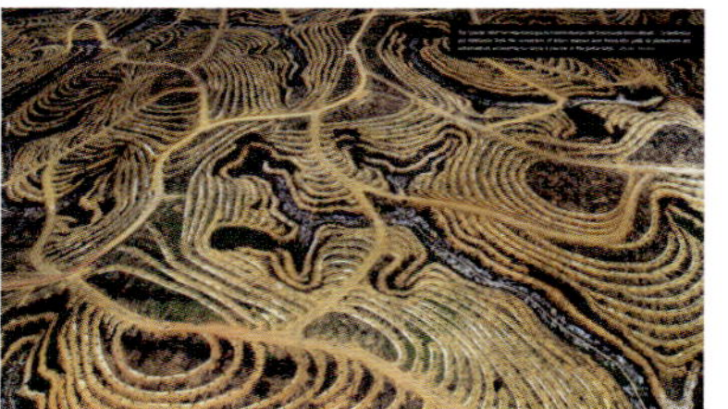

Palm Oil Plantation: Increasing demand for biofuels is linked to increased deforestation as native habitat is converted to palm oil production. Indonesia; image © Yann Arthus-Bertrand.

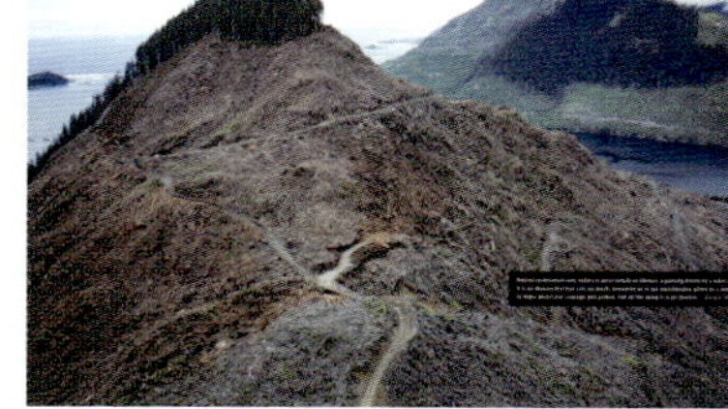

British Columbia Clear-cut: Sometimes called the Brazil of the North, Canada has not been kind to its native forests. Image of clear-cut logging on Vancouver Island; © Garth Lenz.

WILDLIFE LOST

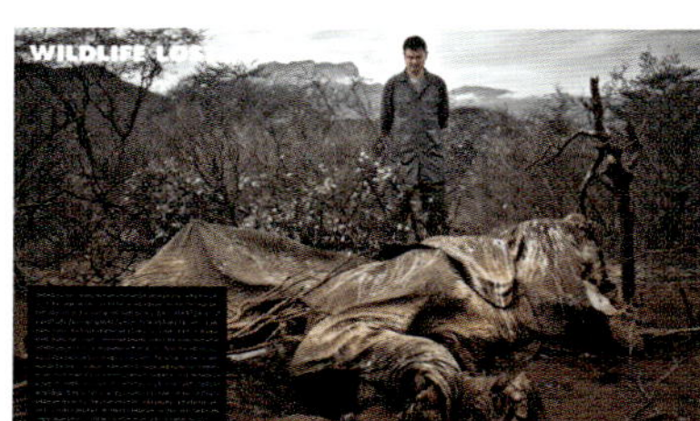

Dead Elephant: Basketball star Yao Ming comes face-to-face with a poached elephant in Northern Kenya; © Kristian Schmidt/Wild Aid.

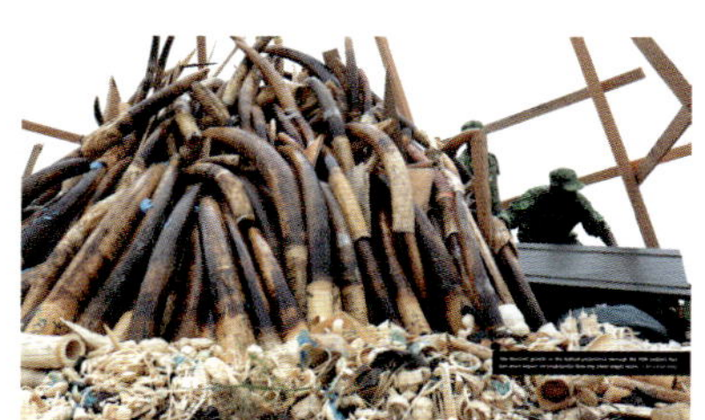

Pile of Tusks: Soldiers arrange a pyre of elephant tusks and thousands of pieces of worked ivory, preparing to burn ivory stocks corresponding to roughly 850 dead elephants, in Libreville, Gabon, Africa, 2012; © Joel Bouopda Tatou/AP.

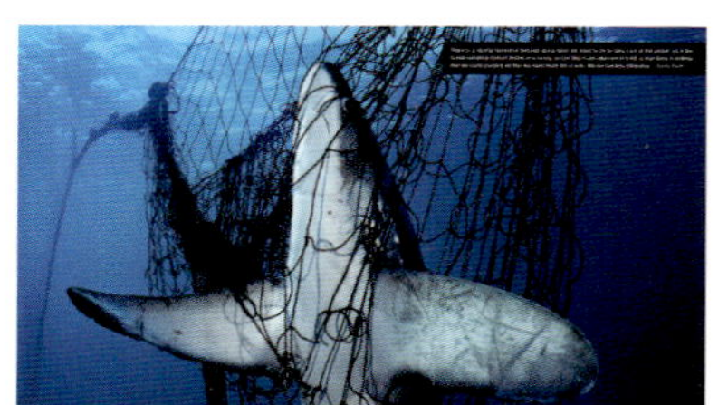

Shark in Net: Thresher shark killed in gill net, Mexico; © Brian Skerry/National Geographic.

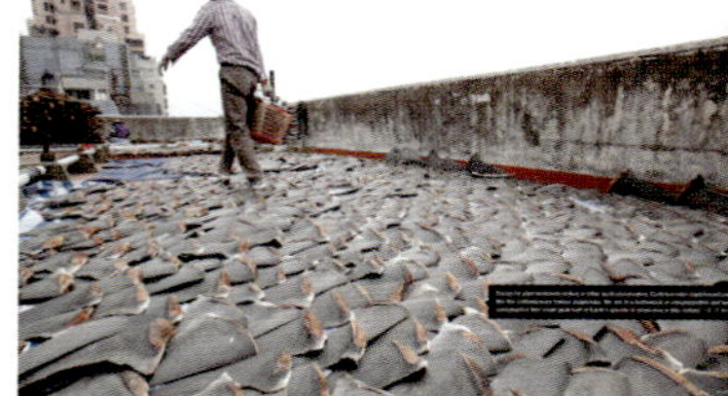

Shark Fins: Steps in the recipe for shark fin soup: A worker collects pieces of shark fins dried on the rooftop of a factory building in Hong Kong; © Kin Cheung/AP.

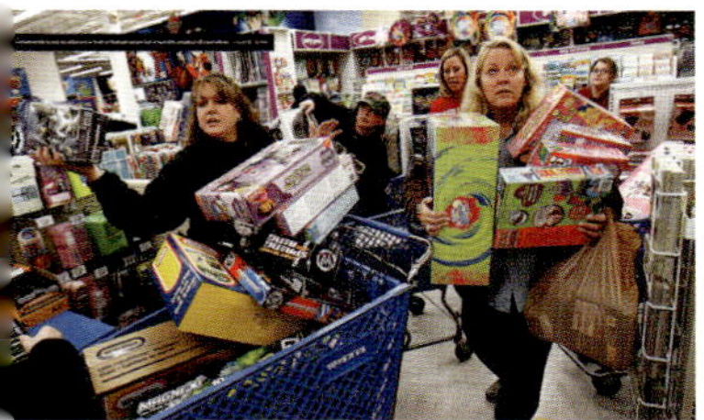

Shopping Mania: In the USA, "Black Friday" bargains rev shoppers into a consumption frenzy; © Tom Pennington/Getty Images.

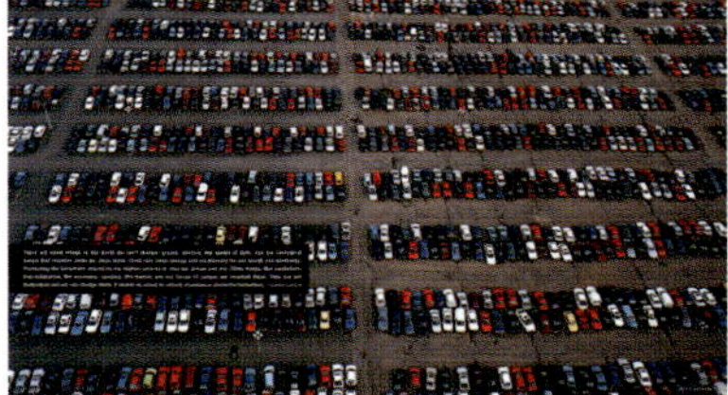

Cars: In the overdeveloped world, car culture and consumer culture are joined at the bumper. At an auction site in Sandwich, Kent, UK, thousands of cars are lined up to sell; Alamy.

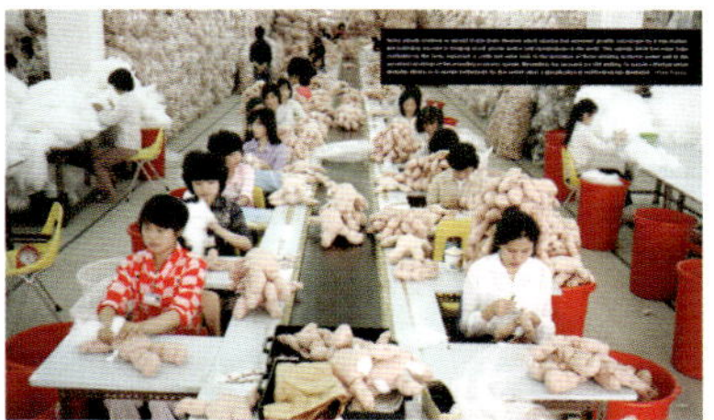

Assembly Line: In a globalized economy, resource extraction and production may occur a world away from where products are marketed and sold. Here, factory workers stuff Cabbage Patch Dolls in Shenzhen, China; Wally McNamee/Corbis.

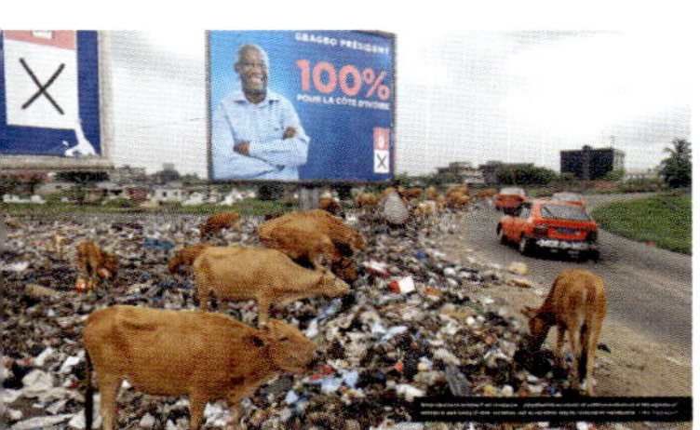

Garbage Cows: Watched over by a billboard depicting former President Gbagbo, cows graze on garbage in Abidjan, Ivory Coast, Africa; © Schalk van Zuydam/AP.

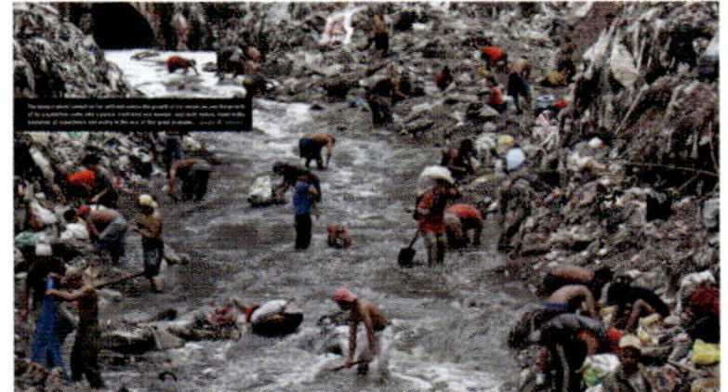

Sorting Garbage: To eke out a living, people search for scrap metal in contaminated water at the bottom of the biggest trash dump (known as "the mine") in Guatemala City, Guatemala; © Rodrigo Abd/AP.

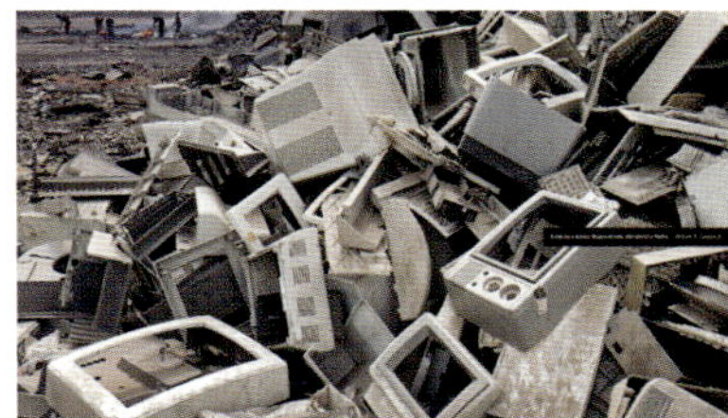

Computer Dump: Massive quantities of waste from obsolete computers and other electronics are typically shipped to the developing world for sorting and/or disposal. Photo from Accra, Ghana; © Peter Essick.

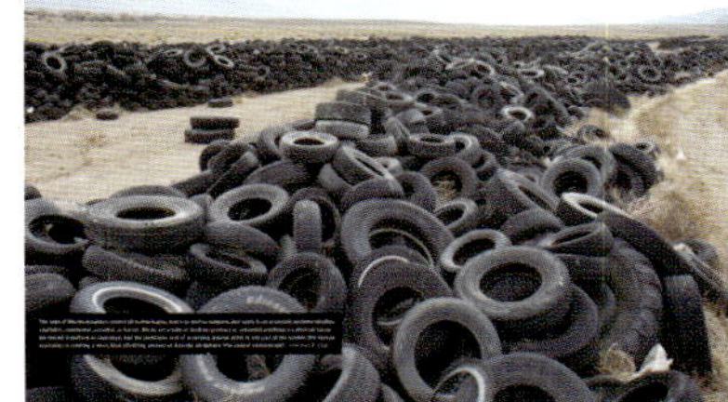

Tire Dump: End of the road for these tires is a desert dumping ground in Nevada, U.S.A.; © Daniel Dancer.

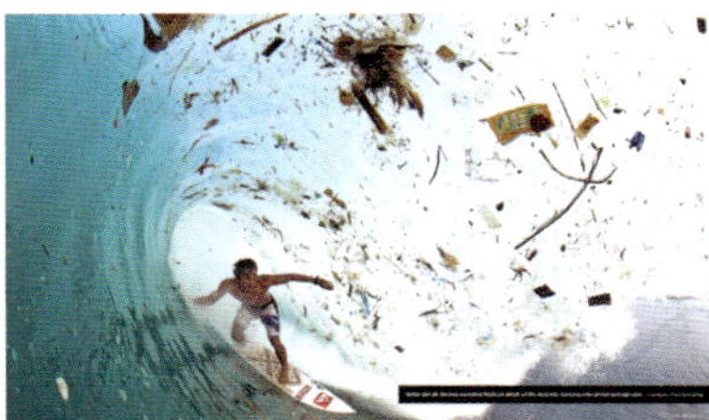

Trash Wave: Indonesian surfer Dede Surinaya catches a wave in a remote but garbage-covered bay on Java, Indonesia, the world's most populated island; © Zak Noyle.

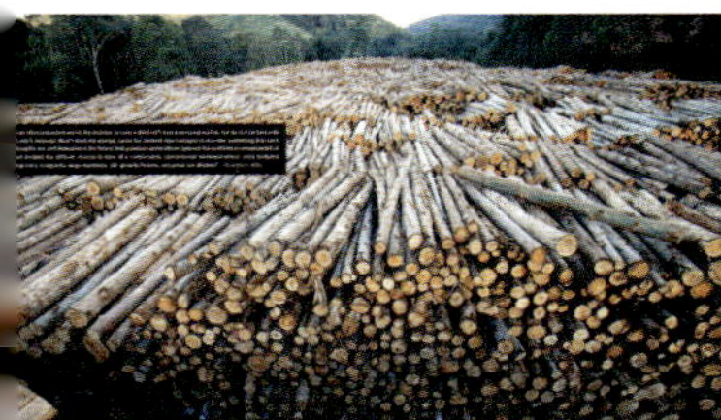

Stacked Logs: Nonnative eucalyptus plantations have displaced an estimated 5 million acres of native forest in Brazil; National Geographic.

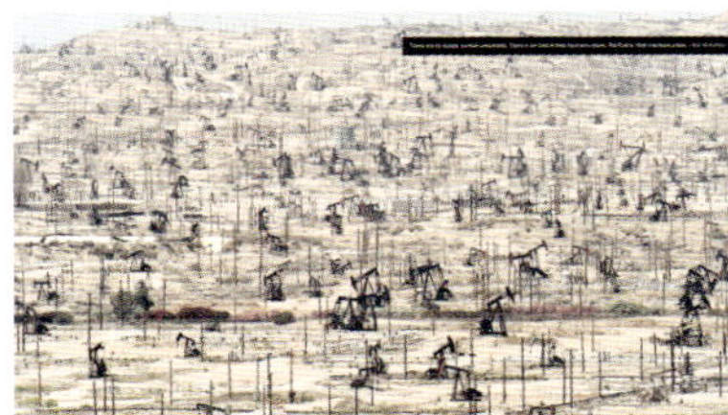

Oil Wells: Depleting oil fields are yet another symptom of ecological overshoot; Kern River Oil Field, California, U.S.; © Mark Gamba/Corbis.

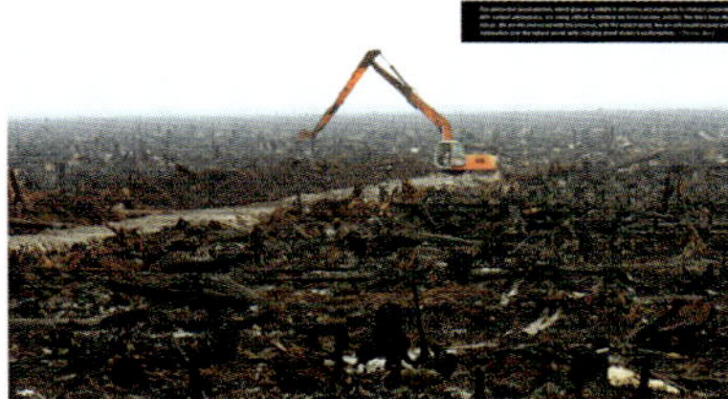

Devastation: Excavator making drainage canal in recently cleared and burned peatland rainforest, Indonesia. The area is being cleared for plantation establishment; © Oka Budhi/Greenpeace.

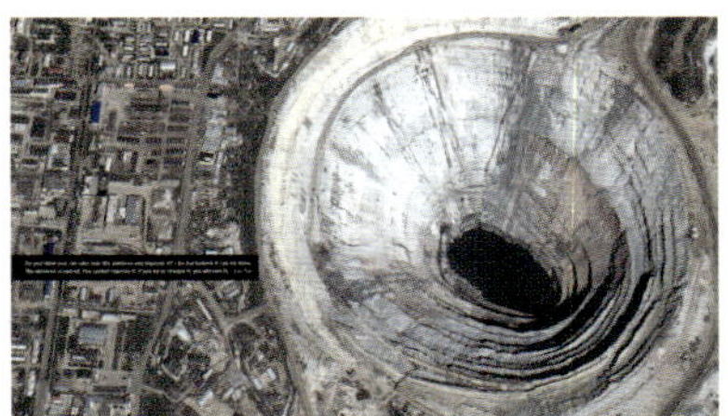

Big Hole: The Mir Mine in Russia is the world's largest diamond mine; Google Earth/2014 Digital Globe.

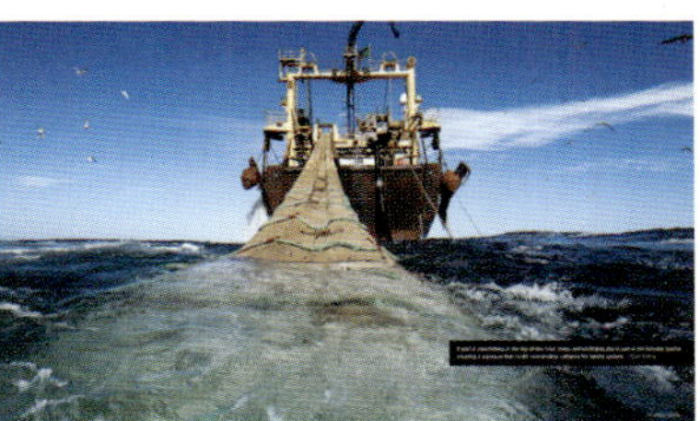

Ship Dragging Net: A 120-meter-long pelagic trawler fishes off the coast of Mauritania, Atlantic Ocean, to support the ever-growing demand for fish protein in the world diet; © Christian Asulnd/Greenpeace.

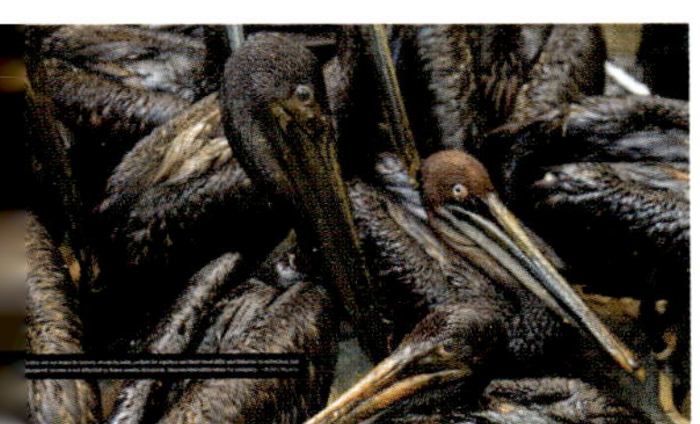

Oily Pelicans: Brown Pelicans, covered in oil from BP's 2010 Gulf of Mexico oil spill, huddle together in a cage at the International Bird Rescue Research Center in Buras, Louisiana; © Lee Celano/Reuters.

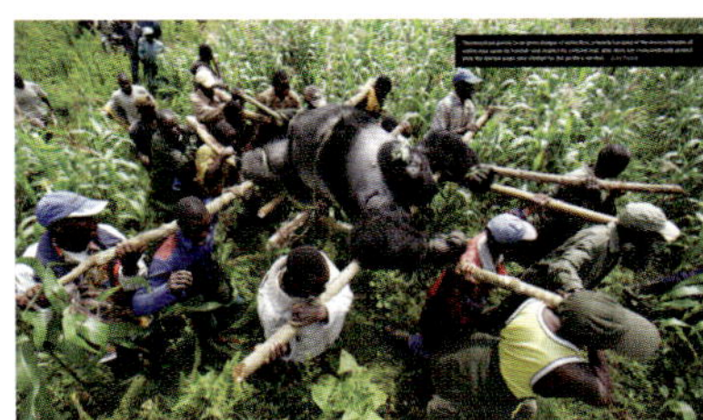

Dead Gorilla: Conservation rangers from an anti-poaching unit work with locals to evacuate the bodies of four mountain gorillas killed in mysterious circumstances; 2012, Virunga National Park, Eastern Congo, Africa; © Brent Stirton/Getty Images.

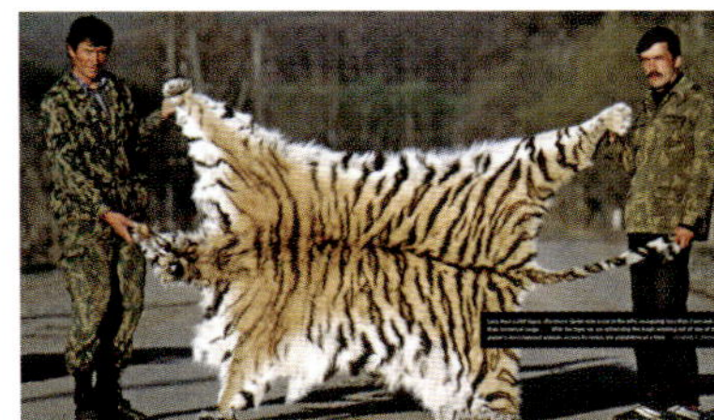

Tiger Skin: Siberian tiger skin recovered from poachers, Siberia, Russia; © Steve Morgan/Photofusion.

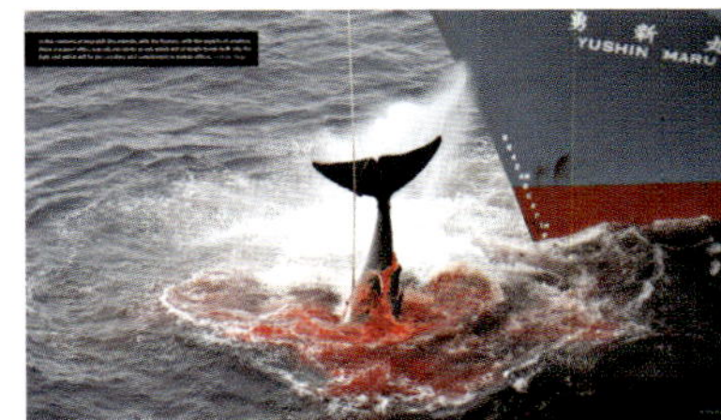

Whaling Ship: The Yushin Maru, a ship in a Japanese whaling fleet, takes a whale in the Southern Ocean off Antarctica. © Kate Davidson/Corbis.

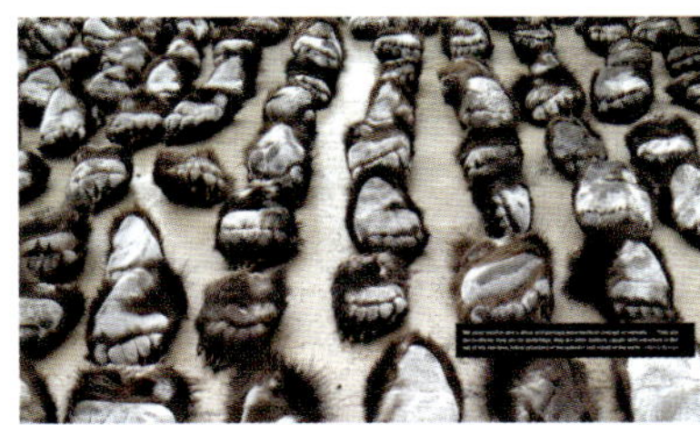

Paws: Trafficking wildlife parts is big business, and contributes to the pressure on various imperiled species. Two Russians were arrested for smuggling 213 bear paws into China at a China–Russia land border, 2013; Reuters.

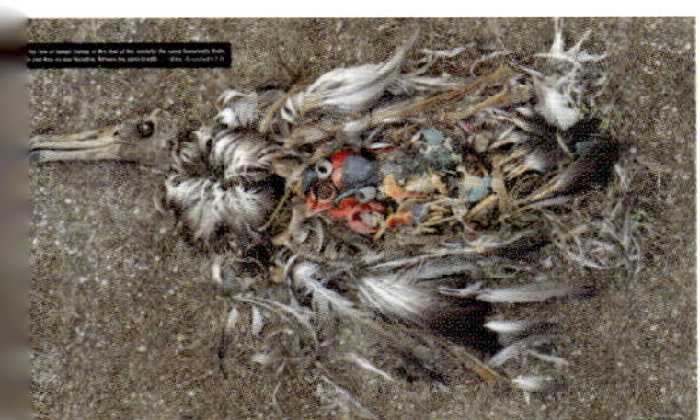

Dead Bird: On Midway Island, far from the centers of world commerce, an albatross, dead from ingesting too much plastic, decays on the beach; it is a common sight on the remote island; Chris Jordan.

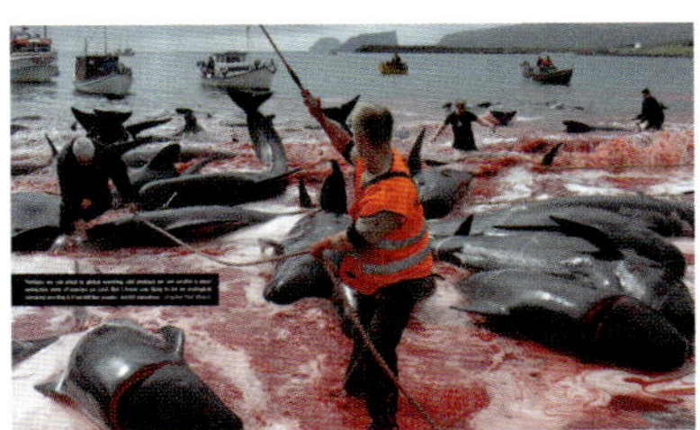

Bloody Whales: In the Faroe Islands, residents participate in a "traditional" whale hunt; 2012; © Andrija Ilic/Reuters.

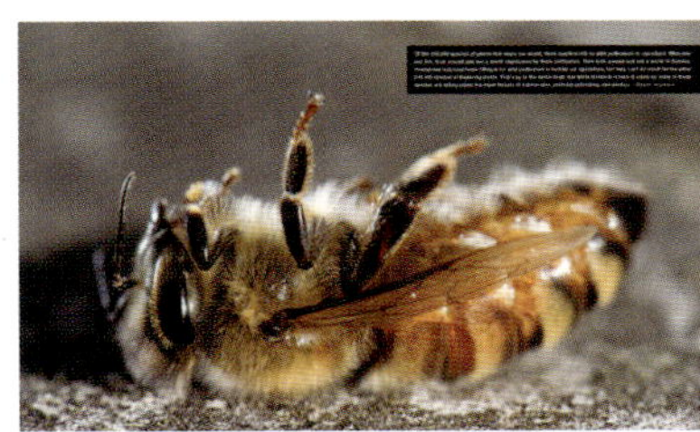

Dead Bee: Colony collapse disorder, a plague affecting domesticated honeybees, has gotten much attention. The decline of native bee and other pollinator populations is similarly alarming; © Robin Loznak/ZUMAPRESS.com.

ENERGY BLIGHT

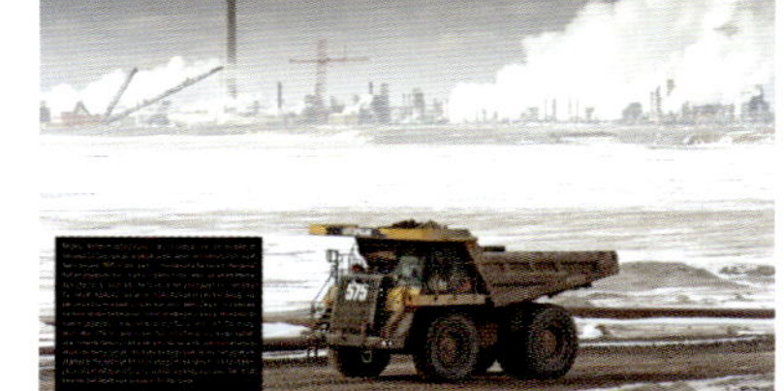

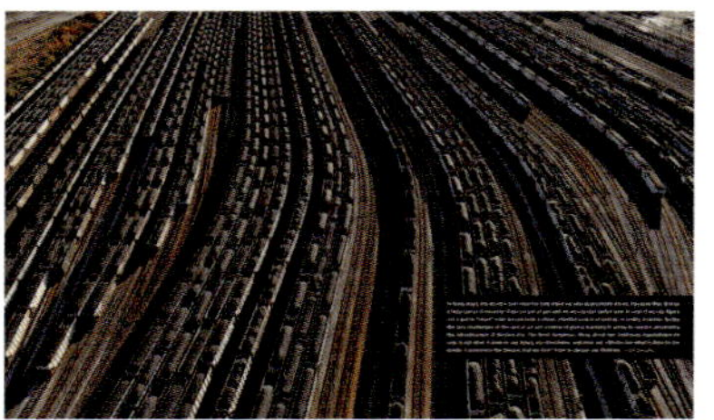

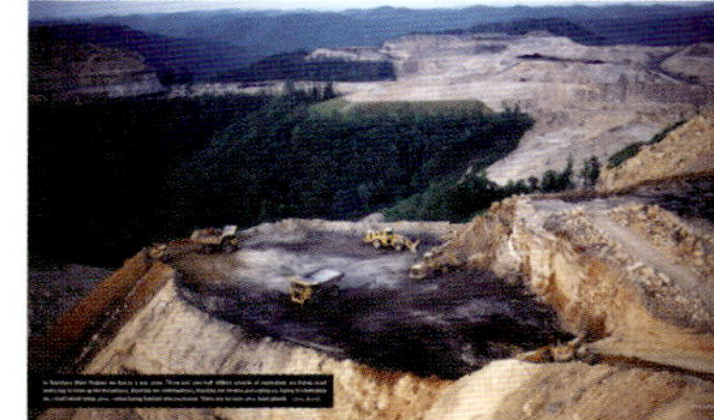

Big Truck: Massive haul trucks support surface mining operations in the tar sands region of Alberta, Canada, one of the largest known deposits of unconventional (in this case bitumen) oil resources; Photo © Garth Lenz.

Coal Trains: Railcars line up to fill waiting ships, Lamberts Point Coal Terminal, Norfolk, Virginia; © Robert M. Kendrick/ National Geographic.

Mountaintop removal: Radical strip-mining in Appalachia (U.S.A.) has decapitated more than 500 mountains; © Robert M. Kendrick/National Geographic.

FOUL WATER

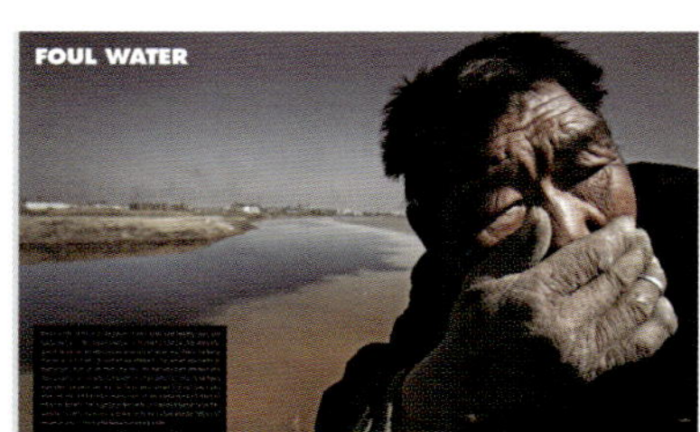

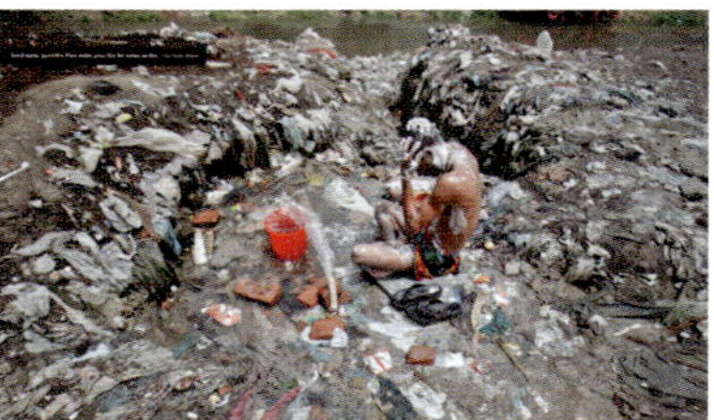

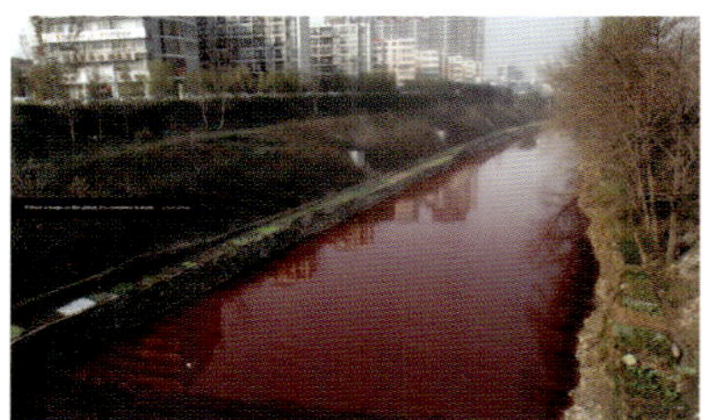

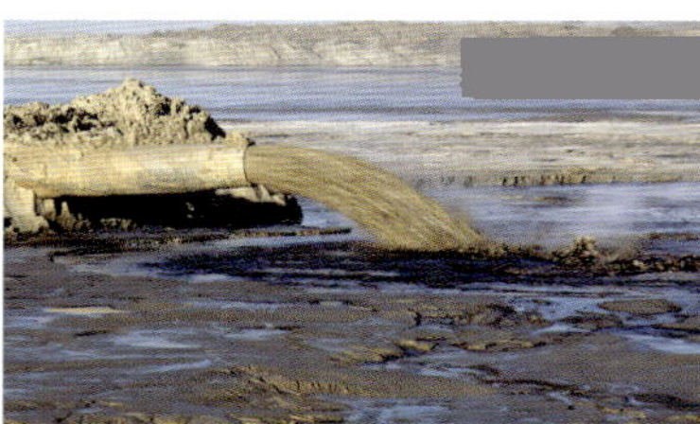

Man Covering His Mouth: A shepherd by the Yellow River cannot stand the smell, Inner Mongolia, China; © Lu Guang.

Man Bathing: A large percentage of the global population has limited access to clean water, and climate change is predicted to increase water scarcity. Here a man uses a broken water pipe in Noida slum, Uttar Pradesh, India, for bathing; © Parivartan Sharma/Reuters.

Red River: The Jian River became apocalyptic in character when red dye was dumped into the storm water pipe network, reportedly by two illegal dye workshops in Luoyang, Henan province, China; STR/AFP/Getty Images.

Drain Pipe: Tar sands-related tailings ponds are among the largest toxic impoundments on Earth and lie in unlined dykes mere meters from the Athabasca River; indigenous communities downstream are fearful of being poisoned by toxic seepage into the food chain. Alberta, Canada; © Garth Lenz.

DARKENING SKIES

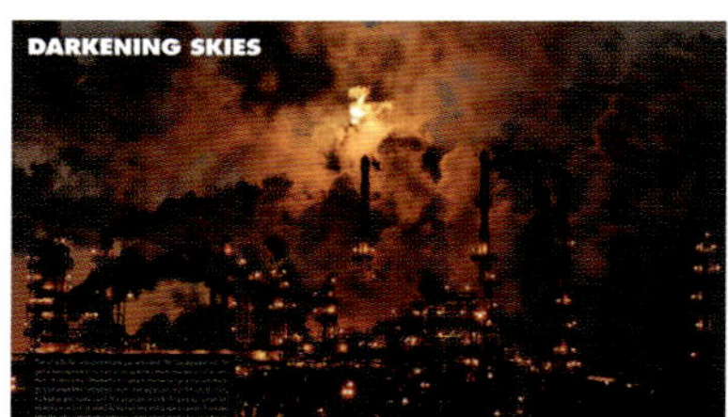

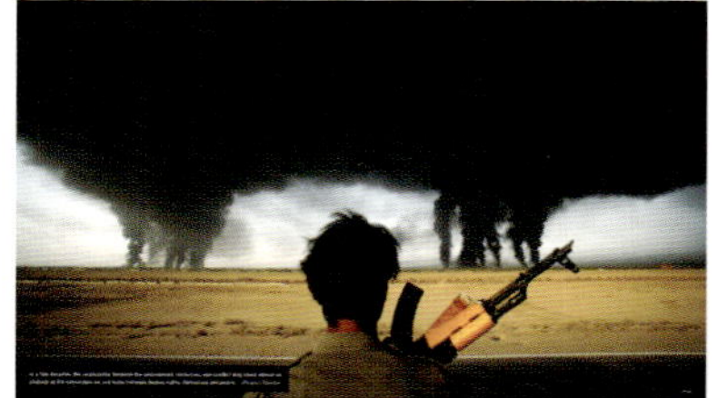

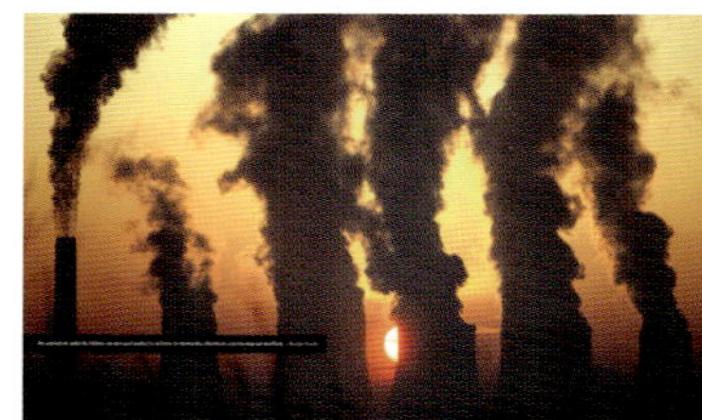

Refinery at Night: Petro-Canada's Edmonton Refinery and Distribution Centre, Edmonton, Alberta, Canada; © Dan Riedlhuber/Reuters.

Guns and Oil: An Iranian soldier watches as smoke billows from multiple burning oil refineries in Abadan, Iran; © Henri Bureau/Corbis.

Smokestacks: Birthplace of the coal-powered Industrial Revolution, England still burns coal to keep the lights on. Nottinghamshire power station; © Chris Knapton/Alamy.

CLIMATE CHAOS

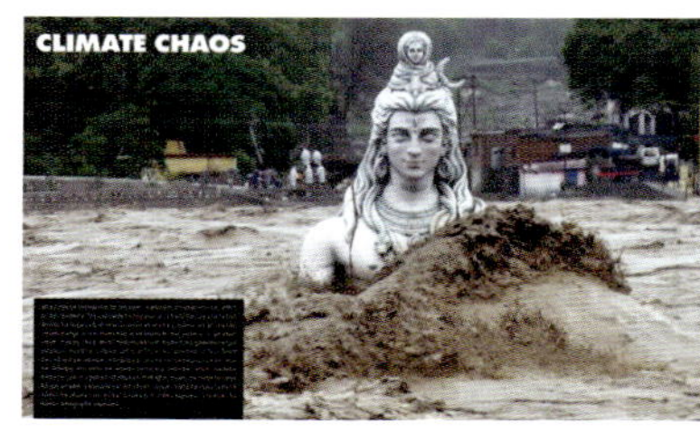

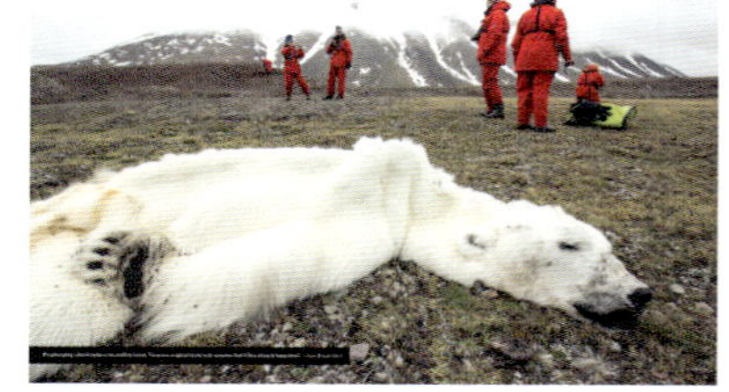

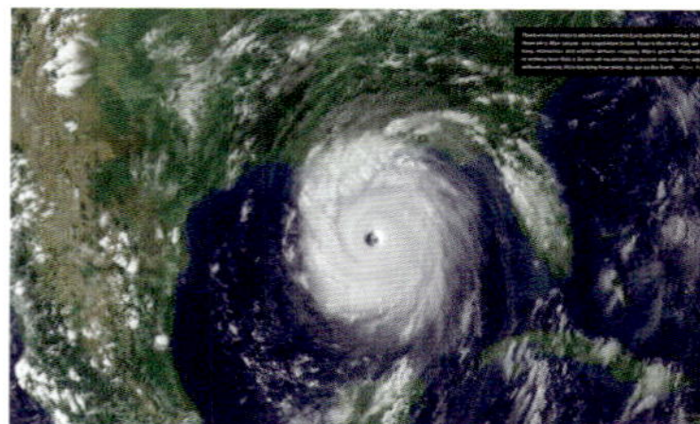

Shiva Flood: A submerged idol of Hindu Lord Shiva stands in the flooded Ganges River in Rishikesh, Uttarakhand, India; AP.

Dead Polar Bear: The western fjords on Svalbard, Norway, that normally freeze in winter, remained ice-free all season. This bear headed north, looking for suitable sea ice to hunt on. Finding none, it eventually collapsed and died; Photo © Ashley Cooper.

Storm from Space: One of the most powerful and disruptive storms in U.S. history, Hurricane Katrina (2005) strikes land; NASA.

PARABLE REDUX

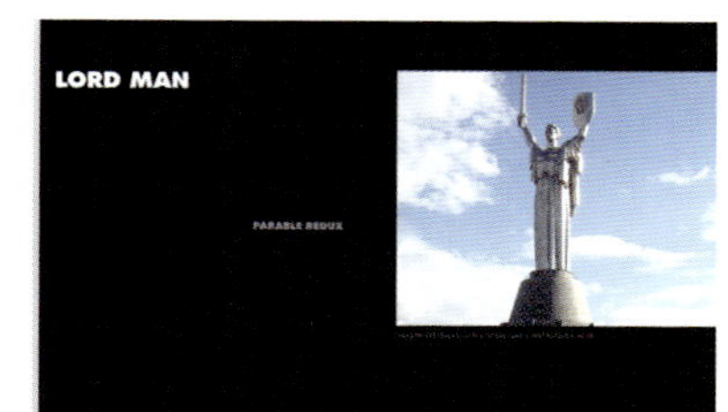

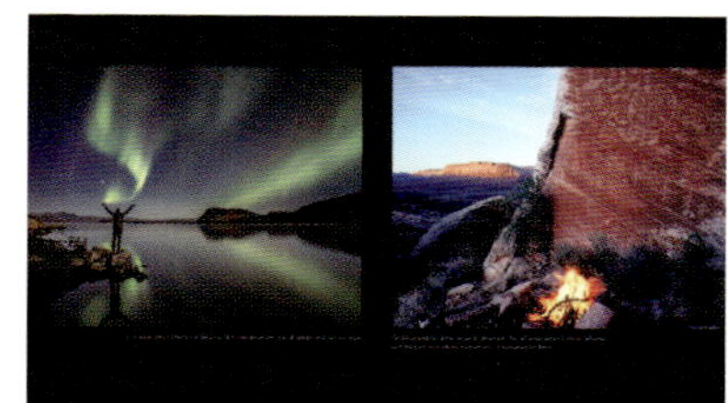

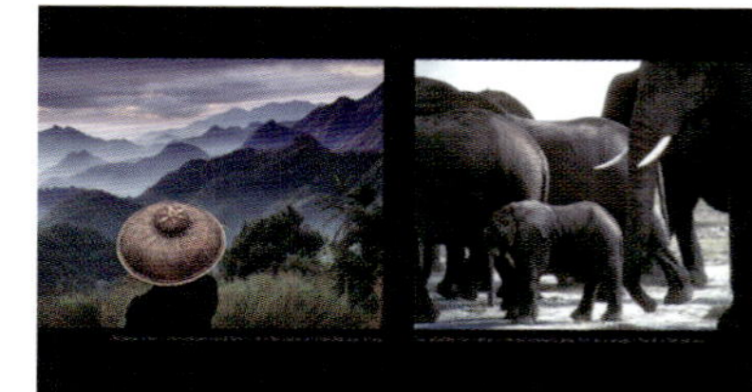

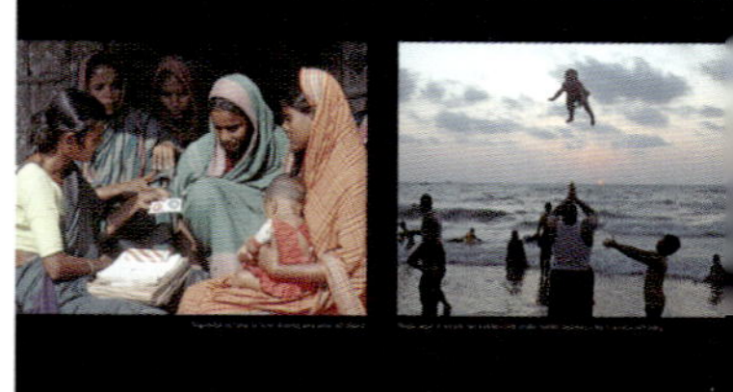

Statue: Motherland Monument, Kiev, Ukraine; © Douglas Tompkins.

Northern Lights: Woman enjoying the view of the Northern Lights, Lake Thingvellir, Iceland; Arctic-Images/ Corbis.

Rock Art: Ancient artwork adorns the rock walls of Canyonlands National Park, Utah, U.S.A.; © Daniel Dancer.

Mountain View: Greeting the morning in the Cordilleras Mountains, Philippines; © Per-Andre Hoffmann/Alamy.

Charismatic: Elephant family in Masai Mara National Park, Kenya, Africa; © Daniel Dancer.

Empowered Women: Women sharing information about the effective use of condoms, family planning clinic, Mobarakdi Village, Bangladesh; Getty Images.

Baby in the Sky: A Palestinian family enjoys playing on the beach at Gaza; © Majdi Fathi/NurPhoto/Sipa USA.

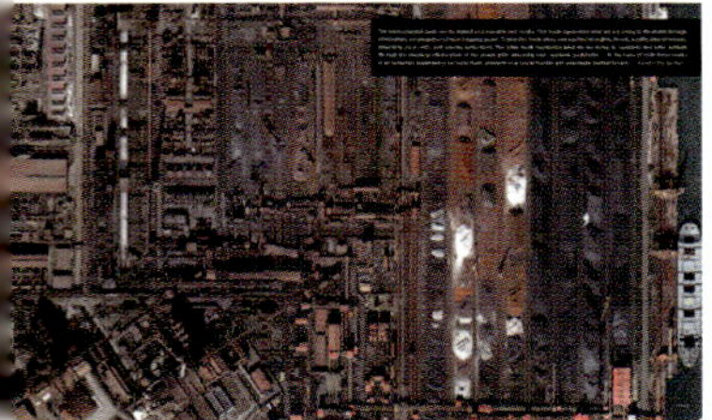

Tokyo Aerial: Major importers of energy resources like Japan must devote large parts of the landscape to energy-related processing and transport; Google Earth/2014 Digital Globe.

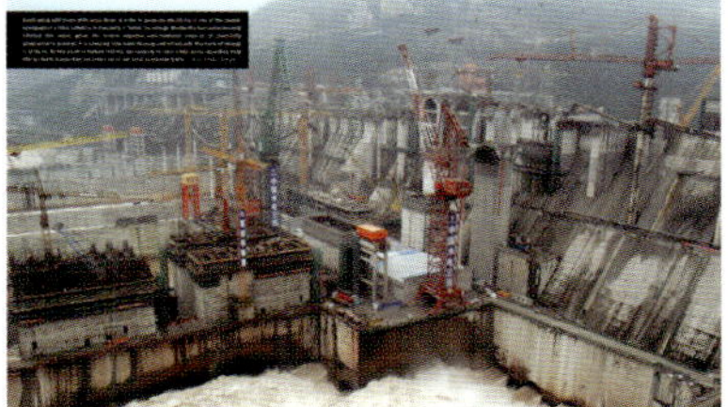

Megadam: The Xiangjiaba Hydropower Station on the Jinsha River, Sichuan province, China. China has been building large dams on its major rivers at a rate unmatched in human history; Corbis.

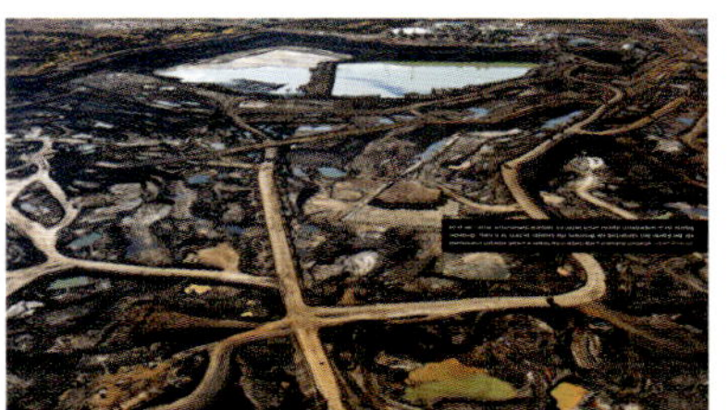

Toxic Landscape: Aerial view of the tar sands region, where mining operations and tailings ponds are so vast they can be seen from outer space; Alberta, Canada; © Garth Lenz.

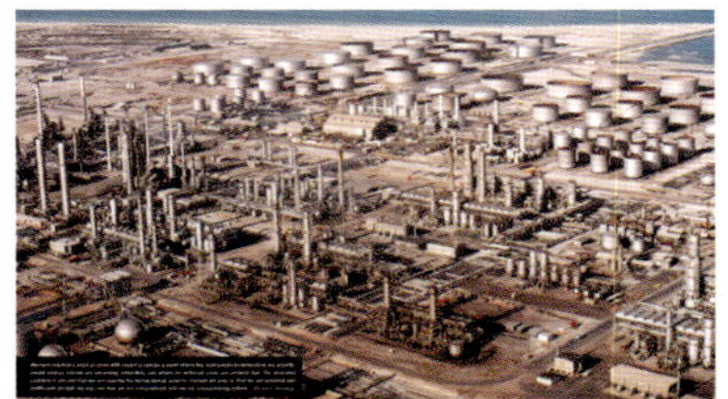

Oil Refinery: Necessary infrastructure in the globalized petroleum economy, oil refineries (such as this one in Saudi Arabia) also are significant sources of air pollution; Alamy.

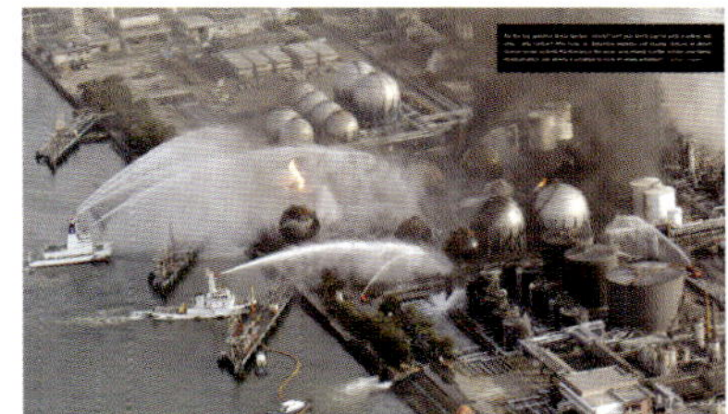

Nuclear Meltdown: The 2011 accident at the Fukushima Daiichi Nuclear Station in Japan galvanized the world's attention and again highlighted the risks of nuclear power. As of 2013 it was reported that the damaged plant was still leaking radioactive water into the Pacific Ocean; Mainichi Newspapers/AFLO.

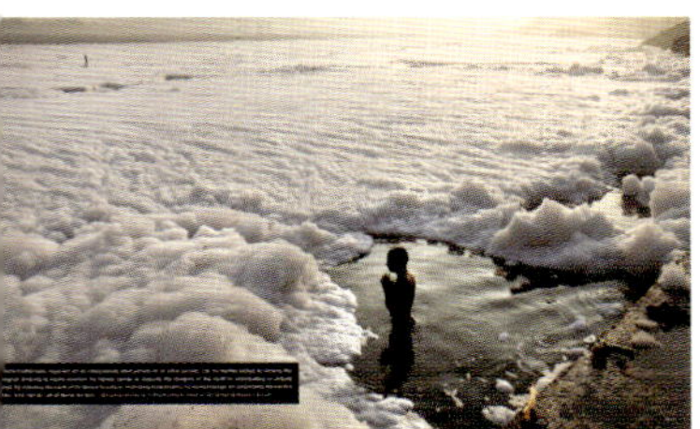

uds River Prayer: A Hindu man offers rayers after a dip in the Yamuna River, urrounded by industrial effluent, New elhi, India; © Manish Swarup/AP.

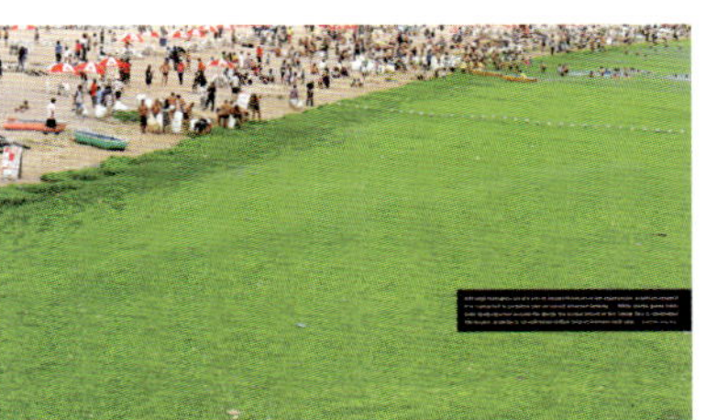

Algae Beach: Likely linked to fertilizer runoff, algae blooms along the coastline of Qingdao, Shandong province, China, are among the largest ever recorded. Bathers, apparently, are not deterred from swimming in the algae-fouled waters; Reuters.

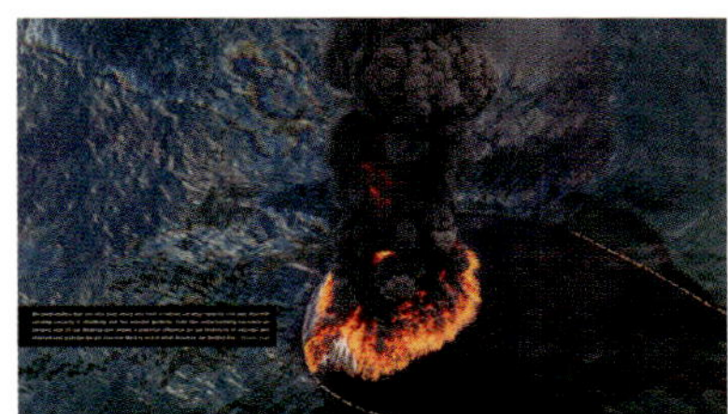

Oil Spill Fire: Aerial view of an oil fire following the 2010 Deepwater Horizon oil disaster, Gulf of Mexico; © Daniel Beltra.

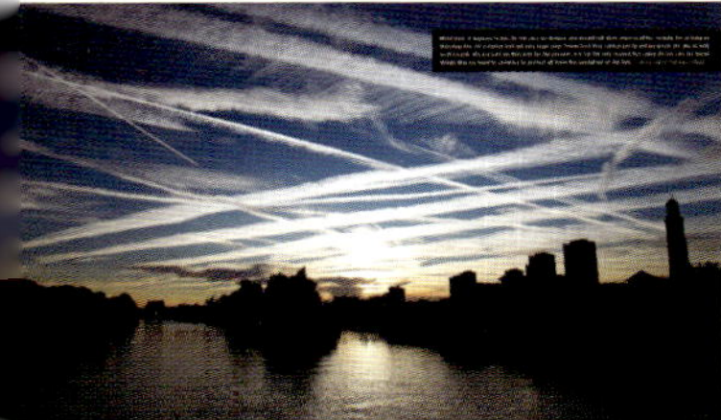

rplane Contrails: Globalized ansportation networks, especially ommercial aviation, are a major ontributor of air pollution and eenhouse gas emissions. Photo of ontrails in the west London sky over the ver Thames, London, England; Ian Wylie.

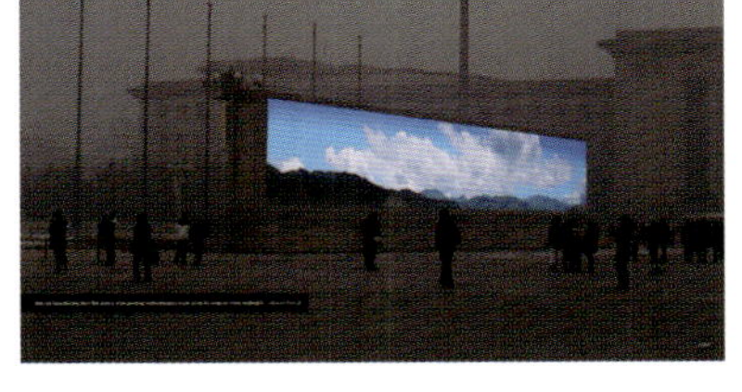

Blue Sky Sign: The ultimate irony—a giant LED screen in Tiananmen Square depicts an image of blue sky during dangerous levels of air pollution on January 23, 2013, in Beijing, China; © Feng Li/Getty Images.

AFTERWORD

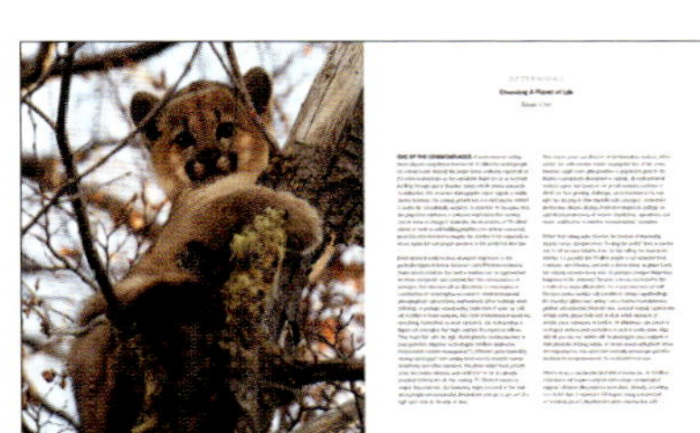

Puma Kitten: A right-sized human population would leave room for all creatures to flourish. Puma kitten in the future Patagonia National Park, Chile; © Chantal Henderson.

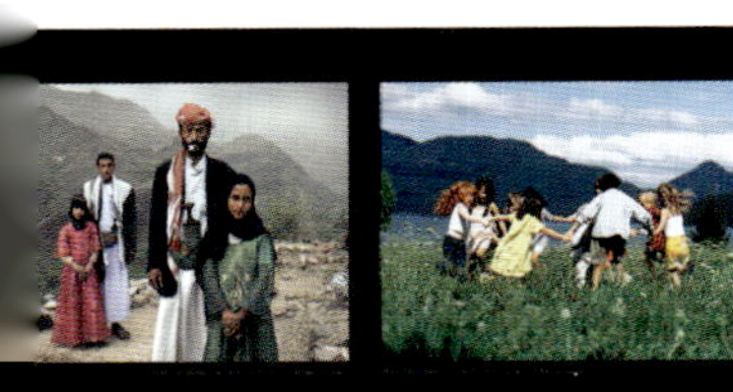

orm Damage: Survivors walk near hicles and floating debris after pertyphoon Haiyan devastated cloban City in central Philippines, ovember 10, 2013; Erik de Castro/Reuters.

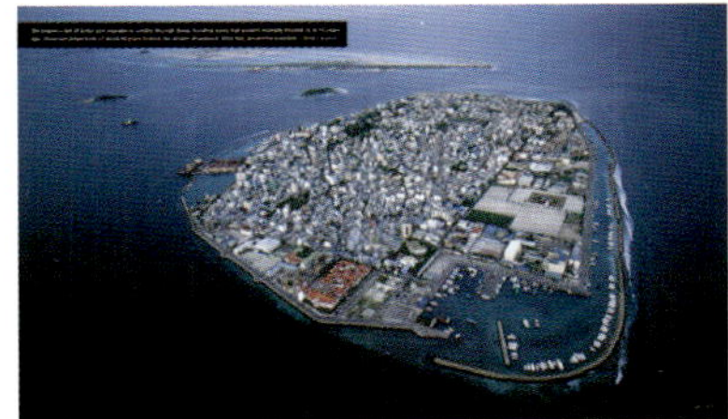

Shrinking Island: One of Earth's most vulnerable nations to climate change, the Maldive Islands are severely threatened by rising sea levels; © Peter Essick.

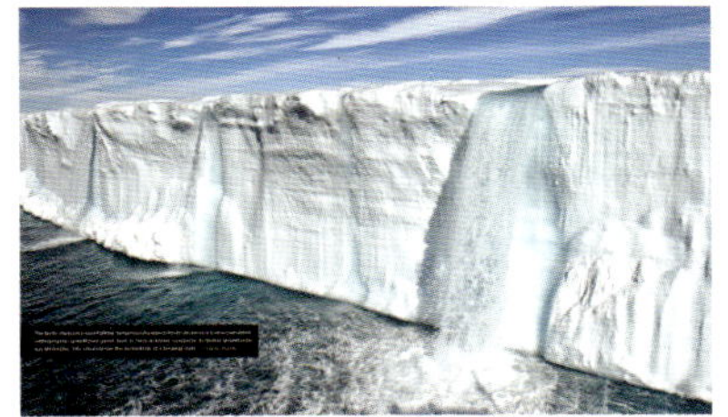

Ice Waterfall: In both the Arctic and Antarctic regions, ice is retreating. Melting water on icecap, North East Land, Svalbard, Norway; © Cotton Coulson/Keenpress.

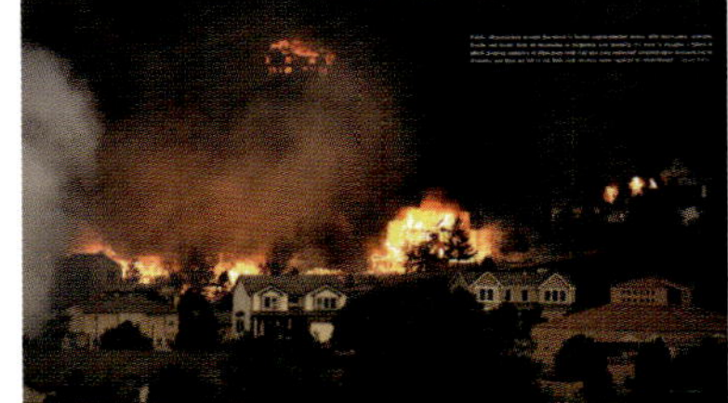

Fire: More frequent and more intense wildfires (such as this one in Colorado, USA) are another consequence of a warming planet; © R.J. Sangosti/Denver Post.

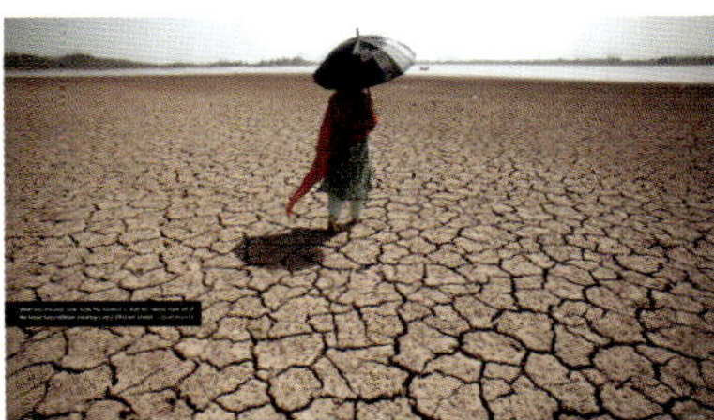

Dry Lake: A woman walks on a dry bank of a dam next to a lake that provides water to Islamabad, Pakistan, June 2012; © B.K. Bangash/AP.

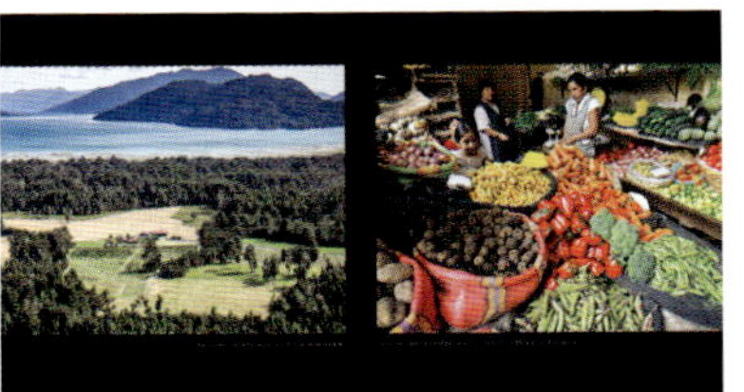

ild Brides: Tahani, 8, is seen with husband Majed, 27, and her former ssmate Ghada, 8, and her husband in jjah, Yemen, July 26, 2010; Stephanie Sinclair.

ds: All children should be safe, be ed, and have the opportunity to perience nature's beauty; Corbis.

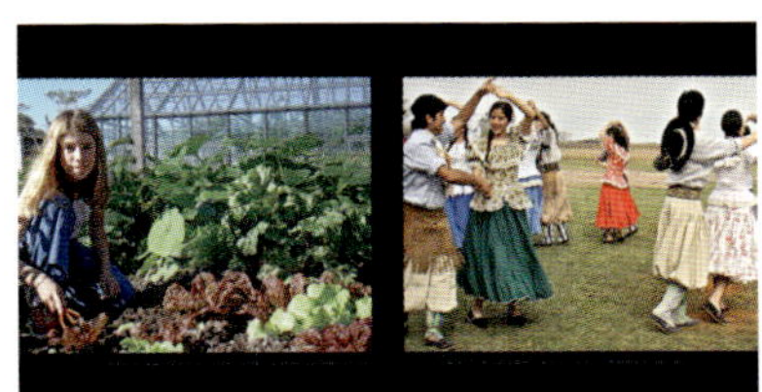

Farm: Model of eco-localism—a small subsistence farm near the village of Hornopiren in southern Chile; © Douglas Tompkins.

Local Produce: Women vendors work behind their vegetable display at a marketplace in Chosica, Peru, near Lima; Alamy.

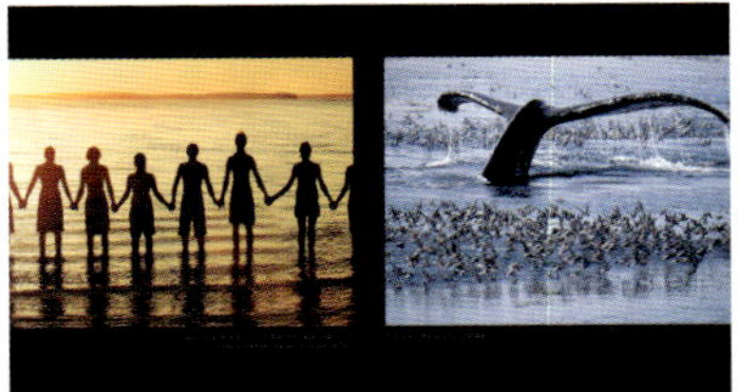

Garden: School-related gardening programs engage children to grow and eat organic food; © Douglas Tompkins.

Dancing: Place-based cultures produce distinctive regional customs. Corrientes province, Argentina; © Douglas Tompkins.

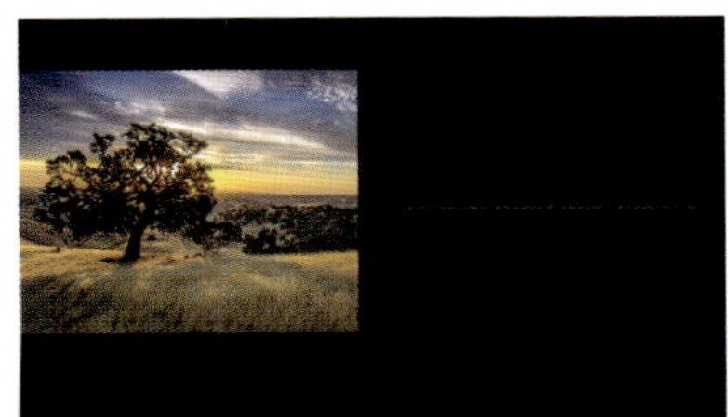

Holding Hands: A wild Earth is a beautiful Earth. California sunset; © Kristy-Anne Glubsh/Design Pics/Corbis.

Whale Tail: Fluke of humpback whale and red-necked phalaropes in Frederick Sound, S.E. Alaska, U.S.A.; © Ron Sanford/Corbis.

Tree of Life: Clouds over an oak tree on a ridge at sunrise, U.S.A.; Marc Crumpler/Getty Images.

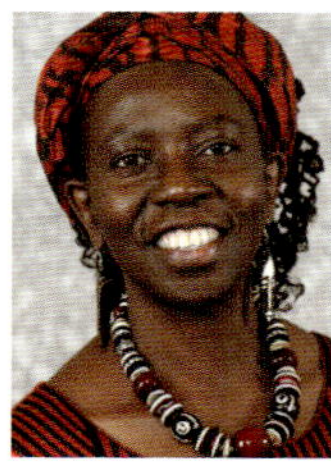

Musimbi Kanyoro, a native of Kenya, is president and CEO of the Global Fund for Women. She is a leading champion for human rights, the health of women and girls, and social-change-centered philanthropy. Kanyoro's international experience in the field of sexual and reproductive health and rights spans more than three decades. Prior to her current position, she was the director of the Population and Reproductive Health program of the David and Lucile Packard Foundation.

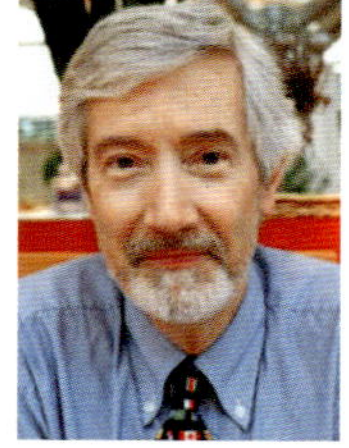

William Ryerson is founder and president of Population Media Center and also serves as CEO of the Population Institute in Washington, D.C. He has a forty-year history of working in the field of reproductive health, including two decades of experience adapting the Sabido methodology of social change communications to various cultural settings worldwide.

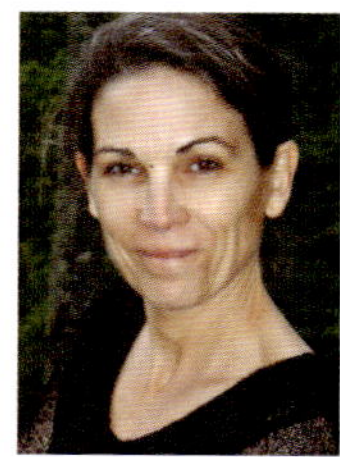

Eileen Crist teaches in the Department of Science and Technology in Society at Virginia Tech. She is author of *Images of Animals: Anthropomorphism and Animal Mind* and coeditor of *Gaia in Turmoil, Life on the Brink: Environmentalists Confront Overpopulation*, and *Keeping the Wild: Against the Domestication of Earth*.

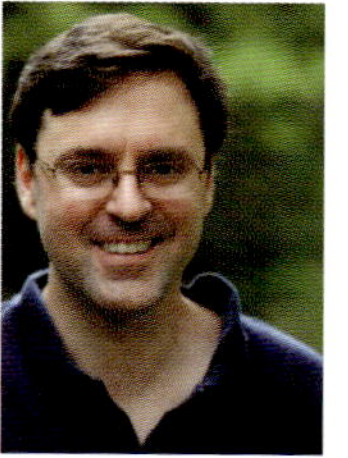

Tom Butler is editorial projects director of the Foundation for Deep Ecology and president of Northeast Wilderness Trust. His books include *Wildlands Philanthropy* and *ENERGY: Overdevelopment and the Delusion of Endless Growth*.

Visit the Global Population Speak Out website to add your voice and lend support to a worldwide campaign of activists, population and development professionals, and ordinary citizens concerned about the enormous size and rapid growth of the human population—and how these issues affect both the future of people and the ability of other species to flourish. "Speak Out" is the planet's leading population advocacy platform, bridging continents, cultures, and languages to bring together a global community of motivated and concerned citizens. Right now, future human population dynamics are being decided. Your voice, your spirit, and your creativity can be the difference between a stabilized and more sustainable human population by mid-century or one that is still rapidly expanding, unsustainably, by the millions.

Join us: populationspeakout.org.

Population Media Center (PMC) is a nonprofit, international nongovernmental organization, which strives to improve the health and well-being of people around the world through the use of entertainment-education strategies, such as serialized dramas on radio and television, in which characters evolve into role models for the audience for positive behavior change. Founded in 1998, PMC has over 15 years of field experience using the Sabido methodology of behavior change communications, positively affecting more than 50 countries around the world.

www.populationmedia.org

The Population Institute (PI) provides essential leadership to promote voluntary family planning and reproductive health services and increase awareness of the social, economic, and environmental consequences of rapid population growth. Founded in 1969 and based in Washington, D.C., PI works to educate policy makers, the media, and the general public about population issues. PI also recruits and trains tomorrow's population activists, and national membership networks to address population issues. The Institute promotes both international and U.S. support for voluntary family planning programs and supports full legal, political, economic, and social equality for women, including sexual and reproductive rights.

www.populationinstitute.org

COLOPHON

Volume Editor/Writer

Tom Butler

Contributors

Musimbi Kanyoro, William Ryerson, Eileen Crist

Art Director

Douglas Tompkins

Book Design

Roberto Carra

Photo Researcher

Daniel Dancer

Editorial Assistance

Mary Elder Jacobsen, Joe Bish, Ryan Buresh, Robert Walker

Proofreading

Mary Elder Jacobsen, Evan Morris

Print Coordination

Meghan Wright-Martin, Beatrice Cheung, Amanda Pfeifer

Production

Usana Shadday, Miney Tan

© 2015 Foundation for Deep Ecology
ISBN: 978-1-939621-23-8

All rights reserved under International and Pan-American Copyright Conventions.
No part of this book may be reproduced in any form or by any electronic or mechanical means without permission in writing from the Foundation for Deep Ecology.

Foundation for Deep Ecology
1606 Union Street, San Francisco,
California 94123
(415) 229-9339
www.deepecology.org
www.tompkinsconservation.org

Printed and Distributed by Goff Books, an imprint of ORO Editions.
www.goffbooks.com
info@goffbooks.com

Printed in China on paper from sustainably managed forests as certified by the Forest Stewardship Council.

8 BILLION

7 BILLION

6 BILLION

5 BILLION

4 BILLION

3 BILLION

2 BILLION

1 BILLION